Atomistic Mechanisms in Beam Synthesis and Irradiation of Materials

MATERIALS RESEARCH SOCIETY
SYMPOSIUM PROCEEDINGS VOLUME 504

Atomistic Mechanisms in Beam Synthesis and Irradiation of Materials

Symposium held December 1–2, 1997, Boston, Massachusetts, U.S.A.

EDITORS:

J.C. Barbour
Sandia National Laboratories
Albuquerque, New Mexico, U.S.A.

S. Roorda
Université de Montreal
Montreal, Quebec, Canada

D. Ila
Alabama A&M University
Normal, Alabama, U.S.A.

M. Tsujioka
Sumitomo Electric Industries, Ltd.
Itami, Hyogo, Japan

Materials Research Society
Warrendale, Pennsylvania

Single article reprints from this publication are available through
University Microfilms Inc., 300 North Zeeb Road, Ann Arbor, Michigan 48106

CODEN: MRSPDH

Published by:

Materials Research Society
506 Keystone Drive
Warrendale, PA 15086
Telephone (724) 779-3003
Fax (724) 779-8313
Website: http://www.mrs.org/

Library of Congress Cataloging in Publication Data

Atomistic mechanisms in beam synthesis and irradiation of materials :
    symposium held December 1–2, 1997, Boston, Massachusetts / editors,
    J.C. Barbour, S. Roorda, D. Ila, M. Tsujioka
        p.cm.—(Materials Research Society symposium proceedings ;
        ISSN 0272-9172 ; v. 504)
        Includes bibliographical references and index.
        ISBN 1-55899-409-2
        1. Materials—Effect of radiation on—Congresses. 2. Ion
    bombardment—Congresses. 3. Thin films—Congresses. 4. Surface
    chemistry—Congresses. I. Barbour, John Charles, 1958- II. Roorda, S.
    III. Ila, D. IV. Tsujioka, M. V. Series: Materials Research Society symposium
    proceedings ; v. 504
TA418.6.A85   1999   99-17074
620.1'1228—dc21   CIP

Manufactured in the United States of America

CONTENTS

*Invited Paper

*Invited Paper

*Invited Paper

*Invited Paper

PREFACE

This proceedings volume represents 70% of the papers presented December 1–2, 1997, in Symposium KK, "Atomistic Mechanisms in Beam Synthesis and Irradiation of Materials," at the 1997 MRS Fall Meeting in Boston, Massachusetts. The symposium focused on understanding the atomistic processes that occur in metals, ceramics (including glasses), and polymers exposed to energetic beams for the purpose of synthesizing or modifying a material. The work presented, in 47 oral and 60 poster presentations over two days, was similar in focus to that traditionally presented in previous years in Symposium A. However, this symposium tried to give greater emphasis to work discussing atomistic approaches for modelling and understanding experimental results. The organization of this symposium, with the large number of poster presentations, was intended to create a smaller meeting atmosphere with greater one-on-one interaction between the participants. The enthusiastic participation in both the oral and poster sessions was testimony to the continued interest in beam-based materials research.

The science presented in this symposium fell into the following categories: (i) surface and thin-film microstructures driven by ion-beam processing, including mechanisms driving preferential orientation in single-crystal and polycrystalline formation; (ii) comparisons and contrasts in the mechanisms involved with high-energy ion-beam irradiation and high fluence-rate effects; (iii) identification of mechanisms in beam-assisted synthesis of materials; and (iv) atomistic modelling to describe improved mechanical and optical properties of coatings and layers treated with ion beams. The papers in this volume have been divided in accordance with the topical divisions used for the oral presentations.

J.C. Barbour
S. Roorda
D. Ila
M. Tsujioka

November 1998

ACKNOWLEDGMENTS

The organizers wish to thank all of the contributors and participants, who by their enthusiastic debate were responsible for the success of this ion-beam symposium. We also thank the reviewers for their time and hard work to carefully ensure the quality of the papers in this proceedings. Finally, we thank the invited speakers for their excellent presentations which provided the backbone for this meeting:

> Sebania Libertino
> Marcel Toulemonde
> Robert C. Birtcher
> Isao Yamada
> David M. Follstaedt
> Carmine A. Carosella
> Richard F. Haglund, Jr.

The following organizations and companies made this symposium possible through their generous support:

> Alabama A&M University,
>   Center for Irradiation of Materials
> High Voltage Engineering Europa B.V.
> National Electrostatics Corp.
> Sumitomo Electric Industries Ltd.

MATERIALS RESEARCH SOCIETY SYMPOSIUM PROCEEDINGS

MATERIALS RESEARCH SOCIETY SYMPOSIUM PROCEEDINGS

Volume 488— Electrical, Optical, and Magnetic Properties of Organic Solid–State Materials IV, J.R. Reynolds, A. K-Y. Jen, L.R. Dalton, M.F. Rubner, L.Y. Chiang, 1998, ISBN: 1-55899-393-2

Volume 489— Materials Science of the Cell, B. Mulder, V. Vogel, C. Schmidt, 1998, ISBN: 1-55899-394-0

Volume 490— Semiconductor Process and Device Performance Modeling, J.S. Nelson, C.D. Wilson, S.T. Dunham, 1998, ISBN: 1-55899-395-9

Volume 491— Tight–Binding Approach to Computational Materials Science, P.E.A. Turchi, A. Gonis, L. Colombo, 1998, ISBN: 1-55899-396-7

Volume 492— Microscopic Simulation of Interfacial Phenomena in Solids and Liquids, S.R. Phillpot, P.D. Bristowe, D.G. Stroud, J.R. Smith, 1998, ISBN: 1-55899-397-5

Volume 493— Ferroelectric Thin Films VI, R.E. Treece, R.E. Jones, S.B. Desu, C.M. Foster, I.K. Yoo, 1998, ISBN: 1-55899-398-3

Volume 494— Science and Technology of Magnetic Oxides, M. Hundley, J. Nickel, R. Ramesh, Y. Tokura, 1998, ISBN: 1-55899-399-1

Volume 495— Chemical Aspects of Electronic Ceramics Processing, P.N. Kumta, A.F. Hepp, D.N. Beach, J.J. Sullivan, B. Arkles, 1998, ISBN: 1-55899-400-9

Volume 496— Materials for Electrochemical Energy Storage and Conversion II—Batteries, Capacitors and Fuel Cells, D.S. Ginley, D.H. Doughty, T. Takamura, Z. Zhang, B. Scrosati, 1998, ISBN: 1-55899-401-7

Volume 497— Recent Advances in Catalytic Materials, N.M. Rodriguez, S.L. Soled, J. Hrbek, 1998, ISBN: 1-55899-402-5

Volume 498— Covalently Bonded Disordered Thin–Film Materials, M.P. Siegal, J.E. Jaskie, W. Milne, D. McKenzie, 1998, ISBN: 1-55899-403-3

Volume 499— High–Pressure Materials Research, R.M. Wentzocovitch, R.J. Hemley, W.J. Nellis, P.Y. Yu, 1998, ISBN: 1-55899-404-1

Volume 500— Electrically Based Microstructural Characterization II, R.A. Gerhardt, M.A. Alim, S.R. Taylor, 1998, ISBN: 1-55899-405-X

Volume 501— Surface–Controlled Nanoscale Materials for High–Added-Value Applications, K.E. Gonsalves, M-I. Baraton, J.X. Chen, J.A. Akkara, 1998, ISBN: 1-55899-406-8

Volume 502— In Situ Process Diagnostics and Intelligent Materials Processing, P.A. Rosenthal, W.M. Duncan, J.A. Woollam, 1998, ISBN: 1-55899-407-6

Volume 503— Nondestructive Characterization of Materials in Aging Systems, R.L. Crane, S.P. Shah, R. Gilmore, J.D. Achenbach, P.T. Khuri-Yakub, T.E. Matikas, 1998, ISBN: 1-55899-408-4

Volume 504— Atomistic Mechanisms in Beam Synthesis and Irradiation of Materials, J.C. Barbour, S. Roorda, D. Ila, 1998, ISBN: 1-55899-409-2

Volume 505— Thin–Films—Stresses and Mechanical Properties VII, R.C. Cammarata, E.P. Busso, M. Nastasi, W.C. Oliver, 1998, ISBN: 1-55899-410-6

Prior Materials Research Society Symposium Proceedings available by contacting Materials Research Society

Part I

Defects and Modelling

DEFECT EVOLUTION IN ION IMPLANTED Si: FROM POINT TO EXTENDED DEFECTS

Sebania Libertino[a], Janet L. Benton[b], Salvatore Coffa[c], Dave J. Eaglesham[b].

[a] INFM and Dipartimento di Fisica, Università di Catania, C.so Italia, 57, I-95130 Catania, Italy.
[b] Bell Labs, Lucent Technologies, 700 Mountain Avenue, Murray Hill, NJ 07974.
[c] CNR-IMETEM, Stradale Primosole, 50, I-95122, Catania, Italy

Abstract

Several recent experiments assessing the role of impurities (C, O), dopants (P, B) and clustering on defect evolution in ion implanted Si are reviewed. Deep level transient spectroscopy measurements were used to analyze the defect structure in a wide range of ion implantation fluences (1×10^8–5×10^{13} cm^{-2}) and annealing temperatures (100-800 °C). By using substrates with a different impurity content and comparing ion implanted and electron irradiated Si samples, many interesting features of defect evolution in Si have been elucidated. It is found that only a small percentage, 4-16 % depending on ion mass, of the Frenkel pairs generated by the beam escape direct recombination and is stored into an equal number of room temperature stable vacancy- (V-) and interstitial-type (I-) defect complexes. Identical defect structures and annealing behavior have been measured in ion implanted (1.2 MeV Si, 1×10^8–1×10^{10}/cm^2) and electron irradiated (9.2 MeV to fluences between 1 and 3×10^{15}/cm^2) samples in spite of the fact that denser collision cascades are produced by the ions. The O and C content of the substrate plays a major role in determining the point defect migration, the room temperature stable defect structures and their annealing behavior. Annealing at temperatures up to 300 °C produces a concomitant reduction of the I- and V-type defect complexes concentration, demonstrating that defect annihilation occurs preferentially in the bulk. At temperatures above 300 °C, when all V-type complexes have been annealed out, ion implanted samples present a residual I-type damage, storing 2-3 I per implanted ion. This unbalance is not observed in electron irradiated samples and it is a direct consequence of the extra implanted ion. The simple point defect structures produced at low ion fluence (1×10^8–1×10^{11} /cm^2) anneal at ~ 550 °C. At higher fluences (~ 10^{12}-10^{13} /cm^2) and for annealing temperatures above 500 °C the deep level spectrum is dominated by two signatures at E_V+0.33 eV and at E_V+0.52 eV that we have associated to Si interstitial clusters. Impurities (C, O and B) play a role in determining nucleation kinetics of these defects, but they are not their main constituents. The dissolution temperature of these clusters indicates that they might store the interstitials that drive transient enhanced diffusion phenomena occurring in the absence of extended defects. Finally, at higher implantation fluence, a signature of extended defects is observed and associated to the presence of {311} defects detected by transmission electron microscopy analyses.

Introduction

Ion implantation is widely used in the processing of VLSI Si devices. However, ion beam damage is responsible of several phenomena, such as transient enhanced diffusion (TED) [1] and extended defects formation [2], which severely hamper our ability to fabricate sub-micron devices. In order to understand and model these processes, a full comprehension of how ion implantation induced defects form and evolve as a function of the implantation dose and thermal processing is required.

In spite of the widespread use of ion implantation in Si, most of our current knowledge of defect structure and annealing behaviour still relies on the detailed investigations performed [3, 4, 5, 6] during the past 20 years on electron-irradiated silicon. In these studies, the structure and annealing behavior of point defects such as divacancy (VV), oxygen vacancy, (O_iV) and carbon-oxygen (C_iO_i), were assessed, using photoluminescence, deep level transient spectroscopy (DLTS), and electron paramagnetic resonance analyses. In the last few years, the purity of the Si substrate has increased considerably, and it is now possible to grow wafers with an oxygen and carbon content comparable or lower than the dopant concentration. Since impurities are effective traps for Frenkel pairs generated by the beam, it is reasonable to expect a different evolution of defects in these pure materials [7]. Moreover, although recent DLTS studies [8,9] show that the same defect structures are produced in Si by both ions and electrons, several issues certainly need to be explored in more detail. A significant difference is, in fact, expected between the damage evolution in ion implanted and in electron irradiated Si. It is known that the defect profile for ion implanted samples is not spatially uniform, as in the electron irradiated case, but the damage is mainly localised in a well defined peak at the end of range. Also the density of collisional events is orders of magnitude higher for ions than for electrons. As a result, room temperature annihilation and defect clustering [10] can be greatly favored in ion implanted samples.

Furthermore, while electron irradiation produces equal numbers of vacancies and self-interstitials, ion implantation produces an imbalance between vacancies and interstitials caused by the presence of the extra-incorporated ion. This last observation has noticeable consequences and provides the basis for the +1 model, [11, 12] which is used to describe interstitial clustering, and all the related phenomena occurring for high fluence and high temperature annealing. In this model, it is assumed that extensive recombination of vacancies and interstitials occurs until only the extra ion is left.

The interstitial excess caused by the extra ion presence drives the Transient Enhanced Diffusion (TED) of dopants. Indeed, detailed TEM analyses and B diffusion experiments have demonstrated that release of the interstitials stored into {311} extended defects, drive B enhanced diffusion. The number of interstitials stored in these defects equals the total number of implanted ions, providing a confirmation of the "plus one" model [13].

However, TED occurs even without the formation and dissolution of extended defects [14], as in the case for low dose implants, or in early stages of the annealing process, before the formation of extended defects. To explain these phenomena, small Si interstitial clusters, not detectable by TEM, have been proposed as the interstitial source for TED. The experimental identification of these clusters could give new inputs to the modeling [15] of transient enhanced diffusion phenomena. The new generation of simulation tools for the design of sub-micron devices, in fact, requires an accurate scientific description of defect-defect and defect-impurity interaction, from simple to extended defects during the various implantation and annealing processes.

Deep Level Transient Spectroscopy (DLTS) has been traditionally used in the low implantation fluence regime and considered unable to provide information in the fluence regime where extensive defect clustering and extended defects formation is observed. In this paper we show that DLTS can be effectively used to monitor defect evolution from simple point defects to extended defects. The aim of this work was to characterize ion implantation damage as a function of the impurity content of the material and in a wide dose and annealing temperature range. Ion implantation induced damage can be divided, based on recent experimental investigations [16], into four groups: *interstitial (I-) and vacancy (V-) type point defect pairs*, obtained in samples implanted to low fluences, $\leq 1\times10^9$ Si/cm^2 and stable from room temperature to 300 °C; *interstitial point-like defects*, obtained after annealing at temperatures in the range 300-500 °C for implantation doses of 1×10^9-1×10^{11} Si/cm^2; *interstitial clusters*, formed after implantation doses $\geq 1\times10^{12}$ Si/cm^2 and annealing temperatures ≥ 550 °C; and *extended defects*, found after implantation at doses $\geq 2\times10^{13}$ Si/cm^2, and annealed for temperatures $\geq$ 680 °C. Experimental results for each of the four groups are presented and discussed.

Experimental

We used p-type (doped with 7×10^{15} B/cm^3) and n-type (doped with either 5×10^{13} P/cm^3 or with 7×10^{14} P/cm^3) Si wafers. Both epitaxial (epi, with an O and C content lower than 1×10^{15} /cm^3) and

4

Czochralski (CZ, with [O]= $7 \times 10^{17}/cm^3$ and [C] $\leq 10^{16}/cm^3$) Si samples were analyzed in order to assess the effect of impurities on defect evolution. Samples were implanted with 1-3 MeV He, to fluences in the range $1 \times 10^8 - 8 \times 10^9/cm^2$ or with 145 keV-1.2 MeV Si to fluences in the range $1 \times 10^9 - 5 \times 10^{13}/cm^2$ or irradiated with 9.2 MeV electrons to a fluence of 1 - $3.5 \times 10^{15}/cm^2$. Thermal anneals at temperatures of 100-680 °C for 30 min-15 hr were performed in flowing He. Finally, Schottky barriers were formed on the samples by evaporating either Ti on p-type Si or AuPd on n-type Si after a dip in HF. In some of the DLTS analyses p^+-n junctions, obtained by B implantation and high temperature diffusion, have been used. The samples were characterized by DLTS measurements using Bio-Rad DL4600 or DL8000 Spectrometers.

Interstitial- and vacancy-type point defect pairs

The point defects (I, V) generated by electron irradiation or ion implantation can migrate through the Si lattice until they either recombine according to the relation I + V = $\emptyset$ or are stored in room temperature stable defect complexes. Fig. 1 schematically summarizes what is known on the formation of the principal room temperature defect complexes. A vacancy will migrate until it either finds another vacancy, forming a divacancy, VV, or it pairs with impurities to create V-type point defect structures such as oxygen-vacancy, OV, or phosphorus-vacancy, PV [6, 8].

Silicon self-interstitials create, by the Watkins' exchange mechanism [17], interstitial impurities such as boron interstitial, B_i, and carbon interstitial, C_i. The favored branch of the reaction depends on to the ratio between B and C. These interstitial species diffuse at room temperature until they are trapped by impurities and form interstitial-impurity pairs such as carbon-carbon (C_s-C_i), carbon-oxygen (C_iO_i), carbon-phosphorous (C_i-P_s) if C_i is generated [5, 18]. Instead, if B_i is generated in suitable concentration the formation of boron related defect complexes such as boron-oxygen, B_iO_i, boron-carbon, B_iC_s and boron-boron, B_iB_s, [19, 20, 21] will take place. These simple point defect pairs have been widely characterized and all of them anneal at temperatures below 450 °C. The wealth of published information on the identity, introduction rates, annealing behavior, and DLTS signatures of vacancy and interstitial related point defects in electron irradiated Si provides a firm foundation for a

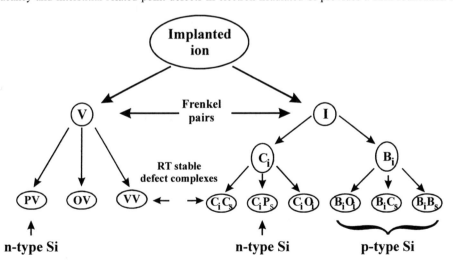

Figure1: *Room temperature stable defect complexes generated by ion implantation with a (light ion) or by electron irradiation in Si.*

comprehensive study of the introduction and evolution of implantation-induced defects in Si.

a. Comparison between electron irradiation and ion implantation in the as-implanted state

In order to apply the knowledge acquired in the last 20 years on electron irradiated samples to io implantation it is fundamental to compare directly the damage introduced with the two processes. I this study we used Si ions because they do not introduce deep levels in Si and because their collisio cascade is diluted. The comparison was performed using the same substrate, thus fixing the impuritie (C and O) and dopant concentrations.

The results of the direct comparison between electron irradiation (9.2 MeV, $3.5 \times 10^{15}/cm^2$) and io implantation (1.2 MeV Si, $1 \times 10^9/cm^2$) are reported in Fig. 2 (*a*) for n-type and (*b*) for p-type samples.

In n-type epitaxial Si, (low C, low O content) the spectra are dominated by V-type defects tha introduce deep levels in the upper half of the band gap, such as the oxygen-vacancy, OV E(0.17 eV the doubly negative charge state of the divacancy, $VV^=$ E(0.22 eV), the singly negative charge state c the divacancy, VV^- E(0.41 eV), and the phosphorous-vacancy, PV E(0.46 eV), normally observed i DLTS after electron irradiation [3]. The spectra of electron irradiated (solid line) and ion implante (dashed line) Si are almost identical, demonstrating that the same defect structures are introduced. N measurable differences arise as a consequence of the denser ion collision cascade.

The impurity content of the substrate plays a major role in determining the defect structure of a: implanted or as-irradiated Si samples. In silicon with a high concentration of carbon (dot dashed line i Fig. 2*a*), the signal related to P_sC_i, E(0.3 eV), is clearly visible [5].

In p-type Si, Fig. 2*b*, the DLTS spectru of a CZ Si sample irradiated with electron reveals the signatures [4] of the carbc oxygen pair (C_iO_i, at E_V+0.36 eV) and of th positively charged divacancy (VV^+, E_V+0.21 eV) and, again, the spectrum aft irradiation (dot dashed line) and ic implantation (not shown) are the same. On more, the impurity content plays a major rol In epitaxial Si samples the C_iO_i concentratic is much lower than in the CZ sample and tw additional defect signatures are present: th boron-carbon pair (B_iC_s, at E_V+0.27 eV) a a B-related defect that contains a sing interstitial [7] (H_1, at E_V+0.45 eV). The lar differences in the DLTS spectra betwe epitaxial (solid line) and CZ (dashed dott line) Si samples spectra can be explain observing that the C_iO_i formation is large suppressed in epitaxial Si due to the lo concentration of both O and C. In this samp B concentration is higher than concentration and it successfully compet with carbon in storing the self interstiti generated by the irradiation. Self-interstit will produce, by the Watkins replaceme mechanism (see Fig. 1), interstitial B ato

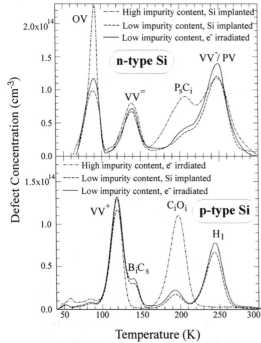

Figure 2: DLTS spectra of ion implanted or electron irradiated (a) n-type and (b) p-type Si samples with different impurity content.

B_i) which are then immobilized into defect complexes, as B_iC_s.

These results imply that the dense ion collision cascade and the incorporation of the extra ion do not produce any measurable effect in the as-implanted defect structure. Hence, it is only the impurity content of the material, which is determining the defect structure in the as-implanted state [7].

b. Room temperature migration of interstitials and vacancies

Ion implantation damage is not uniform in all the sample thickness as in the electron irradiation case, but it is well localized at the end of range of the implanted ion. This characteristic of ion implantation can be used to monitor room temperature recombination and migration of vacancies and interstitials [22, 23]. For this study Si substrates with different C and O impurity concentration, were implanted with helium. Helium has a very low atomic mass (4 a.m.u.) and creates a very diluted collision cascade in Si. This minimizes the direct clustering effect within the single ion cascade. Moreover, a very low implantation dose, such as 5×10^8 He/cm^2, avoids overlapping between different ion track collision cascades. In this study, we used n-type Si and determined the depth profiles of impurity-point defect pairs (OV and P_sC_i) and divacancies.

Depth profiles of the divacancy (●) and oxygen-vacancy (■) defect complexes for a Czochralski Si sample implanted with 1 MeV He, are reported in Fig. 3. The OV concentration in this sample is very high confirming the efficiency of O as a trap for migrating vacancies. Indeed, the OV profile plotted in Fig. 3 has been divided by a factor 1.5 to allow the direct comparison with the VV⁻ profile. The initial generation of I-V pairs as calculated by Transport of Ion In Matter, TRIM, [24] is also reported as a solid line. The simulated generation curve has been divided by a factor of 22 in order to allow comparison. Important information can be drawn from the figure. A large percentage of interstitials and vacancies generated by the beam anneals during the implant, or immediately after the implantation process. Such extensive recombination occurs within the single cascade, because the implantation dose is well below the threshold to have overlapping between cascades. The concentration of electrically active defects is only ~ 16 % of the total concentration of defects generated by the beam. In spite of the extensive recombination, the depth profile precisely mirrors the Frenkel pair distribution predicted by TRIM. The point defects that escape recombination cannot migrate long distances due to the formation of complexes with other vacancies or pairs with O or P atoms and this effect confirms, once again, that impurities are very efficient traps for point defects.

The relative importance of defect clustering and impurity trapping can be changed by varying the impurity content in the material and/or changing the mass of the implanted ion [10, 23]. The role played by impurities and dopants on the migration properties of point defects becomes evident when He implants (3 MeV, 5×10^8 cm^{-2}) are performed on substrates having very low impurities and dopant concentrations. In particular, we compared float zoned and epitaxial Si substrates both doped with 5×10^{13} P/cm^3. The resulting OV depth concentration profiles are reported in Fig. 4.

The reduction in the sample impurity content, in particular from 7×10^{17} O/cm^3 of the CZ samples (see Fig. 3) to 2×10^{16} O/cm^3 of the FZ or ~ 10^{15} O/cm^3 of the epi samples, results in a significant broadening of

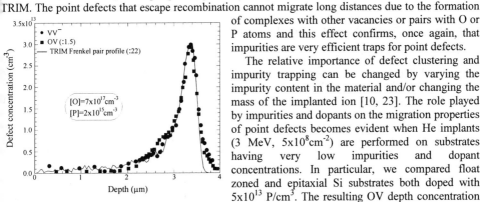

Figure 3: Depth profiles of VV (●), and OV (■) for a CZ sample (4 Ωcm, [O]=7x10^{17} /cm^3 and [C]=1x10^{16} /cm^3) implanted with 1 MeV He, 5x10^8 /cm^2. The profiles are compared with a TRIM simulation of the implant profile (solid line).

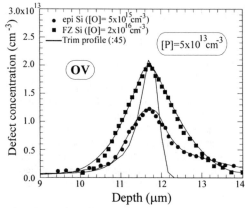

Figure 4: *Depth profiles of OV in samples implanted with 3 MeV He, 5x10⁸ /cm² on epi Si (of 60 Ωcm, [O] ≤ 5x10¹⁵ /cm³ and [C] < 1x10¹⁶ /cm³, •); on FZ Si(60 Ωcm, [O] ~ 1x10¹⁶ /cm³ and [C] ≤ 1x10¹⁶ /cm³, ■). The solid line is a TRIM simulation of the generated Frenkel pairs.*

the OV depth profile (Fig. 4) with respect to the initial defect profile calculated by TRIM (solid line). In a highly pure material, vacancies are free to migrate long distances before they either recombine with interstitials or create stable defect complexes by clustering with other vacancies or pairing with impurities such as O, causing the observed broadening. It should be observed that the experimental depth profiles show a symmetric broadening with respect to the TRIM simulation; moreover, the peak of the defect distribution is deep, ~ 12 μm. Hence, any preferential defect recombination at the surface can be ruled out. A similar depth dependence is found for the VV depth profile, suggesting that for a light ion such as He most of the VV are not formed directly in the collision cascade but result from clustering of two migrating vacancies. [22, 23]. A similar behavior has been observed for interstitial-type defects.

The strong reduction of the sample impurity content also causes a strong reduction in the concentration of vacancy-type defects stable at room temperature. While a value of ~ 16% is observed in CZ Si, the total concentration of V stored in V-type defects is reduced to ~ 3% and ~ 2.6 % for FZ and epi Si, respectively. It occurs because vacancies and interstitials can migrate for a longer distance before being trapped and the probability of annihilation is increased. The experimental results strongly suggest that the room temperature defect migration is in this case a trap limited process where free migrating vacancies are trapped by O and a trapping radius of ~ 1 Å has been measured [23].

c. Annealing behavior of I- and V-type defect complexes

Role of the impurity content

A reduction in impurity content can also change the annealing rate and, hence, the thermal stability of room temperature stable defects. In particular, annealing of defect complexes is the result of either thermal dissociation or annihilation by point defects released by the dissociation of less stable defect structures. The dependence of the annealing temperature on the impurity content is known for some defect complexes, such as the VV [25] and the B_iC_s [20]. We demonstrated [7] that also the C_iO_i complex dissolution temperature is dependent on the impurity content of the sample. The isochronal anneal of the C_iO_i pair for three electron irradiated samples with different impurity concentrations is reported in Fig. 5. The epitaxial (•) and the CZ Si (■) used in this study have a lower carbon concentration, $[C] \leq 10^{16}$ /cm³, than the CZ (▲) used in previous electron irradiation studies [20], [C] = 6x10¹⁶ /cm³. For this high C sample we also report the annealing behavior of the C_i (★) which was not detected in our other epitaxial and CZ samples. The annealing curve of the C_iO_i complex in this high C CZ sample shows the behavior typically reported for this defect. At 100 °C its concentration increases as the C_i diffuses and pairs with the O, present in high concentration, producing more C_iO_i. The concentration of C_iO_i remains stable up to 250 °C and next decreases slightly in the range 250-350 °C (probably due to annihilation by free vacancies released by the dissolution of divacancies). Finally, complete dissolution of the C_iO_i occurs at 400 °C. The stability of the C_iO_i defect is a consequence of

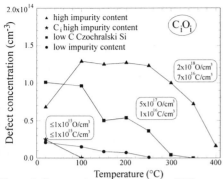

both the high binding energy of the two impurity atoms and of the strong back reaction which competes with dissociation. Indeed, in Si with high O concentrations, when a C_iO_i is dissociated the released C_i will have a strong chance to pair again with O.

In epitaxial Si, having a low concentration ($\leq 10^{15}/cm^3$) of both C and O, the concentration of C_iO_i pairs is greatly reduced since, as shown in Fig. 2b, most of the interstitials are expected to be stored into different, B_i related, complexes. B_i-related complexes dissociate at T < 400 °C [7] releasing free B_i. These B interstitials can recombine with vacancy type defects, releasing free vacancies which then produce the dissolution of the C_iO_i defects. In addition, the dissociation of a C_iO_i complex is not balanced by a back-reaction, because the released C_i has little probability of pairing again with O in the epitaxial material. Complete annealing of C_iO_i is achieved at 250 °C. The annealing temperature of the C_iO_i in the low C

Figure 5: *Concentration of carbon-oxygen (C_iO_i) complexes, as obtained from DLTS, is reported as a function of annealing temperature for epitaxial Si (●) and CZ Si (■). In addition the annealing behavior of C_iO_i (▲) and carbon-interstitial C_i (★) defects in a CZ sample with a higher C content [20] are also reported. All the samples were irradiated with 9.2 MeV electrons to a fluence of $3.5 \times 10^{15}/cm^2$.*

CZ is intermediate to that of the high C CZ and of the epitaxial Si. Once more C and B compete for the capture of interstitials and the formation of stable defects. The presence of B_i stored in room temperature stable defects will again result in a different annealing behavior of the C_iO_i complex.

Role of the extra implanted ion

The concentration of *V- and I-type point defect pairs* (i.e. VV, C_iO_i,) can be measured by DLTS. Hence, the number of I and V stored in these defects can be counted [26, 27]. We have analyzed the

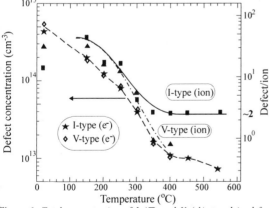

annealing behavior of V-type and I-type point defects in ion implanted and electron irradiated Si samples. We determined the total concentration of vacancies by adding the concentration of the VV (multiplied by a factor two since two vacancies participate to the VV formation) and of the OV. This last contribution is the only one derived from n-type Si spectra since the OV signature is not detectable in p-type samples. The total interstitial concentration is obtained by adding the concentrations of C_iO_i, B_iC_s and H_1. The results are reported in Fig. 6 for both I-type (■) and V-type (▲) defects, calculated for ion implanted samples isochronally (30 min) annealed at temperatures in the range 100-550 °C, in 50 °C steps. The lines, solid for interstitials and dotted and dashed for

Figure 6: *Total concentration of I (■) and V (▲),stored in defect complexes, as a function of annealing temperature for epi Si implanted with 1.2 MeV Si, $1 \times 10^9/cm^2$ and total concentration of I (★) and V (◊) stored in defect complexes for electron irradiated epi Si with $3.5 \times 10^{15}/cm^2$, 9.2 MeV.*

vacancies, are only guides to the eye. The right hand scale of the plot represents the number of defects per implanted ion, N, calculated as:

$$N = \frac{C_d x}{\Phi} \qquad (1)$$

where C_d is the average defect concentration in the region probed during the DLTS measurements, x is the thickness of this region (~ 1 μm) and Φ is the implanted fluence (1×10^9 /cm^2, in Fig. 6).

Although each ion produces ~ 2500 Frenkel pairs (as calculated by TRIM [24]), only ~ 60 per ion escape recombination and form room temperature stable defects. Thermal annealing at temperatures up to 300 °C produce a concomitant annealing of I and V-type defects. Indeed, an equal number of V-type and I-type defects remain although their absolute concentration has been reduced by approximately one order of magnitude. This confirms that, as expected from Monte Carlo simulations [15], most of the defect recombination occurs in the bulk, thus maintaining the balance between interstitials and vacancies. If surface annihilation of I or V had been dominant, the balance between vacancies and interstitials would not have been maintained.

At temperature above 300 °C all of the vacancy-type clusters have been annealed out, but a measurable number of interstitials (~ 3 per implanted ion) is still present in the samples. We suggest that this unbalance between interstitial and vacancy-type defects is due to the extra incorporated ion [11], and it only becomes experimentally detectable when most of the vacancies have recombined. This hypothesis is confirmed by the observation that this imbalance was not observed in electron irradiated samples where I- ($\star$) and V-type ($\Diamond$) defects continue to anneal concomitantly. Despite the fact that the two samples have the same as implanted spectrum (see Fig. 2b), after annealing at 350 °C the difference in the two samples is experimentally detectable.

Interstitial point-like defects

It is now interesting to apply the defect counting method we developed to samples implanted a higher fluences. Implantation at high fluences results in the deterioration of the junction characteristic and in a defect concentration comparable or higher than the doping concentration. It is not possible to perform DLTS measurements in the as-implanted samples. However, after annealing at 400 °C, most o the damage recovers and DLTS measurement are again possible. Is this region, at annealing temperatures of 350-500 °C after all the vacancies anneal, *interstitial point-like defect* form. These point defects store the interstitia "excess" caused by the presence of the extra ion.

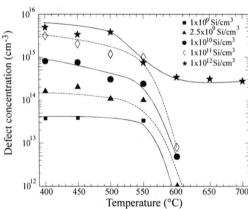

Figure 7: *Residual defects concentration as a function of the temperature for epitaxial Si implanted with 1.2 MeV Si at doses of: $1 \times 10^9 cm^{-2}$ (■), $2.5 \times 10^9 cm^{-2}$ (▲), $1 \times 10^{10} cm^{-2}$ (●), $1 \times 10^{11} cm^{-2}$ (◊), $1 \times 10^{12} cm^{-2}$ (★).*

In Fig. 7 we summarize the results o defect counting on samples that have been annealed for 30 min at temperatures in th range 400-700 °C.

Si samples implanted with 1.2 MeV Si a fluences of 1×10^9 cm^{-2} (■), 2.5×10^9 cm (▲), 1×10^{10} cm^{-2} (●), 1×10^{11} cm^{-2} (◊), 1×10 cm^{-2} (★). The solid and dashed lines are use as a guide to the eye. For low fluences (up t 1×10^{11} cm^{-2}) the defect concentration increas roughly linearly with ion fluence so that, th number of interstitial per ion, calculate according to Eq. (1) remains approximately constant (~ 3) [16, 28]. At temperatures above 550 °C the residual damage is fully recovered.

In contrast, for doses $\geq 1 \times 10^{12}$ Si/cm^2 (★), the defect concentration increase much less than linearly, suggesting that, probably as a consequence of interstitial clustering, the simple defect counting procedure used for low fluence samples, is no longer valid. Upon further annealing at temperatures ≥ 550 °C, the residual damage is still present in the samples, in contrast with lower doses. This result suggests that in this case more stable structures are formed. Furthermore, as detailed in the next paragraph, the shape of the DLTS peaks suggest that defect clusters are formed.

Interstitial clusters

DLTS characterization of ion implanted Si at a dose of 1×10^{12} Si/cm^2 provides evidence of interstitial clusters. The same defect structures were found for a wide dose range, $1 \times 10^{12} - 5 \times 10^{13}$ Si/cm^2 annealing temperatures, 550-800 °C, and times, 10min-15h, regardless of the impurity content and the doping level of the sample. The last evidence suggest that impurities, O or C or B, are not the main constituents of these defects [16].

A typical DLTS spectrum is reported in Fig. 8 for an epitaxial Si sample implanted with 1.2 MeV Si to a fluence of 1×10^{12}/cm^2 and annealed at 600 °C 30 min (solid line), and similar spectra are seen in CZ Si samples. Two defect signatures labeled in figure as B_1, at $E_V + 0.33$ eV and B_2, at $E_V + 0.52$ eV (B-lines) have been detected.

The features of the DLTS peaks provide a confirmation that the B-lines signals are related to a complex defect structure. In fact, simulation of a point-like defect peak (dashed line) having the same capture cross section and the same activation energy as the cluster signals is shown for comparison. The measured peaks are much broader than expected. To successfully reproduce the experimental data a simulation assuming that a narrow defect band of 20 meV (S factor), rather than a single level, present in the Si band gap, was performed [29]. The result, shown as a dotted line in Fig. 8, well describes the experimental data. Broad DLTS peaks have been previously associated [30] with complex defects, and are a result of a distribution in activation energies for carrier emission from these defects. Further studies are in progress to determine how the S factor evolves during annealing and to relate it to a specific cluster feature.

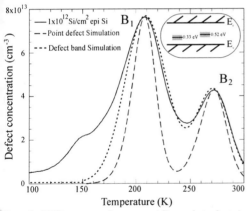

Comparison of the depth profiles of the B-line defects with the Frenkel pair profile generated by the implanted ion provides additional information about the nature of the clusters. We monitored the depth profile distribution of a point-like interstitial defect, C_iO_i (●) and of a cluster, B_1 (■) and compared them with the Frenkel pair and the extra ion distributions simulated by MARLOWE [31], a Monte Carlo code. The results are reported in Fig. 9(a) and (b). The experimental profile of C_iO_i, reported in Fig. 9a precisely mirrors the simulated Frenkel pair profile, while the extra ion profile is narrower and slightly deeper in depth. On the other hand, the B_1 depth profile,

Figure 8: DLTS spectrum of a p-type epi Si sample implanted with 1.2 MeV Si to a fluence of 1×10^{12}/cm^2 and annealed at 600 °C 30min (solid line). Comparison with simulation of the DLTS spectrum for point-like defects (dashed line) and of defect band (dotted line) suggest that the defects B_1 and B_2 are clusters.

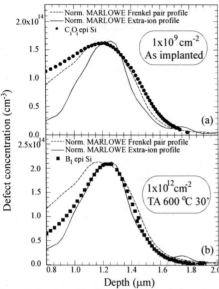

Figure 9: Depth profiles, in p-type epi Si samples, of:
(a) C_iO_i for 1.2 MeV Si, $1x10^9$ cm^{-2}, as
implanted; (b) B_1 for 1.2 MeV Si, $1x10^{12}$ cm^{-2}
annealed at 600 °C for 30 min. The lines are
the Extra ion profile (solid line) and the
Frenkel pair (dashed line) simulations both
calculated with MARLOWE.

Fig. 9b, reaches its maximum at ~ 1.3 μm, which is
the position of the implantation end of range for 1.2
MeV Si, and precisely mirrors the simulation profile
of the extra implanted ion (solid line, divided by a
factor 100).

The evidence that defect clusters form in the
end of range region, which is the region that
experiences the maximum super-saturation of
interstitials, their independence from the
concentration of O, C or B, and the cluster nature of
the defect signatures, support the conclusion that the
B-line defects identified by DLTS are Si interstitial
clusters. Annealing kinetics of these defects are
consistent with measurements of TED at low doses,
below the threshold of extended defect formation.
[32].

Finally, although the interstitial cluster defects
do not contain large numbers of C or O, we have
found that the presence of these impurities in the Si
sample do modify the kinetics of defect formation;
B-lines are formed at lower doses in CZ Si [16]. It
can be speculated that impurity atoms stabilize small
interstitial clusters, preventing their dissolution until
the arrival of other interstitials whereas in epitaxial
Si the clustering process occurs only when the
interstitial super-saturation is large enough to allow
the rapid formation of big clusters that will survive
the annealing process.

Extended defects

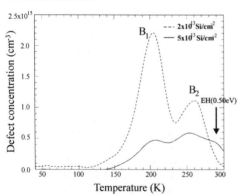

Figure 10: DLTS Spectra of residual damage measured
after 145 keV Si ion implantation at $2x10^{13}$ Si/cm^2
(dashed line) and $5x10^{13}$ Si/cm^2 (solid line) on p-
type Si and annealing treatment at 680 °C 1 hr.

At high implantation fluences, $\geq 5x10^{13}$
Si/cm^2, and annealing temperatures ≥ 680 °C 1hr
a new electrical signature is found in the residual
damage spectrum (solid line) reported in Fig. 10.
Beside the B-lines signatures, still present for
these high fluences, a shoulder, marked with an
arrow in Fig. 10, is visible in the high
temperature part of the spectrum. The measured
activation energy of this peak is at $E_V+0.50$ eV
(EH). In Fig. 10 we also report the DLTS
spectrum of a sample implanted at a lower
fluence ($2x10^{13}$ Si/cm^2) which underwent the
same annealing process. TEM analyses show that
{311} defects are present in the $5x10^{13}$ Si/cm^3
sample but not in the $2x10^{13}$ Si/cm^2. Detailed
DLTS studies of EH evolution as a function of
the annealing time [27] compared with TEM

analyses on the same samples show that the signal observed in DLTS is an electrical signature of the {311} defect. Moreover, such defects show logarithmic capture kinetics, typical of DLTS markers of extended defects [30]. It is interesting to observe that while B-line concentrations increase almost linearly with increasing ion implantation dose from 1×10^{12} Si/cm^2 (see Fig. 8) to 2×10^{13} Si/cm^2 (dashed line, Fig. 10), these signals abruptly decrease as soon as extended defects are formed. These results show that B-lines formation is inhibited [27] by extended defect formation because the two defects compete in storing the interstitial excess.

Conclusions

The comparison of defect evolution in ion implanted and electron irradiated Si samples provided several insights on the mechanisms of damage accumulation and annealing in crystalline Si. The main results can be summarized as follows: at low implantation fluences and annealing temperature below ~ 350 °C no difference is observed in the defect structure and annealing behavior of ion implanted and electron irradiated samples. The impurity content of the substrate (O, C) strongly influences point defect migration, room temperature stable defect structures and their annealing behavior. Annealing produces a concomitant reduction of I- and V-type defect complexes, demonstrating that defect recombination occurs preferentially in the bulk.

For annealing above 350 °C of ion implanted samples, the residual damage is given by I-type defects storing the interstitial excess caused by the extra implanted ion (~ 3 interstitials per ion)

For higher implantation fluences (10^{12}-10^{13} Si/cm^2) and annealing temperatures (550-700 °C) the residual damage produces two well defined DLTS signatures at E_V+0.33 eV and E_V+0.52 eV, which have been associated with interstitial clusters produced by the agglomeration of the excess implanted interstitials.

At high implantation fluences, $\geq 5 \times 10^{13}$ Si/cm^2, and annealing temperatures, ≥ 680 °C 1hr, extended defects, such as {311} have been observed by TEM and correlated to a new electrical signature, at E_V+0.50 eV, found in the DLTS spectra. Extended defects compete with small clusters in storing the interstitial excess.

Acknowledgments

We would like to acknowledge L. Pelaz and G. Gilmer for many useful discussions, D. C. Jacobson and A. Marino for the ion implantations, P. G. Fuochi and M. Lavalle for the electron irradiation and A. Spada for the important collaboration in the realization of the DLTS simulation program. The expert technical assistance of N. Parasole is also acknowledged. This work was partially supported by Progetto Finalizzato MADESS II and Progetto 5% Microelettronica.

References

1. N. E. B. Cowern, K. T. F. Janssen, H. F. F. Jos, J. Appl. Phys. 68, 6191 (1990).
2. K. S. Jones, S. Prussin and E. R. Weber, Appl. Phys. A 45, 1, (1988).
3. L. C. Kimerling, in Radiation Effects in Semiconductors, ed. by N. B. Urli and J. M. Corbett (Inst. Of Phys. Conf. Ser. 31, London 1977), p. 221.
4. M. T. Asom, J. L. Benton, R. Sauer and L. C. Kimerling, Appl. Phys. Lett. 51, 256 (1987).
5. J. L. Benton, M. T. Asom, R. Sauer and L. C. Kimerling, Mat. Res. Soc. Symp. Proc. 104, 85 (1988).
6. G. D. Watkins and J. W. Corbett, Phys. Rev. 121, 1001 (1961).
7. S. Libertino, J. L. Benton, D. C. Jacobson, D. J. Eaglesham, J. M. Poate, S. Coffa, P. G. Fuochi, and M. Lavalle, Appl. Phys. Lett. 70 (22), 3002 (1997).

8. B. G. Svensson, C. Jagadish, A. Hallèn and J. Lalita, Nucl. Instr. Meth. B106, 183 (1995).
9. J. Lalita, N. Keskitalo, A. Hallèn, C. Jagadish and B. G. Svensson, Nucl. Instr. Meth. 120, 27, (1996).
10. B. G. Svensson, B. Mohadjeri, A. Hallèn, J. H. Svensson and J. W. Corbett, Phys. Rev. B 43, 2292 (1991).
11. M. Giles, J. Electr. Soc. 138, 1160 (1991).
12. K. Seshan and J. Washburn, Rad. Eff, 37, 147, (1978).
13. P. A. Stolk, H.-J. Gossmann, D. J. Eaglesham, D. C. Jacobson, C. S. Rafferty, G. H. Gilmer, M. Jaraiz, J. M. Poate and T. E. Haynes, J. Appl. Phys. 81, 6031, (1997).
14. N. E. B. Cowern, G. F. A. van de Walle, P. C. Zalm and D. W. E. Vandenhoudt, Appl. Phys. Lett. 65, 2981, (1994).
15. M. Jaraiz, G. H. Gilmer, J. M. Poate, T. D. de la Rubia, Appl. Phys. Lett. 68, 409 (1996).
16. J. L. Benton , S. Libertino, S. Coffa and D. J. Eaglesham, Mat. Res. Soc. Symp. Proc. 469, 193 (1997).
17. G. D. Watkins, Phys. Rev. B 12, 5824 (1975).
18. C. A. Londos and J. Grammatikakis, in Phys. Stat. Sol. (a) 109, 421, 1988.
19. P. J. Drevinski, C. E. Caefer, S. P. Tobin, J. C. Mikkelsen, Jr., and L. C. Kimerling, in Mat. Res. Soc. Symp. Proc., Vol. 104, ed. by M. Stavola, S. J. Pearton, and G. Davies (Materials Research Society, Pittsburgh, PA, 1988), p. 167.
20. P. J. Drevinski, C. E. Cafaer, L. C. Kimerling and J. L. Benton, Proceeding of the International Conference on the Science and Technology of Defect Control in Semiconductors, ed. by K. Sumino (Elsevier Science, North Holland, Amsterdam, 1990) p. 341.
21. P. J. Drevinski and H. M. De Angelis, Thirteenth International Conference on Defects in Semi-conductors, ed. by L. C. Kimerling and J. M. Parsey, (The Metallurgical Society of AIME, Warrendale, PA, 1985), p. 807.
22. S. Coffa, V. Privitera, F. Priolo, S. Libertino and G. Mannino, J. Appl. Phys. 81, 1639 (1997).
23. S. Libertino, S. Coffa, V. Privitera and F. Priolo, Mat. Res. Soc. Symp. Proc. 438, 65 (1997).
24. J. P. Biersack and L. G. Haggmark, Nucl. Instr. Meth. 174, 257 (1980).
25. L. C. Kimerling, Inst. Phys. Conf. Ser. 31, 221 (1977).
26. S. Libertino, J. L. Benton, D. C. Jacobson, D. J. Eaglesham, J. M. Poate, S. Coffa, P. G. Fuochi and M. Lavalle, Appl. Phys. Lett. 71, 389 (1997).
27. J. L. Benton , S. Libertino, P. Kringhøi, D. J. Eaglesham, J. M. Poate and S. Coffa, J. Appl. Phys. 82, 120 (1997).
28. S. Libertino, J. L. Benton, S. Coffa, D. C. Jacobson, D. J. Eaglesham, J. M. Poate, M. Lavalle and P. G. Fuochi, Mat. Res. Soc. Symp. Proc. 469, 187 (1997).
29. P. Omling, L. Samuelson and H. G. Grimmeiss, J. Appl. Phys. 54, 5117, (1983).
30. J. R. Ayres, S. D. Brotherton, J. Appl. Phys. 71, 2702, (1992).
31. M. T. Robinson and J. M Torrens, Phys. Rev. B 9, 5008, (1974).

IN-SITU DLTS MEASUREMENT OF METASTABLE COUPLING BETWEEN PROTON-INDUCED DEFECTS AND IMPURITIES IN n-Si

K. Kono, N. Kishimoto and H. Amekura
High Resolution Beam Research Station
National Research Institute for Metals
1-2-1 Sengen, Tsukuba, Ibaraki, 305, Japan, kono@nrim.go.jp

ABSTRACT

Deep centers related to proton-induced defects and impurities in n-Si were investigated using an in-situ DLTS system. The measurement system was installed in the beamline of NRIM cyclotron. Floating zone (FZ) or Czochralski (CZ) Si was irradiated at a flux of about 30 nA/cm^2. The total dose was varied between $1.0 \sim 7.2 \times 10^{13}$ ions/cm^2. Samples were irradiated at 200 K or 300 K. Their defect concentrations were determined by in-situ DLTS. Quantitative relationship between oxygen and phosphorus concentration and defect creation rates were presented. Isothermal annealing at 250 K was conducted for a CZ-Si sample and the result indicated the meta-stable nature of the carbon defect complex.

INTRODUCTION

Deep levels related to radiation-induced defects in Si have been extensively studied, up to now. As is well known, deep levels cause the deterioration of Si device performance. Most of the irradiation experiments were conducted using electron irradiation. The nature of the deep centers (such as energy levels or recovery temperature) was well established [1,2]. The reported defects can be summarized as follows. The vacancy-oxygen complex (V-O) has an energy level of 0.16 eV below the conduction band edge, E_c, and recovers at 620-670 K [3]. The energy level of divacancy is E_c-0.23 eV (V-V^{2-}) or E_c-0.39 eV (V-V$^-$) and recovers at 550-610K [4]. Phosphorus-vacancy complex (P-V) has an energy level of E_c-0.43 eV and recovers at 370-420 K [3]. Carbon reacts with defects and creates various deep centers in Si and they show meta-stability [5-8]. Similarly, hydrogen atoms implanted with low-energy protons form deep centers [9,10].

In-situ DLTS is useful to evaluate evolution of such meta-stable centers by irradiation. High-energy protons create collision cascades and possibly give rise to more complex defects, compared to electrons. High-energy protons enable to extract the effect of proton-induced defects without implanted hydrogens.

In this paper, the real-time evolution of deep centers in n-type Si under 17 MeV proton irradiation has been studied with an in-situ DLTS device which is installed in the beamline. Experiments were focused on: (1) interactions between impurities and defects, (2) metastability of carbon-related complexes. For the former, FZ-Si was used as well as CZ-Si to vary the oxygen concentration. For the latter aspect, isothermal annealing was conducted to study the evolution of the carbon-related complex.

EXPERIMENTAL PROCEDURES

Sample preparation

Samples were cut from commercially available Si wafers. A CZ-type Si of resistivity ~10 Ω·cm (phosphorus concentration: 7×10^{14} cm^{-3}) was used. An FZ-type Si wafer had resistivity 3-5 Ω·cm (phosphorus concentration: 3×10^{15} cm^{-3}). Phosphorus concentration was determined by the C-V characteristics. Samples were cut at rectangular shape. Gold was vacuum-evaporated to form a Schottky diode on the front surface. An ohmic contact was formed on the other surface by nickel plating.

Irradiation and measurement

A sample was set within the irradiation chamber of the NRIM cyclotron. A proton beam of 17 MeV was used to irradiate the sample. A He-gas circulator and a precise temperature controller were used to keep the sample temperature constant. In-situ DLTS measurement was conducted right after irradiation or after post-irradiation annealing. The proton beam of 17 MeV has a projectile range of 1.7 mm in Si, according to SRIM calculation [11]. This projectile range is so large that protons do not remain within the sample. This also guarantees a uniform depth profile of defect concentration.

The beam current density of proton was about 30 nA/cm^2 and the total fluence was varied from 1.0×10^{13} ions/cm^2 (5.8×10^{-8} dpa) to 7.2×10^{13} ions/cm^2 (4.2×10^{-7} dpa). The conversion relation used was 1.0 μA·s/cm^2 (6.24×10^{12} ions/cm^2) = 3.6×10^{-8} dpa for 17 MeV proton. The proton irradiation was conducted either at 200 K or 300 K.

All the DLTS measurements were conducted under majority carrier injection. The Schottky diode was biased with a reverse voltage of 4 V and the injection pulse was 1 ms of 4 V which canceled the reverse bias. The delay time was varied from 1 ms through 256 ms.

DLTS measurements were conducted right after irradiation or after annealing. When the samples were irradiated at 200 K, the upper limit for the temperature scan was fixed to 200 K to avoid spontaneous annealing of deep centers.

After proton irradiation of CZ-Si at 200K, an isothermal annealing below room temperature was conducted at 250 K. It was known from a previous experiment that carbon-related defects change quickly at 300 K [12]. Annealing temperature was chosen to be 250 K to slow down the transformation process.

EXPERIMENTAL RESULTS

In order to study the effect of oxygen concentration, an FZ-type sample was irradiated as well as CZ-samples. From atomic absorption analysis, oxygen concentration of CZ-Si was determined to be 2×10^{18} cm^{-3}. To eliminate oxygen overestimation in atomic absorption analysis, relative oxygen density between the CZ- and FZ-Si was determined by Fourier Transform Infrared (FT-IR) measurement. It was concluded that FZ-Si has about 10 times lower oxygen concentration than the CZ-Si.

A typical DLTS spectrum of CZ-Si is shown in Fig. 1, where 17 MeV protons were irradiated at 300 K up to 4.4×10^{13} ions/cm^2. Four types of deep centers (E1, E2, E3, E4) were observed in this spectrum. These levels were assigned to the respective states, in comparison with the published data of the deep levels by electron or proton irradiation

16

[1,8]. The trap E1 is located at 0.12 eV below the conduction band and corresponds to complex of substitutional carbon and Si interstitial $(C_s\text{-}Si_i\text{-}C_s)$ [8]. The E2 trap is at 0.18 eV below the conduction band and corresponds to a vacancy oxygen (V-O) cluster [1]. The E3 is 0.25 eV deep and corresponds to the doubly negative divacancy $(V\text{-}V^{2-})$ [1]. The E4 peak is the sum of vacancy phosphorus cluster (V-P) and singly negative divacancy $(V\text{-}V^-)$ which is 0.41 eV below the conduction band. As far as the energy positions are concerned, there appeared the same peaks for the protons as those for electrons.

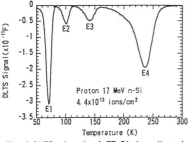

Fig. 1 DLTS signal of FZ-Si irradiated at 300 K. Beam current was about 30 nA/cm².

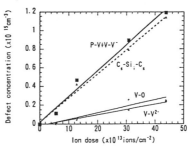

Fig. 2 Defect concentration against proton dose. Defect concentration was determined from peak intensity of DLTS spectra and C-V characteristics.

Figure 2 shows the defect density for the FZ-Si, plotted against proton dose. Irradiation temperature was kept at 300 K. Phosphorus concentration of the FZ-Si sample is 3×10^{15} cm^{-3}. The defect concentration was determined from the peak intensity. The transient capacitance change was compared to the capacitance at the same reverse biased condition. The concentrations of all the deep centers are quantitatively proportional to the irradiation dose. From Fig. 2, linear relationship between irradiation dose and defect concentration was confirmed and defect-creation rate was derived. The defect-creation rate calculated is for the *primary* defects (V,I) which are directly created by collision. The detail is summarized in Table 1.

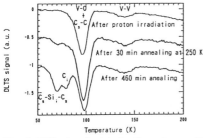

Fig. 3 DLTS signal with isothermal annealing at 250 K. Sample (CZ-Si) was irradiated at 200 K by 17 MeV proton at the dose of 1.3x10¹³ ions/cm².

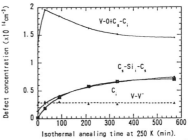

Fig. 4 Change of defect concentration by isothermal annealing at 250 K.

Change of DLTS spectrum after isothermal annealing at 250 K is shown in Fig. 3. Phosphorus doped Si of CZ type was irradiated at 200 K. The total dose was 1.2×10^{13} ions/cm². In the post-irradiation measurement, no obvious deep centers related to carbon were observed. The change of transient capacitance response at 200 K became smaller after post-irradiation annealing. This suggests the existence of carbon- interstitial complex such as C_i-P_s [8]. The recovery of defects is not attributable to the change of either P-V or V-V, because their recovery temperatures are much higher [3,4]. The measurement following isothermal annealing showed the emergence and the transformation of the meta-stable deep centers related to carbon.

Figure 4 shows change in the concentration of the deep centers against isothermal annealing time at 250 K. In-situ measurements were conducted before and after annealing. No obvious carbon-related centers were observed in the measurement following irradiation. After 30 min annealing, a C_s-C_i (0.18 eV) peak emerges accompanying small peaks of C_i (0.16 eV) and C_s-Si_i-C_s (0.12 eV). As annealing time increases, the peak related to C_s-C_i disappears. At the same time, peaks C_i and C_s-Si_i-C_s increase. The sum of number densities, C_s-C_i, C_i and C_s-Si_i-C_s, is almost constant except the first 30 min annealing. This shows that C_s-C_i clusters are transformed into C_i and C_s-Si_i-C_s.

DISCUSSION

During proton irradiation, point defects such as Si interstitials (I) and vacancies (V) are primarily created. This primary process is not sensitive to the irradiation temperature or impurity concentration. The vacancies and interstitials quickly react with each other and with sinks and impurities. Subsequently, the defects remain in the form of clusters (residual defects), i.e., V-O, V-V, V-P, etc. around room temperature. These reaction processes are sensitive to the temperature and impurity concentration. The in-situ DLTS measurement detects the residual defects.

Defect-creation rates of residual defects are summarized in Table 1. The defect-creation rates are estimated as a ratio against the defect-creation rate of *primary* defects. The creation rate for E4 was dissolved into two components of divacancy and P-V, by analyzing C-t characteristics and by curve fitting.

Table I: Defect creation rate for FZ- and CZ-Si.

Irradiation temperature	300 K		200 K	
Type of Si	FZ	CZ	FZ	CZ
P-V (0.43 eV)	6.9×10^{-3}	1.9×10^{-3}	N/A*	N/A*
V-V$^-$ (0.39 eV)	3.5×10^{-3}	1.9×10^{-3}	N/A*	N/A*
V-V^{2-} (0.25 eV)	4.8×10^{-3}	2.4×10^{-3}	2.9×10^{-3}	1.4×10^{-3}
V-O (0.18 eV)	1.9×10^{-3}	7.2×10^{-3}	7.2×10^{-3}	4.5×10^{-3}
C_i (0.16 eV)	0	6.5×10^{-4}	0	0
C_s-Si_i-C_s (0.12 eV)	9.4×10^{-3}	0	7.2×10^{-3}	0

*Defect-creation rate for P-V or V-V is not available for irradiation at 200 K because DLTS scan was ranged between 50 K and 200 K to avoid spontaneous annealing of the defects.

The differences in creation rates for each sample and irradiation temperature arise from reactions of primary defects and impurities. Therefore, it is necessary to consider the elementary processes.

The reactions are as follows:

$$V+I \;\rightleftharpoons\; \emptyset \qquad (1)$$
$$V+S \;\rightleftharpoons\; S \qquad (2)$$
$$I+S \;\rightleftharpoons\; S \qquad (3)$$
$$V+O \;\rightleftharpoons\; V\text{-}O \qquad (4)$$
$$V+V \;\rightleftharpoons\; V\text{-}V \qquad (5)$$
$$V+P \;\rightleftharpoons\; V\text{-}P \qquad (6)$$
$$Si_i+C_s \;\rightleftharpoons\; C_i \qquad (7)$$
$$2C_s+Si_i \;\rightleftharpoons\; C_s\text{-}Si_i\text{-}C_s \qquad (8)$$
$$C_s+C_i \;\rightleftharpoons\; C_s\text{-}C_i \qquad (9),$$

where S stands for sinks, e.g., the surface. The symbol "I" stands for the Si interstitial Si_i.

Those equations are used to evaluate the competitive processes between the reactions to form clusters. It is noted that FZ-Si has only one-tenth oxygen concentration as much as CZ-Si. Phosphorus concentration for samples were: FZ-Si 3×10^{15} cm^{-3}, CZ-Si 7×10^{14} cm^{-3}. Most defects annihilate through the reactions shown in Eqns. (1) through (3). Only a small portion of vacancies and interstitials survive the annihilation. The number of surviving vacancies is fewer than the phosphor or oxygen atoms in our experiment. Thus those reactions compete witheach other. From Table 1, the followings are pointed out: First, creation rate for divacancy is two times larger for FZ-Si than for CZ-Si. Therefore, the high oxygen concentration hinders formation of the divacancy because of the competitive processes (4) and (5). On the other hand, FZ-Si has more than four times higher phosphorus concentration than for CZ-Si. This suggests that the competition between (5) and (6) are not so much as (4) and (5). Second, linear dependence for the defect concentration against proton dose was confirmed. Third, disappearance of the carbon C_i peak in CZ-Si irradiated at 200 K may be attributed to the different carbon complex (C_i-X, where X is an impurity). This idea is supported by the fact that the transient change of capacitance at 200 K decreased after 30 min annealing at 250 K. Fourth, both CZ- and FZ-Si irradiated at 200 K show a lower creation rate. This can be understood as a decrease in vacancy mobility at a lower temperature. Finally, the defect-creation rate is compared to the result obtained by photoconductivity measurement [13]. The total number of residual defects was experimentally obtained to be 0.03 against the primary defects. The ratio for the DLTS measurement was 0.02 for FZ-Si and 0.01 for CZ-Si. Although the experimental methods of photoconductivity and DLTS are quite different, the both results show good quantitative agreement.

An isothermal annealing was conducted at 250 K. Without annealing, deep levels which corresponded to the C_i, C_s-Si_i-C_s and C_s-C_i were not observed. The spectra in Fig. 3 show a decrease in the transient response at 200 K during 30 min annealing. This suggests that C_i-X forms right after the irradiation. After annealing for 30 min, they dissolve and become C_i or C_s-C_i (9). Some react with Si and create Si_i (inverse of (7)). Those Si_i atoms react with C_s and form C_s-Si_i-C_s (8). As the annealing time increases, C_s-C_i dissolves and C_i and C_s-Si_i-C_s increase. This is why the sum of the concentrations of C_s-C_i, C_i and C_s-Si_i-C_s is almost constant. Thus, changes of meta-stable coupling between carbon and defects were directly observed by the in-situ DLTS measurement.

In-situ DLTS measurement was conducted for 17 MeV proton-irradiated CZ- and FZ-Si to study the interaction between impurities and proton induced defects. Relationship between oxygen or phosphorus concentration and proton dose was studied. The higher oxygen concentration in CZ-Si was found to hinder formation of divacancy.

The meta-stable nature of the carbon related complex was studied by isothermal annealing at 250 K after irradiation. Mutual transformation of meta-stable structures (C_i, C_s-C_i and C_s-Si_i-C_s) was successfully observed by virtue of the in-situ DLTS measurement.

REFERENCE

[1] A. Hállen, D. Fenyo, B.U.R. Sundqvist, R.E. Johnson and B.G. Johnson, J. of Appl. Phys. 70 (6), 3025 (1991)

[2] B.G. Svensson, C. Jagadish, and J.S. Williams, Phys. Rev. Lett. 71 (12), 1860 (1993)

[3] J.W. Walker and C.T. Sah, Phys. Rev. B 7 (10), 4587 (1973)

[4] A.O. Evwaraye and Edmund Sun, J. Appl. Phys., 47 (9), 3776, (1976)

[5] L.W. Song, B.W. Benson and G.D. Watkins, Appl. Phys. Lett. 51 (15), 1155 (1987)

[6] G. Ferenczi, C.A. Londos, T. Pavelka, M. Somogyi and A. Mertens, J. Appl. Phys. 63 (1), 183 (1988)

[7] A. Chantre and L.C. Kimerling, Appl. Phys. Lett. 48 (15), 1000 (1986)

[8] T. Asom, J.L. Benton, R. Sauer and L.C. Kimerling, Appl. Phys. Lett. 51 (4), 256 (1987)

[9] B. Holm, K. Bonde Nelsen and B. Bech Nielsen, Phys. Rev. Lett., 66 (18), 2360 (1991)

[10] K. Irmscher, H. Klose and K. Maass, J. Phys. C: Solid State Phys. , 17, 6317 (1984)

[11] J.F. Ziegler, J.P. Biersack and U. Littmark, The Stopping and Range of Ions in Solids, (Pergamon Press, New York, 1985), Chap. 8

[12] K. Kono, N. Kishimoto, H. Amekura and T. Saito, Mat. Res. Soc. Proc. 442, 287 (1997)

[13] H. Amekura, N. Kishimoto, K. Kono and T. Saito, Mat. Sci. Forum, 196-201, 1159 (1995)

METASTABLE AMORPHOUS SiO$_2$ CREATED BY ION BOMBARDMENT

Koichi Awazu*, Sjoerd Roorda, and John L.Brebner
Groupe de recherche en physique et technologie des couches minces (GCM), Département de Physique, Université de Montréal, P.O. Box 6128, Station "Centre-Ville", Montréal, Québec H3C 3J7, Canada.
*Permanent address, Electrotechnical Laboratory, 1-1-4, Umezono, Tsukuba 305, Japan

Structural changes in thin amorphous SiO$_2$ (a-SiO$_2$) films induced by ion bombardment were examined with FT-IR and ESR techniques. 10MeV H$^+$, 15MeV He$^+$, 2MeV Li$^+$, 4MeV C$^+$ and 30MeV Si^{6+} ions, which traverse the 200nm thick of SiO$_2$ layer without chemical reaction with SiO$_2$, were used as ion beams with dosage of 10^{12}-10^{16}cm^{-2}. A decrease in the frequency of one peak in the infrared absorption spectrum from 1078cm^{-1} to 1041cm^{-1} and an increase of a second peak from 805cm^{-1} to 830cm^{-1} were simultaneously observed with cumulative ion dose. The concentration of E' centers has a maximum as a function of dosage in the region of 1-5x10^{19}cm^{-3}. We propose that a reduction of the Si-O-Si bond angle followed by displacement of oxygen is induced with ion bombardment.

INTRODUCTION

It is technologically as well as scientifically important to understand ion-beam-induced phenomena in silica glass (a-SiO$_2$). Yet, we still know very little about the structural changes in a-SiO$_2$ induced by MeV ions. For example, structural defects in latent tracks created with MeV ions remain unidentified. In this paper, structural changes in a-SiO$_2$ were examined with infrared (IR) spectroscopy and electron spin resonance (ESR) using various kinds of MeV ion beams. The results are discussed in term of densification and oxygen vacancies.

EXPERIMENT

SiO$_2$ films were formed on (100) silicon wafers by thermal oxidation at 1000°C for 3 hours. Commercial silicon wafers are usually doped with B or P and also contain oxygen impurities. To avoid oxygen impurity and dopants in silicon wafers, we used pure silicon wafers having high resistance and fabricated by the Floating Zone method and suitable for FT-IR and ESR. This specially made silicon wafer was fabricated by Shin-Etsu Semiconductor Co. LTD.. Oxidation was performed to obtain ≈200nm thick oxide film. Ion bombardment was performed at room temperature and at a residual pressure below 1×10^{-4}Pa. 10MeV H$^+$, 15MeV He$^+$, 2MeV Li$^+$, 4MeV C$^+$, 20MeV Si^{4+} , and 30MeV Si^{6+} beams were generated with the Université de Montréal 6MV Tandem accelerator. Electron spin resonance (ESR) measurements were carried out with a standard X-band Brucker ESP-300E spectrometer at room temperature. Absolute spin concentrations were obtained by double numerical integration of the experimental first-derivative spectra. The optimum power needed to obtain the spectra was 20μW. Relative

Mat. Res. Soc. Symp. Proc. Vol. 504 © 1998 Materials Research Society

error of the ESR signal intensity is within 30%. Infrared (IR) absorption was measured with a Fourier transform IR spectro-photometer using light at normal incidence. Isochronal annealing was performed at 350, 500, 600, and 800°C for 2 hours at a residual pressure below 1×10^{-4}Pa to avoid the effect of oxidation by residual air.

RESULTS

Fig. 1 shows the IR absorption spectra of an a-SiO$_2$ film before and after irradiation with 2MeV Li$^+$. There are two first order absorptions in the IR absorption spectra at 1078cm^{-1} and at 805cm^{-1}, which were labeled ω_4 and ω_3, respectively [1]. A decrease of frequency of ω_4 and an increase of frequency of ω_3 were observed to occur simultaneously with increase of dosage.

Fig.2 shows the frequency shift of ω_4 in the IR absorption spectra as a function of dosage. The closed circle stands for the frequency peak at 1078cm^{-1} of a virgin sample as occurring at a dose of 10^{12}cm^{-2}. No change of the IR-spectrum contour was observed after

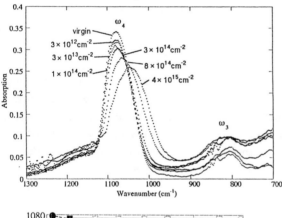

Fig.1 IR spectra of a-SiO$_2$ films before and after irradiation of 2MeV Li$^+$ ions. The two absorption peaks are labeled ω_3 and ω_4.

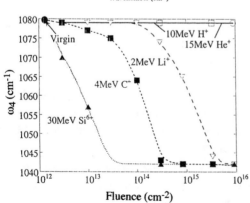

Fig.2 Frequency of ω_4 as a function of cumulative fluence of ion bombardment.

irradiation with either 10MeV H+ or 15MeV He+. In case of 2MeV Li+, 4MeV C+ and 30MeV Si^{6+} ions, in contrast, the frequency shift starts at dose of about 1×10^{14}cm^{-2}, 1×10^{13}cm^{-2} and 1×10^{12}cm^{-2}, respectively. The frequency shifts were saturated at 1042cm^{-1} and no more change was observed with increase of dosage.

Fig.3 shows the IR spectra of the irradiated samples as a function of annealing temperature. The spectrum contour did not change with 350°C treatment for 2 hours. A frequency shift started from 500°C treatment and the peak frequency with 800°C treatment came back to almost the same position as virgin sample.

Fig.4 presents the concentration of E' centers as a function of cumulative dosage. In the case of 10MeV H+ and 15MeV He+, the E' defect concentration was below the detection limit ($<1\times10^{18}$cm^{-3}). We found that the E' center concentration as a function of cumulative dose had a maximum at 3×10^{14}cm^{-2}, 1×10^{14}cm^{-2} and 1×10^{13}cm^{-2} for 2MeV Li+ , 4MeV C+, and 30MeV Si^{6+} ions, respectively.

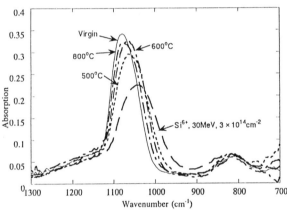

Fig.3 IR spectra of a-SiO with the heating temperature as a parameter. All thermal anneals were performed under residual pressure of 1×10^{-4}Pa and for 2 hours.

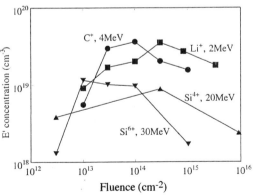

Fig.4 E' concentration as a function of cumulative fluence of ion bombardment.

DISCUSSION

The FT-IR spectra of a-SiO$_2$ exhibit two principal absorption bands connected to particular vibration modes: the bending vibration at about 805cm^{-1} (ω_3) and the asymmetric stretching mode at about 1078cm^{-1} (ω_4). In a central-force idealized continuous random network theory, the frequencies ω_3 and ω_4 are determined by the Si-O-Si bond angle [2]. The angular frequencies of the calculated modes as a function of bond angle are given by

$$\omega_3{}^2 = (\alpha/m_O)(1 + \cos \theta) + (4\alpha/3m_{Si}) \tag{1}$$

$$\omega_4{}^2 = (\alpha/m_O)(1 - \cos \theta) + (4\alpha/3m_{Si}) \tag{2},$$

where α is the central force constant, and m_O and m_{Si} are the atomic weights of O and Si, respectively.

Fig.5 shows the relation between ω_3 and ω_4 as determined from the IR spectra. Frequency ω_3 was increased and frequency ω_4 was decreased with increasing fluence. These results imply a decrease of θ from Eqs. (1) and (2). The solid line denotes the relation between ω_3 and ω_4 in a-SiO$_2$ at a range of densities, as has been reported in Ref [1]. Ion bombardment data in low dosage region corresponds to the data of densified silica. In contrast, ω_3 in high dosage region of ion bombardment is smaller than ω_3 in high pressure region of densified silica (left and upper side). We assume that the frequency shift can be explained by densification of silica induced by ion bombardment in the region of low dosage.

We now discuss what is happening in addition to densification as a function of dosage. As shown in Fig.4, the E' center concentration induced with each ion beam has maximum as a function of dosage. We notice that the dosage which gives a maximum peak of E' concentration also gives a frequency ω_4 of about 1060cm^{-1}. For example, in the case of 2MeV Li$^+$, 4MeV C$^+$ and 30MeV Si^{6+}, the maximum of E' concentration and 1060cm^{-1} value for ω_4 occur at fluence of about 3×10^{14}cm^{-3}, 1×10^{14}cm^{-3} and 1×10^{13}cm^{-3}, respectively. We do not think that the decrease of E' concentration in the high dosage region implies a decrease of the concentration of defects. The maximum concentration of E' centers for each ion is located in the range of $1-4\times10^{19}$cm^{-3} ($\approx$0.05-0.2mol%). When the E' center concentration exceeds 0.05-

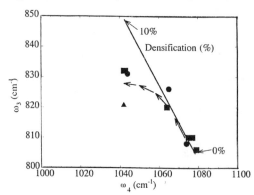

Fig.5 Relation between ω_3 and ω_4 in a-SiO$_2$ with ion bombardment. Arrows indicate increasing ion dose. Solid line denotes the relation in densified silica reported in Ref. [1].

0.2mol%, some E' centers transform to non-paramagnetic centers by interaction with one another (E'- E' interaction) . If we would implant a small amount of oxygen into the damaged layer, we would find an increase in paramagnetic center.

The IR spectrum of a-SiO$_2$ irradiated and annealed at 800°C is almost the same as that of unirradiated a-SiO$_2$. It is unlikely that residual oxygen acts as the terminator of defects because the residual pressure during all annealing procedure was below 1×10^{-4}Pa. We assume that this recovery is responsible for two things, the relaxation of densification on the one hand, and that knock-on oxygen produced by ion bombardment diffuses and compensates vacancies such as E' center, on the other.

It was found that IR spectrum was not further affected with very high dosage. It seems that a sort of metastable structure is reached in SiO$_2$. We speculate that the saturation of the ω_4 peak shift is due to a thermal equilibrium reaction. The densified structure induced by ion bombardment is simultaneously annealed with ion bombardment. Oxygen displaced by ion bombardment is also expected to undergo dynamic annealing.

It is not obvious which aspect of the ion - solid interaction causes the damage. For the ion - energy combinations investigated here, similar damage levels are reached by scaling the ion dose with a factor of 10 (Si to C or C to Li) or 100 (Si to Li). The nuclear stopping, electronic stopping, and momentum each are different by a factor of about 3 (Si to C or C to Li) or 9 (Si to Li). No single parameter can be identified, nor is it clear whether a combination of the above parameters plays a role or whether other factors such as dynamic annealing come into play as well. Nevertheless, it is remarkable that the evolution of the damage with increasing ion dose is similar for all three ions in spite of the fact that the nature of the damage changes from compaction (at low fluences) to predominantly oxygen vacancies (at higher fluences until saturation).

CONCLUSIONS

The initial response of pure a-SiO$_2$ to irradiation with multi-MeV ions is a compaction and a decrease in Si-O-Si bond angles, as induced from the concerted frequency shifts of two absorption at 1078cm^{-1} and 805cm^{-1}. Continued irradiation leads to saturation in the density of E' centers as measured by ESR while at the same time the frequency shifts indicate damage other than compaction. The additional damage is thought to consist of oxygen vacancies, those being responsible for the E' signal. This would be consistent with the observation that annealing at temperatures of up to 800°C suffices to restore the network almost completely to the unirradiated state. The E' density saturates at roughly 0.1 mol%, and at ion dose where this occurs, the IR absorption bands shifting any further, indicating that a quasi-equilibrium (or metastable state) has been reached between the damage introduced by additional ions and dynamic annealing. It was found that heavier and more energetic ions require lower dosage to reach a certain level than do light or less energetic ions but no simple scaling law could be established.

REFERENCES

[1] R.A.B.Devine, J.Vac.Sci.Tech., A6, 3154 (1988).

[2] P.N.Sen and M.F.Thorpe, Phys.Rev.B 15, 4030 (1977).

ACKNOWLEDGMENTS

It is a pleasure to acknowledge the expert assistance of P.Bérichon and R.Gosselin with the operation of the tandem accelerator, of M.Andre for ESR measurement and of A.Ait-Ouali for FT-IR measurement. Thermal oxidation was performed by K.Watanabe in Hitachi Central Laboratories.

MICROSTRUCTURE OF ULTRA HIGH DOSE SELF IMPLANTED SILICON

X.F. Zhu*, J.S. Williams*, D.J. Llewellyn* and J.C. McCallum**
* Department of Electronic Materials Engineering, Research School of Physical Sciences and Engineering, Australian National University, Canberra, ACT 0200, Australia
** School of Physics, University of Melbourne, Parkville, 3052, Australia

ABSTRACT

This study has investigated the microstructure of ultra high dose ($\sim 10^{18}$ cm^{-2}) self implantation into Si. Implants have been carried out into both (100) Si and pre-amorphised Si as a function of implant temperature between liquid nitrogen temperature and 350°C. Results show that high dose implantation into completely amorphous Si (a-Si) produces layers which regrow quite well during subsequent solid phase epitaxy. In contrast, implantation into crystalline Si (c-Si) or part amorphous/part crystalline Si can lead to rich and varied microstructures at elevated temperatures, even extending to porous-like structures in some cases. Strong dynamic annealing and agglomeration of points defects in c-Si is thought to be responsible for such behaviour.

INTRODUCTION

Implantation of self ions into both Si [1] and Ge [2] leads readily to amorphisation under conditions where implantation induced defects are not particularly mobile. In Ge, continued bombardment to doses in excess of 10^{16}Ge cm^{-2} can lead to an intriguing porous structure [2,3] thought to arise from vacancy accumulation in amorphous Ge. Such a process has not been observed in Si. However, irradiation of crystalline Si (c-Si) can lead to substantial agglomeration of point defects (at elevated temperatures) to form, for example, local interstitial-based clusters and loops [4] and vacancy excesses leading to voids [5]. Nanocavities can also be formed in c-Si by H or He bombardment to high dose followed by annealing [6,7]. It is interesting to establish whether such structures can form in amorphous Si (a-Si) and whether continued self ion bombardment to very high doses can, under any circumstances, lead to nucleation of phase transformations to more dense forms of Si [8] (from interstitial clusters) or to porous Si (from irradiation induced voids). Indeed, in an early study [9] we observed an odd irradiation-induced structure for self ion irradiation of Si (to doses beyond 10^{17} cm^{-2}) in which the Si appeared "black" to the eye and porous-like in an SEM. It was impossible to establish how this odd structure formed but it was thought to arise from impurity incorporation (e.g. oxygen) under high dose implantation at elevated temperatures. In the current study we set out to investigate high dose self ion irradiation of both c-Si and pre-amorphised Si as a function of implant temperature in an attempt to resolve some of the above issues.

Mat. Res. Soc. Symp. Proc. Vol. 504 © 1998 Materials Research Society

EXPERIMENTAL

In the present experiments, Cz Si (100) wafers (both p-type and n-type) of 1-10 Ωcm where irradiated with 40 keV Si ions to doses up to 10^{18}cm^{-2} at temperatures from liquid nitrogen to 350°C. Half of the samples had previously been "amorphised" by implantation with 100 keV Si ions to a dose of 3 x 10^{15}cm^{-2} at room temperature prior to the high dose 40 keV implant. In most cases the dose rates were kept at or below 10μA and the pressure in the implant chamber (housing a liquid nitrogen cooled cryoshield) was always $\leq$ 1 x 10^{-7} torr. Following implantation, the samples were analysed by 2 MeV He$^+$ Rutherford backscattering and channeling (RBS-C) and scanning electron microscopy (SEM) with selected samples analysed by cross-section transmission electron microscopy (XTEM). Most samples were annealed on a "hot block" attached to a time resolved reflectivity (TRR) apparatus to examine the kinetics of solid phase epitaxy of amorphous layers which was thought to be a simple and sensitive way to determine major structural changes/inhomogeneities in the high dose amorphous Si.

RESULTS AND DISCUSSION

In all cases where the 40 keV, 10^{18}Si cm^{-2} implants were carried out into continuously amorphous Si, even at elevated temperatures up to 250°C, the amorphous Si behaved quite normally during subsequent solid phase epitaxy. In Fig.1 we show an example (RBS-C spectra) for high dose irradiation into a-Si at 175°C. The typical a-Si thickness is around 2200-2600Å after implantation. After annealing on the TRR apparatus, essentially perfect epitaxy takes place, as confirmed by XTEM analysis [10]. Indeed, the regrowth rate is close to that for low dose amorphous layers. In some cases, for 25-250°C irradiations, a retardation in growth rate is observed, which peaks at around a factor of 2 slower regrowth at about the end of range of 40 keV Si in Si (around 950Å). Details of these slight regrowth rate effects will be treated elsewhere [10], but we note here that our measured sputtering coefficient for 40 keV Si into Si is close to 0.8, leading to a slight "deposition" of Si for a 10^{18}cm^{-2} implant.

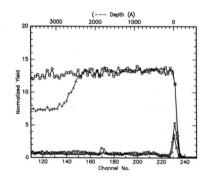

Fig.1
RBS-C spectra for 40 keV, 1 x 10^{18}Si cm^{-2} implanted into a-Si at 175°C: random (O), as-implanted (+), annealing 600°C (✗) and virgin (Δ).

The conclusion from the above results is that no *major* structure rearrangements occur with completely amorphous Si during high dose implantation even at elevated temperatures. This implies that a-Si is quite homogenous as has been found for low dose implanted a-Si [11]. It may well be that irradiated a-Si is quite "plastic" and flows to resist defect accumulation and local stress-concentration centres [12]. However, such behaviour is not the same in c-Si irradiated to high doses at elevated temperatures where continuous a-Si layers are not formed. In such cases, a complex microstructure develops which continuously changes with increasing dose. We illustrate a typical microstructure for 40 keV irradiation to 10^{18}Si cm^{-2} at 175°C into c-Si by the RBS-C spectra and XTEM micrograph in Figs.2 and 3, respectively. After implantation, the RBS-C spectrum (Fig.2) indicates a near-surface layer which may be amorphous (to about 700Å) and a heavily disordered layer extending to about 1800Å. The corresponding XTEM micrograph indicates that the surface layer is not completely amorphous, containing pockets of c-Si, rich in point defect agglomerations and some extended defects. The deeper c-Si layer contains mainly loops and extended defects of interstitial character [10]. In Fig.2 we also show the RBS-C spectrum after annealing at 600°C on the TRR apparatus. Clearly, the near-surface amorphous material has recrystallised but the resultant c-Si material is very defective. The 600°C anneal has not altered the deeper layer appreciably.

We wish to emphasise that small changes in the implantation conditions (dose rate, temperature) lead to quite major changes in the microstructure for elevated temperature implantation in c-Si. Clearly, defect annihilation and agglomeration processes, including build up of interstitial (and presumably vacancy) clusters, nucleation of amorphous layers and beam-induced crystallisation [13] are very sensitive to implantation parameters and to defect mobility and interactions in c-Si. Such sensitivity to implant parameters sometimes led to microstructural non-uniformities across the implanted region of a single sample and, in extreme cases, to visible colour changes across the sample. We illustrate below a particularly

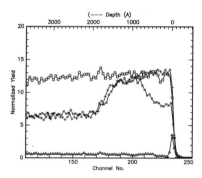

Fig.2
RBS-C spectra for 40 keV, 1 x 10^{18}cm^{-2} Si implanted into c-Si at 175°C: random (O), as-implanted (+), annealed 600°C (×) and virgin (Δ).

Fig.3
XTEM micrograph corresponding to as-implanted case in Fig.2; two arrows mark the wafer surface position.

striking example of non-uniformity across a wafer.

Fig.4 shows RBS-C spectra from two regions on a sample irradiated with 40 keV Si ions at 175°C to 1 x 10^{18}Si cm^{-2}. This sample had been originally "amorphised" with 100 keV, 3 x 10^{15} cm^{-2} ions at room temperature but parts of this implant (see Fig.6 below) were not initially amorphous to the surface. From one part of the sample (open triangle spectrum), the 40 keV implant gave a continuous amorphous structure (~ 2600Å) similar to that in Fig.1. This layer recrystallised normally as before. However, an adjacent region of the sample appeared "black" to the eye, at first sight indicative of contamination. No evidence of carbon or other contamination was found and the RBS-C spectrum (crosses) is also shown in Fig.4. The "disordered" layer appeared much thinner in this case. Subsequent optical microscopy and SEM (shown in Fig.5) revealed that the "black" region of the sample was extremely rough and crater-like, thus giving rise to absorption of visible light.

It is puzzling as to how such a layer developed during implantation, particularly in view of our previous observation that high dose implantation into completely amorphous Si does not appear to lead to anomalous microstructures. A strong clue to this behaviour relates to the fact that our initial amorphisation scheme was not always sufficient in amorphising the Si completely to the surface. This is illustrated in Fig.6, where changing the dose rate of the 100 keV, 3 x 10^{15}Si cm^{-2} implant, at nominally room temperature, from 7.5 μA cm^{-2} to 15μA cm^{-2} resulted in changes from a continuous a-Si layer to a buried layer. Indeed, non-uniformities can exist across a single implant and we believe this to be the case for the sample containing the "black" region. We suggest that the "black" region of the sample originally contained a crystalline region at the surface. It may be that beam-induced crystallisation and defect accumulation during 175°C, 40 keV Si irradiation in this region led to build up of a near-surface vacancy excess and surface expansion. Once a rough, non-planar surface is exposed during irradiation (Fig.5) sputtering is high and the crater-like layer is eroded. This may explain the thinner layer observed by RBS-C in Fig.4. Thus,

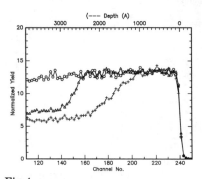

Fig.4
RBS-C spectra from two regions of a non-uniform 40 keV, 1 x 10^{18}Si cm^{-2}, 175°C implant into nominally a-Si: random (O), normal a-Si region (Δ) and "black" region (✛).

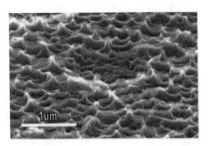

Fig.5
A typical SEM micrograph from a typical "black" region of a. sample implanted at 250°C.

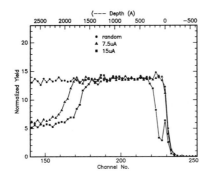

Fig.6
RBS-C spectra showing the effect of small changes in dose rate (beam current) for 3×10^{15}Si cm^{-2}, 100 keV, room temperature implants.

we have observed quite anomalous crater-like microstructures in originally part-amorphous/part crystalline samples subjected to irradiation at elevated temperatures. We suggest that defect accumulation (vacancy excesses) in the crystalline regions may account for the observed behaviour. The appearance of a "Si-black" region in elevated temperature implants was not an isolated occurrence and was found in four separate occasions for irradiations between 175°C and 250°C. The phenomenon clearly warrants further investigation.

CONCLUSION

Implantation at high doses into a completely amorphous Si layer, even at elevated temperatures, does not lead to anomalous microstructures in a-Si. Indeed, the amorphous Si appears to be quite homogeneous. However, irradiation of c-Si or part crystalline/part amorphous Si can lead to a rich variety of microstructures during high dose elevated temperature implantation as a result of defect agglomeration in c-Si. Indeed, we have observed a crater-like, possibly porous Si structure in some cases which may originate from defect (vacancy) agglomeration in near-surface c-Si layers.

ACKNOWLEDGEMENTS

The authors gratefully acknowledge Dr Roger Heady for the SEM micrograph and Mr Frank Brink for his assistance in TEM operation at the ANU EM Unit.

REFERENCES

1. G.L. Olson and J.A. Roth, *Mat.Sci. Reports* **3** (1988) 1.

2. B.R. Appleton, O.W. Holland, J. Narayan, O.E. Schow III, J.S. Williams, K.T. Short and E.M. Lawson, *Appl.Phys.Lett.* **41** (1982) 711.

3. H. Huber et.al. *Nucl.Instrum.Meth*. (in press 1997).

4. D.J. Eaglesham et.al, in Microscopy of Semiconducting Materials, Oxford, 1995, ed. A.G. Cullis and A.E. Stanton-Bevan [Inst. Phys. Conf. Ser. **146**, 451 (1995)].

5. O.W. Holland, L. Xie, B. Nielson and D.S. Zhou, *J.Elec.Mat*. **25** (1996) 99.

6. J. Wong-Leung, E. Nygren and J.S. Williams, *Appl.Phys.Lett.* **68** (1995) 417.

7. S.M. Myers, D. Follstaedt and D.M. Bishop, Mat.Res.Soc.Symp.Proc. **316** (1994) 33.

8. J.Z. Hu, L.D. Merkle, C.S. Menoni and I.L. Spain, *Phys.Rev.* **B34** (1986) 4679.

9. J.S. Williams et.al. Mat.Res.Soc.Symp.Proc. **27** (1984) 205.

10. X.F. Zhu and J.S. Williams, to be published.

11. D.L. Williamson, S. Rooda, M. Chicoine, R. Tabti, P.A. Stolk, S. Acco and F.W. Saris, *Appl.Phys.Lett.* **67** (1995) 226.

12. S. Rooda, L. Cliche, M. Chicoine and R.A. Masut, *Nucl.Instrum.Meth.* **B106** (1995) 80.

13. J.S. Williams, Trans.Mat.Res.Soc.Jpn. **17** (1994) 417.

NUMERICAL MODELING AND EXPERIMENTAL MEASUREMENTS OF PULSED ION BEAM SURFACE TREATMENT

Michael O. Thompson* and T.J. Renk**
* Dept. of Materials Science, Cornell University, Ithaca, NY 14853, mot1@cornell.edu
** Sandia National Laboratories, Dept 9521, Albuquerque, NM 87185, tjrenk@sandia.com

ABSTRACT

Pulsed ion beam treatment of materials provides an attractive alternative to pulsed laser processing for near surface modification of semiconductors, metals and polymers. The transfer of energy to the sample occurs by electronic and nuclear stopping over depths extending to several microns depending on ion species and voltage. A numerical code for modeling the melt and solidification behavior of materials under ion beam processing has been developed. The code and parameter extraction procedures were validated experimentally by comparing simulations with experimental measurements of the melt duration in silicon. The sensitivity of the melt behavior to variations in the beam properties was also investigated. The quiescent nature of the melt was confirmed by measurements of the diffusion of arsenic in the melt. These results demonstrate that simulations of the ion beam treatment can quantitatively match experimental results with no adjustable parameters.

INTRODUCTION

Pulsed laser treatment of materials has been extensively researched over the last fifteen years. Following nanosecond irradiation, impurities can be incorporated into semiconductors at levels limited only by the lattice stability [1], morphological stability [2], and segregation effects [3]. For metals, supersaturated alloy compositions [4] and non-equilibrium crystalline and glassy phases have been produced [5]. Today, pulsed excimer lasers are used industrially to drill fine holes in polyimide and for low temperature processing of Si devices [6]. Finally, fundamental materials properties including, for example, impurity diffusion and segregation models [7], and homogeneous nucleation [8] have also been measured during the transient melt induced by pulsed lasers.

For materials processing, the laser is primarily a convenient source of heat that can be delivered to the near surface region in a sub-microsecond time scale. As such, many of the properties of a laser are actually deleterious in materials processing. The coherence of the light requires extensive efforts at homogenization to avoid interference effects. Metals generally have a high reflectivity leading to poor coupling of energy and large spatial variations in fluence on non-homogeneous surfaces. In addition, the large absorption coefficient (10^6 cm^{-1}) of most metals limits the processing depth to only several hundred nanometers due to ablation and surface damage. In contrast, In ceramics, it is often difficult to even get sufficient absorption to induce melt before the onset of damage. Semiconductors, such as silicon, are generally more amenable to laser processing, but melt depths and durations are limited to approximately 500 nm and 300 ns, respectively, due to the metallic nature of the liquid phase.

Pulsed ion beams are an attractive alternative to laser processing since they are not subject to these same limitations [9]. In a pulsed ion beam, ions accelerated to potentials of 100-1000 kV impinge on a material surface held in moderate vacuum (10^{-4} Torr). With current densities of 50-200 A/cm^2 and pulse durations of 100-1000 ns, these beams deliver 0.1-10 J/cm^2 of energy directly into the near surface. Depending on the ion species and voltage, the energy is delivered over depths ranging from 10 nm (low energy or heavy ions) to several microns (protons at high energy). The total energy deposition is independent of surface preparation and properties, although the specific energy deposition may vary with phase composition. Since energy is delivered in depth, it also possible to melt and process layers extending to nearly 10 microns without surface damage. In contrast to conventional ion implantation, however, the total number of ions delivered in a single ion beam pulse (typically 10^{12}-10^{13} cm^{-2}) is insignificant from a metallurgical point, and usually insignificant for the electrical properties as well. With the recent development and refinement of the MAP (Magnetically confined Anode Plasma) ion diode, the incident ion species can now be selected and reproducible operation over a large number of experiments is possible.

Mat. Res. Soc. Symp. Proc. Vol. 504 © 1998 Materials Research Society

There are however, several new complications arising from the use of ion beams. In particular, the formation of the ion plasma in the MAP diode is complex and yields numerous ion charge states and some beam contamination (usually protons and carbon). Additionally, each of the charge states is accelerated to a different energy by the diode. Moreover, travel from the diode gap to the sample position temporally separates the beam due to differences in the time of flight. Finally, each species and charge state delivers energy over a distinct range governed by the energy and ion dependent stopping power. While the list of species present in the beam is constant from shot to shot, the fraction of each ion may change substantially.

For quantitative study of materials processing, it is critical that an accurate simulator of the melt and solidification behavior be developed. For routine irradiation, only the temporal behavior of the ion voltage and the total ion current are measured. In this paper, we discuss a process for extracting the ion species and energies from these measurements. These parameters are then coupled into a numerical code that calculates the temperature and phase changes in a one-dimensional layered structure. To verify the accuracy of the species extraction and heat flow simulations, we have performed direct measurements of the melt duration of silicon under pulsed ion irradiation, and of arsenic diffusion in the molten phase.

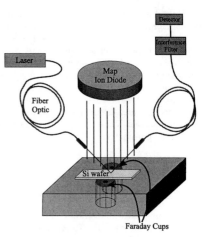

Figure 1: Schematic diagram of the experimental setup for measuring melt duration and incident ion currents.

EXPERIMENTAL

Pulsed ion beam irradiation of test samples were carried out at Sandia National Laboratories using the RHEPP pulsed power accelerator coupled to a MAP ion diode. All of the experiments discussed in this work used nitrogen as the diode gas and operated at a peak accelerating voltage of 800 kV. Details of the RHEPP accelerator and ion diode are given elsewhere [10]. For these conditions, fluences were 0.5 to 3.0 J/cm^2 over approximately 100 cm^2.

Samples scribed from $\langle 100 \rangle$ silicon single crystal wafers were mounted on a test stand equipped with a fiber coupled reflectivity measurement setup (Fig. 1). A 5 mW HeNe laser, coupled through a 62.5 μm diameter fiber, was focussed onto a 0.5 mm diameter spot at the center of the sample using a GRIN (GRaded INdex) lens. The laser light was nominally unpolarized at the sample. Reflected light was collected with a second GRIN lens and coupled into a 200 μm fiber. An interference spike filter was inserted before a silicon avalanche detector to eliminate stray light from the diode firing (minimal) and from ablation induced plumes (significant at high fluences). All signals were digitized and stored for analysis.

Since the fluence from the ion diode varied over a centimeter spatial scale, two Faraday cups were mounted directly adjacent to the Si samples. These two cups, separated by 1 cm, provided both a measurement of the ion current and an estimate of the variation of the beam fluence near the sample. Variations of up to 20% were observed between the two

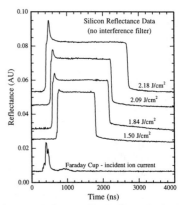

Figure 2: Reflectance from a Si sample showing melt duration for a variety of fluences between 1.5 and 2.2 J/cm^2.

cups. For each experiment, the fluence was assigned as the average of the two cups with error bars extending to include both measurements. Because the ion beam tended to "peak" on axis, it is possible that the actual fluence could be above either Faraday cup measured fluence. This spatial variation constitutes the largest source of error in these experiments.

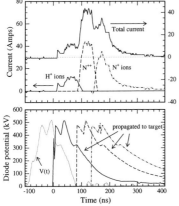

Figure 3: Total ion fluence from the Faraday cup. The contributions from protons, N^+ and N^{++} are resolved. Lower graph shows the propagated voltage for each of the ion species to the sample position

Upon melting, Si transforms from a semiconductor to a metal with an increase in reflectivity from approximately 35% to 65% at 632 nm. Reflectance traces for a series of irradiations between 1.5 and 2.2 J/cm^2 are shown in figure 2. The reflectance shows the characteristic jump at the onset of melting, followed by a plateau while the surface is molten, and finally the drop back to the crystalline level as the melt front approaches the surface. The sharpness of the final drop reflects the melt duration variation over the laser probe diameter (for a 1 m/s solidification velocity, the reflectivity should drop in 10 ns for a uniform melt). At the highest fluences (above 2-3 μs melt duration), some ablation occurs as evidenced by a sharp spike in the reflectance at melting and visible roughening of the surface. (An interference filter was normally used to block this ablation plume.) Significant damage, evidenced also by structure in the plateau region of the reflectivity traces themselves, did not occur until the melt duration approached 4-5 μs.

The total ion current, as measured by one of the Faraday cups, is shown in figure 3 for a sample irradiated at approximately 2 J/cm^2. With the nitrogen beam, three distinct ion components, based on arrival time, could be resolved in the ion current; protons and two charge states of nitrogen (N^+ and N^{++}). The time origin was arbitrary chosen as the arrival of protons at the target (57.3 cm from the diode for this example). Doubly charged nitrogen ions arrived 85 ns later, followed finally by N^+ ions 135 ns after the protons. The ion voltage, measured at the diode itself, was propagated for each species to the target as shown in lower half of figure 3. The low energy wraparound (low energy ions produced early that arrive after high energy ions produced later) was truncated assuming that the high energy ions constituted the majority of the beam. This low energy component is estimated to be no more than 5% of the total ion flux. Note also that N^{++} ions were actually accelerated to 1 MeV energies.

The ion current was split between the three species as follows. All current before the first arrival time of N^{++} was assigned to protons. At the N^{++} arrival, the proton current was linearly reduced to zero over 10 ns, while the N^{++} current was simultaneously ramped up. The fluence for each individual Faraday cup was then determined by integrating the voltage-current traces. These data also formed the input to the numerical modeling code described below.

This procedure was repeated for a large series of shots to determine the fluctuations in species

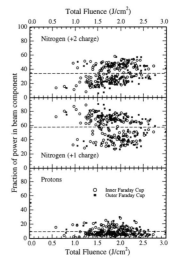

Figure 4: Distribution of total beam fluence between the three species. On average, the beam is 34% N^{++}, 57% N^+ and 9% H^+.

distribution, as shown in figure 4. Of the total fluence, protons never constituted more than 30% of the power and on average only 10%. However, the split between the N^+ and N^{++} varied substantially. On any given shot, either N^+ or N^{++} was the majority component (almost randomly); the split apparently depends on minor variations in the diode behavior. From a practical standpoint, though, the split between the N^+ and N^{++} species affects only the distribution of energy in depth in the sample but not the total fluence calculation. Further, as shown below, simulations indicate that the melt behavior does not depend strongly on the ion species either since both N^+ and N^{++} are stopped within the molten zone.

SIMULATIONS

A numerical simulation code developed to model pulsed laser processing of materials was modified [11] to handle pulsed ion beams. This code follows heat transport, melt, and solidification phase changes in a non-homogeneous 1-D layered structure using an enthalpy-conserving algorithm. Included also is support for full temperature dependence of the thermal conductivity and specific heat, arbitrary phase selection rules, and velocity dependent interface undercooling and overheating.

The energy absorption module of the code was modified to follow multiple ion species and currents through the layered structure. Within each layer, the ion energy is reduced by the energy dependent stopping power (dE/dx) with the enthalpy transferred to the lattice assuming immediate thermalization. Stopping powers were determined from TRIM calculations [12] for the appropriate targets and incident ions. Specific ion voltages and currents, as determined from the extraction process described above, were used for all simulations.

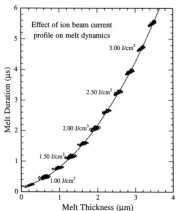

Figure 5: Effect of the beam profile (ion species distribution) on the melt duration and maximum melt depth.

Because of the multiple species and time dependent voltage, the temperature-depth-time dependence is significantly more complex than that observed during laser heating. The first ion specie to arrive, protons, has a low stopping power and hence provides minimal heating over several microns. Doubly charged nitrogen arrives 85 ns later and, because of the high energy and long range, also only heats the surface region. Since the N^+ has the highest specific energy deposition (J/cm^3), surface melting, except at the highest fluences, does not occur until shortly after the arrival of these ions at 135 ns. Once molten, however, the specific energy deposition profile is essentially "washed out" by the melt front propagating to a 1-3 μm depth. For deep melts, solidification requires nearly 5 μs and thermal diffusion distances over this time scale are much greater than all but the proton range.

The effect of varying the distribution of ion species (as observed experimentally) was investigated by simulating melts for a wide range of beam profiles as a function of fluence (Fig. 5). As anticipated above, the melt duration is essentially independent of the specific split, varying by less than ±100 ns even for 5 μs melt durations. The maximum melt thickness, more dependent on the preheating of the deep regions, varied by ±75 nm; the variation though was essentially independent of the actual melt depth.

A direct comparison of the experimentally observed melt durations and the calculated melt durations is shown in figure 6. For each measurement, the error bars represent the full range of the experimentally determined fluence (two Faraday cups) and the experimentally determined melt duration. The error in the melt duration is insignificant on this scale. At each fluence, a simulation was performed using the average of the beam profiles from the two Faraday cups, shown as the filled squares. The dashed line represents the duration versus fluence for an "average" beam. As discussed above, there is little difference between individual simulations and the average behavior.

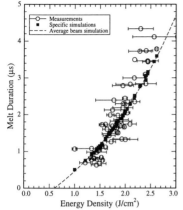

Figure 6: Comparison of measured melt duration and simulations. Filled squares are simulations specific to each fluence.

For fluences up to 2.0 J/cm², although there is considerable scatter in the data, there is no statistically significant difference between the experimental measurements and the simulated melt duration. This is remarkable given that there are no adjustable parameters in either the measurements or the simulations. With the exception of some ambiguity in the ion species determination (which does not affect the results in any case), all of the thermodynamic and experimental parameters are directly measured. This result effectively validates both the simulation codes and the experimental parameter extraction.

At higher fluences, there is consistent deviation of the experimental data, with measured melt durations exceeding the calculated values by as much as 40%. This cannot be fully explained at this time. At fluences above the ablation or damage threshold, one would normally expect the experimental duration to be shorter than simulated since a fraction of the incident fluence would be removed by the ablated material. Thermal transport properties are also not expected to change significantly since, although the surface is roughened at high fluences, the material remains a single crystal. We believe that this deviation is most likely due to the spatial variation of the beam fluence. The long melt durations were not intentionally selected, but rather arose as occasional fluctuations from a lower average beam behavior. These "hot shots" are more likely to include large spatial variations, and probably more peaked behavior along the axis.

ARSENIC DIFFUSION

The total melt duration during pulsed beam processing is governed primarily by heat flow into the substrate. For silicon, the high thermal conductivity of the liquid ensures that the molten layer rapidly becomes isothermal and thus the solidification behavior is dependent only on the thermal conduction of the solid phase. However, given the high fluences and deep melts involved in pulsed ion beam processing, the liquid may not remain quiescent (unmixed) during the melt. Convection in the liquid would modify diffusional mixing in Si, and potentially affect the validity of thermal simulations in metal alloys. A direct measurement of the mixing of arsenic during the melt was thus undertaken to determine if convective mixing was induced by the ion beam.

Single crystal Si wafers were implanted with 100 keV As ions to a total dose of 8×10^{15} ions/cm². The projected range at this energy is only 58 nm and all of the arsenic remains within 100 nm of the surface. Samples were irradiated by the pulsed ion beam at fluences between 0.5 and 2.2 J/cm² with the melt duration directly measured. Rutherford Backscattering Spectrometry (RBS) measurements of the arsenic profile (Fig. 7) were fit to the error function diffusion equation, yielding an estimate of the total diffusion distance ($\sqrt{Dt}$) for each sample. The preliminary data from four samples are shown in Fig. 7c. Because of the extremely long melt durations (compared to laser experiments), the diffusion distances are substantial and determination of the $\sqrt{Dt}$ value is simplified. A linear fit through these points yields an average liquid phase diffusion constant for arsenic in silicon of $2.19 \pm 0.20 \times 10^{-4}$ cm²/s. The value obtained from similar experiments (although with shorter melt durations) using pulsed laser irradiation is 2.32×10^{-4} cm²/s [13]. We conclude from these data that even for melts of 2.5 μs (approaching the ablation threshold), the liquid remains essentially quiescent and only diffusional mixing is important.

SUMMARY

Quantitative measurements and simulations have been performed for pulsed ion beam irradiated silicon. For the pulsed ion beam, all of the experimental conditions can be readily measured,

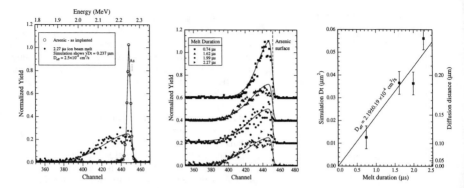

Figure 7: RBS measurements of arsenic diffusion in silicon following a 2.27 μs melt. Solid curves are fits of the diffusion profile to an error function diffusion form. Measured diffusion length (squared) as a function of the melt duration. Best fit yields a diffusion coefficient of $2.19 \pm 0.20 \times 10^{-4}$ cm^2/s.

including ion species and total fluence, leaving no adjustable parameters for numerical heat flow simulations. One-dimensional simulations were compared with experimentally measured melt durations and show extremely good agreement for melt durations up to 2.5 μs. Although the beam profile and ion species distribution varied from pulse to pulse, these variations are almost irrelevant for studies involving melt and solidification. The slight differences in volume energy deposition become insignificant compared to thermal diffusion of heat over the molten period.

Measurements of the liquid phase diffusion of arsenic during pulsed ion beam irradiation were also performed. The diffusion constant determined from these experiments is consistent with measurements obtained during laser irradiation. This strongly suggests that the liquid, even for long melt durations, remains quiescent and convective mixing can be neglected.

These results for silicon, where experimental verification is possible, provide strong evidence for the general validity of ion beam simulations. This is critical since direct measurements during ion beam processing of metals are, at best, extremely difficult. The validity of the simulation code and the experimental parameter extraction processes permit simulations to be used reliably for understanding pulsed beam treatment of metals, ceramics and polymers.

This work was supported by the United States Department of Energy under Contract DE-AC04-94AL85000. Sandia is a multi-program laboratory operated by Sandia Corporation, a Lockheed Martin Company, for the United States Department of Energy.

REFERENCES

[1] P. S. Peercy, M. O. Thompson and J. Y. Tsao, Appl. Phys. Lett. **47**, 244 (1985).

[2] D. E. Hoglund, M. J. Aziz, S. R. Stiffler, M. O. Thompson, J. Y. Tsao and P. S. Peercy, J. Crys. Growth **109**, 107 (1991) .

[3] A. G. Cullis, H. C. Webber, J. M. Poate and A. L. Simons, Appl. Phys. Lett. **38**, 800 (1981).

[4] P. M. Smith and M. J. Aziz, *Acta Metall. Mater.* **42**, 3515 (1994).

[5] P. Mazzoldi, L. F. Dona dalle Rose and D. K. Sood, *Radiation Effects* **63**, 105 (1982).

[6] P.M. Smith, P. G. Carey and T.W. Sigmon, Appl. Phys. Lett. **70**, 342 (1997)

[7] L. M. Goldman and M. J. Aziz, *J. Mater. Res. Proc.* **35**, 257 (1985).

[8] S. R. Stiffler, M. O. Thompson and P. S. Peercy, Phys. Rev. B **43**, 9851-5 (1991)

[9] R. Fastow, Y. Maron and J. W. Mayer, Phys. Rev. B **31**, 893 (1985); S. A. Chistjakov et al., *Physics Letters* **131**, 73 (1988).

[10] J. B. Greenly, M. Ueda, G. D. Rondeau and D. A. Hammer, J. Appl. Phys. **63**, 1872 (1988).

[11] M. J. Uttormark, Ph.D Thesis, Cornell University, 1992.

[12] TRIM90, J. F. Ziegler, IBM T. J. Watson Research Center, Yorktown Heights, NY.

[13] J. A. Kittl, M. J. Aziz, D. P. Brunco and M. O. Thompson, J. Crys. Growth **148**, 172-182 (1995).

MODELING COLLISION CASCADE STRUCTURE OF SiO_2, Si_3N_4 and SiC USING LOCAL TOPOLOGICAL APPROACHES

C. Esther Jesurum*†, Vinay Pulim†, Linn W. Hobbs‡and Bonnie Berger*†

* Department of Mathematics †Laboratory for Computer Science
‡Department of Material Science and Engineering
Massachusetts Institute of Technology, Cambridge, MA 02139, USA

Abstract

Many crystalline ceramics can be amorphized within high-energy collision cascades whose overlap leads to global structural amorphization. Because the structural rearrangements amount to topological disordering, we have chosen to model these rearrangements using a topological modeling tool as an alternative to molecular dynamics simulations. We focus on the tetrahedral network compounds SiO_2, Si_3N_4, and SiC, each compound comprising corner-sharing tetrahedral units, because they represent increasingly topologically constrained structures. SiO_2 and SiC are easily amorphized experimentally, whereas Si_3N_4 proves very difficult to amorphize. In this model, we consider the tetrahedron as the base unit, whose identity is largely retained throughout. In a collision cascade, all bonds in the neighborhood of a designated tetrahedron are broken, and we reform bonds in this region according to a set of local rules appropriate to crystalline assembly, each tetrahedron coordinating with available neighboring tetrahedra (insofar as is possible) in accordance with these rules. We generate fairly well connected amorphized structures for SiO_2, but run into underconnected networks for Si_3N_4 and SiC which are irreparable without rebreaking and reforming primary tetrahedral bonds. The resulting structures are analyzed for ring content and bond angle distributions for comparison to the crystalline precursors.

1 Introduction

Amorphizability is governed topologically by available structural freedom f, which represents the difference between available degrees of freedom for and constraints on each atom [1]. Values for SiO_2, Si_3N_4 and SiC are, in order, 0, -1.5 and -3, which indicates that such freedom should be marginally available in {4,2}-connected (4 tetrahedral vertices, each shared by two tetrahedra) SiO_2, but much less available in {4,3}-connected Si_3N_4 and highly unavailable in {4,4}-connected SiC. Structural freedom criteria hence predict that SiO_2 should be relatively easy to amorphize by any means, Si_3N_4 much less so, and SiC very difficult. Irradiation with energetic medium- to heavy-weight ions results in collision cascades within which overall structural amorphization can be effected through direct impact amorphization or cascade overlap to achieve a critical defect density [2]. The cascade phenomenon represents an ideal platform for topological modeling of amorphization, since the violent cascade displacements are largely uncorrelated and represent atomic randomization followed by reassembly into chemically-preferred coordination polyhedra whose relative dispositions are then likely to be governed by the same sorts of local rules that also apply to long-range ordered crystalline assembly. While the conventional view is that epitaxial recrystallization will take place partially (or completely above a critical temperature) from the cascade boundary, we have explored first an alternative view in which topological disorder can propagate in the reassembly if sufficient local options are available; only when propagation of topological disorder is

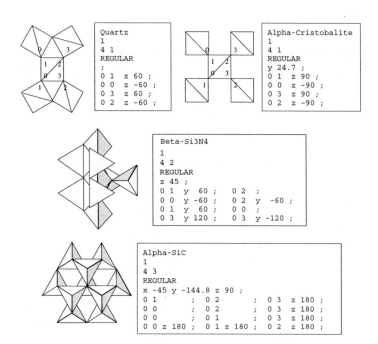

Figure 1: Local rules for the assembly of ideal quartz (top left), α-cristobalite (top right), β-Si$_3$N$_4$ (middle) and α-SiC (bottom).

frustrated will propagation of long-range order from the periphery present the only option. Our second approach uses epitaxial recrystallization as the basic method for reconnection of the collision cascade. We analyze resulting structures for *local clusters* [3] compared to the crystalline precursor clusters as well as for underconnection and bond angle distribution.

2 Local-Rules based modeling

We simulate the generation of crystals using *local rules*. The viability of this approach has been demonstrated for the six known network silicas as well as α and β forms of Si$_3$N$_4$ and SiC [4]. The approach (first described in [5]) is to use the local information of one tetrahedron (or more than one for some, more complicated structures) to iteratively "grow" arbitrarily large structures. A rules file will contain the following information (illustrated in Figure 1): The first line in the rules file defines the number of topologically inequivalent tetrahedra (in this case only one, referred to as type "0," because all tetrahedra in each of the structures illustrated have equivalent environments), the next two lines specify the regular tetrahedral unit, with 4 vertices each sharing 1, 2 or 3 other tetrahedra, and the fourth line indicates the orientation of the initial tetrahedron in terms of rotations about the x, y and z axes with respect to a canonical orientation inscribed within a unit cube with sides parallel to these axes. The last four lines specify the tetrahedron type (type "0" in all cases), vertex number and rotations for each neighboring tetrahedron bonded to (in succeeding lines) vertices 0, 1, 2 and 3 of the initial tetrahedron; the rules files for Si$_3$N$_4$ and SiC have respectively two or three rotation sets, corresponding to the two or three additional tetrahedra respectively sharing the vertex.

The assembly operation begins with the initial tetrahedron in its chosen orientation, and new tetrahedra are introduced, rotated and bonded. After being added, the first-neighbor tetrahedra become, in turn, initial tetrahedra for still further additions, and so forth. Local topology–the environment around a tetrahedron in tetrahedral networks–can be described using irreducible closed circuits along the connected network known as *primitive rings*. All tetrahedra belonging to the set of primitive rings passing through a tetrahedron in the network comprise the *local cluster* of that tetrahedron. The local clusters and primitive ring content for crystalline silica, Si_3N_4 and SiC network polymorphs have been determined and reported elsewhere [4]. Quartz is comprised of 6-rings and 8-rings (dominated by 8-rings), cristobalite comprises only 6-rings; β-Si_3N_4 is dominated by 6-rings, but with substantial numbers of 3- and 4-rings, while the two SiC polymorphs have only 3-and 4-rings in equal number.

Large models comprising 2000 tetrahedra of each of three crystalline structures (ideal quartz, β-Si_3N_4 and α-SiC) were assembled prior to cascade disordering. During erection of the initial models and during subsequent reconnection after cascade disordering, all networks were optimized using springs inserted between vertices which exerted forces and torques on connecting tetrahedra. The elastic energy of these springs was globally minimized after each tetrahedron addition or reconnection step; additional routines monitored and rejected tetrahedral intersections. Global optimization associated with each connection was continued at least until the spring segment for that connection was less than half the Si-X distance.

3 Modeling collision cascade disorder

We simulate collision cascade disorder in three stages. In the first stage, all the bonds connecting tetrahedra are broken and the crystal boundary is expanded. In the second stage, the now isolated tetrahedra are "disturbed" by being rotated through randomly selected angles $\leq 20°$. The boundary expansion is then relaxed as these "excited" tetrahedra are next allowed to reform connections, to each other and to boundary tetrahedra, according to a chosen set of local rules, *e.g.* a rule set for the same or another crystalline polymorph or a modified rule set. Both interior and boundary are settled using the optimization scheme described above. Innovations to the second stage are currently being investigated that would guarantee less residual topology from the precursor structure. One way to do this is to add repulsive forces which will serve to space out the excited tetrahedra.

3.1 Results of cascade simulations

As a first example, connections between 400 central tetrahedra were destroyed in a 2000-tetrahedra model of ideal quartz and the structure regrown using α-cristobalite rules; the choice of a high-temperature polymorphs (α-cristobalite) on which to base the reconstruction rule set was deliberate, since the temperature at which reconstruction is actually effected within a cooling cascade is expected to be high. The result is an amorphous reconstructed volume, substantially bonded (71%, Table 1) both internally and to the crystal boundary, and with little tetrahedral distortion (as evidenced by short remanent optimization springs).

We performed analogous cascade simulations for β-Si_3N_4 and α-SiC regrown using the same rules as the precursor structure. Whereas all SiO_2 cascades, reported here and elsewhere [4] were fairly well re-connected and exhibited little distortion, the regrowth volumes in β-Si_3N_4 and α-SiC were more poorly connected (Table 1), exhibiting a larger fraction of underconnected tetrahedra (39% and 44% respectively vs. 27% for SiO_2), both at the boundary and in the interior; amorphized regions for β-Si_3N_4 and α-SiC also exhibited large optimization springs, corresponding to large tetrahedron distortions. Further optimization failed to improve these results and in fact, in some cases, caused the structure to suffer more tetrahedral distortion and intersection.

Precursor/Regrow	Si-X-Si Angles θ for Precursor/Regrow	Si-X-Si Angles θ in Cascade	Percent Underconnection
Ideal Quartz/ α-Cristobalite [a]	155.6/145.3	145.5	28.75
Ideal Quartz/ α-Cristobalite [b]	155.6/145.3	152.12	26.75
β-Si$_3$N$_4$/β-Si$_3$N$_4$	119.8 (avg.)	120.2	38.6
α-SiC/α-SiC	109.5	115.6	43.9

[a] Original model of regrowth.
[b] Recrystallization model of regrowth.

Table 1: Average bond angles and underconnection in the cascade amorphization of SiO$_2$, Si$_3$N$_4$ and SiC.

Local clusters incorporating tetrahedra residing at the periphery mostly retained their precursor topologies, but those in the cascade interior developed different ring configurations. The local clusters in cascade-amorphized ideal quartz, while having enough variation to be considered amorphous, still resembled quartz topology (dominated by 8-rings). In contrast, those in the interior of the cascades in β-Si$_3$N$_4$ and α-SiC (shown in Figure 2) were not at all Si$_3$N$_4$- or SiC-like and exhibited large distortions and energetically unfavorable remanent springs.

Figure 3 plots the distributions of Si-X-Si inter-tetrahedral angles θ for all three compounds, which are considerably broadened and shifted from the corresponding angles (Table 1) for β-Si$_3$N$_4$ (119.8° average) and α-SiC (109.5°). These broad distributions are further evidence of poor reconnection. For example, if we consider SiC to be comprised of [CSi$_4$] tetrahedra, an average Si-C-Si angle of 115 indicates large distortion. The distribution for the ideal-quartz cascade is more narrow and shows an average bond angle of 145.5° close to the regrowth structure (155.6° for ideal-quartz, 145.3° for α-cristobalite).

3.2 Modeling recrystallization reconnection

We are currently investigating an alternate view of cascade regrowth using recrystallization at the periphery as the basic method. In this technique, we regrow the structure from the periphery (outside in) in a similar fashion to the way we normally grow crystals from the inside out, directly from the rules files. In essence, we remove all the tetrahedra from the interior and then we put them all back one at a time. The only important detail to note is that we use a gradient of rules based on distance to the periphery; that is, nodes closer to the boundary use rules that are similar to the precursor rules and nodes closer to the center use rules similar to the regrow rules. The simulator achieves this by creating two hybrid rules sets, one which is a mix of precursor/regrowth rules in a 2:1 ratio favoring the precursor and another which is a 2:1 mix favoring the regrowth rules.

Results for SiO$_2$: When performing this recrystallization simulation of cascade reconnection, naturally, if one uses regrowth rules identical to precursor rules, one obtains the original structure. With this in mind, we performed an experiment using regrowth by recrystallization on ideal quartz regrown with α-cristobalite, and the results were distinguishable from the experiment using ordinary regrowth. The underconnection was improved (see Table 1) and the average Si-O-Si bond angle in the cascade was closer to the precursor. Presumably, with a larger cascade region, the interior would contain many more nodes influenced by the regrowth rules and therefore, even more cristobalite-like structure. In our small cascade the interior local cluster (see Figure 2) was still quartz-like but was less dominated by 8-rings than the original cascade.

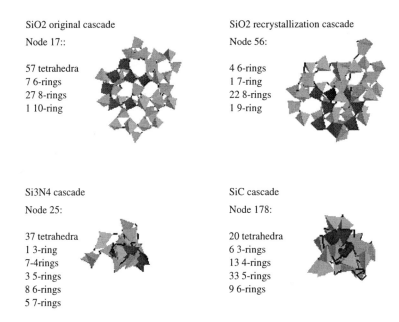

SiO2 original cascade

Node 17::

57 tetrahedra
7 6-rings
27 8-rings
1 10-ring

SiO2 recrystallization cascade

Node 56:

4 6-rings
1 7-ring
22 8-rings
1 9-ring

Si3N4 cascade

Node 25:

37 tetrahedra
1 3-ring
7-4rings
3 5-rings
8 6-rings
5 7-rings

SiC cascade

Node 178:

20 tetrahedra
6 3-rings
13 4-rings
33 5-rings
9 6-rings

Figure 2: Interior local clusters for the cascade amorphized SiO_2, Si_3N_4 and SiC.

4   Discussion and Conclusions

It seems that the prediction offered by structural freedom analysis accounts for the inability to satisfactorily resolve the cascades for Si_3N_4 and SiC. The procedure in which we disturb the tetrahedra is not really possible since it occasions overlaps which are too difficult to resolve in overconstrained structures. Perhaps the elastic energy that remains is what drives recrystallization in Si_3N_4. The explanation of the relative ease of amorphization in SiC can be accounted for by the allowance of antisite disorder as is confirmed by current research [6]. We believe that a model which allows for the breaking up of tetrahedral subunits with preferential bonding as in current kinetics simulations [7] will be more fruitful in determining the local structure of amorphized SiC.

Future enhancements of cascade simulation will involve repulsive forces between near-tetrahedra inversely proportional to their distance from the center of the cascade. This will serve to simulate the greater disorder far form the periphery of the cascade. There is also evidence that an additional reconnection stage may reduce residual underconnection.

Acknowledgments

This work was funded by the Office of Basic Energy Science, U. S. Department of Energy through grant DE-FG02-89ER45396, to whom two of us (LWH, CEJ) are much indebted for support. Computational facilities were used through the assistance of the Advanced Research Projects Agency under contract N0014-95-1-1246 which funds the Computational Biology Group in LCS, and for providing one of us (VP) with an research assistanceship supported by Department of Energy grant DE-FG02-95ER25253.

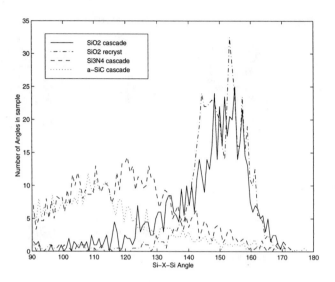

Figure 3: Si-X-Si bond angle distribution for cascade amorphization experiments in Si_3N_4, SiC and SiO_2 (two techniques).

References

[1] P.K. Gupta and A.R. Cooper, "Topologically disordered networks of rigid polytopes," J. Non-Cryst. Solids 123 (1990) 14.

[2] W.J. Weber, "Radiation-induced defects and amorphization in zircon," J. Materials Research 5 (1990) 2687-2697.

[3] C. S. Marians and L. W. Hobbs, "Local structure of silica glasses," J. Non-Cryst. Solids 119 (1990) 269.

[4] L.W. Hobbs, C.E. Jesurum, V. Pulim and B. Berger, "Local topology of silica networks," Phil. Mag. (1998, in press); C.E. Jesurum, V. Pulim and L.W. Hobbs, "Topological modeling of amorphized tetrahedral ceramic network structures," J. Nuclear Mat. (1998, in press).

[5] C.E. Jesurum, V. Pulim, L.W. Hobbs and B. Berger, "Modeling of topologically-disordered tetrahedral ceramic network structures using local rules", In: G. Matthews and R. Williams, editors, *13th International Conference on Defects in Insulating Materials*, Materials Science Forum, 239-241 (Trans-Tech Publications, Switzerland, July 1996) 37-40.

[6] R. Devanathan, W. Weber and T. Diaz de la Rubia, "Atomistic simulation of defect production in β-SiC," In *Materials Research Society Symposium Proceedings* (Dec. 1997, in press).

[7] R. Schwartz, "A multi-threaded simulator for the kinetics of virus shell assembly," SM thesis, M.I.T., Cambridge, MA, 1996.

ATOMISTIC SIMULATION OF DEFECT PRODUCTION IN β-Sic

R. Devanathan*, W. J. Weber*, and T. Diaz de la Rubia**
* Pacific Northwest National Laboratory, Richland WA 99352, ram@pnl.gov
** Lawrence Livermore National Laboratory, Livermore, CA 94550

ABSTRACT

The process of defect formation and the threshold energies for Si and C displacements along various crystallographic directions in cubic silicon carbide (β-SiC) have been examined using molecular dynamics simulations. A combination of Tersoff and first-principles potentials was used to model the inter-atomic interactions. The lowest threshold energies for C and Si displacements were found to be 28 and 36 eV, respectively. These displacement threshold energies show excellent agreement with the results of recent first-principles calculations in SiC and with experimental observations. Simulation of a 10 keV Si cascade yielded values of about 0.1 ps for the cascade lifetime and about 3.5 for the ratio of the number of surviving C defects to Si defects. Anti-site defects were found on both Si and C sublattices. These defects may play an important role in the amorphization of SiC by energetic particle irradiation.

INTRODUCTION

Silicon carbide in the form of SiC/SiC composites is a candidate for structural applications in fusion reactors. In addition, SiC is the leading candidate material for semiconductor devices in high-power, high-frequency, high-temperature, and high-radiation applications for which conventional Si and GaAs-based devices are unsuitable. There are potential applications for such devices in the military (more electric aircraft and combat vehicles), commercial power generation (electrical switching systems), and petrochemical industries (direct process monitoring). The recent commercial availability of device-quality wafers has provided a much needed boost for SiC device technology. In the applications discussed above, the material is likely to undergo atomic displacement damage as a result of neutron irradiation in a fusion reactor or ion-implantation during semiconductor processing. An understanding of the production of point defects by energetic beams and their subsequent evolution is needed to realize the potential of this technologically useful material.

Several experimental studies have been performed to examine the effects of energetic particle irradiation in SiC [1,2]. These studies have shown that SiC can be amorphized below 250 K by energetic ions or electrons at a dose of about 0.6 displacements per atom. However, the small time and distance scales associated with radiation damage events has precluded a complete understanding of the process of defect production. The details of this process can be obtained by molecular dynamics simulations performed with realistic inter-atomic potentials. Such simulations can enable us to interpret experimental results and tailor the properties of the material to suit the demands of the application. Previous efforts to simulate displacement events in SiC have been hampered by the lack of reliable potentials [3]. In the present work, a new inter-atomic potential scheme has been used to calculate displacement threshold energies (E_d), and model a 10 keV Si cascade in SiC. The results are compared to first-principles calculations and experimental observations to test the validity of the approach used in the present work.

DETAILS OF THE SIMULATIONS

The molecular dynamics simulations were performed using a modified form of the MDCASK code [4] on a massively parallel Cray Research Inc. T3D supercomputer at Lawrence Livermore National Laboratory. The inter-atomic interactions were modeled using the Tersoff potential [5] for distances greater than 1.4 Å, a realistic repulsive potential [6] for distances less than 0.5 Å, and a combination of the two potentials for intermediate distances. This scheme has been discussed in detail elsewhere [7]. The calculations of E_d were performed using 8000 atoms in the microcanonical ensemble (constant energy, volume, and number of atoms). The initial temperature of the cell was 150 K. A larger simulation cell containing 192,000 atoms was used for the simulation of a 10 keV Si cascade in SiC. In this case, the volume and number of atoms were kept constant and the temperature of the cell was controlled by coupling the atoms in the 4 outermost layers to a heat reservoir at 300 K. The bottom layer was kept static and a damping force was applied to the boundary layers to prevent the energy leaving the cell from re-entering at the opposite boundary as a result of periodic boundary conditions. The simulation cell was equilibrated, initially, for 2 ps. Then, an atom was given a kinetic energy of 20-120 eV for the displacement events and 10 keV for the cascade simulation. The configurations and velocities were recorded periodically and analyzed using a Silicon Graphics Inc. O2 workstation to understand the creation and evolution of atomic-level damage in SiC.

RESULTS AND DISCUSSION

Table 1 compares the E_d values for C and Si atoms along various crystallographic directions to the corresponding values calculated in an independent first-principles molecular dynamics study [8] using local-orbital local-density approximation (LDA) method (Fireball96). The Fireball96 method has been shown to provide an accurate description of point defect properties in Si [9].

Table 1. Displacement threshold energies in β-SiC

Atom	Direction	Present work (eV)	Fireball96 [8,9] (eV)
C	[1 0 0]	31±1	28.5
C	[1 1 0]	38±1	38.5
C	[1 1 1]	28±1	27.5*
C	[$\bar{1}\,\bar{1}\,\bar{1}$]	71±1	
Si	[1 0 0]	36±1	
Si	[1 1 0]	71±1	
Si	[1 1 1]	113±1	
Si	[$\bar{1}\,\bar{1}\,\bar{1}$]	39±1	36.9

* A recent full plane-wave LDA pseudopotential calculation [8] yielded a C displacement energy of 14.6 eV along [1 1 1], which is much less than the values from the present work and Fireball96.

It is evident from table 1 that the present results are in excellent agreement with the results of the Fireball96 calculation. In addition the lowest values obtained in the present work for E_d, 28 and 36 eV, respectively, for C and Si, are in good agreement with the corresponding experimental estimates of 20 and 40 eV [10]. The present calculations also show that the displacement threshold energy surface in SiC is highly anisotropic. E_d varies from about 28 to 71 eV for C, and 36 to 113 eV for Si. This may explain the large scatter in the E_d values determined by various experimental techniques [10]. The present calculations clearly show a large difference between E_d values along [1 1 1] and [$\bar{1}$ $\bar{1}$ $\bar{1}$], which reflects the difference in the atomic packing between these two directions. The agreement between the present results and the independent first-principles calculation on the one hand and experimental results on the other suggests that the combined Tersoff-repulsive potential provides a good description of displacement cascade events in β-SiC.

Fig. 1 shows the evolution of a 40 eV Si displacement event along [001] in SiC. This energy is above the threshold value for displacement along [001]. For clarity, only a portion of

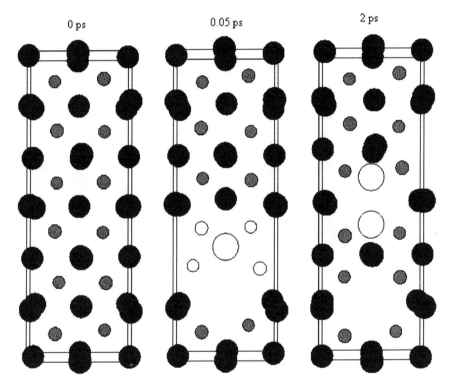

Figure 1. Three stages in a 40 eV [0 0 1] Si displacement event in β-SiC. The Si atoms are shown as large black circles and the C atoms as small gray circles. The white circles are Si and C 'defects' with potential energies at least 2 eV above the corresponding perfect crystal values.

the simulation cell containing 8000 atoms is shown. At 0 ps, a Si atom near the bottom of the figure was given an energy of 40 eV along the upward [001] direction. At 0.05 ps, the Si atom was displaced from its lattice site and in the process had imparted a high potential energy (at least 2 eV above the corresponding perfect crystal value) to a ring of carbon atoms along its path. At a time of 2 ps after the start of the event, the C atoms returned to a low energy state, but the Si primary knock-on atom (PKA) had formed a stable [001] dumb-bell interstitial defect leaving behind a vacancy at its original site. These defects can be clearly seen in fig. 1.

Fig. 2 shows a 10 keV Si displacement cascade along [4 11 $\overline{95}$] at the maximum extent of the damage (0.094 ps) and in its final state (9.4 ps). For the sake of clarity only a portion of the simulation cell measuring (87.2x87.2x130.8 Å) is shown. At 0.094 ps, the number of defects produced in the cascade reaches a maximum and the defects are vacancies and interstitials. With the passage of time, many of these defects recombine creating replacements and anti-site defects, and the number of surviving defects decreases. The range of the PKA was found to be about 90 Å and it came to rest within 0.16 ps of cascade initiation. The mixing parameter Q was found to be about 1.7 $Å^5$/eV which is much less than values of 12 and 48 $Å^5$/eV reported for a 5 keV Si cascade in Si [11] and a 10 keV Au cascade in Au [12].

0.094 ps   9.40 ps

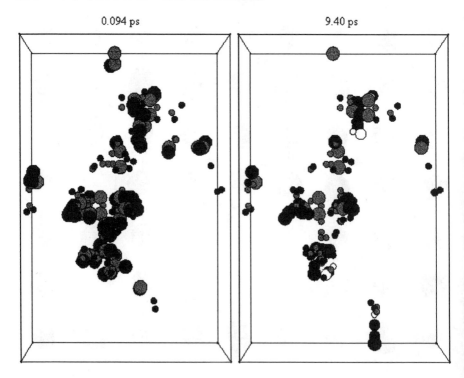

Figure 2. A 10 keV Si cascade in β-SiC close to the [0 0 $\overline{1}$] direction shown at 0.094 ps (maximum extent of damage) and 9.4 ps (final state of damage) after primary knock-on initiation. The Si atoms are shown as larger circles. Black, dark grey, light grey, and white represent, interstitials, replacements, vacancies, and anti-site defects, respectively.

Fig. 3 shows the number of vacancies and anti-site defects of Si and C in a 10 keV Si cascade in β-SiC as a function of time. The number of interstitials was found to be nearly the same as the number of vacancies of the same species and so is not shown in the figure. The ratio of surviving C vacancies to Si vacancies was found to be about 3.5. This higher value can be attributed to the higher displacement threshold energy for Si than for C, and to the occurrence of twice as many replacements on the Si sublattice as on the C sublattice. The Si vacancies and interstitials created in the cascade are less stable and annihilate in larger numbers than the C defects. Nearly 40% of the displaced Si atoms recombined with Si vacancies, thereby reducing the number of surviving Si defects. The peak in the number of vacancies at about 0.1 ps corresponds to the maximum extent of damage shown in fig. 2. The number of defects saturated within about 0.5 ps indicating that the cascade lifetime is of the order of 0.1 ps in β-SiC which is considerably shorter than the value of 2 ps in Si [11]. The shorter cascade lifetime, lower mixing efficiency, and more 'diffuse' cascade damage in SiC compared to Si is consistent with previous observations [11] that amorphization within the ion track, by quenching of a molten cascade region, does not occur in SiC. These differences between the damage states in SiC and Si offer insights into the greater radiation resistance of SiC.

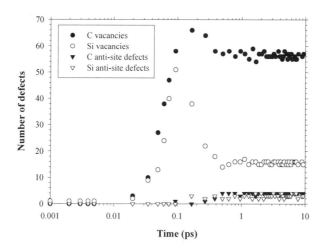

Figure 3. The number of point defects of Si and C in a 10 keV Si cascade as a function of time.

Since amorphization does not occur within the ion-track in SiC, it must occur by the gradual accumulation of lattice defects. It is evident from fig. 3 that stable anti-site defects form on both the C and Si sublattices. This has important implications for the amorphization of SiC. Previous experimental [13] and simulation [14] studies have shown that chemical disorder plays an important role in the amorphization of binary intermetallic compounds. Hobbs et al. [15] have suggested that β-SiC should be difficult to amorphize in the absence of chemical disorder based on considerations of network topology. The fact that SiC is readily amorphized by particle

49

irradiation below 250 K has been attributed to anti-site disorder [15]. The accumulation of anti-site defects changes the topology of SiC, increases the structural freedom, and makes it easier to amorphize the material.

CONCLUSIONS

The occurrence of displacement damage in β-SiC has been successfully modeled using a combination of Tersoff and first-principles repulsive potentials. The lowest displacement energy (E_d) was found to be 28 eV for C and 36 eV for Si. The threshold energy surface was found to be highly anisotropic which explains the large scatter in experimentally determined values of E_d. The values of E_d calculated in the present work are nearly identical to those obtained independently by accurate first-principles calculations. In addition, the E_d values show good agreement with experimental observations thereby validating the interatomic potential scheme employed in the present work. The mixing efficiency was much lower, the cascade lifetime much shorter, and the damage more diffuse in a 10 keV Si cascade in SiC compared to previous reports for cascades in Si. Stable anti-site defects formed by recombination of defects in the cascade. The accumulation of these anti-site defects may be responsible for the amorphization of SiC.

ACKNOWLEDGEMENTS

This work is sponsored by the U. S. Department of Energy, Division of Materials Sciences, Office of Basic Energy Sciences under Contract DE-AC06-76RLO 1830 (PNNL). Work performed at LLNL was funded by the Office of Fusion Energy, U. S. Department of Energy under Contract W-7405-Eng-48.

REFERENCES

1.   W. J. Weber and L. M. Wang, Nucl. Instrum. Methods B **106**, p.298 (1995).
2.   H. Inui, H. Mori, A. Suzuki, and H. Fujita, Philos. Mag. B **65**, p.1 (1992).
3.   H. Huang, N. M. Ghoniem, J. K. Wong, and M. I. Baskes, Modell. Simul. Mater. Sci. Eng. **3**, p.615 (1995).
4.   T. Diaz de la Rubia and M. W. Guinan, J. Nucl. Mater. **174**, p.151 (1990).
5.   J. Tersoff, Phys. Rev B **39**, p.5566 (1989); ibid. **49**, p.16349 (1994).
6.   K. Nordlund, J. Keinonen, and T. Mattila, Phys. Rev. Lett. **77**, p.699 (1996).
7.   R. Devanathan, T. Diaz de la Rubia, and W. J. Weber, J. Nucl. Mater. (1998) (in press).
8.   W. Windl, T. J. Lenosky, J. D. Kress, and A. F. Voter, Mat. Res. Soc. Symp. Proc. **490** (1998) (in press).
9.   W. Windl, T. J. Lenosky, J. D. Kress, and A. F. Voter, Nucl. Instrum. Methods B (1998) (in press).
10. S. J. Zinkle and C. Kinoshita, J. Nucl. Mater. (1997) (in press).
11. T. Diaz de la Rubia, M. J. Caturla, and M. Tobin, Mat. Res. Soc. Symp. Proc. **373**, p.555 (1995).
12. M. Ghaly, R. S. Averback, and T. Diaz de la Rubia, Nucl. Instrum. Methods B **102**, p.51 (1995).
13. D. E. Luzzi and M. Meshii, Res Mechanica **21**, p.207 (1987).
14. R. Devanathan, N. Q. Lam, P. R. Okamoto, and M. Meshii, Phys. Rev. B **48**, p. 42 (1993).
15. L. W. Hobbs, A. N. Sreeram, C. E. Jesurum, and B. A. Berger, Nucl. Instrum. Methods B **116**, p.18 (1996).

DEFECT PRODUCTION DURING ION-ASSISTED DEPOSITION OF MOLYBDENUM FILMS STUDIED BY MOLECULAR DYNAMICS

PETER KLAVER AND BAREND THIJSSE
Laboratory of Materials Science, Delft University of Technology, Rotterdamseweg 137,
2628 AL Delft, Netherlands

ABSTRACT

Low energy argon-ion assisted growth of thin molybdenum films (≈ 60 Å) has been studied by molecular dynamics simulations. The effects of a single ion impact are discussed, but more particularly we consider film growth from a manufacturing viewpoint and examine the properties of the completed films. Results for ion-beam assisted deposition are compared with those for unassisted growth (i.e. physical vapor deposition). Surface morphology, defect generation, argon incorporation, and the various responsible atomic mechanisms are discussed.

INTRODUCTION

One of the methods to modify the microstructural properties of thin films is to assist the film-deposition kinetics by concurrent low-energy ion bombardment [1–3]. To apply this technique effectively it is important to understand the processes taking place at or just below the surface, because surface morphology, adatom mobility and the generation and interaction of defects are significant factors for many thin film properties. However, the collision events following the impact of even a single ion on an ideally flat surface are quite complicated and depend critically on local impact conditions [4], so that computer simulation methods are virtually indispensable for understanding the atomic-scale mechanisms playing a role in a more complex situation such as the evolution of a growing film.

In the present work we report on the effects of argon ions of up to 250 eV on the growth of molybdenum films (≈ 60 Å) on molybdenum (110) and (100) substrates. This specific system was chosen to be studied by molecular dynamics simulations, because these films are also being manufactured in our group and studied experimentally by thermal helium desorption spectrometry (THDS) [5]. Our primary interests are the morphology and the defect state of the films. This paper is a follow-up of earlier work, in which somewhat less realistic interaction potentials were used [6]. We emphasize experimental rather than idealized conditions, i.e. argon ion impacts on non-flat surfaces and properties of completed films. Due to the limited scope of this paper, the trapping and desorption of helium ions, essential for the THDS experiments, and comparisons between simulations and experiments will be reported elsewhere.

COMPUTATIONAL METHOD

The simulations were conducted using the velocity-Verlet method with a variable integration time step such that no atom experiencing a force moves more than 0.02 Å in a single step. To improve accuracy, accelerations are taken into account in testing an atom's impending displacement against this limit. Periodic boundary conditions are applied in the in-film directions. Mo atoms and Ar ions arrive from random positions well above the surface (positive z-axis). The atoms in the two lowest planes are treated in a special way to simulate a large substrate, allowing the system to lose the excess energy and momentum transferred by the incoming particles [6]. The substrate is held at 300 K.

The Mo-Mo interaction is described by the Johnson-Oh EAM-potential [7], which yields a cohesion energy of 6.82 eV/atom, a bcc-fcc energy difference of 0.075 eV/atom, a bcc lattice constant of 3.147 Å, and a relaxed vacancy formation energy of 3.01 eV. The spacing between the first two planes of the (110) surface is contracted by 6.5%; the (100) surface exhibits damped oscillatory distortions, beginning with a 2.2% expansion and followed by a 1.5% contraction. Between 1.6 Å and 2.1 Å the EAM-potential is smoothly joined to the Molière potential with Firsov screening length [8], which is used at small interatomic separations. The Ar-Mo and Ar-Ar interactions are represented by the Molière-Firsov potential only. Omitting an attractive part is justified by the temperatures considered in this work (300 K and above) and by the fact that Ar always finds itself in high potential-energy situations. Since only low-energy processes are studied, the system could be kept to a modest size. The (110) surface orientations are studied in a box with an xy-base of 44.1 × 44.5 Å^2 (280 atoms per plane), (100) surfaces in a 44.1 × 44.1 Å^2 box (196 atoms per plane).

The growth conditions are as follows. The kinetic energy of the impinging Mo atoms is 0.17 eV, or 2000 K, a typical value for vapor deposition; the arrival rate is one atom per 1.4 ps. The depositions start on a substrate of four or more atomic planes. In some cases, when the simulation box is nearly filled, the system is rapidly cooled to 0 K, and the atoms are "pushed down" over a certain length through the bottom of the box. The remaining atoms in the box are then brought to 300 K, and the deposition is resumed. This makes it possible to study film evolution up to 60 Å. Several conditions were exactly copied from our experiments: Ar ions arrive along the substrate normal, Mo atoms along a 15° off-normal direction making a 13° angle with the (100) direction in the base plane. Also, most low-energy Ar beams are intentionally contaminated by 9% neutral Ar atoms of 250 eV. *The energies of these beams are indicated by an asterisk*, e.g. 100* eV. The ion-to-atom arrival ratio γ is 0.1.

As in many molecular dynamics simulations of materials, time is the most limiting factor. The deposition rate in this work is nearly 6 × 10^9 higher than in typical experimental conditions (1 Å/s). This implies that virtually no thermally activated processes can be studied. For instance, if 15 ps is taken as a typical time interval in the growth simulations (say, between ion impacts) and 10 Å as a typical diffusion length, a simple calculation based on a Debye frequency of 1 × 10^{13} s^{-1} shows that only diffusion processes with an activation energy of less than 50 meV can be observed at 300 K. Experimentally, however, when this time interval is longer by a factor 6 × 10^9, processes of up to 0.6 eV can take place. This means that in the simulations low-energy phenomena such as surface diffusion and, possibly, self-interstitial mobility will not be represented realistically. To make up for this, several configurations have been annealed at 2000 K for 4 ns to allow processes up to a 1.3 eV activation energy to be studied separately. Clearly the fast ballistic effects in the collision cascades do not suffer from the time limitation.

In this work the surface topology of the growing films is characterized by a single roughness parameter R, defined as the RMS variation of the z-coordinates of the surface atoms. Surface atoms are defined as atoms with 3–10 neighbors within a distance of 3.8 Å, halfway between the 2nd and 3rd nearest neighbor distances in bcc Mo. (For the (001) surface the range is 3–9.)

RESULTS

Nanoscale events following the impact of an energetic argon ion

More than 150 impacts of a 250 eV argon ion on a 40 Å thick (110)-grown film were studied to get detailed insight in the atomic mechanisms taking place. Averaged over these impacts the following data are obtained. The Ar ion sputters 0.46 Mo atoms out of the film and has an

Fig. 1. Top view of 40 Å thick (110) Mo film after 250 eV argon ion impact. Greyscale denotes total number of lost and gained neighbors as a result of the impact.

52

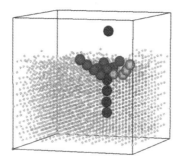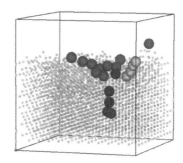

Fig. 2. Mo (100) film just before (left) and 1 ps after (right) the impact of a 250 eV argon ion. Four groups of Mo atoms, shown as large spheres, undergo a replacement collision sequence. The bottom one creates a self-interstitial, the other three each create an adatom. The argon ion becomes trapped in the four-fold vacancy cluster that it creates. Small spheres shown are also Mo atoms.

8% probability of being incorporated in the film. The incorporation is substitutional in the lattice: the ion creates its own vacancy (V) or vacancy cluster (V_n) up to n=4, at an average depth of $\approx$ 7 Å below the surface. As a result of the impact 20 Mo atoms are displaced by one or more nearest neighbor distances, 11 bulk atoms lose so many neighbors that they become surface atoms (adatom creation), and 10 surface atoms gain so many neighbors that they become bulk atoms. Fig. 1 shows an example of these surface activities. Replacement collision sequences (RCS's) involving 3 or more Mo atoms are very common: on average 2.0 RCS's occur, with an average length of 4.1 atoms. Many end on the surface, creating an adatom; some are directed downward, creating a self-interstitial. The longest RCS found involves 14 atoms. Fig. 2 illustrates some of these processes. (Note that Fig. 2 was taken from a series of impacts on a (100)-film.)

The sputter yield at a (100) film surface is 0.63, significantly higher than the value 0.46 for the (110) film. The reason is the columnar structure of the (110) surface (see below), making it less probable for detached Mo atoms to leave the film. Ion channeling effects play no significant role.

100 eV ions cause less violent effects: no vacancy clusters are formed, almost no RCS's take place and only 10 Mo atoms are displaced. The trapping probability of a 100 eV ion is on the order of 0.3%, which shows that argon trapping from mixed beams of 91% 100 eV + 9% 250 eV is largely due to the 250 eV ions.

The impact of a 25 eV ion is even hardly noticed (2 displaced atoms).

<u>Film Morphology</u>

The upper four curves in Fig. 3 show the (110)-surface roughness as a function of film thickness for four different deposition conditions. It is seen that ion assistance retards surface roughening: more ion energy per deposited Mo atom leads to a slower increase of the

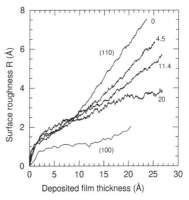

Fig. 3. Surface roughness as a function of nominal film thickness for several deposition conditions. For the four (110) films the ion energy per Mo atom is indicated in eV: (0)=no ion assistance, (4.5)=25*eV γ=0.1, (11.4)=100*eV γ=0.1, (20)=100 eV γ=0.2. The (100) film is grown without ion assistance.

roughness. Only after R has reached a value of 2.4 Å (one monolayer) the deposition conditions begin to make a difference. This shows that certain minimum height variations on the surface are *necessary* for the smoothing action of the ion energy and momentum transfer to become effective. It can be concluded from Fig. 3 that 100 eV ions do have a considerable effect, more so than in the case of trapping, mentioned above. The reason is that in this case the displacement processes are much more directly associated with the surface and hence can be set into motion with lower energies.

What we actually observe in Fig. 3 is the onset of columnar growth. This is illustrated in Fig. 4 for the 100* eV (γ=0.1) case. The development of columns with open boundaries is clearly seen. Once begun to appear, these boundary regions can no longer be filled completely. The effect of ion assistance is to increase the minimum film thickness at which the columns begin to develop.

Annealing such a film for up to 4 ns at 2000 K reveals surface diffusion. About 300 atoms with low coordination numbers begin to diffuse over the surface with a mean activation energy of 1.2 eV. With time, more and more atoms find a position with higher coordination. The most striking effect is a strong reduction in "nano-roughness", which we measure by the number of atoms that are only four- or five-fold coordinated. In terms of larger-scale features, the columnar structure is only negligibly affected (Fig. 5). Therefore, *if* the magnitude of the nano-roughness would be a significant factor for the onset of columns during growth, this effect could be an artefact of our unrealistically high deposition rate, which in effect bypasses diffusion. However, detailed analysis shows that the nano-roughness is almost constant during the first 30 Å, i.e. in the thickness range where the columns begin to develop (this can also be seen in the bottom two frames of Fig. 4). Apparently then, the nano-roughness does not play an important role in the onset of columnar growth. From this it seems safe to conclude that the columns would have developed also when a much more realistic (lower) deposition rate would have been used.

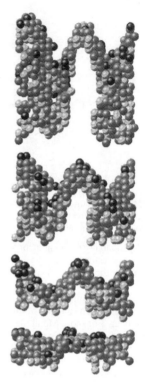

Fig. 4. Columnar growth developing during deposition of a (110) film. Shown are only the surface atoms for four film thicknesses. The nominal film thickness increases by 15 Å between two successive frames (bottom to top). Greyscale indicates number of neighbors.

Films grown in (100)-orientation have a quite different morphology (as can already be seen in Fig. 2). The surface roughness is significantly smaller (Fig. 3, bottom curve). No columnar structure develops. However, an interesting phenomenon was observed in (100) films deposited from directions other than 15° off-normal. For 0° and for 30° off-normal deposition a large void of the size of some tens of vacancies developed in the film, in both cases approximately 30 Å above the substrate. On continued deposition the surface closed and re-assumed its normal flatness (Fig. 6). Even after detailed inspection of the deposition runs we found no apparent reason or particular event that had caused these voids to develop, nor do we understand the significance of the incident vapor direction. We are left with the somewhat unsatisfactory explanation that subtle statistical fluctuations in the vapor flux, combined with the high binding energy of molybdenum, are responsible for these large defects; more work is needed in this area. Post-deposition annealing and intentional attempts to break up a void by directed ion impacts revealed that these voids are quite stable. The void changed shape somewhat, but did not collapse even under 250 eV ion bombardment.

Ion assistance during (100) deposition and annealing of these films has effects similar to those in the (110) case. There is only an insignificant effect on the morphology.

<u>Defects</u>

Vacancies and vacancy clusters are built into the growing films in various concentrations. Unassisted deposition produces defective films with concentrations up to order 1%. Clusters may be as large as V_5, with V_{12} as an exception. Ion assistance of 25* eV has only a small effect on the vacancy content, but 100* eV clearly reduces the concentration. In (100) and (110) films we found a vacancy concentration on the order of 1×10^{-3}, implying that on average the net effect of a 100* eV ion is to remove about 0.1 vacancy. This is despite the vacancies and vacancy clusters created by the 250 eV ions that are part of the 100* eV beam (Fig. 2).

The concentration of built-in 100* eV argon ions is approximately the same as the vacancy concentration (1×10^{-3}), which implies a 1% trapping probability. Furthermore, 1–2 Mo self-interstitials are produced per 100* eV ion. Argon self-sputtering has been observed, but as a rare event.

In all cases of film growth the first few deposited atomic planes contained no vacancies.

CONCLUSIONS

Molecular dynamics studies of the deposition of Mo films on Mo substrates under Ar ion-assistance up to 250 eV show large differences in morphology between films grown on (110) and (100) substrates. (110) films gradually develop a columnar structure after a few initial epitaxial atomic planes. Ion beam assistance postpones this effect. Contrary to this, (100) films stay relatively flat. However, under certain incident vapor angles, large, stable voids may grow into a (110) film.

Replacement collision sequences following an ion impact are a frequent source of self-interstitials and adatoms, but the activities taking place at the surface are more varied than this. Post-deposition anneals reveal the "long-term" effects of surface diffusion and yield a value of 1.2 eV for the activation energy. However, the complexity of the evolving surface geometry makes it unrealistic to compare the

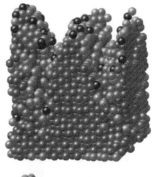

Fig. 5. Upper part of a (110) Mo film of 60 Å thickness before (top) and after (bottom) annealing at 2000 K for 2.1 ns. Greyscale indicates number of neighbors. The "nano-roughness", as evidenced by the atoms that are only four- or fivefold coordinated (black), strongly reduces upon annealing. The larger-scale surface features are unaffected.

Fig. 6. Cross section of growing (100) Mo film, at 40 Å thickness, showing large void, over which the surface is about to close. Greyscale indicates potential energy. Only the upper part of the film is shown.

diffusion data with known properties of well-defined crystallographic surfaces, such as diffusion coefficients.

Ion beam assistance reduces the concentration of grown-in vacancies and vacancy clusters, which without assistance is on the order of 1%. A 100* eV beam (a mixture of 91% 100 eV and 9% 250 eV) with an ion-to-atom ration of 0.1 reduces this concentration to 1×10^{-3}. On the other hand, about 1% of the 100* eV ions become entrapped in the film. These are predominantly the 250 eV ions, which find a resting place in the vacancies or vacancy clusters that they create themselves.

ACKNOWLEDGMENTS

This work is part of the research program of the Stichting voor Fundamenteel Onderzoek der Materie (Foundation for Fundamental Research of Matter), and was made possible by financial support from the Nederlandse Organisatie voor Wetenschappelijk Onderzoek NWO (Netherlands Organization for Scientific Research).

REFERENCES

1. F.A. Smidt, Int. Mater. Reviews **35**, 61 (1990).
2. S. Esch, M. Breeman, M. Morgenstern, T. Michely, and G. Comsa, Surf. Sci. **365**, 187 (1996).
3. G.S. Was, D.J. Srolovitz, Z. Ma, and D. Liang, Mat. Res. Soc. Symp. Proc. **441**, 311 (1997).
4. H.A. Atwater, Solid State Phenomena **27**, 67 (1992).
5. J. van der Kuur, E.J. Melker, T.P. Huijgen, W.H.B. Hoondert, G.T.W.M. Bekking, A. van den Beukel, and B.J. Thijsse, Mat. Res. Soc. Symp. Proc. **396**, 587 (1996); J. van der Kuur *et al.*, this conference.
6. A. Robbemond and B.J. Thijsse, Nucl. Instr. Meth. B **127/128**, 273 (1997)
7. R.A. Johnson and D.J. Oh, J. Mater. Res. **4**, 1195 (1989).
8. W. Eckstein, Computer Simulation of Ion-Solid Interactions (Springer, Berlin, 1991).

ARGON–SURFACE INTERACTIONS IN IBAD–DEPOSITED MOLYBDENUM FILMS

JAN VAN DER KUUR, BAS KOREVAAR, MARTIN POLS, JACQUELINE VAN DER LIN-
DEN, AND BAREND THIJSSE
Delft University of Technology, Laboratory of Materials Science, Rotterdamseweg 137, 2628
AL Delft, Netherlands

ABSTRACT

Argon incorporation and defect creation were studied experimentally. Direct desorption measurements have been used to establish the argon implantation profile. A projected range and
range straggle of 0.8 and 3.5 Å were found. Argon is incorporated at substitutional positions.
The creation rate of defects by argon was studied by helium desorption spectrometry. A net
creation rate of $(0.7 \pm 0.4) \times 10^{-3}$ vacancy/argon atom was found. Ion assisted deposition at elevated substrate temperatures shows that all incorporated argon acts as helium trap. Argon fluence
variations show an effective cross-section for self sputtering of 31 Å^2, a trapping probability of
6.5%, and a maximum achievable argon concentration of 4×10^{-3}.

INTRODUCTION

Ion beam assisted deposition (IBAD) is a technique commonly used for manipulating the
composition or microstructure of thin films, to obtain films with specific properties like improved
hardness, adhesion, optical, electrical or magnetic characteristics. Most phenomena have been
observed for energy-to-atom ratios ranging between 1 and 100 eV/atom [1], by applying ion
beams with energies exceeding 100 eV. A consequence of the ion assistance is that, depending
on the beam energy, part of the ions will be incorporated in the film. This may affect electrical
and other properties considerably. In this paper, we present thermal desorption measurements of
incorporated argon in thin molybdenum films deposited under argon ion assistance. In addition
to direct desorption measurements, thermal helium desorption spectrometry (THDS) is applied
as a tool to determine the number of defects created by the argon bombardment in the films.

EXPERIMENTAL

The films are deposited in a UHV apparatus (1×10^{-10} mbar), described in more detail
elsewhere [2]. The system consists of a deposition chamber for producing IBAD films connected
to a chamber in which the films are analyzed by THDS. The initial substrate was a polished,
annealed molybdenum disk of 2 mm thickness and 10 mm diameter. Thereupon all films were
deposited in succession, each film having been annealed to 2000 K before the next film was
deposited. It was confirmed that the stack of layers kept functioning as a reproducible substrate.
The molybdenum atoms originate from a 3 kW electron beam evaporation source and arrive at
15° off–normal incidence, with a typical deposition rate of 1 Å/s. A Kaufman source (normal
incidence) is used for Ar^+ ion assistance.

The THDS technique has been described more extensively elsewhere [3, 4]. THDS begins
with 'decorating' the defects in the film with helium as probe particles, using an ion beam with
a fluence Φ. After thermalization and diffusing through the film, some of the helium probe
particles are trapped by defects. Next, the trapped helium is thermally excited until it dissociates
from the various traps. This is done by increasing the sample temperature T linearly with time

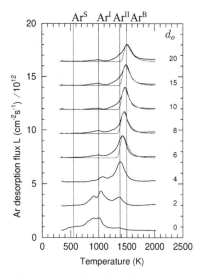
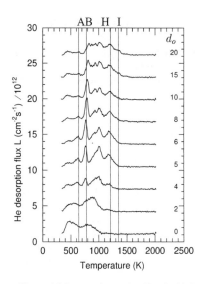

Figure 1: Argon and helium desorption spectra of a polycrystalline molybdenum substrate implanted with 3×10^{14} atoms/cm^2 argon ions of 250 eV after which overlayers of various thicknesses (d_o in Å) have been deposited, as indicated in the figures. The helium decoration energy is 100 eV, and $\Phi = 1 \times 10^{14}$ he/cm^2. The heating rate is $\beta = 40$ K/s. The black curves are measured, the grey curves are calculated.

(to 2000 K), using a 2.5 kV electron beam. During the temperature ramp the helium desorption flux L is monitored as a function of T, using a quadrupole mass spectrometer. The L vs. T data form the 'thermal desorption spectrum': a collection of peaks, each originating from the dissociation of helium at a particular energy, i.e. the dissociation of He from a particular type of defect in the material. In general the helium/defect dissociation energy increases with increasing trap size. Simultaneous with the helium desorption signal, the desorption of argon incorporated in the film during the ion assistance is measured by alternately sampling (at 40 Hz) the argon and helium channels of the mass spectrometer.

RESULTS

The purpose of the first experiment is to establish the implantation profile of argon in the polycrystalline molybdenum substrate, as a model for understanding incorporation of argon in IBAD films. The second experiment is aimed at the determination of the defect creation rate by a 250 eV argon beam. Based on this information, the argon and defect concentration of IBAD films deposited at various substrate temperatures will be discussed. Finally, the implantation and argon self-sputtering probability is determined by measuring the argon incorporation for various argon fluences.

The implantation profile of 250 eV argon ions in molybdenum

The dissociation energy of argon atoms from their traps depends on the trapping depth under the surface. The shallower an argon atom is incorporated in the substrate, the more effectively the presence of the surface can lower its dissociation energy, for example as a result of surface contraction. This implies that desorption peaks at higher temperatures correspond to argon incorporated deeper in the layer. By depositing an overlayer that is known to contain few additional defects, one effectively changes the average distance between the incorporated argon and the surface in a controlled manner. This leads to a corresponding change in the relation between the distinct desorption peaks originating from different distances from the surface, which makes determination of the depth profile possible. Figure 1 shows the argon desorption spectra for the polycrystalline substrate implanted with 3×10^{14} atoms/cm^2 argon ions of 250 eV, and subsequently covered by overlayers various thicknesses (d_o). The peaks will be referred to as indicated in the figure. This notation has been introduced in previous work on argon irradiation of molybdenum [5]. We will first focus on the behavior of the Ar^B peak.

When d_o is increased, the area of the Ar^B peak increases while the peaks at lower temperatures virtually disappear. Apparently, the influence of the surface has disappeared when d_o exceeds 6 Å. Further increase of the overlayer thickness only yields a shift of the maximum of the Ar^B peak. This is caused by diffusional delay owing to the increased average distance to the surface. Fitting the diffusion behavior by a computer code HOP [6] yields the attempt frequency and the activation energy corresponding to the diffusion process. The grey curves in the figure show a very close match between the results of these simulations and the experiment. The corresponding activation energy is 4.16 eV, and the attempt frequency is $\nu = 3.5 \times 10^{16}$ Hz, or $D^0 = 2.8 \times 10^{-4}$ m^{2}s^{-1}. These values are in fair agreement with the values reported elsewhere [7] for self diffusion in molybdenum ($4.0 - 5.7$ eV). This indicates that the Ar^B peak originates from argon incorporated at substitutional positions in the molybdenum lattice, and that diffusion requires vacancy assistance, as is the case for self diffusion in molybdenum [7]. The value of ν is high in comparison with the Debye frequency, due to entropy effects [8].

All desorption peaks found at lower temperatures merge into the Ar^B peak when the overlayer thickness exceeds 6 Å. This means that these other desorption peaks also originate from argon in substitutional positions, because deposition of an overlayer only changes the proximity of the surface, but does not provide enough energy to change the trapping positions of the argon atoms. (Note that the small residue of the Ar^I peak at $d_o > 6$ Å is caused by parallax between the argon and molybdenum beams. An aperture defining the exposed substrate area inhibits complete overlap of both beams.)

The origins of the Ar^S and Ar^I peaks are demonstrated by the lowest spectra in figure 1. When d_o is increased from 0 to 2 Å, the Ar^S peak almost completely disappears, while the Ar^I peak increases slightly. A 2 Å overlayer ($\sim$ 1 monolayer) causes disappearance of argon positions in the first atomic layer below the surface. This leads to the conclusion that Ar^S is located in the first atomic layer, which also is corroborated by the small dissociation energy. It also demonstrates good surface coverage. This implies surface mobility, since without mobility one would expect 63% coverage based on Poisson statistics. This is not in agreement with surface diffusion data obtained at high temperatures [7], but does agree with the observation of a Zone II microstructure in molybdenum films grown at low temperatures [9]. Simultaneous with the disappearance of S-positions in the first atomic layer, the number of positions in the second atomic layer increases by the same amount, and decreases by the number of argon positions in the former second atomic layer. This leads to a net increase of the Ar^I peak. Since it is clear that Ar^S is located in the first atomic layer, and that the Ar^B peak originates from positions below 6 Å from the surface, the Ar^I positions have to be located at substitutional positions in the second and third atomic layer.

When d_o is increased, an Ar^{II} peak appears in the left flank of the Ar^B peak. The position of the peak maximum of the Ar^{II} peak is independent of the overlayer thickness, which rules out the possibility of thermal vacancy assistance during diffusion. At this point, it has to be realized that

250 eV argon bombardment not only causes incorporation of argon, but also creates vacancies and interstitials in the substrate. Other experiments (not shown) indicate the collapse of vacancy clusters at 1350 K, leaving mobile monovacancies in the substrate. A likely explanation is that this sudden excess concentration of monovacancies leads to rapid desorption of a part of the substitutional argon in the layer at 1380 K.

Approximation of the depth profile with a truncated Gaussian, using the relation between the areas of the Ar^I, Ar^{II} and Ar^B peaks following from the analysis described above, yields a projected range and a range straggle of 0.8 and 3.5 Å, respectively. These values are in agreement with values predicted by the computer code MARLOWE, as shown e.g. in [10].

Mono- and polyvacancies created by 250 eV argon bombardment

Argon ions with energies over 80 eV are capable of creating vacancies and interstitials in molybdenum [10]. Helium desorption spectrometry is employed to establish the amount and types of defects created by the argon bombardment. Figure 1 shows the helium spectra corresponding to the experiment described in the previous section. The defects are decorated by implanting helium ions with an implantation energy of 100 eV, which is below the threshold energy for vacancy creation in bulk molybdenum (250 eV). The helium fluence is 1×10^{14} He/cm^2, and the heating rate is $\beta = 40$ K/s. The peaks will be referred to as indicated in the figure [5].

The A and B peaks originate from trapping of helium atoms by substitutional argon atoms at Ar^B positions, i.e. argon at interstitial positions further than 6 Å below the surface. It has been shown that the A-peak corresponds to one helium atom bound to a substitutional argon atom, while the B-peak corresponds to multiple (up to 5) helium atoms bound to one substitutional argon atom [5]. The fact that the A and B-peak originate specifically from Ar^B positions will be demonstrated conclusively in the next section. When the overlayer thickness is increased, the B-peak increases relative to the A-peak. This means that the average number of trapped helium atoms per argon atom increases correspondingly, which implies an increased trapping probability. This does not automatically mean that the number of substitutional argon atoms increases correspondingly. The simple fact of putting an overlayer on top of the substrate itself already increases the trapping probability. Correction for this effect yields that despite this, the number of traps increases with the overlayer thickness, which is in correspondence with the observed increase of the Ar^B peak.

The H-peak at 1200 K has been ascribed to monovacancies containing a single helium atom, and the I-peak has been associated with vacancy clusters in the molybdenum lattice [5]. The behavior of the H-peak resembles the behavior of the Ar^B peak. Both peaks grow when the overlayer thickness increases. Monovacancies created by the argon beam will especially be formed in the first atomic layers. Apparently, superficial monovacancies manifest themselves at lower temperatures than the H-peak, which is presumably caused by the same mechanism as suggested for the Ar^B peak. HOP simulations show that the combined area of the H and I-peaks corresponds to a defect concentration of $(5 \pm 3) \times 10^{-4}$. This corresponds to a net defect creation rate of $(0.7 \pm 0.4) \times 10^{-3}$ vacancy/argon.

The presence of single-filled monovacancies is demonstrated by the H-peak. When the helium decoration fluence is sufficiently high, part of the the monovacancies will be multiple filled, giving rise to a series of peaks between 800 and 1000 K [5]. The distribution and area of the peaks observed in the present spectra, however, cannot be explained by this mechanism alone. We will show in the next section that the broad band of peaks between the B and H-peaks is caused by incorporated argon.

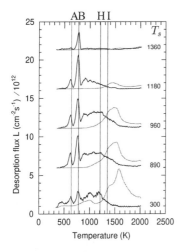

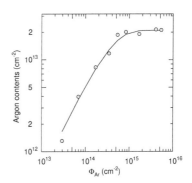

Figure 2: Helium (black) and argon (grey) desorption spectra of a 50 Å Mo IBAD film ($E = 250$ eV, IAR $= 0.02$) deposited at various substrate temperatures T_s. $\Phi = 1 \times 10^{14}$ He/cm^2 and $\beta = 40$ K/s.

Figure 3: Incorporation of 250 eV argon in a polycrystalline molybdenum substrate.

Incorporation of argon in IBAD layers deposited at elevated temperatures

Increasing the substrate temperature T_s implies increasing the surface mobility. When the surface mobility is sufficiently high, formation of vacancies in the layer is prevented. On the other hand, argon incorporation is expected to persist up to higher temperatures since the part of the argon that is implanted sufficiently deep can only escape through vacancy assisted diffusion. Consequently, by ion beam assisted deposition at elevated temperatures the effects of incorporated argon on the helium desorption spectrum can be isolated.

Figure 2 shows the helium and argon desorption spectra corresponding to a 50 Å molybdenum layer deposited at various substrate temperatures T_s. The ion-to-atom arrival ratio (IAR) is 0.02, and the argon ion energy is 250 eV. Deposition at 1360 K shows unambiguously that the A and B peaks originate from substitutional argon atoms at positions below 6 Å, since argon at shallower positions in the layer (Ar^l) can not be incorporated at this temperature.

The measured argon concentration and the trap concentration as determined by the helium trapping probability are both 2×10^{-4} at a substrate temperature of $T_s = 1180$ K and $T_s = 1360$ K. At these temperatures formation of mono- and polyvacancies during deposition is not expected, since several experiments [11, 12] indicate mobility of vacancies at 550 K. Apparently, all incorporated argon acts as helium trap. This leads to the conclusion that the band of peaks between 800 and 1200 K also originates from argon in substitutional positions, which confirms the suggestion made above about the origin of the band.

Trapping and self sputtering probability of argon

When the argon fluence is increased, the amount of incorporated argon also increases. Simultaneously, the self sputtering probability of incorporated argon increases owing to the increased probability for an incoming argon atom to encounter an incorporated argon atom. By observing the amount of incorporated argon as function of argon fluence, the trapping and the effective cross-section for self sputtering can by estimated. Figure 3 shows the argon contents of a poly-crystalline molybdenum substrate for various argon fluences. The solid line has been fitted to the data points, based on the following simple model for trapping and self sputtering:

$$N_{tr}(\Phi_{Ar}) = \frac{p}{\sigma} \left(1 - e^{-\sigma \Phi_{Ar}}\right) \tag{1}$$

where N_{tr} is the number of trapped argon particles per unit area, Φ_{Ar} is the argon fluence, p is the trapping probability, and σ is the effective cross-section for argon self sputtering. Based on this fit, the trapping probability is $p = 6.5\%$, and $\sigma = 31 \text{ Å}^2$. Using the maximum implantation depth of 8 Å, it follows that saturation occurs when the argon concentration is 4×10^{-3}.

CONCLUSIONS

Argon is incorporated in thin molybdenum films grown under 250 eV argon bombardment. Direct desorption measurements have been used to establish the argon implantation profile. A projected range and range straggle of 0.8 and 3.5 Å were found. Argon is incorporated at sub-stitutional positions. The creation rate of defects by argon was studied by helium desorption spectrometry. A net creation rate of $(0.7 \pm 0.4) \times 10^{-3}$ vacancy/argon atom was found. Ion assisted deposition at elevated substrate temperatures shows that all incorporated argon acts as helium trap. Argon fluence variations show an effective cross-section for self sputtering of 31 Å², a trapping probability of 6.5%, and a maximum achievable argon concentration of 4×10^{-3}.

ACKNOWLEDGEMENTS

This work is part of the research program of the Stichting voor Fundamenteel Onderzoek der Materie (Foundation for Fundamental Research of Matter), and was made possible by financial support from the Nederlandse Organisatie voor Wetenschappelijk Onderzoek NWO (Netherlands Organization for Scientific Research).

REFERENCES

1. F.A. Smidt, International Materials Reviews **35**, 61–128 (1990).

2. J. van der Kuur, E.J.E. Melker, T.P. Huijgen, W.H.B. Hoondert, G.T.W.M. Bekking, A. van den Beukel, and B.J. Thijsse, Mat. Res. Soc. Symp. Proc. **396**, 587 (1996).

3. A.A. van Gorkum and E.V. Kornelsen, Vacuum **31**, 89 (1981).

4. A. van Veen, A. Warnaar, and L.M. Caspers, Vacuum **30**, 109 (1980).

5. A. van Veen, W.Th.M. Buters, G.J. van der Kolk, L.M. Caspers, and T.R. Armstrong, Nucl. Instrum. Methods **194**, 485–489 (1982).

6. W.H.B. Hoondert, *Trapping and diffusion of noble gas atoms in some off-stoichiometric ceramics studied by thermal desorption spectrometry*, Ph.D. thesis, Delft University of Technology, 1993.

7. H. Mehrer (ed.), *Landolt-Börnstein*, vol. 26, Springer-Verlag, 1990.

8. D.L. Smith. *Thin-Film deposition*. McGraw-Hill, Inc., 1995.

9. O.P. Karpenko, J.C. Bilello, and S.M. Yalisove, J. Appl. Phys. **76** (8), 4610 (1994).

10. H.A. Filius and A. van Veen, Radiat. Eff. Defects Solids **108** 1, (1989).

11. R.H.J. Fastenau, L.M. Caspers, and A. van Veen, Phys. Stat. Sol. A **34** 227 (1975).

12. K. Maier, M. Peo, B. Saile, H.E. Schaefer, and A. Seeger, Philos. Mag. A **40** (5), 701 (1979).

SIMULATION OF INTERSTITIAL CLUSTER MOBILITY AND CLUSTER MEDIATED SURFACE TOPOLOGIES

B. D. WIRTH*, G. R. ODETTE*, D. MAROUDAS**, and G. E. LUCAS**
*Department of Mechanical and Environmental Engineering, University of California, Santa Barbara, CA 93106
**Department of Chemical Engineering, University of California, Santa Barbara, CA 93106-5080

ABSTRACT

Molecular statics and molecular dynamics (MD) simulations based on the embedded atom method (EAM) are used to model the energetics and mobility of tightly bound clusters of self-interstitial atoms (SIA) in bcc iron. Single and clusters of SIA are directly produced in displacement cascades generated in neutron and high energy charged particle beam irradiations. The clusters are composed of <111> split dumbbells and crowdions with binding energies in excess of 1 eV. Clusters containing specified 'magic' numbers of SIA can be described as perfect prismatic dislocation loops with Burgers vector b=a/2<111>; however, the core region is extended compared to an isolated edge dislocation and the loops are intrinsically kinked. As the loops grow, SIA occupy successive edge rows, with minimum energy cusps found at the magic numbers corresponding to filled hexagonal shells. The SIA clusters are highly mobile, undergoing rapid one-dimensional diffusion on their glide prism. The activation energy for glide diffusion is less than 0.3 eV and the corresponding mechanism is related to easy motion of the intrinsic kinks. The kinks, which are preferentially observed on the hexagonal corners, propagate around the loop periphery, resulting in stochastic increments of glide. Image drift forces bias cluster motion near free surfaces and the consequential annihilation of clusters at the surface produces islands bounded by hexagonal ledges. The potential effect of islands modifying surface topology is also discussed. While these simulations are specific to iron, similar behavior is expected for other cubic alloys.

INTRODUCTION

The structure and migration characteristics of self-interstitial clusters has significant impact on the micro/nanostructural evolution under neutron irradiation. Self-interstitial atoms (SIA) and their clusters are formed in neutron and charged particle induced displacement cascades [1-3][1]. Despite its importance in understanding the kinetics of micro/nanostructural evolution during irradiation [4,5], quantitative understanding of the structural, energetic, and transport properties of point defects and point-defect clusters in body-centered cubic (bcc) iron is still lacking.

In this paper, a comprehensive study is presented of the structure, energetics and dynamics of small SIA clusters in bcc iron, with up to 92 atoms (N ≤ 92), using the Finnis-Sinclair type many-body potential, as modified by Calder and Bacon [2,6]. Through a variant of the simulated annealing method we determine the structural characteristics and cluster formation energetics. Our structural relaxation scheme involves isothermal-isobaric Metropolis Monte Carlo (MC) simulation in conjunction with conjugate gradient quenching [7]. In all of the simulations reported here, periodic boundary conditions are applied to supercells that contain from 2000+N to 54000+N atoms. The different supercell sizes have been used to ensure that interactions of the cluster with its periodic images are negligible. The dynamics of SIA clusters has been studied by molecular dynamics (MD) simulation at temperatures of 560 K, corresponding to typical irradiation temperatures encountered in nuclear reactor pressure vessels. The MC and MD are carried out using a modified version of the MOLDY code [8]. A number of the MD simulations have been performed with SIA cluster sinks present in the simulation cell, such as internal free surfaces of different orientation to determine the effect of image forces in biasing cluster motion.

---

[1] A few key references are cited representative of a larger literature.

RESULTS AND DISCUSSION

Self-Interstitial Cluster Structure and Energetics

The structure of SIA clusters in iron has been a matter of some confusion and controversy. The lowest energy configuration of an isolated SIA is a <110> split dumbbell. Two metastable configurations are very near to the ground state energy, the <111> split dumbbell has a formation energy only 0.11 eV higher, while the so called crowdion <111> SIA orientation is only 0.15 eV higher than the <110> split dumbbell. Crowdions are defined as arrays of atoms in the close packed directions, where m atoms share m-1 lattice sites. If m is even, the center of mass of the array is located at a lattice site and the configuration is a <111> split dumbbell; if m is odd, the center of mass is located symmetrically between two lattice sites and the configuration is the <111> crowdion. Note, that the <111> crowdion and split dumbbell configurations are separated by a translation of a/4<111> and, indeed the crowdion is the saddle point configuration for migration of the <111> split dumbbell [5]. The atomic displacements around the lattice sites decrease with distance from the SIA center-of-mass.

Clusters of SIA with <110> orientations have very high energies associated with extensive strain field overlap. However, when the SIA in clusters are in the <1̄11> crowdion/split dumbbell orientation, positive and negative dilation centers align in much lower energy configurations. Minimum energy configurations are found when the clusters are made up of mixtures of <111> split dumbbells and <1̄11> crowdions. The centers of mass of the SIA lie in an extended (>2) set of {110} planes. Note in this view of the clusters, the SIA maintain their individual identity, and the clusters have a three dimensional character. Figure 1 shows [11̄1] (Figure 1a) and [001] (Figure 1b) projections of clusters containing 37 and 91 SIA respectively (special magic numbers which are discussed below). The circles are the center of mass of the extended arrays of atoms in the <111> direction; filled circles correspond to crowdions while the open circles are <111> split dumbbells. The number of (220) planes occupied by the clusters is shown in Figure 1c. The 37 and 91-member SIA clusters occupy eight and six (220) planes, respectively.

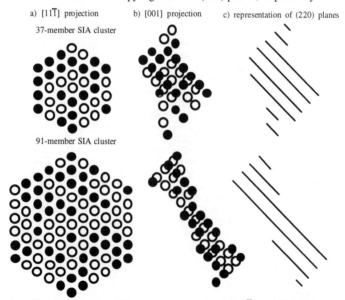

a) [11̄1] projection   b) [001] projection   c) representation of (220) planes

37-member SIA cluster

91-member SIA cluster

Figure 1 -- 37 and 91-member SIA cluster structure in a) [11̄1] and b) [001] projection. Filled and open circles are the center of mass of <111> crowdions and split dumbbells, respectively. c) Additional (220) lattice planes occupied by the clusters.

However, the [11$\bar{1}$] projection shown in Figure 1a resembles a prismatic dislocation loop, bounded by a/2<111> edge dislocations. A Burger's loop yields the expected a/2<111> vector. Hence, it is useful to consider the configuration of an isolated edge dislocation that would form each of the six edges of a prismatic loop. An edge dislocation is composed of two extra partial planes of atoms that maintain the ABAB stacking sequence of (220) planes in a bcc crystal. In the prismatic configuration, this would be a quasi two dimensional (compact) perfect loop, e.g., without a stacking fault. Thus, the major difference between a perfect prismatic edge dislocation loop and the SIA cluster is the much larger number of {110} planes occupied in the latter. Note that the number of planes decreases with loop size. Based on a simple linear extrapolation, the SIA cluster would be expected to approach a nominal perfect loop configuration at minimum sizes on the order of 270 SIA. This corresponds to a loop diameter of about 4.7 nm.

Indeed, the alternative descriptions can be unified, within a more general framework of dislocation structures by considering the SIA cluster to be a kinked, and in some cases highly kinked, prismatic loop. It may seem strange that the ground state energy for such a loop is intrinsically kinked; however, this is conceptually not that surprising if one considers the strong self-interaction energies, perhaps, akin to the symmetrically kinked segments in dislocation dipoles. The perfect hexagonal loop character occurs only with the filling of successive hexagonal shells surrounding a central SIA with a total of 7, 19, 37, 61, 91, ..., 1+6Σ_2(n-1) SIAs. Cluster sizes in between the magic numbers can be considered to contain a jog.

Figure 2 shows the cluster energy as a function of size. In all cases the SIAs, which have a formation energy of 4.76 eV in the lowest energy <110> split dumbbell orientation, are tightly bound to the clusters with energies in excess of 1 eV/SIA (Figure 2 inset). Note that the cluster energy is minimized and the binding energy maximized at the cluster sizes corresponding to filled hexagonal shells, the so-called magic clusters. We have compared the formation energy of the magic clusters with a continuum elasticity expression for the self-energy of a prismatic hexagonal dislocation loop given by Hirth and Lothe [9] as:

$$E = \frac{6\mu b^2 L}{4\pi(1-\upsilon)}[\ln(\frac{L}{\rho}) + C]$$

(1)

where μ is the shear modulus, ν is Poisson ratio, b is the burger's vector magnitude, L is the hexagonal side length, ρ is the cutoff radius, and C is a constant which includes dislocation-dislocation interactions and the dislocation core energy. Using nominal values of the shear

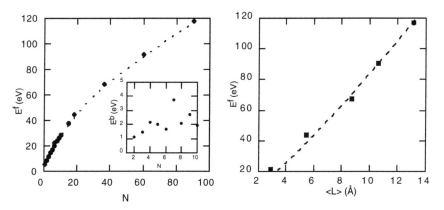

Figure 2 -- SIA cluster formation and binding (inset) energy with cluster size N.

Figure 3 -- Formation energy with hexagonal loop edge size <L>, fit provided by eq (1).

modulus and Poisson ratio for bulk bcc iron and the recommended value of ρ [9] provides good agreement with SIA cluster formation energetics, as shown in Figure 3, fit to a core energy of 0.4 eV/Å. This core energy is consistent with other computer simulation studies using pair potentials to describe the edge dislocation in iron having a range of core energies between 0.4 to 0.6 eV/Å [10].

Self-Interstitial Cluster Mobility

A number of MD simulations performed at a temperature of 560 K have revealed a common migration mechanism for SIA clusters over the size range $3 \leq N \leq 91$. The most salient feature of the migration process is that it occurs as a fully one-dimensional random walk with a very low activation energy. Within the timescale of the simulations there are no indications that the cluster will reorient to another <111> direction, or that there is any substantial change in cluster configuration. Further details of the description of the mechanism depends on the viewpoint adopted to describe the cluster. Viewed as identifiable cluster SIAs the migration mechanism is the collective local disassociation and reassociation of SIAs by a sequence of expansion and contraction of the number of occupied {110} planes, causing what might be described as an amoebae-like motion. These local expansion-contraction or disassociation-reassociation processes tend to start at the cluster corners and edges.

Indeed, the other view is that the cluster is an intrinsically kinked, hexagonal perfect dislocation loop, where highly extended cluster locations are taken to be the kinks. In this view the easy motion of the loop on its glide prism is associated with the easy propagation and re-formation of the kinks. While it is difficult to conceive of small clusters as loops (mostly cores), the agreement with loop energetics shown in Figure 3 is strong evidence that this is the case. Indeed, even at very small sizes ($N \leq 11$) an approximate $N^{2/3}$ scaling is observed; and the apparent excess energy of non-magic number clusters (compared to values from fits to the magic number loop energies with filled hexagonal rows) ranges from a total of about 0.5 to 2.5 eV averaging about 1.5 eV. This is reasonably consistent with the energy of edge dislocation jogs of around 1 eV [9]. The association of SIA clusters with jogged and kinked loops is very powerful, since it provides a link between the physics of defect clusters to understanding based on a much more developed field of dislocation mechanics and dynamics.

We have not yet quantified the mobility of the clusters/loops. However, a 37-member SIA cluster at 560 K placed in a simulation supercell of 35189 atoms had a net displacement in excess of 5.5 nm in 180 ps. As shown in Figure 4, a 91-member cluster had a net displacement of (x_{cl}) about 6 nm in 135 ps. Using the relation for one-dimensional random walk $x_{cl}^2 = 2D_{cl}t$ and $D_{cl} = D_{cl}^o \exp(-E_{cl}^m / kT)$, yields estimates of E_{cl}^m of 0.04, 0.15 and 0.26 eV for D_{cl}^o of 10^{-6}, 10^{-5} and 10^{-4} m^2/s, respectively. We have previously found that the activation barrier for <111> migration of a <111> SIA to be 0.04 eV [5].

Thus, even an upper bound estimate suggests that the cluster/loops are highly mobile. There are at least two significant conceptual consequences of SIA cluster migration on the micro/nanostructural evolution which occurs under neutron irradiation. First, the directionality of SIA diffusion may limit vacancy-interstitial recombination during cascade aging (a direct or near hit between a mobile SIA cluster and a stationary vacancy cluster is required for recombination). Second, the high mobility of SIA clusters separates the time scales over which the fate (recombination or diffusion away from the cascade) of interstitial features is determined relative to vacancies. In this case, cascade vacancy aging processes can be decoupled from cascade interstitial aging processes [11].

We have also performed SIA cluster migration simulations at 560 K in the presence of cluster sinks, such as vacancy loops and internal surfaces. In the former case, a four-member SIA cluster annihilated with a 10-member faulted vacancy loop of a/2[110] orientation initially located 2.9 nm from the SIA cluster. In spite of the fact that the <111> trajectory of the SIA cluster intersected only the edge of the vacancy loop, strong loop-loop interactions led to rapid diffusional drift of the SIA cluster, resulting in complete annihilation of the interstitials. Other studies have focused on the dynamics of a 37-member SIA cluster in the presence of several low index ({100},{110} and {111}) internal free surfaces. Part of the motivation for these simulations is to determine the effect of drift bias due to the image forces on the SIA cluster migration and the behavior of the cluster, such as re-orientation, near the sink.   In the case of {110} and {111} surfaces oriented parallel to the glide prism, the clusters showed no evidence of changing either migration mechanisms or orientation within the 100 ps of the MD simulations.

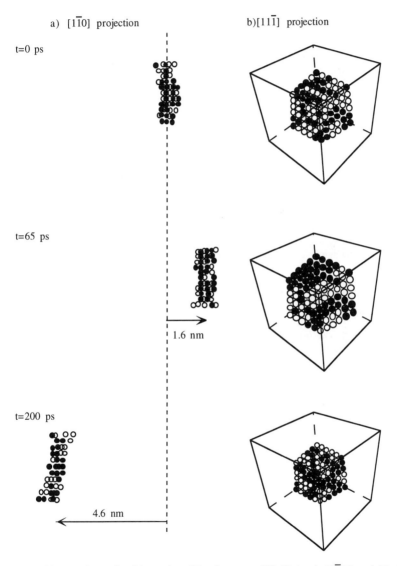

a) [1$\bar{1}$0] projection   b)[11$\bar{1}$] projection

t=0 ps

t=65 ps

1.6 nm

t=200 ps

4.6 nm

Figure 4 -- MD snapshots of a 91-member SIA cluster at 560 K in a) [1$\bar{1}$0] and b) [11$\bar{1}$] projection at 0, 65 and 200 ps. Filled and open circles are the center of mass of <111> crowdions and split dumbbells, respectively.

However, {100} and {110} free surfaces which intersect the glide prism attract the SIA clusters over short distances. Upon reaching the surface, the clusters reconstruct slightly to epitaxially form distorted hexagonal islands, which persist during further simulation of approximately 10 ps. Note, over longer times the island would be expected to reconfigure to form a lower energy (by a Wulff-type reconstruction) island. The creation of such surface features under ion bombardment may influence crystal growth and surface chemical reaction rates (e.g., ion assisted processes). Measurement of surface topologies using atomic force microscopy methods may also be a useful approach to characterize SIA cluster formation and migration properties.

It should be noted that these results must be considered as qualitative and may be influenced by other processes not modeled (e.g., SIA cluster trapping/detrapping interactions with solutes and other microstructural features) as well as details of the simulation (e.g., the interatomic potential). Further, the interaction of SIA clusters with internal sinks such as dislocations and grain boundaries is a key issue that is yet to be resolved. Finally, while these simulations are specific to iron, similar behavior is expected for other cubic alloys.

CONCLUSIONS

In summary, we have presented the results of a comprehensive atomistic-scale MD and MC study of the structural and migration characteristics of SIA clusters in bcc iron using a Finnis-Sinclair type interatomic potential. SIA clusters form as extended, prismatic hexagonal loops with a Burgers vector of $a/2<111>$. We have shown that these SIA clusters are consistent with highly kinked prismatic dislocation loops with energies that are in good agreement with continuum elasticity theory, assuming a nominal dislocation core energy of 0.4 eV/Å. The SIA clusters are highly mobile, with diffusional migration energies < 0.3 eV, undergoing easy one-dimensional glide on hexagonal {110} slip prisms, in association with the intrinsic presence and easy motion of kinks. The SIA cluster migration mechanisms and orientations do not change in the presence of free surfaces, although image forces bias cluster migration near the intersecting surface sinks. The mobile clusters initially form hexagonal islands upon intersecting a surface. The effect of cluster migration on cascade aging processes and island formation in mediating surface topologies and reaction sites in cubic, metallic materials has been noted.

ACKNOWLEDGMENTS

Fruitful discussions with G. E. Beltz (UCSB) and R. E. Stoller (ORNL) are gratefully acknowledged. This work was supported by the US Nuclear Regulatory Commission under contract number NRC-04-94-049.

REFERENCES

1. M. T. Robinson, J. Nucl. Mater. **216** (1994) 1.
2. A. F. Calder and D. J. Bacon, J. Nucl. Mater. **207** (1993) 25.
3. W. J. Phythian, R. E. Stoller, A. J. E. Foreman, A. F. Calder and D. J. Bacon, J. Nucl. Mater. **223** (1995) 245.
4. G. R. Odette, in IAEA Technical Report Series, Vienna, in press.
5. B. D. Wirth, G. R. Odette, D. Maroudas and G. E. Lucas, J. Nucl. Mater. **244** (1997) 185.
6. M. W. Finnis and J. E. Sinclair, Philos. Mag. A **50** (1984) 45.
7. D. Maroudas and R. A. Brown, Phys. Rev. B **47** (1993) 15562.
8. M. W. Finnis, MOLDY 6, UKAEA Harwell Laboratory, AERE R-13182 (1988).
9. J. P. Hirth and J. Lothe, Theory of Dislocations, 2nd edition, Krieger Publishing, Malabar, FL 1992.
10. J. O. Schiffgens and K. E. Garrison, J. Appl. Phys., **43** no. 8 (1972) 3240.
11. G. R. Odette and B. D. Wirth, "A Computational Microscopy Study of Nanostructural Evolution in Irradiated Pressure Vessel Steels", J. Nucl. Mater., in press.

ADVANCES IN ION AND LASER BEAM TECHNOLOGY: ACHIEVEMENTS OF JAPANESE GOVERNMENT AND UNIVERSITY PROJECTS

Isao Yamada
Ion Beam Engineering Experimental Laboratory, Kyoto University,
Sakyo, Kyoto 606, Japan   i-yamada@kuee.kyoto-u.ac.jp

ABSTRACT This paper reviews recent R&D activity in advanced beam technologies funded by MITI and JST. A large-scale national project known as AMMTRA (Advanced Material-Processing and Machining Technology Research Association) funded by MITI has made a significant contribution to industrial applications of beam processing by developing high density ion and laser beam equipment. A subsequent program, the ACTA (Advanced Chemical Processing Technology Research Association) project involves development of multi-beam deposition systems for metal and dielectric materials formation including several systems which combine ion and laser beams, ion beams and CVD and plasma, and laser beams with sputtering. NEDO Proposal Based Project and JST (The Japan Science and Technology Corporation)-Exploitation and Application Study Project are fundamentally aimed to transfer technology from university to industry. Research in gas cluster ion beam technology is currently being conducted in the following areas: (i) shallow ion implantation for 0.1 μ m PMOS junction formation, (ii) high rate etching and smoothing for Si, diamond, SiC and metal oxide films and (iii) thin film depositions for optical filter and transparent conductive film. The gas cluster ion beam processes are discussed in comparison with traditional ion beam processing methods which are presently limited by available atomic and molecular ion beams.

INTRODUCTION

The history of ion beam processing of materials dates back at least as far as 1907 when J.J.Thomson confirmed, for the first time, the existence of a positive ray by reproducing an experiment done by Goldstein in 1885 [1].This experiment also pointed out that there exist surface modification processes such as implantation, etching and coating. According to Wegmann [2], the first ion implantation experiment was done by Rutherford in 1906 when he bombarded aluminum foil with α particles. Since a beginning in the year 1960, industrial applications of atomic and molecular ion beams have progressed extensively and now involve a wide range of processes. Ion beams are particularly indispensable as doping technology for semiconductor device fabrication.

The advancement of ion beam technology is strongly related to equipment development. During the last few years, extensive development has been made in advanced ion beam equipment and processing technologies, particularly for processing of materials other than semiconductors. A large-scale Japanese national project named AMMTRA (Advanced Material-Processing and Machining Technology Research Association) was conducted during the period 1986-1993 and made significant contributions to industrial applications by developing large scale ion beam systems [3,4]. Following this project, during the period 1990-1996, the ACTA (Advanced Chemical Processing Technology Research Association) project was conducted to establish new processes for material synthesis and modification by means of combined physical and chemical processing [5].

Quite new and unique approaches have also been started in the US. One well-known technology is called PIII (Plasma Immersion Ion Implantation) [6] or PSII (Plasma Source Ion Implantation) [7]. In Japan, research and development of a three dimensional ion implantation technique which is similar to PIII was started in 1995 as a NEDO project at the Ion Engineering Institute Corporation [8].

Since the discovery and subsequent development of ion beam processing, beams of either atomic or molecular ions have been used. The author has proposed a process technology which uses a beam of ions consisting of a few hundreds to thousands of atoms generated from various kinds of gaseous materials. This is referred to as 'gas cluster ion beam processing' [9-11]. In gas cluster ion beam processing, non-linear collisions occurring during the impact of accelerated cluster ions upon substrate surfaces produce fundamentally low energy bombarding effects at very high density. These bombarding characteristics can be applied to (i) shallow ion implantation, (ii) high yield sputtering and smoothing, (iii) surface cleaning and (iv) low temperature thin film formation. These characteristics could extend new application fields which would not be possible by traditional ion beam processing. Research of gas cluster ion beam processing is now going on in different industrial fields such as electron devices, optical devices and mechanical components under a NEDO consortium program and under a program of the Japan Research Development Corporation (JST) [12].

Continuing demands for new and improved materials and structures needed to support the progress of high technology industries are pushing ion beam technology further towards its limits in terms of ion energy, ion dose, deposited energy density and processing temperature and wafer size. Equipment capable of meeting requirements of very low energy, very high energy, high current, and large target area for many different applications is desired. This paper reviews recent progress of advanced beam processes which will be needed to meet future requirements for new technology and novel materials.

LARGE SCALE NATIONAL PROJECTS UNDER THE NEDO PROGRAM

Ion beam technology promises to provide important new materials for use in diverse engineering fields. However, there are many examples where potentially useful industrial applications have not yet been realized even though extensive successful laboratory results are available. A key barrier to be overcome is often that results obtained on a laboratory scale still require substantial research and development using large-scale equipment. The issues of production feasibility, yield, cost and other important factors must be taken into account before useful industrial applications can be established. In order to overcome these problems, development of much larger scale equipment is necessary.

The national research and development program popularly known as the large-scale project of NEDO carries out technological R&D activities in various industrial fields in close cooperation between government research institutes and industrial and academic organizations. Two of these projects, the AMMTRA project and the ACTA project are summarized below:

AMMTRA Project

As has already been reported in numerous international conferences and articles, the AMMTRA project included development of a high-current metal ion beam system (Nissin Electric Co. Ltd., Al, Si ions:100mA), an integrated low-energy ion beam system (ULVAC Japan Ltd., Ar ion:100eV, 1mA), a high-current large-sheet ion beam system (ULVAC Japan Ltd., N_2 ion :100keV,250mA), an ionized multiple-beam system with high deposition rate (Mitsubishi Elec. Corp., Al, Bi, Sr, Cu, Ca beams: deposition rate 1μ m/min on 200 mm ϕ substrate area), a focused gas-phase ion beam system (JEOL Ltd., He ions:10nm beam diameter, 10pA, 100keV), and a high-energy ion beam system (Hitachi Ltd., N ion: 4MeV, >1mA). Concepts and schematics of the five advanced industrial-scale ion beam systems are shown in Fig. 1. The project was successfully completed in 1993 [3,4].

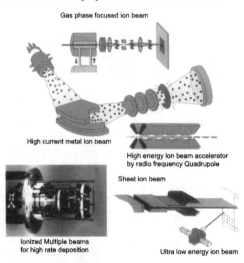

Gas phase focused ion beam

High current metal ion beam

High energy ion beam accelerator by radio frequency Quadrupole

Sheet ion beam

Ionized Multiple beams for high rate deposition

Ultra low energy ion beam

Fig.1   AMMTRA R&D project for large scale ion beam equipment development.

ACTA Project

The ACTA project was started by 21 industrial organizations, 6 government institutions and 1 US institute in December 1990 as one of several large scale projects funded by NEDO. The ACTA project has focused on setting the targets for super-precise, super-fine manufacturing systems and super-qualified materials synthesis and surface modifications. The purpose and intended technological breakthroughs are very ambitious and the technological approach is also new and original. The aim is to develop innovative process technologies to impart to materials high grade sensing functions, active chemical reactions, high performance magnetism, strong anti-corrosion and heat resistance, high toughness or biological adaptability [13].

Unique processes which have been developed under the ACTA project have been basically combined operations. They have included (i) interface modification processes in ion beam assisted film deposition for thick film coatings (Nissin Electric Co., Ltd. and Sumitomo Electric Industries Ltd.) [14, 15], (ii) combined CVD and ion beam processes (Ishikawajima-Harima Heavy Industries Co., Ltd.) [16], (iii) multi-source plasma CVD processes (Denki Kagaku Kogyo Co., Ltd.) [17], (iv) a sputter deposition system combined with ion beam and laser beam irradiation (Matsusita Electric Industrial Co., Ltd.) [18], (v) a combined high temperature gas beam and high velocity molecular beam process and (vi) molecular beam and laser beam combined processes. Detailed results from these have been reported [19]. Some of the related ion and laser processes and their results are discussed below.

In ion beam assisted deposition for thick film coatings, special attention has been paid to interface formation. Strong adhesion characteristics have been accomplished by controlling interface layers of deposited Si and Ni films on WC-Co substrates [14]. A combined ion beam irradiation and vapor deposition method has been developed and applied. A very thick layer of 400μ m of Ni-TiN nano-composite film was formed on the above mentioned interface layer by a newly developed filtered vacuum arc-evaporation system. The film is a graded Ni-TiN nano-composite material. Figure 2 shows the concept of the equipment. The films are expected to be used as cutting tool materials with characteristics of both extreme hardness (>Hv 19.6 Gpa) and very high fracture toughness (>10MN/m$^{2/3}$) superior to those of conventional ceramic-metal composite materials [15]. Figure 3 shows the characteristics of high toughness and high hardness materials developed for cutting tool applications. The area of composite materials combines good hardness and high toughness, and, as indicated in this figure, represents the focus of new research.

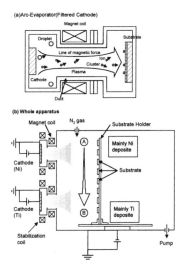

Fig.2 Schematic diagrams of (a) filtered cathode arc-evaporator and (b) complete apparatus – from ref.[15].

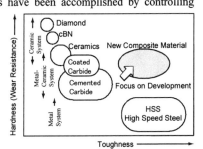

Fig.3 Characteristics of several high hardness materials and specification of areas of interest under this project – from ref.[15].

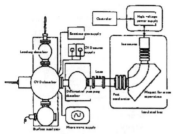

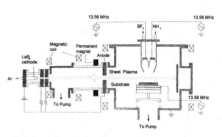

Fig.4 Schematic diagram of combined CVD and ion beam system – from ref.[16].

Fig.5 Schematic diagram of combined plasma and CVD system – from ref.[17].

Combined plasma CVD and ion beam processing has been developed for making ceramic insulating films which have strong adhesion and high electrical insulation characteristics at high temperature. The project objective is to form a very highly resistive ($> 10^{12}\Omega \cdot$ cm) and highly adherent thin film of aluminum oxide for use at an operating temperature of 800℃ [16]. Aluminum oxide films have been deposited on nickel based super alloy (Inconel 718) by thermal CVD and plasma and ion assisted CVD. A process combining CVD and ion beam system has been developed. In this process, ion beams are generated in a conventional ion source and are mass analyzed by a sector-type magnet system and are then accelerated to a differentially pumped chamber connected to the CVD system. Figure 4 shows the construction of this system. Plasma is generated in the CVD chamber by microwave excitation. In the CVD process the source material is (iso-$C_3H_7O)_3$Al and substrates are Inconel 718. Kr ions are accelerated at 10kV towards the CVD chamber. The insulating films will be used for fabrication of sensors applied on turbine blades.

A multi-source plasma CVD system has been developed to form SiO_x, SiN_x, and SiC_x thin films at high deposition rate (1.7 μ m/min) [17]. As shown in Fig.5, plasma is formed by an

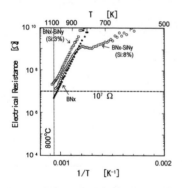

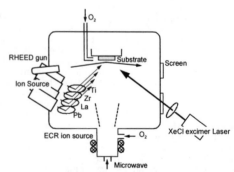

Fig.6 Electrical resistance dependence on temperature of NBx film an dBNx-SiNy composite films deposited by combined plasma and CVD system – from ref.[17].

Fig.7 Schematic diagram of combined laser and sputtering system – from ref.[18].

electron beam emitted from a LaB_6 cathode in a plasma chamber with confined magnetic field. The plasma is introduced through the anode slit attached to the CVD chamber. A sheet plasma is produced by a magnetic field and introduced above a substrate holder. Figure 6 shows the electrical resistance dependence on temperature of BN_x films and BN_x-SiN_y composite films. SiH_4 and NH_3 gases were separately introduced through two plasma generators above the substrate. The films thus produced show very high resistance of $10^7 \Omega$ at 800℃.

A system for sputter deposition combined with laser and ion beam irradiation, shown in Fig.7, has been used to produce high quality ferroelectric thin films with perovskite structure at a (low) substrate temperature of 415℃ [18]. As well, high quality films have been produced on room temperature substrates by use of excimer-laser irradiation during film deposition. This technique can be applied to form various kinds of IR sensors, strain gauges, etc.

NEDO PROPOSAL BASED PROJECT

In addition to the large scale research and development projects, quite new and unique procedures have been proposed and carried out by funding from a Proposal-Based Advanced Industrial Technology R&D Program of NEDO. One such project, referred to as '3 dimensional ion implantation' is a plasma pulse type ion process [6,7] which was developed under a

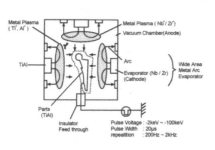

Fig.8   Schematic diagram of plasma pulse type ion process – from ref.[8].

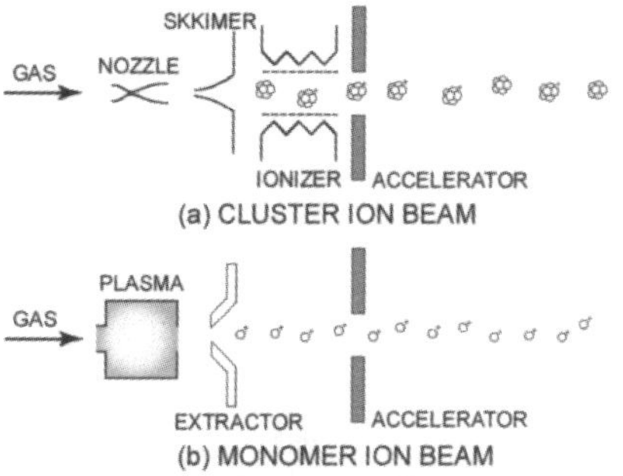

Fig.9   Comparison of cluster ion beam and traditional ion beam system

collaboration of Doshisha University and the Ion Engineering Institute Corporation. Figure 8 shows the schematic of the designed equipment. Equipment development is currently being carried out [8].

Since the discovery and subsequent development of ion beam processing, beams of either atomic or molecular ions have been used. Materials processing using mass-analyzed gas cluster ion beams, proposed in ref [9] represented a novel development in this field. A comparison of gas cluster ion beam system and a

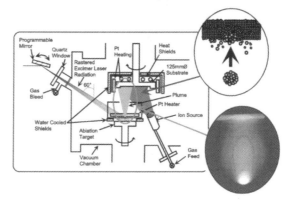

Fig.10   Gas cluster and laser combined thin film deposition system

traditional one is shown in Fig.9. This project has enabled   the study of material modification by large gas cluster bombardment. Surface modification of ceramics such as Al_2O_3 has been demonstrated [19]. In addition to the fundamental R&D, a NEDO Consortium R&D project has been started to fabricate ultra-high-quality transparent conductor films by applying a combined process of gas cluster ion beam and laser beam. The project objective is to form very low electrical resistivity, high optical transmission films at low substrate temperature ($<10^{-5}$ $\Omega \cdot$ cm at $100°C$) [20]. The concept of the deposition equipment is shown in Fig. 10 [21]. The equipment essentially consists of combined gas cluster and vapor deposition by e-beam evaporation or laser ablation.

JST PROJECTS FOR GAS CLUSTER ION BEAM PROCESSING

R&D of gas cluster ion beam processing has been supported by JST since 1989 in order to proceed toward possible industrial applications. Figure 11 summarizes fundamental characteristics of typical application fields. Possible industrial applications are being investigated by Kyoto University Ion Beam Engineering Experimental Laboratory in collaboration with several industrial partners. They include (i) shallow ion implantation (Fujitsu Research Laboratory Ltd. for ULSI application), (ii) atomic scale surface smoothing and low damage processes for metals, dielectrics, super conductors and diamond film surfaces (Matsushita Research Lab. for diamond surface finishing., Mitsubishi Materials Co. for SOR and X-ray Lithography , Adachi-Shin Ind. Co., Ltd. for non-spherical plastic lens mold surface smoothing, Japan Pillar Corp. Ltd., for SiC surface smoothing for SOR mirrors) (iii) very high yield sputtering and etching processes. (iv) high quality thin multi-layer film coatings for reliable and durable optical filters (Adchi-Shin Ind. Co., Ltd.). Typical results are described below.

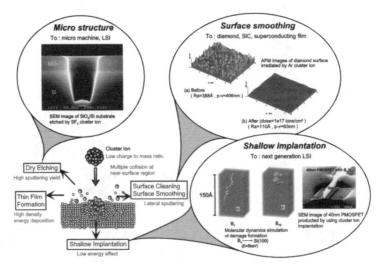

Fig.11   Characteristics of gas cluster ion beam and typical application areas

Semiconductor doping: shallow implantation

To test whether a larger cluster can make truly shallow junctions, molecular dynamics simulations of B cluster implantation into Si substrates for various cluster sizes have been performed [22]. Figure. 12 shows the result of a 5 keV boron implantation into Si(100) substrates with 10 atom clusters and with one atom after an elapsed time of 0.3 ps. The results show that the range of cluster implanted atoms is much shallower than monomer ion implantation for the same acceleration energy. Figure. 12 also shows that an entirely different kind of damage profile is created by cluster ion implantation. The first clusters that have been demonstrated for shallow junction formation were $B_{10}H_{14}$ ions [23, 24]. $B_{10}H_{14}$ ion implantation for p-type source/drains junctions was performed at an acceleration energy of 3 keV to a dose of $2x10^{14}$ions /cm² then annealed at 1000 ℃ for 10 sec. A junction depth of 20 nm has been achieved. For S/D extensions, $B_{10}H_{14}$ ion implantation at 2keV has been carried out to a dose of $1x10^{12}$ions/cm² followed by annealing at 900 ℃ for 10 sec. A 7-nm ultra-shallow junction without transient enhanced diffusion

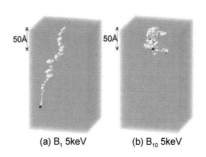

(a) B, 5keV   (b) B₁₀ 5keV

Fig.12 Molecular dynamics simulation of (a)B cluster implantation (10 atoms) and (b)monomer boron implantation into crystalline Si(100), after an elapsed time of 0.3 ps.

(TED) and thermal diffusion (TD) has been achieved. TED and TD are considered to be the most important issue in dealing with shallow junction formation in traditional ion implantation technology [25]. A typical dopant profile implanted at 3keV is shown in Fig. 13. For p-type junctions, such a shallow junction has not yet been achieved by the traditional ion implantation method due to implant channeling which results in deep implant. A comparison of B profiles obtained by B, BF_2 and $B_{10}H_{14}$ implantation at the same energy (5keV), is shown in Fig. 14. The result suggest that cluster ion implantation is one of the solutions for shallow implantation. Moreover, to obtain the same boron dose in the substrate, the ion dose of decaborane, for example, was ten times lower.

0.1nm PMOSFETs have been fabricated to demonstrate $B_{10}H_{14}$ cluster implantation for shallow source/drain formation. Devices were made using $B_{10}H_{14}$ implants at 2keV with a dose of 10^{12} ions/cm² for the source/drain extension regions. For the deeper source/drain region, where electrical contact was made to the junctions, $B_{10}H_{14}$ was implanted at 30keV to a dose of

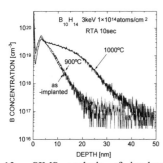

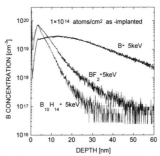

Fig.13 SIMS analysis of implanted B concentration after implantation of $B_{10}H_{14}$ at 3KeV to a dose of $10x10^{13}$ions/cm² before and after 10s rapid thermal annealing.

Fig.14 SIMS analysis of implanted B profiles by B, BF_2 and $B_{10}H_{14}$ at 5keV. Ion doses for B and BF_2 is 10^{14}cm⁻², ion doses for $B_{10}H_{14}$ are 10^{13}cm⁻².

2×10^{14} ions/cm^2. Dopant activation at 900 and 1000℃ for 10s was used for the source/drain extension and for the deep source/drain regions, respectively. The highest drive current (0.40 mA/μ m (@I_{off} of 1nA/μ m and V_d= -1.8V) shows 15% improvement compared with published data [26,27]. A low S/D series resistance R_{sd} of 760 ohm-μ m is achieved even if a high sheet resistance (>20kohm/sq) is used for the extension regions (due to the diminished extension length). The smallest PMOSFET with a L_{eff} of 38 nm has been demonstrated for the first time. Figure 15 shows the SEM image of the device having a poly-Si gate length of 40 nm after sidewall removal [28].

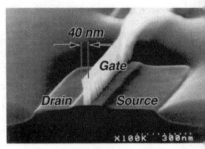

Fig.15 SEM image of PMOSFET having a 40nm gate

High rate etching and surface smoothing

Ion-enhanced dry etching techniques are widely used for a variety of device fabrication operations because of their high spatial resolution. Recent requirements for etching are very difficult to implement because of smaller dimension fabrication. Ion bombardment during etching processes creates considerable long range damage in the substrate surface, and it becomes increasingly important to understand the possible damage to the material induced by the processing. A number of experiments performed in conjunction with theoretical simulations have shown, for example, that damage extends to a depth of more than 1000 Å for incident ions having energies of about 300eV. It is claimed by Hu [29] that fortuitous channeling of ions causes penetration deep into the material. Another mechanism that would introduce defects even deeper into the substrate might be that of defect diffusion.

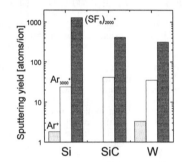

Fig.16 Comparison of sputtering yields for Si, SiC and W by SF6 cluster ion beam with the size of 2000 at an acceleration voltage of 20kV.

Gas cluster ion beam etching is characterized by a high sputtering rate and by a surface smoothing effect at low energy ion bombardment. Cluster ion beam etching produces different bombarding effects than traditional monomer ion bombardment in etching rates, surface smoothing and damage creation. Figure 16 shows a comparison of sputtering yields for Si, SiC and W substrates bombarded by SF_6 cluster ion beam with a size of 2000 at an acceleration voltage of 20 kV. Sputtering yield is higher when the bombardment is by cluster ion than by Ar monomer ion beam. When SF_6 is used as a cluster ion beam, physical sputtering is observed on a gold substrate which does not react with SF_6. However, chemical sputtering of the W substrate has been observed in this case. Reactive ion sputtering by SF_6 cluster ion beam for W substrate was much higher than the physical sputtering on Au. The effects on resultant surface roughness after etching are also quite different. Physical sputtering by Ar cluster ion beam produced a smooth surface, and this continued as the bombardment continued. However in case of SF_6 surface roughness could not be reduced. In Fig.17, AFM images of W and Au substrate surfaces bombarded at different incident angles are shown . Bombardment by normal incident beam has smoothed the surface, whereas bombardment at an inclined angle did not produce the same effect.

Potential applications of cluster ion beam sputtering and smoothing have also been demonstrated in metals, Si, CVD diamond, SiC and YBCO film substrates: (i) Surface smoothing of metal has been applied to non-spherical plastic lens mold. The mold is too small and too complicated for surface smoothing to be performed by mechanical etching. Gas cluster

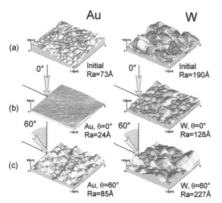

Au   W

(a) 0° Initial Ra=73Å   0° Initial Ra=190Å

(b) 60° Au, θ=0° Ra=24Å   60° W, θ=0° Ra=128Å

(c) Au, θ=60° Ra=85Å   W, θ=60° Ra=227Å

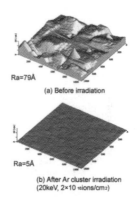

Ra=79Å

(a) Before irradiation

Ra=5Å

(b) After Ar cluster irradiation (20keV, 2×10 ¹⁶ions/cm²)

Fig.17 SF$_6$ cluster (cluster size 2000) ion bombardment on gold and tungsten, as shown. a)before; b-c)after bombardment at normal incident beam and at 60° respectively. (SF$_6$ cluster ion beam irradiation at 20kV to a dose of $7x10^{15}cm^{-2}$)

Fig.18 AFM images of YBCO thin film a)before and b)after Ar cluster ion beam irradiation at 20kV to a dose of $2x10^{16}cm^{-2}$

ion beam etching could make the surface smoother even when scratch-type traces are present. Surface roughness of a few nanometers could be easily obtained. (ii) A very high rate etching of Si by SF$_6$ cluster ion bombardment was applied for Si substrates. Sputtering yield of 1300 /cluster-ion has been achieved by SF$_6$-cluster beam with the size of 2000 at 20keV. (iii) Surface smoothing of CVD deposited diamond and SiC substrates has been performed for X-ray lithography mask formation. (A diamond membrane is expected to X-ray mask membranes because of its high Young modulus (559-1054GPa) and small thermal expansion coefficient $(0.8x10^{-6})$). Surface smoothing of diamond by cluster ion beam etching was performed using Ar or Ar and O$_2$ mixture at 20keV. A transparent thin film of $2\,\mu$ m CVD diamond with surface roughness of 5 nm has been obtained. A synchrotron orbital radiation (SOR) mask is being fabricated at the moment. (iv) Surface smoothing of superconducting yttrium -barium copper oxide (YBCO) film is being performed in order to fabricate SQUID circuits on its surface [30].

No effect was observed on the surface smoothness due to grain boundaries or composition. Figure 18 shows the surface roughness of the YBCO films before and after bombardment by Ar cluster ions of the size of 2000 at 20 keV

Thin film formation

Cluster ion assisted deposition for thin multi-layer filters and for transparent conductive film formation are on-going programs. For thin multi layer deposition of TiO$_2$ and SiO$_2$, oxygen gas cluster ion beam is irradiated

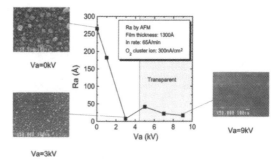

Va=0kV

Va=3kV

Va=9kV

Ra by AFM
Film thickness: 1300Å
In rate: 65Å/min
O$_2$ cluster ion: 300nA/cm²

Transparent

Fig.19 Surface roughness and transparency vs. acceleration voltage. Grain structure is shown inset.

during Ti or Si deposition. It is expected that this technique will produce films with a high packing density and smoother interface structure. Work is in progress in this direction. Transparent optical conductive film has been deposited with In metal vapor and O_2 gas cluster at different acceleration energies. Figure 19 shows the effect of acceleration energy on the film's structure and transparency. With increasing energy, the number of unoxidized metal grains decreases and film transparency improves. This indicates that bombardment by ion clusters promotes the reaction between metal and oxygen and enhances film formation.

CONCLUSIONS

The present status of advanced beam processing and their on-going projects was reviewed. Demands and prospects for future beam processing were also presented. Several ion beam systems which allow for industrial production scale have been developed as Japanese large scale project. Due to recent rapid developments in high technology, especially in the electronic industry, fabrication processes allowing smaller lateral and vertical dimensions and better reliability are urgently required. Laser and ion combined processes will undoubtedly be used as a new processing technique in this field. Cluster ion beam processing, which is fundamentally different from traditional ion beam processes due to non-linear collisions, constitutes a significant step in advanced ion beam processing. Its contribution to this field is in its ability to produce features and devices which could not have been realized by traditional ion beam processing

Acknowledgments

The author wishes to thank the members of the AMMTRA and ACTA projects for their kind cooperation and for providing materials used in this report. He also appreciates long-term support given by the Japan Science and Technology Corporation for R&D program for gas cluster ion beam processing .

References

1. A.Hellenmans and B.H.Bunch: *The Timetable of Science*(Simons & Schuster, New York, 1988).
2. J.F.Zigler Ed., *Handbook of Ion Implantation technolgy* (North Holland, Amsterdam 1988)p.1.
3. I.Yamada, *Mater. Res. Soc. Symp. Proc., Vol.268, "Materials Modification by Energetic Atoms and Ions,"*, S.Granbowswki, S.A Barnett, S.M. Rossnagel and K.Wasa Eds., 1992, (Pittsburgh, 1996) pp.261-272.
4. I.Yamada, *Surf. and Coat. Technol.* 51,(1992) 514-521.
5. ACTA Final Report (Industrial Technology Council of the Ministry of International Trade and Industry) 1997.
6. J.R.Conrad, J.Radtke, R.A.Dodd and F.J.Worzala, *J.Appl.Phys.* 62 (1987) 4591
7. N.W.Cheung, *Nucl. Instr. and Meth.* B 55 (1991) 811.
8. M.Inoue, Y.Suzuki, T.Takagi, *Nucl. Instr. and Meth.* B121 (1997) 1.
9. I.Yamada, Proc. 14th Symp. on Ion Sources and Ion-Assisted Technology, Tokyo, (The Ion Engineering Society of Japan, Tokyo,1991) pp.227-235.
10. I.Yamada and J.Matsuo: *Mater. Res. Soc. Symp. Proc., Vol. 396, "Ion Solid Interactions for Materials Modification and Processing"*, Eds. D.B.Pocker, D.Illa, Y-T.Cheng, L.R.Harriott and T.W.Sigmon, 1995, (Pittsburgh, 1996) pp.149-154.
11. I.Yamada and J.Matsuo: *Mat. Res. Soc. Symp. Proc., Vol. 427, "Advanced metallization for Future ULSI"*, K.N.Tu, J.M.Poate, J.W.Mayer and L.C.Chen Eds., 1996 (Pittsburgh, 1997) pp265-276.
12. The name of this project is *Japan Science and Technology Corporation, Exploitation and Application Study Program: "Gas cluster ion beam processing for high performance surface treatment".*
13. Proc. ACT International Symposium, Sponsored NEDO, ACTA (Advanced Chemical Processing Technology Research Association,) 1996, October ,Tokyo.

14. Y.Murakami, N.Kuratani, S.nishiyama, O.Imai, K.Ogata, *Nucl. Instr. and Meth.,* B121 (1997) 212.
15. M.Irie, H.Ohara, A.Nakayama, N.Kitagawa, T.Nomura, *Nucl. Instr. and Meth.,* B121 (1997) 133.
16. H.Nakai, H.Kuwahara, J.Shinohara, T.Kawaratani, T.Sassa, Y.Ikegami, *Nucl. Instr. and Meth.,* B112 (1996) 280.
17. R.Nonogaki, S.Yamada, T.Wada, Nuclear Instruments and Methods, B121 (1997) 121.
18. I.Kanno, S.Hayashi, R.Takayama, H.Sakakima, T.Hirao, Nucl. Instr. and Meth., B112 (1996) 125.
19. I.Yamada, R&D Report for NEDO, FY 1995 (contract no.A0076).
20. NEDO Consortium program, 1997, Project office: Management office :Osaka Science and Technology Center, Utsubo, Osaka Japan: "R&D for ultra -high quality transparent conductive film fabrication" .
21. Epion Corp. Catalogue (Bedford Mass. USA)
22. T.Aoki, N.Shimada, D.Takeuchi, J.Matsuo, Z.Insepov and I.Yamada in *Proc. Solid State Devices and Materials,* (1996) p. 49-54.
23. K.Goto, J.Matsuo, T.Sugii, H.Minakata, I.Yamada and T.Hisatugu: *IEDM Tech. Dig.,* p.435, 1996
24. I.Yamada, J.Matsuo E.C.Jones, D.Takeuchi, T.Aoki, K.Goto and T.Sugii: *Mat. Res. Soc. Symp. Proc., Vol. 438, "Materials Modification and Synthesis by Ion Beam Processing,* D.Alexander, B.Park, N.Cheung, and W.Skorupa Eds., 1996 (Pittsburgh, 1997) pp 368-374.
25. K.S. Jones and G.A. Rozgonyi, in *Rapid Thermal Processing Science and Technology,* R. B. Fair ed., (Boston: Academic Press, 1993) 123.
26. M.Rodder, Q.Z. Hong, M.Nandakumar, S.Aur, J.C.Hu and I-C. Chen, *IEDM Tech. Dig.,* p.563, 1996
27. M.Bohr, S.S.Ahmed, S.U.Ahmed, M.Bost, TR. Ghani, J. Greason, R. Hainsey, C.Jan, P. Packan, S. Sivakumar, S. Thompson and S. Yang, *IEDM Tech. Dig.,* p.847, 1996
28. K.Goto, J.Matsuo, Y.Tada, T.Tanaka, Y.Momiyama, T.Sugii, and I.Yamada: *IEDM Tech. Dig.,* 1997, in press.
29. E.L.Hu, C-H Chen and D.K.Green, *J.Vac.Sci. Technol.* B14 (1996) 3632.
30. W.C.Chu, Y.P.Li, J.R.Liu, J.Z.Wu, S.C.Tidrow, N.Toyoda, J.Matsuo and I.Yamada, *Appl.Phys Let.,* 1997, In press.

MOLECULAR DYNAMICS SIMULATION OF FULLERENE CLUSTER ION IMPACT

Takaaki AOKI, Toshio SEKI, Masahiro Tanomura, Jiro MATSUO, Zinetulla INSEPOV, Isao YAMADA
Ion Beam Engineering Experimental Laboratory, Kyoto University, Sakyo Kyoto 606-01 Japan, t-aoki@kuee.kyoto-u.ac.jp

ABSTRACT

In order to interpret the projection range and to reveal the mechanism of damage formation by cluster ion impact, molecular dynamics simulations of a fullerene carbon cluster (C_{60}) impacting on diamond (001) surfaces were performed. When the kinetic energy of C_{60} is as low as 200eV/atom, C_{60} implants into the substrate deeper than a monomer ion with the same energy per atom because of the clearing-way effect. The kinetic energy of the cluster disperses isotropically because of the multiple-collision effect, and then a large hemispherical damage region is formed. When the energy of the cluster is as high as 2keV/atom, the cluster dissociates in the substrate, and then cascade damage is formed like in a case of a monomer ion impact. The projection range of incident atoms becomes similar to that of the monomer with the same energy per atom. However, the number of displacements of C_{60} is larger than the summation of 60 monomer carbons. The displacement yield of fullerene is 4 to 7 times higher than that of monomer carbon. This result agrees with the measurement of the displacements made on sapphire substrates with C_{60} and C_2 irradiation.

INTRODUCTION

A cluster consists of tens to several thousands of atoms, and cluster ion irradiation shows different non-linear effects from that of monomer ion irradiation. Damage formation, high yield sputtering and lateral sputtering by cluster ion irradiation have been previously observed [1-4]. When a cluster impacts on a solid surface with its high density of particles, the collision zone acquires higher energy and a larger number of collisions occur between the cluster and surface atoms compared to a monomer impact. Molecular dynamics (MD) simulation is one of the effective methods which is widely used for studying the impact processes of various cluster ions on solid surfaces. As in a former study [5], we have performed the MD simulation of Ar cluster impacting on a Si (001) surface, and have explained a typical cluster impact process. Multiple collisions between cluster and surface atoms transform the kinetic energy of the cluster to the target's surface layers isotropically. Following the impact, crater-like damage is formed and some atoms are sputtered in a direction lateral to the surface.

Ar gas clusters obtained in our experiment range from 100 to 3000 atoms, and have a broad size distribution. It is desired to use much smaller and well-selected cluster ion beams to do more precise analysis of the irradiation effect of cluster ion.

Mat. Res. Soc. Symp. Proc. Vol. 504 © 1998 Materials Research Society

Fullerene (C_{60}), which is a stable large carbon molecule, enables us to obtain a well-selected cluster ion beam, and is useful to investigate the non-linear irradiation effect of cluster ions by experimental methods [6-8].

In this work we have performed MD simulation of C_{60} impacting on a diamond (001) surface and investigated the range of incident energy in which C_{60} shows non-linear properties in the penetration depth and in stopping power [9]. It has also been suggested [9], that C_{60} irradiation should have superior properties for surface modification compared to monomer ion irradiation. In the present paper, we compare the mechanism of damage formation by C_{60} ion impacts obtained by MD with the experimental results of damage measurement of sapphire irradiated with C_{60} and C_2 ions.

SIMULATION METHOD

In this paper Tersoff's empirical model potential [10] is applied to describe the interaction between carbon atoms. This potential is suitable to describe equilibrium properties of the material. However, the Tersoff model, which is based on the Morse potential, is not valid for simulation of high-energy collisions. In order to simulate collisions of high-energy ions, the 2-body repulsive term of the Tersoff model ($f_R(r)$ in ref. [10]) is replaced by ZBL model [11,12], when the inter-atomic distance (r) is less than 0.4069Å. The difference of the potential energy between two models at this distance is 136.97eV. This energy is added to the ZBL potential at $r < 0.4069$Å to connect between these two potential models smoothly.

Diamond (001) surface was chosen as a target material, with the maximum number of carbon atoms of about 270,000. Periodic boundary conditions are applied to (010) and (100) boundaries, and the atoms of the lowest layer are fixed to retain bulk structure. The initial temperature of the target is 0 K. C_{60} molecule is constructed by locating carbon atoms at the vertices of a truncated icosahedron. For the Tersoff potential model used, the radius of C_{60} equals 3.68Å, which gives a minimum potential energy of -4.05eV. C_{60} is directed at normal direction toward the surface, and C_1 is directed with the incident angle tilted 7° from the surface normal and twisted 30° to (001) direction in order to avoid channeling effect. C_{60} is incident on a diamond surface with energy ranging from 20eV/atom to 3keV/atom and C_1 with energy ranging from 200eV/atom to 2keV/atom. For simulation of C_1, 128 trials have been done and the result is averaged in order to obtain statistical value of penetration depth and displacement yield.

RESULT AND DISCUSSION

Fig.1 shows the snapshots of sixty-four C_1 ions and one C_{60} ion impacting on a diamond (001) surface. The snapshots of C_1 are made by overlapping each one of 64 trials. The large open circles indicate incident carbon atoms and the black dots indicate their trajectories. Gray areas indicate displaced substrate atoms that are defined as the atoms having potential energy 2eV above the bulk state. Both figures show the picture 0.1ps after the impact.

In the case of the incident energy of 200eV/atom, a C_{60} particle impinges into the substrate deeper than carbon monomer ions that have the same incident energy

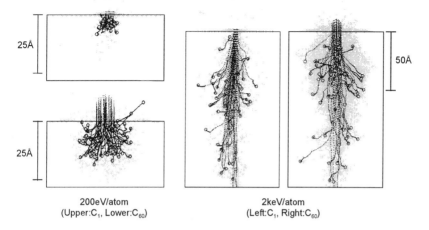

25Å

25Å

50Å

200eV/atom
(Upper:C_1, Lower:C_{60})

2keV/atom
(Left:C_1, Right:C_{60})

Fig.1: Snapshots of sixty-four C_1 ions and one C_{60} ion impacting on diamond (001) surface. Large open circles indicate the incident carbon atoms and black dots indicate theirs trajectory. Gray ones indicate displaced substrate. Each figure shows the picture 0.1ps after impact.

This is caused by clearing-way effect as a consequence of high-density irradiation. The clearing-way effect occurs when one incident atom follows the other atom, which clears the way for the following atoms. The following atom can penetrate into solids without significant energy loss [13]. The implanted atoms cause large number of collisions between each other and the atoms of substrate. Because of this multiple-collision, the kinetic energies of the incident atoms are efficiently transferred to the substrate in the lateral direction compared to the monomer ion impact. Therefore, the displacement region by a C_{60} impact spreads and its shape becomes hemispherical. The clearing-way and multiple-collision effects were first observed in high-density particle irradiation, and are regarded as characteristics of cluster ion impact [5,9].

As the incident energy becomes larger, the lateral dispersion of incident atoms reduces, whereas the penetration depth increases. In the case of the incident energy of 2keV/atom, a C_{60} penetrates into the substrate without interruption, but after embedding itself into the substrate it dissociates to individual carbon atoms, and each incident atom implants into the

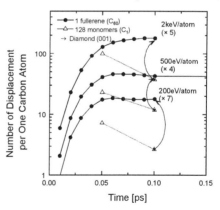

Fig.2: Time dependence of the number of displacement per one carbon atom formed by irradiation of C_{60} and C_1 with the energy of 200eV/atom, 500eV/atom or 2keV/atom.

83

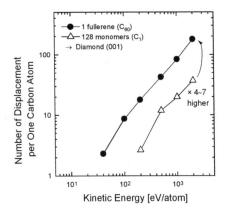
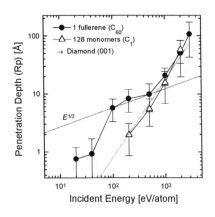

Fig.3: Incident energy dependence of the number of displacements formed by C_{60} and C_1

Fig.4: Incident energy dependence on the penetration depth of implanted carbon atoms of C_{60} and C_1

substrate obeying the collision cascade mechanism. The trajectories of C_{60} become similar to the ones of monomer ion that has same incident energy per atom. The shape of the damaged region in the shallow surface region changes from hemispherical to cylindrical. At the end of the range of the cylindrical displacement, far from the surface, interspersed displacements occur.

Fig.2 shows time dependence of the number of displacements per incident atom by C_1 and C_{60}. The number of displacements by C_1 impact is calculated by averaging the results of 128 trials. At 0.1 ps after impact, about 20, 40 and 200 displacements/atom are formed by C_{60} impact with the incident energy of 200eV/atom, 500eV/atom and 2keV/atom, respectively. Each value is 4 to 7 times higher than those produced by monomer ions because of a non-linear effect of high-density particle irradiation. In the case of monomer ion impact, displacements are formed by knock-on mechanism and tend to result in point defects. In the case of clusters such as C_{60}, the opposite occurs. The kinetic energy of cluster is transferred to the substrate atoms around the cluster by a collective movement and more homogeneously. Hence, the mean potential energy of each displacement by cluster is much less than by monomer ion, and a large number of displacements are formed which actually means that the structure is amorphousized. In each case of monomer ion irradiation, the number of displacements decreases during the period of 0.05 ps to 0.1 ps. On the other hand, preliminary results obtained from MD calculations after 0.5ps for C_{60} at 500eV/atom indicate that the number of displacements per cluster ion impact remains almost constant (i.e. show very small decrease) [data not shown].

Fig.3 and Fig.4 show the energy dependence of the displacement yield, which is the number of displacement induced by one incident atom, and the mean penetration depth, respectively. In the case of energy ranging from 200eV/atom to 2keV/atom, the displacement yields of C_1 and C_{60} are proportional to the incident energy, and the yield of C_{60} is from 4 to 7 times higher than that of C_1. However, the profile of penetration depth of C_{60} becomes much steeper, as the incident energy increases, in

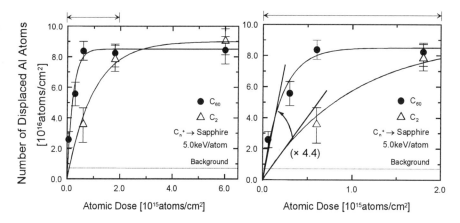

Fig.5: Atomic dose dependence of the number of displaced Al atom in sapphire, irradiated with C_{60} and C_2. The incidence energy of both C_{60} and C_2 is 5keV/atom. The right figure shows the magnified view of the left one.

contrast with that of C_1. When the energy of C_{60} is less than 500eV/atom, C_{60} penetrates into the substrate deeper than C_1 because of the clearing-way effect, and penetration depth is proportional to $E^{1/3}$ (where E is the incident energy), because C_{60} disperses isotropically in all directions by the multiple-collision effect. When the energy is less than 100eV/atom, even C_{60} does not penetrate the substrate and displacements are not induced. On the other hand, as the incident energy increases above 500eV/atom, the lateral dispersion of the momentum of incident atoms reduces, and the penetration depth of C_{60} becomes similar to that of C_1. From the viewpoint of the penetration depth, one C_{60} with energy above 500eV/atom could be treated as 60 individual carbon atoms, but it still shows non-linear property from the viewpoint of damage formation.

We performed experiments of C_{60} or C_2 ion irradiation onto sapphire surface and measured the displacement of Al atoms by Rutherford back-scattering (RBS) spectroscopy [6]. Fig.5 shows the atomic dose dependence of the number of displaced Al atoms in sapphire irradiated with C_{60} and C_2. Both C_{60} and C_2 are incident with the same energy of 5keV/atom. In each case, the number of displacements saturates around 8.5×10^{16} atoms/cm^2 at the atomic dose of 6.0×10^{15} atoms/cm^2. The saturation value is equivalent to the projection range of incident atoms and this result implies that each C_{60} and C_2 ion with incident energy of 5keV/atom have the same projection range. When the number of displacements is not saturated, the displacements increase with increasing atomic dose, and the displacement yield can be calculated. The displacement yield of C_{60} and C_2 are about 400 and 90 respectively, and the ratio of the yield of C_{60} to the one of C_2 is 4.4. This ratio agrees with the results of MD simulation, as shown in Fig.3 and Fig.4.

CONCLUSION

Molecular Dynamics simulations of fullerene impacting on diamond (001) surfaces were performed to investigate non-linear effects in damage formation by cluster ion irradiation. When the incident energy is as small as 200eV/atom, a C_{60} ion implants deeper than monomer carbon with the same incident energy per atom because of the "clearing-way" effect. The kinetic energy of the cluster disperses isotropically through many collisions between cluster and surface atoms. The kinetic energy of C_{60} is delivered to the surface atoms in the shallow surface area, and a large and hemispherical damage region is formed.

When the energy of the cluster increases, their lateral kinetic momenta decrease and their projection range increases. The ratio of lateral dispersion of incident atoms to their projection range also decreases with increasing cluster energy and the shape of the damage region changes from hemispherical to cylindrical. Consequently, the cluster dissociates in the substrate and cascade damage is formed as in the case of a monomer ion. In the case of energy as large as 2keV/atom, the projection range of incident atoms becomes similar to that of monomer with same energy per atom. However, the number of displacements created by fullerene is still larger than the number created by monomer carbon. The displacement yield of fullerene is calculated to be from 4 to 7 times higher than that of monomer. These results are supported by measurements in sapphire of displacement damage induced by C_{60} and C_2 ion irradiation. The results presented in Fig.5 show good agreement in both projection range and displacement yield. These results suggest that C_{60} irradiation will show improved performance for damage formation and surface modification at low dose compared to monomer ions.

REFERENCES

1. I.Yamada, W.L.Brown, J.A.Northby and M.Sosnowski Nucl. Instr. And Meth., B79, p. 223, (1993).
2. I.Yamada, J.Matsuo, Z.Insepov and M.Akizuki, Nucl. Instr. And Meth., B106, p. 165, (1995).
3. D.Takeuchi, K.Fukushima, J.Matsuo and I.Yamada, Nucl. Instr. and Meth. B 121, p. 493, (1997).
4. N.Toyoda, H.Kitani, J.Matsuo and I.Yamada, Nucl. Instr. and Meth. B121, p. 484, (1997).
5. T. Aoki, J.Matsuo, Z.Insepov and I.Yamada, Nucl. Instr. and Meth. B121, p. 49, (1997).
6. M.Tanomura, D.Takeuchi, J.Matsuo, G.H.Takaoka and I.Yamada, Nucl. Instr. and Meth. B121 p. 480, (1997).
7. T.Seki, T.Kaneko, D.Takeuchi, T.Aoki, J.Matsuo, Z.Insepov and I.Yamada, Nucl. Instr. and Meth. B121, p. 498, (1997).
8. T.Seki, in this issue.
9. T.Aoki, T.Seki, J.Matsuo, Z.Insepov and I.Yamada, to be published in J. Mat. Chem. and Phys.
10. J.Tersoff, Phys. Rev. B, B39, p. 5566, (1989).
11. J.P.Biersack and L.G.Haggmark, Nucl. Instr. And Meth. 174, p. 257, (1980).
12. J.F.Ziegler, J.P.Biersack and U.Littmark, The Stopping and Range of Ions in Solids, Pergamon Press, New York, 1985, pp. 321.
13. V.I.Shulga and P.Sigmund, Nucl. Instr. And Meth. B47, p. 236, (1990).

HIGH QUALITY OXIDE FILM FORMATION BY O_2 CLUSTER ION ASSISTED DEPOSITION TECHNIQUE

J. Matsuo[*], W. Qin[*], M. Akizuki[*,**], T. Yodoshi[**] and I. Yamada[*]

[*] Ion Beam Engineering Experimental Laboratory, Kyoto University Sakyo, Kyoto 606-01, Japan
[**] Microelectronics Research Center, SANYO Electric Co., Ltd. 180 Ohmori, Anpachi-cho, Anpachi-Gun, Gifu 503-01, Japan,

ABSTRACT

A new oxide film formation technique using gas-cluster ion beams has been developed. O_2 cluster ions were used to irradiate during the evaporation of metal atoms, and PbO_x and In_2O_3 films were grown. At the acceleration voltages above 5 kV, polycrystalline PbO_x films preferentially oriented to (111) were obtained. A significant smoothing effect was observed with an acceleration voltage as low as 1 kV. An average surface roughness of 0.9 nm was obtained at 7 kV. Oxygen cluster ion beams are also utilized to grow In_2O_3 films, which are widely used as conductive-transparent films in flat panel display. In_2O_3 was deposited on glass or silicon substrates with simultaneous irradiation with an oxygen cluster ion beam. Highly transparent (80%) and low resistivity ($<4 \times 10^{-4}$ Ωcm) films were obtained with 7keV oxygen cluster ion beams. Kinetic energy of above 3keV is necessary to obtain low resistivity films. These results clearly indicate that the kinetic energy of the cluster is effectively used to enhance oxidation on the surface without radiation damage, in spite of the high acceleration voltages.

INTRODUCTION

Oxide films are widely used materials in electrical and optical devices. There is a strong requirement to provide high-quality and low temperature oxide films processing methods. Many different oxidation techniques, such as Pulsed Laser Deposition(PLD), Chemical Vapor Deposition(CVD) and unbalanced magnetron sputtering deposition have been investigated. We have introduced a new technique called "Cluster Ion Assisted Deposition" for oxide film formation. This successive deposition process can use oxygen cluster ion beams, which can transport thousands of atoms per ion with very low energy per constituent atom. For example, an average energy per constituent atom of a 10 keV cluster ion containing 1000 atoms is only 10 eV. As a result, the interactions between the cluster and substrate atoms occur the in near-surface region[1]. Moreover, cluster ions can deposit their energy with a high density within a very localized surface region. Therefore, the irradiation of low-energy gas cluster ions can be expected to enhance the chemical reactions on the substrate surface[2].

We have reported room temperature oxidation by low-energy oxygen cluster ion beams[3]. High-quality SiO_2 films of up to 11 nm thick were formed at room temperature on Si surfaces. A strong enhancement of the oxidation was found using O_2 cluster ion irradiation. This could be the result of high-density energy deposition at near surface regions.

In this paper, O_2 cluster ion-assisted deposition was examined to form oxide films. In this technique, oxygen cluster ion beams are simultaneously irradiated to substrates during evaporation of metals at room temperature. Two different oxide films were studied. One is PbO_x, which is an important material for lead-based ferroelectric films used in nonvolatile memory devices[4]. The other is In_2O_3 films, which is widely used as visibl transparent and conducting film in a variety of optoelectronic devices such as liquid crystal display panels and solar cells[5,6]. Irradiation effects of oxygen cluster ions on oxidation, crystallinity and surface morphology were investigated.

87

2. Experimental

Figure 1 shows a schematic diagram of low energy gas cluster ion assisted deposition system which has been developed at Kyoto University. The detailed technique for generation of gas cluster ion beams and for the cluster-size separation has been described previously[7]. Cluster beams from gas sources, such as O_2, CO_2 and Ar gas, can be formed by adiabatic expansion. High pressure (5 atm.) gas mixture of 30 % He in O_2 is introduced into vacuum through a Laval nozzle, which is cooled down to 150 K by cold N_2 gas in order to obtain high intensity neutral cluster beams[8]. Clusters were ionized by electron bombardment with ionization voltage of 150 V and electron current of 45 mA. Size selection of the cluster ions was performed using retarding potential technique to eliminate monomer ions and small cluster ions. O_2 cluster ions were accelerated up to 9 kV. The mean size of O_2 cluster ions used in this experiment was 2000 and average energy per constituent atom was less than 4.5 eV. High current O_2 cluster ion beams of a current density of 300 nA/cm^2 were obtained on substrates, when acceleration voltages were above 7 kV.

Soda lime glass, Corning #7059 and p-type silicon substrates were mounted on a hexagonal sample holder to deposit six samples in each experiment. The distance between the substrate and the ionizer was 10 cm. In or Pb metal of 99.9 % purity was evaporated from a carbon crucible heated by resistive heater with tantalum radiation shields. Uniformity of the film thickness was about 10 %. The thickness of the PbO_x and In_2O_3 films was approximately 80 nm and 100 nm, respectively. Thickness of the films was measured by contact profiler, and sheet resistance was measured with four point probe. Surface morphology was observed by scanning electron microscopy (SEM), and atomic force microscope (AFM) was used to measure average surface roughness (Ra) of the oxide films. Composition of the film was evaluated by auger electron spectroscopy (AES), Rutherford back scattering spectrometry (RBS) and X ray photoelectron spectroscopy (XPS). Optical transmission spectra of the In_2O_3 films were measured with multipurpose recording spectrophotometer. X-ray diffraction (XRD) and reflection high energy electron diffraction (RHEED) were used to evaluate the crystalline structures.

3. Results and Discussion

3.1 Cluster assisted deposition of PbO_x

Figure 2 shows the acceleration voltage dependence of the average surface roughness of PbO_x films deposited with O_2 cluster ion irradiation. The Ra reduced drastically with increasing the acceleration voltage. A significant smoothing effect of gas cluster ion beams has been reported on various kinds of surfaces, such as Pt, Cu, polycrystalline Si, SiO_2 and Si_3N_4 films[1,9-13]. This smoothing effect can be explained by lateral sputtering, which has been demonstrated both experimentally and theoretically for cluster energy of above several keV[12,13] . However,

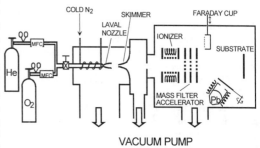

Figure 1   Schematic diagram of the low energy cluster ion-assisted deposition apparatus.

in the present work, the smoothing effect was observed even at an acceleration voltage of 1keV. Another mechanism may be responsible for the smoothing at such low energy.

Figure 3 shows the XRD patterns of samples deposited with O_2 cluster ion irradiation as a parameter of the acceleration voltage (Va). The Pb films deposited without oxygen cluster were polycrystal Pb metal with a (111) preferred orientation, which was consistent with previously reported results[14]. The films deposited with 1keV O_2 cluster ion irradiation produced a similar crystalline structure of Pb as the films deposited without oxygen, but a significant reduction in the surface roughness was observed. As shown in Fig. 3, increasing the acceleration voltage of oxygen cluster ions, the Pb (111) peak disappeared concomitantly with the appearance of a PbO (111) peak in the XRD patterns. The PbO_x films were grown with (111) orientation at an acceleration voltage above 5 kV. According to the RHEED observation, polycrystalline PbO films were obtained. The films deposited with 5 keV O_2 cluster ion irradiation had the same Ra of 1.1 nm within the range of deposition rate from 0.7 to 2.4 nm/min. Ra was not found to depend

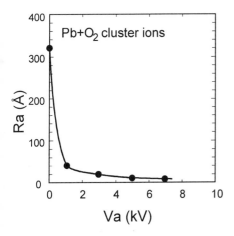

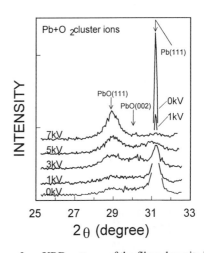

Figure 2   Acceleration voltage dependence of the average surface roughness (Ra) of PbO_x films deposited with O_2 cluster ion irradiation.

Figure 3   XRD patterns of the films deposited with O_2 cluster ion irradiation as a function of the acceleration voltage

on deposition rate, in spite of the increasing intensity of the PbO (111) peak as the deposition rate decreased. Therefore, the surface smoothness is not attributable to the crystalline structure of the films.

The PbO_x films with (111) preferred orientations were grown with decreasing deposition rate from 2.4 to 0.7 nm/min. These results clearly demonstrate that the bombardment of O_2 cluster ions can enhance the crystallization of the films during the deposition. It has been reported that randomly oriented PbO_x films are formed by reactive ion beam sputtering deposition[15]. PbO_x films with a (100) preferred orientation were formed by the oxidation of Pb (111) films[16]. The crystalline structure of PbO_x films, deposited with O_2 cluster ion irradiation, is different from those formed by other techniques.

3.2 Cluster assisted deposition of In_2O_3

Tin doped indium oxide (ITO) films, which are visibly transparent and conducting films, are widely used in a variety of optoelectronic devices. Reactive Magnetron sputtering[17], activated reactive evaporation[18] and chemical vapor deposition[19] have been used to grow ITO. In most of these deposition techniques, high substrate temperature, typically above 200 °C is required to obtain highly transparent and conductive films. ITO films are very sensitive to damage created by ion bombardment, and low energy ion beam is necessary to obtain low resistance films[20]. Therefore, cluster ion assisted deposition technique is suitable for ITO film growth, because cluster ions are basically low energy ions. Average cluster size and maximum ion energy used in this study were 2000 and 10keV, respectively. Therefore, the mean energy of each constituent atom is about 5eV, which is significantly lower than ion energy used in other techniques.

The In/O ratio of the films is shown in figure 4 as a function of acceleration voltage (Va). Indium evaporation rate was kept at 20 nm/min. When O_2 cluster ion energy was above 5 keV, stoichiometric In_2O_3 films were grown at room temperature. The oxygen cluster beam current density was about 100 nA/cm^2 at 3 keV. Transparent films could not be obtained at low acceleration voltage (<3 kV), even when deposition rate was as low as 2 nm/min. When ion current density was kept at 100 nA/cm^2, highly transparent films were easily obtained at 7keV. This result clearly demonstrates that energetic cluster ion has a great capability to oxidize the In surface. High acceralation voltage of above 5keV is necessary for cluster ions to form high quality In_2O_3 films as a result of enhanced chemical reactions. Approximately 10 % of oxygen atoms of the irradiated O_2 cluster ions have been incorporated in the films.

According to SEM observation of In_2O_3 films formed at various acceleration voltage, small particles, likely to be precipitates of metalic In were found below 3 keV. No particle was observed when stoichiometric In_2O_3 films were obtained. AFM measurements clearly shows that very smooth films were formed above 5 keV. The surface roughness of the films was less than 2nm, which is sufficiently smooth to be utilized in optoelectronic devices.

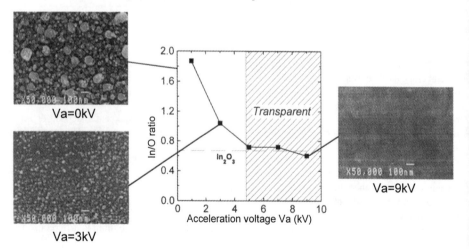

Figure 4   Acceleration voltage dependence of ion current density and In/O ratio

Figure 5 shows the dependence of the In/O ratio on In evaporation rate, resistivity and transparency. These samples were deposited at room temperature with various In deposition rate, while keeping ion current density (300 nA/cm^2) and energy (7 keV) constant. The lowest resistivity of 4.7×10^{-4} $\Omega \cdot$cm with transparency of 80 % was obtained at an In rate of 4.2 nm/min. Comparing with films deposited by other methods at this temperature without post-annealing, this resistivity is the lowest value to our knowledge. Even though the ratio of In deposition rate to O$_2$ cluster current density was changed, stoichiometric In$_2$O$_3$ films were obtained. Resistivity was found to be not sensitive to In evaporation rate for In evaporation rates between 3 nm/min and 6 nm/min. Thus cluster ion assisted deposition technique offers a stabilized process, which is very convenient for industrial applications. The bombardment effect of cluster ions offers a new ion assisted thin film formation technique.

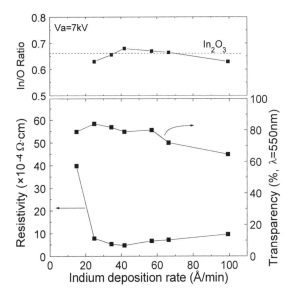

Figure 5   In evaporation rate dependence of resistivity, transparency and In/O ratio

Conclusions

PbO$_x$ films were grown on SiO$_2$ by O$_2$ cluster ion-assisted deposition. The irradiation of the energetic O$_2$ cluster ions enhanced the oxidation, the crystallization and the smoothness of PbO$_x$ films. The polycrystalline PbO$_x$ films formed with O$_2$ cluster ion irradiation, were (111) preferentially oriented at acceleration voltages above 5 kV.

High quality In$_2$O$_3$ films were also obtained by O$_2$ cluster ion assisted deposition at room temperature. It was found that it is necessary to use energetic O$_2$ cluster ions greater than 5 kV to form stoichiometric films. Films with the lowest resistivity of 5×10^{-4} $\Omega \cdot$cm with transmission greater than 80 % have been realized in a wide range of In deposition rates of 3~6 nm/min. The bombardment effect of cluster ions offers a new ion assisted thin film formation technique.

References

1. I. Yamada, W.L. Brown, J.A. Northby and M. Sosnowski, Nucl. Instr. and Meth. B, **79**, p.223(1993).

2. J. Matsuo, N.Toyoda and I. Yamada, J.Vac. Sci. Technol. B, **14**, p.3951(1996).

3. M. Akizuki, J. Matsuo, S. Ogasawara, M. Harada, A. Doi and I. Yamada, Jpn. J. Appl. Phys. **35**, p.1450(1996).

4. J. F. Scott and C. A. Araujo, Science, **246**, p.1400(1989).

5. A. A. Karim, C. Deshpandey, H. J. Doerr and R. F. Bunshah, Thin Solid Films, **172**, p.111(1989).

6. K. L. Chopra and I. Kaur, Thin Film Devices and Applications, Plenum, New York, 1983.

7. J. Matsuo, H, Abe, G.H. Takaoka and I. Yamada, Nucl. Instr. and Meth. B, **99,** p.244(1995).

8. M. Akizuki, J. Matsuo, M. Harada, S. Ogasawara, A. Doi, I. Yamada, Mater. Sci. Eng., **A217-A218**, p.78(1995).

9. I. Yamada, and J. Matsuo, Z. Insepov, D. Takeuchi, M. Akizuki and N. Toyoda, J.Vac. Sci. Technol A, **14**, p.781(1996).

10. I. Yamada and J. Matsuo, Mat. Res. Soc. Symp. Proc., **396**, p.149(1996).

11. M. Akizuki, J. Matsuo, M. Harada, S. Ogasawara, A. Doi, K. Yoneda, T. Yamaguchi, G. H. Takaoka, C. E. Ascheron and I. Yamada: Nucl. Instr. and Meth. B, **99**, p.229(1995).

12. H. Kitani, N. Toyoda, J. Matsuo and I. Yamada, Nucl. Instr. and Meth. B, **121**, p.489(1997).

13. T. Aoki, J. Matsuo, Z. Insepov and I. Yamada, Nucl. Instr. and Meth. B, **121**, p.49(1997).

14. M. Murakami, J. Angelillo, H. C. W. Huang, A. Segmuller and C. J. Kircher, Thin Solid Films, **60**, p.1(1979).

15. J. M. Vandenberg, S. Nakahara and A. F. Hebard, J. Vac. Sci., Technol., **18**, p.268(1981).

16. J. W. Matthews, C. J. Kircher and R. E. Drake, Thin Solid Films, **42**, p.69(1977).

17. R. P. Howson, I. Safi, G. W. Hall and N. Danson, Nucl. Instr. and Meth. B, **121**, p.96(1996).

18. P. Nath and R. F. Bunshah, Thin Solid Films, **69**, p.63(1980).

19. L. A. Rykara, V. S. Salum and I. A. Serbinov, Thin Solid Films, **92**, p.327(1982).

20. Y. Suzuki, F. Niino, S. Okada, M. Yoshida, 1994 SID International Symposium Digest of Technical Paper SID, **xviii+996**, p.943(1994).

SIZE DEPENDENCE OF BOMBARDMENT CHARACTERISTICS PRODUCED BY CLUSTER ION BEAMS

T. SEKI, M. TANOMURA, T. AOKI, J. MATSUO and I. YAMADA
Ion Beam Engineering Experimental Laboratory, Kyoto University, Sakyo Kyoto, Japan

ABSTRACT

Cluster ion beam processes provide new surface modification techniques, such as surface smoothing, high rate sputtering and very shallow implantation, because of the unique interactions between cluster and surface atoms. To understand interactions with cluster and surface, Scanning Tunneling Microscope (STM) observations have been done for single impact traces.

Highly Oriented Pyrolitic Graphite (HOPG) surfaces were bombarded by carbon cluster ions ($Va \leq 300kV$), and large ridges and craters have been observed as a result of single cluster ion impact. The impact site diameters are proportional to the cluster size up to 10 atoms, and increase drastically for cluster sizes above 10. This indicates that non-linear multiple collisions occur only when a local area is bombarded by more than 10 atoms at the same time.

INTRODUCTION

A cluster is an aggregate of a few to several thousands atoms. We have been accelerating cluster ions to various targets. Because a local area is bombarded by many atoms constituting a cluster ion, high-density energy deposition and multiple-collision are realized. Because of the interactions, cluster ion beam processes can produce unusual new surface modification effects, such as surface smoothing, high rate sputtering and very shallow implantation [1-4]. Large impact ridges and craters have been observed on solid surfaces [5,6], when Ar cluster ion beams with a size of more than 100 atoms were irradiated on solid surfaces. Such a large crater, which has not been observed on monomer ion bombardments, is caused by high-density energy deposition to the surface and multiple-collision between cluster and surface atoms. In this work, we investigated the threshold size of clusters to find such a difference between monomer impacts and cluster impacts by means of STM. We believe that the threshold size gives us information to reveal the mechanism of the cluster-surface interaction.

Fullerenes were frequently used as a source of carbon cluster beams. [7-9]. Smaller sizes of Ar clusters, less than 100, can't be generated, but smaller sizes of carbon clusters, up to 70, can be generated as a consequence of cracking of the fullerene. These small clusters are ionized and subsequently accelerated to high energy. In order to observe cluster-surface interactions, it is important to investigate a single trace formed by a cluster impact on a solid surface. Especially the size dependence of the trace diameter is most important, because the cluster size, which is a unique parameter of the cluster ion beam, causes unique interactions which occur between cluster atoms and surface atoms. We have obtained clear images of a single trace formed by a C_n ion impact and investigated the mechanism of the cluster-surface interaction.

EXPERIMENT

Fig.1 shows a schematic of the carbon cluster ion implantation system. Neutral carbon clusters were generated by heating fullerene powder to 400-600°C by an oven in the hot-hollow cathode ion source. After ionization of the neutral clusters by electron bombardment, the cluster ion beam was focused by an eintzel lens. Some of the fullerene molecules were dissociated by electron bombardment. Cluster ions were mass-separated by a 90° magnet, and accelerated up to 400kV. Fig.2 shows the mass spectrum of carbon cluster beam generated from mixed C_{60}, C_{70} powder. Various sizes of clusters up to 70 and some doubly charged ions were generated as a consequence of cracking fullerene. C_n^{2+} beams are generated and doubly charged ions are able to be accelerated to 800keV. Subsequently, the carbon cluster beams were scanned in order to produce a uniform irradiation and deflected at 3° in order to remove neutral beams.

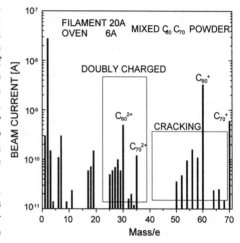

Fig.2 : Mass spectrum of carbon cluster beam generated from mixed C_{70}, C_{60} powder. Various sizes of clusters, of up to 70 atoms, and some doubly charged ions were generated as a consequence of cracking fullerene.

We have used HOPG as a target, because HOPG has the characteristics of high conductivity, chemical stability and an atomically-flat surface. We have first confirmed that the surface of HOPG is sufficiently flat to discuss trace structures of ion impact with an atomic level [5]. An atomically-flat

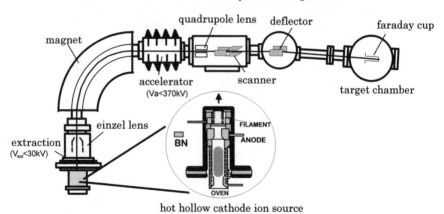

Fig.1 : Schematic of the carbon cluster ion implantation system

surface is obtained by cleaving and is suitable for STM observation in air. We used a mechanically polished Pt-Ir tip for the probe. The conditions used for these observations were as follows: the tunneling current was constant at 0.7nA and the bias voltage was +1.0V. We observed HOPG surfaces irradiated by C_n ions accelerated up to 600keV. Doubly charged ions were used to give an acceleration energy of more than 400keV. The ion doses were varied from 1×10^{10} to 1×10^{11} ions/cm^2. At these low doses, each ion impact trace can be individually distinguished in STM images and they don't overlap.

RESULTS AND DISCUSSION

When an HOPG surface was irradiated by C_{60} ions accelerated to Va=300kV at a dose of 5×10^{10} ions/cm^2, craters and large hills with diameter of about 150Å were observed on this surface [10]. Large crater formations are one of the typical characteristics of cluster ion impact as a consequence of high-density energy deposition and multiple-collision between cluster atoms and surface atoms [5]. The formation of a crater by C_{60} impact indicates that the C_{60} impact process has the same effects. We have reported that the number of defects per incident atoms

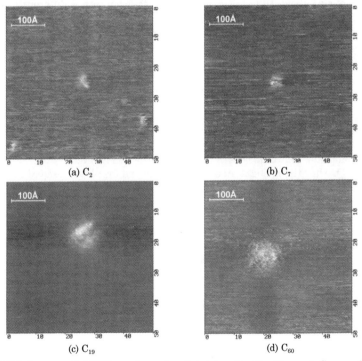

(a) C_2

(b) C_7

(c) C_{19}

(d) C_{60}

Fig.3 STM images of HOPG surfaces irradiated by C_2, C_7, C_{19} and C_{60} accelerated to Va=5kV/atom. Small hills, with diameter of about 50Å, were observed on the surface irradiated by C_2 or C_7. Large hills, with diameter of 100-150Å, were observed on surfaces irradiated by C_{19} or C_{60}.

produced by C_{60} is several times larger than that by carbon monomers at the same velocity [11]. Therefore, non-linear multiple collisions, which are not proportional to sizes, occur when HOPG surface is bombarded by C_{60}.

Fig.3 shows STM images of HOPG surfaces irradiated by C_2, C_7, C_{19} and C_{60} accelerated to Va=5kV/atom. Small hills with diameter of about 50Å were observed on the surface irradiated by C_2 or C_7. Large hills with diameter of 100-150Å were observed on surfaces irradiated by C_{19} or C_{60}. A large difference between cluster sizes 7 and 19 was observed. This indicates that in case of size above 19 the mechanism of trace formation is different from that by small carbon cluster, with a size less than 7.

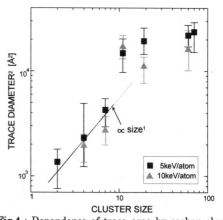

Fig.4 : Dependence of trace area by carbon cluster impact on the cluster size. Carbon cluster was accelerated to Va=5keV/atom or Va=10keV/atom.

Fig.4 shows the dependence of the trace area on the cluster size. The carbon cluster was accelerated to Va=5keV/atom or Va=10keV/atom. The trace areas are proportional to cluster size up to 10 for both energies, and increase suddenly for cluster sizes above 10 atoms. For clusters below 10 atoms, the trace diameter for 5keV/atom is higher than for 10keV/atom. This is due to the fact that nuclear stopping power for 5keV is higher than for 10keV [10]

We propose a model to explain the drastic increase of the trace diameter. Fig.5 shows the proposed model of damage formation by cluster impacts, which is described as follows.

When a monomer ion collides with the surface, the incident atom excites the atoms on an area (S_a). After the incident atom goes through the first layer, excited atoms in S_a lose their energy and the rim of S_a doesn't become damage. Finally, the only damage remaining in each S_a is represented by S_d.

When a cluster ion collides with the surface, the cluster is broken and the atoms constituting the cluster are scattered on an area (S_s). The scattered atoms impinge on the surface with the energy divided among the atoms constituting the cluster. If the cluster size is small $(S_a \times Size < S_s)$, affected areas are isolated and appear as the small hills observed in Figs.3(a) and 3(b). Therefore the damaged area of the first layer (S_1) is,

$$S_1 = S_d \times Size \tag{1}$$

which is proportional to the cluster size. Because the same effects occur in another layer, if STM scan is influenced by the damage area from the first to the n-th layer, the damage area (S) measured by STM is given by

$$S = \sum_{i=1}^{n} a_i S_d Size \qquad (a_i : \text{influence constant})$$

$$= A \times S_d \times n \times Size \qquad (A = \sum_i^n a_i) \tag{2}$$

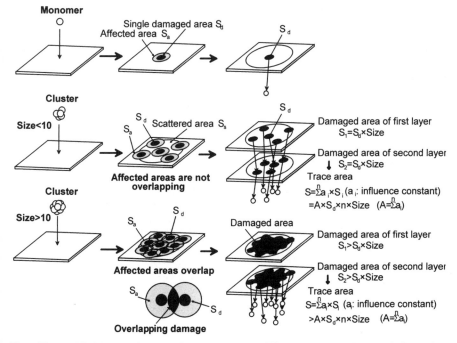

Fig.5 : The model of damage formation by cluster impacts. When cluster ion collides with the surface, atoms constituting cluster are scattered on an area (S_s) and excites the atoms on an area (S_a). If the cluster size is small, affected areas are isolated. If cluster size is large, affected areas overlap and overlapping damage occurs. The magnitude of the damage area is enhanced.

where i is the layer number and a_i is degree by which the STM scan is influenced by the damage area of the i-th layer.

According to this model the trace area is proportional to the cluster size, when the cluster size is small ($S_a \times Size < S_s$). We can say that such damage formation resembles monomer damage formation.

The affected areas start to overlap with increasing cluster size. When the cluster size is larger than the scattered area ($S_a \times Size > S_s$), each area damaged by the atoms of one cluster cannot remain isolated like in the case of small clusters because of its geometrical size, and thus overlapping damage occurs. As multiple collisions in the area are affected by overlap, atoms are excited strongly. Thus, because such strongly excited atoms are disordered, the magnitude of the damage area of the 1-st layer (S_1) is enhanced. S_1 is,

$$S_1 > S_d \times Size \qquad (3)$$

and the damage area (S) measured by STM is given by

$$S > \sum_i^n a_i S_d Size \qquad (a_i : \text{influence constant})$$

$$> A \times S_d \times n \times Size \qquad (A = \sum_i^n a_i) \qquad (4)$$

which is larger than the monomer like damage of the smaller cluster. As a result of this, the trace area increases suddenly, when the affected areas start to overlap. However, the trace area is limited to the scattered area (S_s) and S_s increases slightly with the cluster size. Therefore, when the cluster size is larger, the trace area increases slightly.

Applying this model to the dependence of trace area on cluster size, our results, shown in fig.4, show that monomer like damage formation occurs when cluster size is less than 10 and cluster damage formation occurs when cluster size is 10 or more. According to this model, non-linear multiple collision effects occur only when a local area is instantaneously bombarded by more than 10 atoms, which appears to be the threshold value.

CONCLUSION

We observed large hills on HOPG surfaces irradiated by carbon cluster with size of up to 70. The impact site diameters were found to be proportional to cluster size for clusters of up to 10 atoms and increase discontinuously for cluster sizes above 10 atoms. This can be explained by considering that small affected areas overlap. This indicates that non-linear multiple collision effects occur only when a local area is instantaneously bombarded by more than 10 atoms.

REFERENCE

1. I.Yamada, J.Matsuo, Z.Insepov, M.Akizuki, Nucl. Instr. & Meth. B106 (1995) 165.

2. I.Yamada, W.L.Brown, J,A,Northby and M.Sosnowski, Nucl. Instr. and Meth., B79 (1993) 223.

3. G.H.Takaoka, G.Sugawara, R.E.Hummel, J,A,Northby, M.Sosnowski and I.Yamada, Mat. Res. Soc. Symp. Proc., 316 (1994) 1005.

4. Z.Insepov, M.Sosnowski and I.Yamada, Advanced Materials '93 IV/Laser and Ion Beam Modification of Materials, ed. I.Yamada et al, Trans. Mat. Res. Soc. Jpn. 17 (1994) 1110.

5. T.Seki, T.Kaneko, D.Takeuchi, T.Aoki, J.Matsuo, Z.Insepov and I.Yamada, Nucl. Instr. and Meth., B121 (1997) 498.

6. D.Takeuchi, T.Seki, T.Aoki, J.Matsuo and I.Yamada, to be published in J. Mat. Chem. and Phys. (1998).

7. G.Bräuchle, S.Richard-Schneider, D.Illig, R.D.Beck, H.Schreiber and M.M.Kappes, Nucl. Instr. and Meth., B112 (1996) 105.

8. M.Döbeli, F.Ames, R.M.Ender, M.Suter, H.A.synal and D.Vetterli, Nucl. Instr. and Meth., B106 (1995) 43.

9. A.Hallen, P.Hakanson, N.Keskitalo, J.Olsson, A.Brunelle, S.Della-Negra and Y.Le Beyec, Nucl. Instr. and Meth., B106 (1995) 233.

10. T.Seki, T.Aoki, M.Tanomura, J.Matsuo and I.Yamada, to be published in J. Mat. Chem. and Phys. (1998).

11. M.Tanomura, D.Takeuchi, J.Matsuo, G.H.Takaoka and I.Yamada, Nucl. Instr. and Meth., B121 (1997) 480.

DOES THE LATENT TRACK OCCURRENCE IN AMORPHOUS MATERIALS RESULT FROM A TRANSIENT THERMAL PROCESS?

M. TOULEMONDE[+], Ch. DUFOUR[0], E. PAUMIER[+0] and F. PAWLAK[+]
[+] CIRIL (CEA/CNRS), BP 5133, 14070 Caen Cedex 5 (France),Toulemonde@ganil.fr
[0] LERMAT-ISMRA (ESA 6004 CNRS), 6 Bd du Maréchal Juin, 14050 Caen Cedex (France)

ABSTRACT

Heavy ion irradiations in the electronic stopping power (S_e) regime have been performed in amorphous materials. Latent tracks have been observed in amorphous semiconductors (a-Ge, a-Si) and their radii have been deduced from a phenomenological analysis in an amorphous metallic alloy, in vitreous silica and "polymer" like amorphous carbon. A transient thermal model is developed describing the energy diffusion by the electron gas, by the atomic lattice and the energy exchange between the two subsystems. According to Fick's law, the classical equations of heat flow in the two subsystems (electrons and atoms) are numerically solved in a cylindrical geometry taking into account the temperature dependence of all the parameters. A simulation of annealing of nuclear collisions induced defects in crystalline iron allows to determine a local temperature. Electronic defect creation occurs when S_e increases and becomes larger than a threshold which is correlated with the appearance of a molten phase. Using such a criterion, the radii of latent tracks are reproduced in both a - Ge and a - Si with the same value of the electron-phonon coupling despite large differences in their lattice thermodynamic parameters. Such a model is applied to amorphous metallic alloy $Fe_{85}B_{15}$, vitreous silica and amorphous carbon.

INTRODUCTION

Two models have been proposed in order to explain the appearance of latent tracks induced in matter by the slowing down process of incident ions in the electronic stopping power regime. The first one was the thermal spike proposed by Desauer [1] and reconsidered for metals by Seitz and Koehler [2]. The second one was the ionic spike, proposed by Fleischer et al. [3] that explains that metals are insensitive to the electronic excitation produced by fission fragments irradiation. In both models the key is the high mobility of the electrons in metals. In the ionic spike model for metals the Coulomb repulsion between lattice ions was considered as too quickly screened by the return electrons which inhibits the ionic impulse. In the thermal spike model for metals the electronic energy was considered as spread out in a too large volume to induce a significant increase of the lattice temperature. Now a systematic use of heavy ion accelerators has enlarged the number of materials (metals, semiconductors and insulators) [4-12] which present defect creation induced by heavy ions in electronic stopping power (S_e) regime. Especially all amorphous materials for which the electron mobility is greatly reduced are sensitive [10]. Moreover due to their weak electron mobility they are more sensitive [4-14] than the same materials in their crystalline phase. Hence both models must be reconsidered. In the course of time the ionic spike arises in 10^{-14} - 10^{-13} s while the increase of lattice temperature occurs in 10^{-13} - 10^{-12} s. As the thermal spike appears after the ionic spike, it can anneal all the previous atomic displacements. Then it is necessary to know whether the observed latent tracks might be a consequence of the temperature rise. This time scale scheme is supported by the experimental results following the electronic excitation induced by the high power femtosecond laser pulse irradiations of silicon [15] and GaAs [16]. These experiments have shown that melting arises anyway in a time of $2 \cdot 10^{-13}$ s [15-16] i.e. in a time larger than a characteristic phonon time, even if atomic motions could occur in a shorter time [15]. The lattice temperature increase follows the relaxation of the energy deposited in the electron system by electron-electron and electron-phonon interactions [17,18]. In that case [18] the space and time evolution of the electron and lattice temperatures is governed by a set of coupled non-linear differential equations which describes the energy diffusion in the electron and lattice systems respectively. This description was partially used to quantitatively explain the giant dimensional change in amorphous PdSi [19] and the radii of the latent tracks in a-Ge and a-Si [4,20] induced by heavy ion irradiations. In these two previous calculations analytical solutions [20,21] were used limiting a quantitative comparison with the experiments because the temperature dependence of all the parameters was not taken into account.

The goal of the present work is to determine the radii of the latent tracks in different amorphous materials in a transient thermodynamic model usually called a thermal spike. The two coupled equations are numerically solved in cylindrical geometry using realistic values of the parameters governing the electron energy diffusion and electron-phonon coupling.

PHYSICAL BASIS

Seitz and Koehler [2] established the theory of the temperature spike assuming that the electron gas and atomic lattice were both continuous media where they could write the classical equations of heat flow according to Fick's law. These equations are written in cylindrical geometry :

$$C_e(T_e)\frac{\partial T_e}{\partial t} = \frac{1}{r}\frac{\partial}{\partial r}\left[rK_e(T_e)\frac{\partial T_e}{\partial r}\right] - g(T_e - T_a) + B(r,t) \qquad (1)$$

$$C_a(T_a)\frac{\partial T_a}{\partial t} = \frac{1}{r}\frac{\partial}{\partial r}\left[rK_a(T_a)\frac{\partial T_a}{\partial r}\right] + g(T_e - T_a) \qquad (2)$$

C_i, K_i and T_i are the specific heat, the thermal conductivity and the temperature respectively, i is refering either to the electrons (e) or to the lattice atoms (a). g is the coupling constant describing the electron-phonon interaction [17-18] and $B(r,t)$ is the energy deposited in the electron system during the slowing down of the heavy ions. r is the radial coordinate and t the time. It is assumed, as previously [22], that the energy deposition time τ is equal to 10^{-15} s, the time necessary to slow down the δ-rays electrons [23]. In the present case we assume that $B(r,t) = AD(r)\alpha e^{-\alpha t}$ where $\alpha = 1/\tau$ and A is chosen so that the integral of $B(r,t)$ in space and in time gives the value of the electronic stopping power S_e. $D(r)$ is the spatial energy density distribution as deduced from Waligorski et al [23].

The importance of the initial energy density distribution was clearly evidenced in yttrium garnet [24]. For the same value of S_e, a lower beam energy results in a higher damage cross section. Such an effect was also observed in metals [25-27]. This initial spatial energy distribution on the electrons can be characterized by the radius λ_w of the cylinder in which 66% of the energy is deposited [22]. For the same material λ_w increases when increasing the beam energy [22]. Now for a given beam energy λ_w decreases when increasing the specific mass of the materials. As an example, for a beam energy of 5 MeV/amu λ_w is equal to 4 nm and 8 nm, for iron and crystalline quartz respectively.

In order to take into account the initial spatial energy distribution and the temperature dependence of all the parameters, the two equations are numerically solved. The temperature evolution of the electronic subsystem parameters has been previously described [26, 27]. The thermodynamic lattice parameters are taken from equilibrium measurements. *The only free parameter is the electron-phonon coupling strength g [28,29].*

In the thermal spike model one can introduce the parameter λ which is the mean diffusion length of the energy on the electrons, $\lambda^2 = D_e(T_e)C_e(T_e)/g$, linked to the electron-phonon coupling strength. From previous calculations [30] mean values of D_e and C_e can be deduced: 3.5 cm^{2}s^{-1} and 0.4 Jcm^{-3}K^{-1} respectively leading to $\lambda^2_m = 1.4/g$. This mean λ_m value which corresponds to a mean value of the thermal diffusion length, can be compared to λ_w the spatial initial energy distribution length.

RELATION BETWEEN THE THRESHOLD OF DEFECT CREATION AND THE MELTING TEMPERATURE.

It has been experimentally shown that, due to the electronic excitation, defects created by nuclear collisions are annealed by the electronic excitation [7,31]. Defect annealing can be thermally activated as suggested by Vineyard [32]. Below the electronic stopping power threshold of damage creation, the only defects are due to nuclear collisions. So knowing the initial spatial distribution of nuclear collision defects [33], one can calculate the probability P(r,t) to anneal them, supposing that the thermal atomic jump frequency for the interstitial of a correlated Frenkel pair is

$$P(r,t) = P_0 \exp(-E_a / k_B T_e(r,t)) \qquad (3)$$

where $P_0 = k_B T_D / h$, h is the Planck's constant, E_a the activation energy and T_D the Debye temperature. So the final defect density at a distance r is $N_f(r) = N_i(r) \prod_{t=0}^{\infty} [1 - P(r,t)\Delta t]$. If we sum from r = 0 to r = ∞, we obtain the total number of residual defects D_f. D_i is the total initial number of created defects by nuclear collisions $\left[D_i = \int_0^{\infty} N_i(r)2\pi r dr \right]$ leading to a damage efficiency equal to D_f/D_i. A detailed calculation was performed for iron [33] to predict the damage efficiency evolution [7] taking into account the nuclear defect annealing by S_e and using the g factor previously determined [27]. The results of the calculation with g = 1.44·10^{+12} Wcm^{-3} K^{-1} are presented in figure 1. One can see that a kink appears between 15 and 20 keV/nm as for the experimental results. Applying a factor of 0.40 to the results of the calculation one can very closely follow the experimental results. This normalisation might come from the fact that we have neglected the athermal defect recombination in the cascades [34].

Now damage creation appears when S_e becomes larger than a threshold value for which the model shows that the melting temperature is reached. This criterion will be used in the following and immediately applied to iron (fig 2) assuming the latent track diameter corresponds to the diameter of molten cylinder.

COMPARISON OF THE g VALUE WITH THEORETICAL ESTIMATION .

Now one can compare the determined g value with a theoretical description of the electron-phonon coupling [28,29]. If the lattice temperature is not much smaller than the Debye temperature T_D, the g factor may be approximately expressed as

$$ g = \frac{\pi^2 m_e n_e v^2}{6\tau_e(T_e)T_e} \qquad (4) $$

where m_e is the electron mass, n_e the electronic density, $\tau_e(T_e)$ the electron mean free time between two collisions at temperature T_e and v the speed of sound in the metal, linked to the Debye temperature T_D and the atomic density n_a by $v = k_B T_D / \hbar(6\pi^2 n_a)^{1/3}$. The determination of $\tau_e(T_e)$ is very difficult. To bypass this difficulty, we have related $\tau_e(T_e)$ to the electrical conductivity $\sigma_e(T_e)$ or the thermal conductivity $K_e(T_e)$ of the metal under study and then

$$ g = \frac{\pi^4 (k_B n_e v)^2}{18 L \sigma_e(T_e) T_e} \quad \text{or} \quad g = \frac{\pi^4 (k_B n_e v)^2}{18 K_e(T_e)} \qquad (5) $$

where L is the Lorentz number. As previously [27], the temperature evolution of the g factor will follow the temperature evolution of the thermal or electrical conductivity. The quoted g factor will be determined at 300 K using the measured values of the thermal or electrical conductivity of the metal under study.

Such a definition is in agreement with the fact that the electron-phonon coupling strength is larger for an amorphous metallic alloy since its thermal conductivity is lower than the one of the same material in its crystalline phase [20]. Moreover g is constant since the thermal conductivity of an amorphous material is nearly constant.

Assuming that the electronic stopping power threshold of defects creation is linked to the appearance of a molten phase and using the thermal spike model, we can determine experimental g values for different metals. Table I gives g values deduced from Kaganov's model [28] (column B), from fs laser experiments [18] (column C) and from the electronic stopping power threshold for damage creation by heavy ions [27]. With the values proposed in column B one can deduce that Cu, Au, Nb and W are not sensitive (column D) to heavy ions as shown by experiments [35]. Moreover for Bi and Fe the agreement is satisfactory (column B and D). However, one has to increase the g value for Ti (column B and D). Such a discrepancy is not surprising if we compare the Kaganov's theory to the results obtained using fs laser experiments (column B and C). An agreement could be obtained between experiments and relation 5 if n_e/n_a is chosen between 0.5 and 2. The mean diffusion length λ_m is calculated

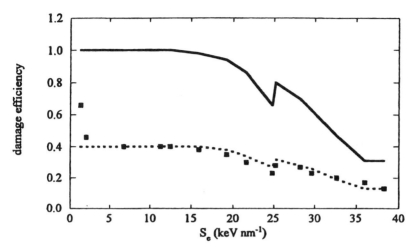

g.1 : theoretical (lines) [33] and experimental (squares) [7] damage efficiency in iron. The solid line
the direct result of the calculation of annealing by S_e of the nuclear defects. The dashed line
rresponds to the calculation for which a factor of 0.4 has been applied.

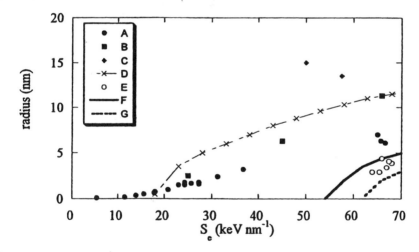

g.2 : latent track radii in a-Fe₈₅B₁₅ and crystalline iron.
A) experimental latent track radii in a-Fe₈₅B₁₅ extracted from in-situ resistance measurements [36].
B) experimental latent track radii in a-Fe₈₅B₁₅ extracted from the incubation fluence [37].
C) experimental latent track radii in a-Fe₈₅B₁₅ extracted from AFM [39].
D) thermal spike calculation for a-Fe₈₅B₁₅ for a beam energy of 10 MeV/amu.
E) experimental latent track radii in crystalline iron extracted from in-situ resistance measurements [7].
F) thermal spike calculation for crystalline iron for a beam energy of 3 MeV/amu.
G) thermal spike calculation for crystalline iron for a beam energy of 20 MeV/amu.

from the g values (column B, C, D) and reported in column E. These values will be compared to those found in amorphous materials. Moreover Bi which is a material with a low melting point and with a low g value has the largest latent tracks. This is due to the fact that the energy even deposited in a large volume overcomes the energy necessary to melt [33].

So for crystalline metallic materials the g definition [29] (equation 5) leads to realistic predictions of metal sensitivity provided that the value of n_e/n_a should lie between 0.5 and 2. Applying this definition to crystalline iron with the determined value of g and taking $n_e/n_a = 1$, the value of the electrical resistivity at 300 K from equation 5 can be calculated : the determined value, $\rho = 11\ \mu\Omega\,cm$, is in agreement with the measured value (9.9 $\mu\Omega\,cm$) on pure iron. This agreement supports the validity of the g determination [28].

Table I
Values of the electron-atom coupling g in 10^{10} Wcm^{-3}K^{-1} for metals quoted in column A, calculated with $n_e/n_a = 1$ [27] in column B, deduced from fs laser experiments [18] in column C and from heavy ion irradiations in column D. Column E gives the range of λ values in nm deduced from g. (* insensitive materials under swift heavy ions [35] in agreement with the thermal spike calculation [30]).

A	B	C	D	E
Cu	12.7	4.8	*	33-54
Au	2.3	2.8	*	71-78
Ti	203	185	809	4.2-8.7
Fe	119		144	9.9-10.8
Bi	20		40	19-26
Cr	~ 220	42		8-18
Nb	40	387	*	6-19
W	~ 30		*	22
Pb	3.8	12.4		34-61

AMORPHOUS MATERIALS : a-Fe$_{85}$B$_{15}$, a-Ge AND a-Si, a-SiO$_2$ AND a-C:H

In order to apply the thermal spike model to amorphous materials one has to know the energy necessary to melt them. It is known that this energy is 20 % lower than the one in the corresponding crystalline phase [20]. Moreover the lattice thermal conductivity of an amorphous material is of the order of 0.05 Wcm^{-1}s^{-1}. Using the relation 4 such a value leads to high electron-atom coupling.

Amorphous metallic alloys : a-Fe$_{85}$B$_{15}$

It has been shown that all amorphous materials [10] are sensitive. a-Fe$_{85}$B$_{15}$ is the only amorphous metallic alloy for which there is a significant number of results [5,6,36,37]. From a modelisation of the resistance evolution versus the fluence the damage cross section σ_p is extracted assuming that the relative saturation resistivity increase is 4% [36]. The chemical etching reveals that the damage is concentrated along the ion path [38]. So it is possible to assume that the increase of resistivity is due to a defect creation along the ion path in a cylinder with a radius R such that $\sigma_p = \pi R^2$. Recently, hillocks of matter with a diameter of 27 nm have been observed on the surface of irradiated samples [39]. Large anisotropic growth is observed after an incubation fluence ϕt_c [36,37] linked to another damage cross section $\sigma_c = 1/\phi t_c$. All the deduced experimental radii are reported in figure 2. The electronic stopping power damage threshold which is extracted from the in-situ resistance measurements [36] and the appearance of a steady state of the anisotropic growth [37]   (S$_e$ = 15 keV/nm) is independent of the modeling of the   experimental data. The value of g that best fits the threshold is g = 4.5 10^{12} Wcm^{-3}K^{-1} (i.e. λ_m = 5.6 nm). Then the molten cylinder radius is plotted versus S$_e$ in fig. 2. It shows a better agreement with damage cross sections deduced from the incubation fluence than with the other experimental results. It must be pointed out that the thermal spike model was used to explain the anisotropic growth [40,41] in such materials.

Amorphous semiconductors a-Ge, a-Si

In a-Ge and a-Si the latent tracks have been observed by transmission electron microscopy [4]. So a direct comparison can be made. Using the same assumption for the energy necessary to melt [21], the g factor has been evaluated to fit the latent track radii (fig.3). The agreement is obtained for g = 7 10^{12} Wcm^{-3}K^{-1} (i.e. λ_m = 4.5 nm) higher than in the case of an amorphous metallic alloy. Using the same electron-phonon coupling for the a-Si it is possible to calculate the latent track radius with a very good agreement.

Amorphous insulators

In the case of insulators the major problem comes from the definition of the parameters governing the energy diffusion. We have followed the model made by Katin et al. [42]. In insulators the charges are localized and there are no free electrons outside the excited region. So it is proposed that hot electrons excited in the conduction bands behave like hot electrons in metals. Then the energy dissipation in the electronic system proceeds via the ionization of bound electrons at the periphery of the excited region. The energy spreading stops when the

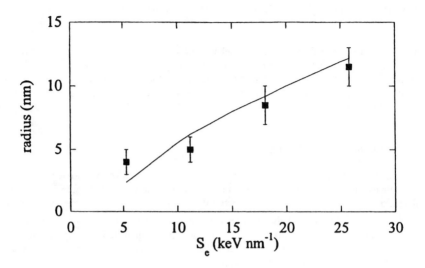

Fig.3 : latent track radii in a-Ge : squares are the experimental data and the line corresponds to
thermal spike calculations for a beam energy of 1 MeV/amu.

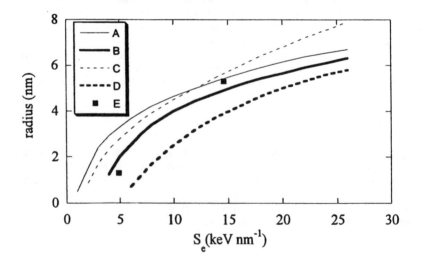

Fig.4 : latent track in vitreous SiO₂ (squares [44]). A and B correspond to the thermal spi
calculation in vitreous SiO₂ for a λ_m value of 2.5 nm and for beam energies of 0.05 MeV/amu (A) a
10 MeV/amu (B). C and D correspond to the thermal spike calculation in SiO₂ quartz for a λ_m val
of 4 nm and for beam energies of 0.05 MeV/amu (C) and 10 MeV/amu (D).

electron energy becomes smaller than the ionization one. Therefore the mean diffusion length λ should increase when the ionization energy decreases. That means that λ should increase when the band gap E_g decreases. Such a description is supported by comparison with experimental results [43]. For the present calculation for insulators it will be assumed that the mean value of $<C_e D_e>$ is equal to 2 [43] leading to $\lambda^2_m = 2/g$.

Vitreous silica : a-SiO₂

The threshold of anisotropic growth due to the electronic stopping power has been determined to be 2 keV/nm with a beam energy of 4 MeV/amu [11]. At the same beam energy latent track radii have been extracted using infrared absorption [44]. Using these two experimental results the λ_m value is equal to 2.5 nm (i.e. $g = 3.2 \ 10^{13}$ Wcm^{-3}K^{-1}). The results of the calculation are reported in fig.4 for two beam energies (0.05 MeV/amu and 10 MeV/amu). It appears that at low beam energy vitreous silica is sensitive to the electronic stopping power at a value as low as 0.6 keV/nm. As an example 1 MeV Ar ions with $S_e = 1.1$ keV nm^{-1} should induce defects from curve A (fig. 4). The calculated radii evolution versus the electronic stopping is also in agreement with chemical etching threshold in a-SiO₂ [12]. Indeed at a value of 4.6 keV/nm the latent track radius is around 3 nm, radius necessary to have an efficient chemical etching in amorphous materials [38] and in crystalline materials [45].

In the same picture the radii evolution in quartz is also presented. Such a calculation [43] with a λ_m value equal to 4 nm (i.e. $g = 1.3 \ 10^{13}$ Wcm^{-3} K^{-1}) closely follows the experimental results [9]. In this material the S_e threshold of swelling [46] appears at 1.8 keV nm^{-1}, quite in agreement with both the damage creation and the calculation [43]. The comparison between crystalline and amorphous materials shows a strange behaviour : at high values of S_e, the latent track radii in vitreous silica begin to be lower than the ones in crystalline SiO₂. This is due to the fact that the lower value of λ_m in a-SiO₂ inhibits the expansion of the energy distribution on the atoms.

Amorphous carbon : the "polymer"-like a-C:D

Amorphous carbon can be considered as the disordered phase of graphite whose g value is of the order of 10^{13} Wcm^{-3}s^{-1} (i.e. $\lambda_m = 3.7$ nm). Taking into account the low λ value for an amorphous material, the latent track radii [47,48] deduced from infrared absorption and nuclear analysis have been fitted. Such a damage corresponds to the breaking of the C-D bonds, leaving radicals which are saturated a long time after [48] by hydrogen. In a first calculation using the previous hypothesis concerning the energy necessary to melt and a lattice specific heat following the Dulong-Petit law it was impossible to fit the results (fig. 5). So the appearance of melting fails to describe the observed effect in a "polymer-like" a-C:D. But the temperature of breaking the hydrogen bonds was previously determined by irradiation of polymers [49] and is equal to 1873 K. So a second calculation is tried assuming that 1873 K is the temperature necessary to break the bonds between carbon and deuterium : the result presented in figure 5 shows good agreement with $\lambda_m = 1.5$ nm (i.e. $g = 1.6 \ 10^{14}$ W cm^{-3}s^{-1}). So in this specific material it seems that it is also a transient thermal process which is responsible for the experimental effects but the physical meaning of the electronic stopping power threshold is different : it corresponds to a bond breaking and not to the appearance of a molten phase.

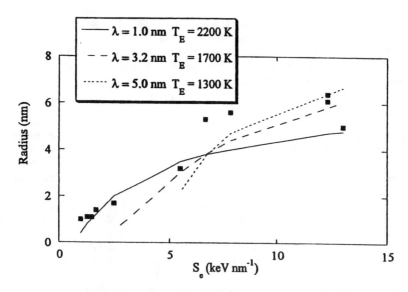

Fig.5 : Polymer-like amorphous deuterated carbon : experimental data (squares [48]), thermal sp calculations for three couple values of $\lambda=\lambda_m$ and T_E (T_E corresponds to the C-D bonds breaking).

CONCLUSIONS

A thermal spike model is able to predict the sensitivity of amorphous inorganic materials assuming that the experimental observations result from the quench of a molten phase. In a-$Fe_{85}B_{15}$ the agreement arises with the latent track deduced from anisotropic growth and not with AFM and in-situ resistance measurements. More experiments are needed to understand this feature. The agreement is quite good for the semiconductors a-Ge and a-Si and for the insulator a-SiO_2. In the case of a polymer-like material a-C:D, this model can explain the latent track radii assuming the energy of C-D bond breaking is $k_BT = k_B1873$ as previously determined.

Comparing the mean diffusion length λ_m of the energy on the electron deduced from the g determination to the radius λ_w of the initial energy density distribution on the electrons, the following classification can be made. In the case of metallic materials $\lambda_w < \lambda_m$ and the energy deposited on the lattice atoms is governed by the energy diffusion in free electron gas. In the case of insulators $\lambda_w > \lambda_m$, the energy deposited on the lattice is governed by the δ-ray electron energy distribution. For the semiconductors both effects are involved in the energy deposition on the lattice. Such a classification is not so sharp in two extreme cases : for cluster beams with high S_e for which λ_m is always larger than λ_w and for ion beams with high velocity for which λ_m could be less than λ_w in some cases.

REFERENCES :

1. F. Desauer, Z. Physik 38, 12(1923)
2. F. Seitz and J.S. Koehler, Sol. St. Phys. 2, 305(1956)
3. R.L. Fleisher, P.B. Price and R.M. Walker, Nuclear Tracks in Solids, University of California Press, (1975)
4. K. Izui and K.S. Furuno Proc, XIth Int. Cong. On Electron Microscopy (Kyoto 1986)1299
5. A. Audouard, E. Balanzat, G. Fuchs, J.C. Jousset, D. Lesueur and L. Thomé, Europhys.Lett. 3, 327(1987)
6. A. Audouard, E. Balanzat, G. Fuchs, J.C. Jousset, D. Lesueur and L. Thomé, Europhys.Lett. 5, 241(1988)
7. A. Dunlop, D. Lesueur, P. Legrand, H. Dammak and J. Dural, Nucl. Instr. Meth. Phys. Res. B90, 330(1994)
8. A. Dunlop, D. Lesueur, J. Morillo, J. Dural, R. Spohr and J. Vetter, C.R. Acad. Sci. Paris 309, 1277(1989)
9. A. Meftah, F. Brisard, J.M. Costantini, E. Dooryhée, M. Hage-Ali, M. Hervieu, J.P. Stoquert, F. Studer and M. Toulemonde, Phys. Rev. B49, 12457(1994)
10. S. Klaumünzer, G. Schumacher, Li Changlin, S. Löffler, M. Rammensee and H.C. Neitzert, Rad. Eff. Def. Sol. 108, 131(1989)
11. A. Benyagoub, S. Löffler, M. Rammensee, S. Klaumünzer and G. Saemann-Ischenko, Nucl. Instr. Meth. Phys. Res. B65, 228(1992)
12. A. Sigrist and R. Balzer, Helv. Phys. Acta 50, 49(1977)
13. M. Toulemonde, J. Dural, G. Nouet, P. Mary, J.F. Hamet, M.F. Beaufort, J.C. Desoyer, C. Blanchard and J. Auteytner, Phys. Stat. Sol. 114, 467(1989)
14. M. Levalois, P. Bogdanski and M. Toulemonde, Nucl. Instr. Meth. Pys. Res. B63, 14(1992)
15. H.W.K. Tom, G.D. Aumiller and C.H. Broto-Cruz, Phys. Rev. Lett. 60, 1438(1988)
16. P. Saeta, J.K. Wang, Y. Siegal, N. Bloembergen and E. Mazur, Phys. Rev. Lett. 67, 1023(1991)
17. P.B. Allen, Phys. Rev. Lett. 59, 1460(1987)
18. S.D. Brorson, A. Kazeroonian, J.S. Moodera, D.W. Face, T.K. Cheng, E.P. Ippen, M.S. Dresselhaus and G. Dresselhaus, Phys. Rev. Lett. 64, 2172 (1990)
19. S. Klaumünzer, Ming-Dong Hou and G. Schumacher, Phys. Rev. Lett. 57, 850(1986)
20. M. Toulemonde, Ch. Dufour and E. Paumier, Phys. Rev. B46, 14362(1992)
21. G. Szenes, Mat. Sc. Forum 97-99, 647(1992)
22. B. Gervais and S. Bouffard, Nucl. Instr. Meth. Phys. Res. B88, 355(1994)
23. M.P.R. Waligorski, R.N. Hamm and R. Katz, Nucl. Tracks Radiat. Meas. 11, 309(1986)
24. A. Meftah, F. Brisard, J.M. Costantini, M. Hage-Ali, J.P. Stoquert, F. Studer and M. Toulemonde, Phys. Rev. B48, 920(1993)
25. Z.G. Wang, Ch. Dufour, B. Cabeau, J. Dural, G. Fuchs, E. Paumier, F. Pawlak and M. Toulemonde, Nucl. Instr. Meth. Phys. Res. B107, 175(1996)
26. Ch. Dufour, A. Audouard, F. Beneu, J. Dural, J.P. Girard, A. Hairie, M. Levalois, E. Paumier and M. Toulemonde, J. Phys. Condens. Matter 5, 4573(1993)
27. Z.G. Wang, Ch. Dufour, E. Paumier and M. Toulemonde, J. Phys. Condens. Matter 6, 6733 (1994) ; 7, 2525(1995)
28. I.M. Kaganov, I.M. Lifshitz and L.V. Tanatarov, Zh. Tekh. Fiz. 31, 273(1956) [Sov. Phys. JET 4, 173 (1957)]

29. Ch. Dufour, Z.G. Wang, M. Levalois, P. Marie, E. Paumier, F. Pawlak and M. Toulemonde, Nucl. Instr. Meth. Phys. Res. B107, 218(1996)
30. Ch. Dufour, E. Paumier and M. Toulemonde, Nucl. Instr. Meth. Phys. Res. B122, 445(1997)
31. A. Iwase, S. Sasaki, T. Iwata and T. Nihira, Phys. Rev. Lett. 58, 2450(1987)
32. G. H. Vineyard, Radiat. Eff. 29, 245(1976)
33. Z.G. Wang, Ch. Dufour, E. Paumier and M. Toulemonde, Nucl. Instr. Meth. Phys. Res. B115, 577(1996)
34. A. Iwase, T. Iwata, T. Nihira, J. Phys. Soc. Jap. 61, 3878(1992)
35. A. Dunlop and D. Lesueur, Rad. Eff. Def. Sol. 126, 123(1993)
36. A. Audouard, E. Balanzat, J.C. Jousset, D. Lesueur and L. Thomé, J. Phys. Condens. Matt. 5, 995(1993)
37. A. Audouard, J. Dural, M. Toulemonde, A. Lovas, G. Szenes and L. Thomé, Phys. Rev. B54, 15690(1996)
38. C. Trautmann, R. Spohr and M. Toulemonde, Nucl. Instr. Meth. Phys. Res. B83, 513(1993)
39. A. Audouard, J. Dural, M. Toulemonde, A. Lovas, G. Szenes and L. Thomé, Europhys. Lett. to be published.
40. Hou Ming-Dong, S. Klaumünzer and G. Schumacher Phys. Rev. B41, 1144(1990)
41. H. Trinkaus and A.I. Ryazanov, Phys. Rev. Lett. 74, 5072(1995)
42. V.V. Katin, Yu. V. Martinenko, Yo. V. Yavlinskii, Sov. Phys. Techn. Lett. 13, 276(1987)
43. M. Toulemonde, J.M. Costantini, Ch. Dufour, A. Meftah, E. Paumier and F. Studer, Nucl Instr. Meth. Phys. Res. B116, 37(1996)
44. M.C. Busch, A. Slaoui, P. Siffert, E. Dooryhée and M. Toulemonde, J. Appl. Phys. 71, 2596(1992)
45. M. Toulemonde, N. Enault, Jin Yun Fan and F. Studer, J. Appl. Phys. 68, 1545(1990)
46. C. Trautmann, J.M. Costantini, A. Meftah, K. Schwartz, J.P. Stoquert and M. Toulemonde, present symposium.
47. F. Pawlak, E. Balanzat, Ch. Dufour, A. Laurent, E. Paumier, J. Perriere, J.P. Stoquert and M. Toulemonde, Nucl. Instr. Meth. Phys. Res. B122, 579(1997)
48. F. Pawlak, Ch. Dufour, A. Laurent, E. Paumier, J. Perriere, J.P. Stoquert and M. Toulemonde, Nucl. Instr. Meth. Phys. Res.131, 135(1997)
49. M.B. Lewis and E.H. Lee, J. Nucl. Mat. 203, 224(1993)

MECHANISMS OF ENERGY TRANSFER IN SOLIDS
IRRADIATED WITH SWIFT HEAVY IONS

G. SZENES,
Department of General Physics, Eötvös University, Muzeum krt 6-8, Budapest, H-1088, Hungary, Szenes@Ludens.elte.hu

ABSTRACT

The damage cross section velocity effect is studied in $LiNbO_3$, $Y_3Fe_5O_{12}$ and SiO_2 α-quartz. The application of our thermal spike model reveals that the efficiency of track formation varies by a factor of two in the range of 2-4 MeV/nucleon. The effect is explained by the varying fraction of energy deposition to the lattice involving lattice ions.

INTRODUCTION

When a swift heavy ion penetrates into a solid, it interacts with the target atoms via elastic nuclear collisions and inelastic electron excitations. In a broad velocity range, the inelastic losses dominate over the elastic ones (electronic stopping regime). During the relaxation of the excited electron system, the energy is transferred to the lattice and various types of defects can be formed. If the lattice is not fully recovered after the passage of the ion, a track is formed. We are interested only in the formation of amorphous tracks in the electronic stopping regime that can be related to a well-defined temperature, the melting point T_m.

Recently, we analyzed amorphous track formation by applying a phenomenological thermal spike model [1] and this model has been applied to all systematic results published to date: magnetic insulators [1,2], mica [3], $LiNbO_3$ [4], high-T_c superconductors (HTCSs) [5] and SiO_2 α-quartz [6]. We concluded that amorphous track formation in insulators by high velocity ions can be scaled and analytically described without adjustable parameters by taking into account only the thermal properties of the target: the average specific heat c and the melting point T_m [1,6].

In the course of the investigation of yttrium iron garnet (YIG) Meftah et al. [7] found that at equal values of the electronic stopping power S_e low velocity (LO) ions are more efficient for track formation than high velocity (HI) ions. Our analysis of this damage cross section velocity effect revealed that the efficiency of energy deposition in the thermal spike g is by about twice higher for LO than for HI irradiation [2]. We showed that this is also valid for $LiNbO_3$ [4] and SiO_2 α-quartz [6]. In this paper we analyze the variation of the damage cross section with the ion velocity in detail and discuss the mechanisms that can be responsible for this dependence.

THEORY

Our thermal spike model predicts that the square of the effective track radius R^2_e is proportional to S_e (linear regime) after an initial logarithmic variation. The following equation was deduced [1]

$$R_e^2 = gS_e/(2.7\pi\rho cT_o) \qquad \text{for} \quad R_e > a(0), \qquad (1)$$

where c, and ρ are the average specific heat and the density, respectively, T_0 is the difference between the melting and the irradiation temperatures, the term gS_e is the fraction of S_e deposited in the thermal spike and g is the efficiency parameter. The parameter a(0) has the physical meaning of the initial width of the assumed Gaussian temperature distribution in the spike [1]. We found that in all insulators studied so far a(0)=4.5 nm for HI irradiations and a(0)≤4.5 nm is expected for LO conditions. The physical process leading to Eq.(1) was discussed in Ref.[4,6]. Such linear behavior has been found for R_e>4.5 nm in YIG [2], $LiNbO_3$ and SiO_2 [6] irradiated with low velocity ions. In Fig.1 we plotted the appropriate experimental data from Refs. [7-11].

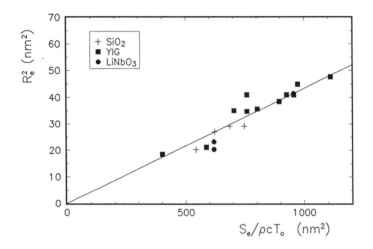

Figure 1. Plot of track sizes according to Eq.(1) for R_e>4.5 nm.

The figure quantitatively proves the validity of Eq.(1), since besides the clear linear behavior, the scaling by ρcT_0 predicted by Eq.(1) works well. This is an important point, because in what follows we base our considerations on this equation.

RESULTS AND DISCUSSION

Meftah et al. found that larger tracks are induced in YIG by low velocity than by high velocity ions [7]. Our analysis of YIG data showed that g_{HI}=0.17 and g_{LO}=0.35 for HI (E≥7.6 MeV/nucleon) and LO (E≤2.2 MeV/nucleon) irradiation, respectively [2]. When the data of $LiNbO_3$ and SiO_2 α-quartz were also included in the analysis a slightly higher value g_{LO}=0.36 was obtained. In the intermediate range (IM) 2.2 MeV/nucleon<E<7.6 MeV/nucleon g_{HI}<g<g_{LO}. In the present paper we study the shape of the g=g(E) curve based on the scaling property offered by Eq.(1). Consequently, only tracks in the linear regime are analyzed in this paper. The a(0) value is not known for IM and LO conditions, therefore the efficiency cannot be determined from tracks belonging to the logarithmic regime.

We evaluate the g values from individual data points according to Eq.(1). In Fig.2 such a g-E plot is shown using the experimental data for $LiNbO_3$ [9,10], SiO_2 [11], $NiFe_2O_4$ [12], besides those of YIG [7,8]. The solid line is to guide the eye. The dashed line shows the g value

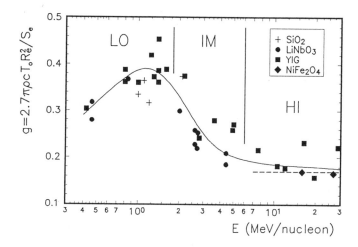

Figure 2. Variation of the efficiency g with the ion velocity. The solid curve is to guide the eye. The dashed line corresponds to the value determined from the initial logarithmic track evolution [2].

determined by the analysis of the initial logarithmic track evolution in various materials [1]. The g values obtained in $LiNbO_3$ and $NiFe_2O_4$ smoothly fit to those represented by the dashed line, while YIG data show considerable scatter.

The g(E) curve is nearly constant in the range of 0.7-2 MeV/nucleon. In this range g≈0.36 and then the efficiency g may decrease again with decreasing ion velocity. However, the irradiation at about E=0.45 MeV/nucleon were made with initial specific beam energies equal to 5-6 MeV/nucleon. This energy was then reduced by applying aluminium degrader foils. At the low energy side of the Bragg curve the actual S_e value is very sensitive even to a small variation of E. Consequently, a small deviation (3-5%) of the foil thickness from the nominal value or a slight tilting or bending of the foils can lead to a considerable difference between the calculated and the true value of S_e. Thus, this low g values require confirmation and these experiments are in progress.

The track data depicted in Fig.2 at high E values are in agreement with our previous analysis of tracks with R_e<4.5 nm [1], where we showed that in the logarithmic regime of track formation g=0.17 for E≥7.6 MeV/nucleon (dashed line in Fig.2). However, we cannot determine the shape of the function g(E) reliably in the IM range. In addition to the experimental error of R_e^2, errors due to the calculation of S_e and of the scaling factors are also possible. Thus, we conclude that the present set of data does not provide reliable information about the separation of the YIG and $LiNbO_3$ curves in the IM range.

The velocity effect indicates a redistribution of the deposited (kinetic and potential) energy with the ion velocity. At high ion velocities either the delayed energy deposition is more significant or a lower fraction of the ion energy is transferred to the target atoms close to the ion path compared to the case of low velocities.

Various models have been proposed to explain how the energy of the bombarding ion is transferred to the target atoms. Toulemonde, Dufour and Paumier assumed that the relaxation of the excited electron system takes place by electron-phonon scattering and S_e is fully converted into the energy of the thermal spike [13]. This 100% efficiency is too high. It is in contradiction with the Monte-Carlo calculations of Bouffard [14] and to our estimates of g, as well. However, T_m cannot be reached in this model when a realistic efficiency is used. Also, the track region is highly defected because of the high ion concentration [15], which makes it implausible that the electron-phonon scattering is the only mechanism.

The excited electrons lose their energy mainly through scattering by ions and neutrals in lattice points. In metallic materials the ion charge is screened very quickly by the conduction electrons. This screening process is much slower in insulators and the excited electrons are less effective for screening. Thus, it is expected that in insulating materials, ion interactions are more important than in solids with metallic conductivity. Previously, we observed no velocity effect in the formation of amorphous tracks in HTCSs [5] where the conductivity is metallic. This lead us to the idea that the velocity effect observed in amorphous track formation may be related to various ion interaction mechanisms. However, we do not deny that a velocity effect may exist in metallic materials when other mechanisms are operative and other type of tracks (different from amorphous ones) are induced by the ion beams.

Gervais and Bouffard applied a Monte-Carlo calculation for the simulation of the interaction of swift heavy ions with solid targets [15]. They found that the density of the primary and secondary holes in SiO_2 bombarded by 10 MeV/nucleon Pb ions is about 10^{22} cm^{-3} even at r=4.5 nm from the trajectory. Recall that for a Gaussian temperature distribution (which is assumed in our model) with a(0)=4.5 nm more than 60% of the thermal energy is confined in this volume. These very high hole density values show that in this volume nearly all atoms are ionized and most of them are multiply ionized.

When the electron-neutral (e-n) and the electron-ion (e-i) scatterings are compared at equal impact parameter values, the energy transfer in a scattering event is higher in the latter case. It is also known that for the Rutherford scattering the cross section $\sigma \sim Z^2$, where Z is the ion charge. On the other hand, if an electron is scattered by a neutral and the polarization effects are neglected, then the Coulomb field has an effect only at small values of the impact parameter when the particle penetrates into the electron cloud of the atom. Thus, besides the higher elementary energy transfer the scattering cross section on ions is also much higher. According to a simple estimate the total energy transfer to the lattice ions $\Delta W \sim Z^3 N_i^2$ where N_i is the ion number density. The contribution of the (e-i) scattering to the relaxation of the electron system becomes comparable or even higher than that of (e-n) scattering when approaching the center of the track. Thus, the (e-i) scattering is an efficient channel of energy transfer in the track region and therefore the energy of the thermal spike is expected to vary with the ion velocity.

An ion-ion interaction mechanism was proposed by Seiberling et al. who assumed that the ionized lattice atoms can gain considerable kinetic energy as a result of their electrostatic interaction [16]. They suggested that this process can lead to the formation of a thermal spike after thermalization and concluded that this was a much faster process than the interaction with the excited electrons. One must check, however, whether the effective lifetime of multiply charged ions is sufficient for acquiring such energy.

In insulators we expect that large positive unscreened charges exist in lattice points close to the trajectory for about 1-10 fs. Seiberling et al estimated that two triply ionized neighboring U atoms can obtain $\varepsilon=10$ eV energy in the lattice of UF_4 due to the Coulomb force [16]. If we calculate the time t necessary to gain this energy we get t= 0.07 ps. This is obviously too high compared to the screening time. The U atoms are heavy, their separation is large which explains the long time interval. To estimate the significance of these parameters, a calculation for UO_2 was made and the results are summarized in Table I.

Table I. Ion interactions in UO_2.

ion pair	ion separation	ε parameters	0.1 eV	1 eV	10 eV
O^{3+}-O^{3+}	0.27 nm	t(fs)	0.72	2.32	8.48
		Δs (nm)	0.001	0.01	0.13
U^{3+}-O^{3+}	0.167 nm	t(fs)	0.39	1.25	4.29
		Δs (nm)	0.001	0.01	0.15
U^{3+}-U^{3+}	0.386 nm	t(fs)	5.73	18	73
		Δs (nm)	0.0015	0.015	0.21

ε - kinetic energy of the ion pair

t - time necessary to acquire ε,

$2\Delta s$ - increase of the ion separation

The data in Table I. show again that the energy transfer to heavy atoms by the ion-ion interaction mechanism requires a long time. When two neighboring oxygen and uranium atoms in UO_2 are triply ionized, the oxygen atom acquires a kinetic energy of 1 eV for t=1.2 fs and 10 eV for about 4 fs. The time t is inversely proportional to the charges of neighboring ions. Thus, even higher energy transfers are possible. This mechanism is important only before screening, while the (e-i) scattering is efficient after screening, as well since the excited electrons can penetrate through the electron cloud.

While heavy atoms are sluggish, light atoms with small separation from neighbors can quickly gain high energies which considerably exceed the thermal energy at the melting point. However, this contribution cannot be responsible alone for the energy of the thermal spike. For LO conditions g=0.36, which means that 36% of the electronic energy losses are converted to thermal energy in the vicinity of the trajectory. Even the total electrostatic energy of the ionized lattice atoms could not cover this energy. In fact, only an atomic rearrangement is possible in a nearly constant volume, thus the reduction of the electrostatic energy is much less compared to the total one.

Both the contributions of the e-i scattering and of the electrostatic interaction of ions both are sensitive to the number density of ions. It is generally accepted that this parameter increases when E is reduced. This can explain the increase of the energy deposited in the thermal spike. Further important parameters are the average ion charge and the mean lifetime of ions. For quantitative comparison with the observed velocity effect further theoretical considerations are necessary.

CONCLUSIONS
We analyzed the amorphous track formation in YIG, LiNbO$_3$ and SiO$_2$ α-quartz and found that the efficiency of energy deposition close to the trajectory is doubled in a narrow range of the variation of E. This suggests that interactions involving target ions are responsible for the effect. We expect that these processes affect many ion-solid interaction phenomena either through the enhanced energy density close to the ion trajectory or through the reduced energy deposition outside this region.

ACKNOWLEDGMENT

This research was partially supported by The National Science Research Fund (OTKA, Hungary) under contracts T014987 and T017344.

REFERENCES

[1] G. Szenes, Phys. Rev. B **51,** 8026 (1995).
[2] G. Szenes, Phys. Rev. B **52,** 6154 (1995).
[3] G. Szenes, Nucl. Instr. Meth. B **107,**) 146 (1996.
[4] G. Szenes, Nucl. Instr. Meth. B **116,** 141 (1996).
[5] G. Szenes, Phys. Rev. B **54,** 12458 (1996).
[6] G. Szenes, Nucl. Instr. Meth. B **122,** 530 (1997).
[7] A. Meftah, F. Brisard, J.M. Costantini, M. Hage-Ali, J.P. Stoquert, F. Studer, and M. Toulemonde, Phys. Rev. B **48,** 920 (1993).
[8] A. Meftah, J.M. Costantini, M. Djebara, N. Khalfaoni, J.P. Stoquert, F. Studer, and M. Toulemonde, Nucl. Instr. Meth. B **122,** 470 (1997).
[9] B. Canut, R. Brenier, A. Meftah, P. Moretti, S. Ould Salem, S.M.M. Ramos, P. Thevenard, and M. Toulemonde, Nucl. Instr. Meth. B **91,** 312 (1994).
[10] B. Canut, S.M.M. Ramos, R. Brenier, P. Thevenard, J.L. Loubet, and M.Toulemonde, Nucl. Instr. Meth. B **107,** 194 (1996).
[11] A. Meftah, F. Brisard, J.M. Costantini, E. Dooryhee, M. Hage-Ali, M. Hervieau, J.P. Stoquert, F. Studer, and M. Toulemonde, Phys. Rev. B **49,**12457 (1994).
[12] M. Toulemonde and F. Studer, Solid State Phenomena **30/31,** 477 (1993).
[13] M. Toulemonde, C. Dufour and E. Paumier, Phys. Rev. B **46,** 14362 (1992).
[14] S. Bouffard, Nucl. Instr. Meth. B **107,** 91 (1996).
[15] B. Gervais, and S. Bouffard, Nucl. Instr. Meth. B **88,** 355 (1994).
[16] E. Seiberling, J.E. Griffith, and T.A. Tombrello, Radiat. Eff. **52,** 201 (1980).

RADIATION EFFECT ON THE VISCOSITY OF THE SIMPLE B_2O_3 GLASSES

A. Barbu and G. Jaskierowicz
Laboratoire des Solides Irradiés (CEA-CNRS), Ecole Polytechnique,
91128, PALAISEAU, cedex , FRANCE.

ABSTRACT

The creep velocity of B_2O_3 glass fibers under irradiation with 2.5 MeV electrons at low flux (around $5 \ 10^{13}$ $e^-cm^{-2}s^{-1}$) has been studied versus temperature. As under very high energy heavy ions irradiation (1.6 GeV argon), the viscosity is drastically reduced below 300°C. An important difference between the two kinds of irradiation is the occurrence of a compaction phenomenon at the beginning of electron irradiation experiments. The results can be understood by assuming two totally different mechanisms : a relaxation driven by melting of the glass along the path of the ions for very high energy heavy ion irradiations and a relaxation driven by individual radiation induced points defects for high energy electron irradiations.

INTRODUCTION

In 1961 Mayer and Leconte [1] reported that the creep velocity of a silica fiber under fast neutron irradiation at 60°C is the same as that at 600°C without irradiation. Despite this very early observation, no direct studies of the effects of irradiation on the viscosity of glasses have been performed until 1989 when we studied the creep rate of a pure B_2O_3 glass fiber irradiated with 1.66 GeV argon ions [2]. We showed that as in fused silica, the viscosity of B_2O_3 is drastically reduced under irradiation. With such high energy heavy ions, the energy is almost totally deposited in electronic excitations. The linear rate of energy deposition dE/dx is extremely large (750 eV/nm) and localised along the path of the ions. We proposed a model based on the fact that at such a very large dE/dx, the glass must be extremely fluid during a very short time in the wake of the ion (thermal spike) [2].
The purpose of this paper is to report new results about the effect of 2.5 MeV electron irradiations on the creep velocity of B_2O_3 glass fibers. With electrons, dE/dx is very small (0.29 eV/nm) compared to swift ions and results in interactions with matrix electrons or atoms with small energy transfers. The electron flux is such that the mean energy deposition per unit volume and per unit time (0.15 eV/nm^3s) has the same order of magnitude as during Ar swift ion irradiations (0.52eV/nm^3s). However in contrast to swift ions where the energy deposition takes place along the path of the ions in a cylinder of radius approximately equal to 20nm, the energy deposition under electron irradiations is homogeneous and the defect created can be considered as isolated.
A study of the viscosity of the B_2O_3 was already reported [3, 4] but it was at very high electron flux giving a mean energy deposition per unit volume and per unit time five orders of magnitude larger (2 10^4 eV/nm^3s). Futhermore, the method was very indirect (study of the coagulation of lead particles performed in situ in transmission electron microscopes) and the viscosity was very low compared to the viscosity obtained under swift ion irradiation.
In this paper we report also an attempt to measure the viscosity of a borosilicate glass under irradiation.

EXPERIMENTAL

B_2O_3 glass fibers have been drawn out from the melt. Their diameter d was 1.5 mm. The specific experimental apparatus built for high energy heavy ion irradiation [2] has been fixed at the end of one of the line of our 2.5 MeV electron Van de Graaff accelerator. It is able to measure viscosity between 10^9 and 10^{14} Ns/m^2. A 12 μm stainless steel window separates the experimental device, in which a 1 bar Helium atmosphere is introduced, from the vacuum of the accelerator line. Only 3 cm of the 10 cm of the fibers was irradiated. The heating of the irradiated part is only due to the energy loss of electrons within the glass. With the chosen diameters of the fibers, the energy is mainly dissipated by convection at the surface and not by conduction along the fiber; the temperature is then almost constant on the 3 cm irradiated. The total loss of energy throughout the fiber does not exceed 0.3MeV. In this experiment, the temperature of irradiation depends on the electron flux that varies from 6.2 10^{12} $e^-cm^{-2}s^{-1}$ to 4.2 10^{14} $e^-cm^{-2}s^{-1}$. A first evaluation of the temperature in the irradiated area is carried out through a sapphire window by means of a radiometric infrared microscope assuming both a transmittance of the window and an emissivity of the sample equal to the unity. The flux is measured using a Faraday cup.

RESULTS AND DISCUSSION

We have shown previously that, under high energy heavy ion irradiation, the elongation of fibers varies linearly with time at any temperature as soon as the irradiation starts [2].
Under electron irradiation, a fresh fiber (not previously irradiated) first shortens for some time and then creeps linearly. Figure 1a shows such a behavior at 130°C, temperature obtained with a flux of 1.3 10^{14} $e^-cm^{-2}s^{-1}$.

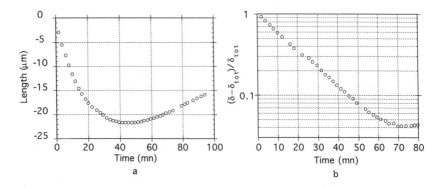

Figure 1: a - Change of the length of a fresh fiber under 2.5 MeV electron irradiation at a temperature of 130°C (corrected value of 160°C), and a flux of 1.3 10^{14} $e^-cm^{-2}s^{-1}$. b - Semi log plot showing the exponential t-dependence of the negative dilation

The irradiation of a crystalline metallic wire (austenitic steel), known to be unable to shorten under irradiation has shown that the shortening of the glass fiber is not an artefact due to the thermal expansion of the whole experimental chamber.

Negative dilatation

The subtraction of the creep elongation to the total elongation gives the true shortening of the fiber. Let us define the negative dilatation as $\delta = (1 - l_o) / l_{irr}$ where l_o is the initial length of the fiber and l its actual length and l_{irr} the length of the irradiated zone. Figure 1b shows that δ follows an exponential law :

$$\delta = \delta_{tot}\left[1 - \exp\left(-\frac{t}{20}\right)\right]$$ where t is the time in minutes and $\delta_{tot} = 10^{-3}$.

Even if we have not performed any density measurement, we can admit that the shortening comes from the well known compaction phenomenon extensively studied in fused silica [5-7]. However in fused silica δ doesn't follow an exponential law, but rather a power law where the exponent is between 3/4 and 1/2 [6].
After 70 mn, the compaction stops and the length of the fiber increases linearly.
When the temperature is subsequently decreased (increased), a new linear strain rate is reached after a transient period of $\approx$ 3mn. It is difficult to know whether any new compaction occurs during the transient period but clearly if it exists it is small.

Viscosity

The temperature (the flux) is increased by steps and the creep velocity $\dot{\varepsilon}$ is given by the slope of curve during the linear regime. The viscosity η is then calculated using

the formula: $\eta = \dfrac{4F}{3\pi d^2\dot{\varepsilon}}$ where F the applied load is 0.5 kg. The change of the fiber diameter d is not considered since for the whole experiment it is smaller than 2%.

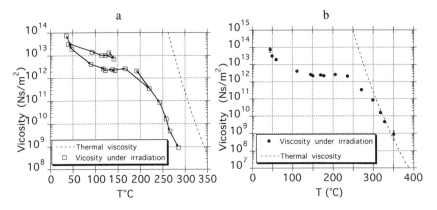

Figure 2 : a) Viscosity under irradiation versus the temperature given by the infrared pyrometer without any correction.
b) Viscosity under irradiation versus the corrected temperature. Only the points corresponding to the increasing temperature steps are plotted.

Figure 2a shows a plot of the viscosity versus temperature given by the infrared pyrometer without any correction. The whole values are obtained on the same fiber.

The arrows give the chronology. The first η value is obtained at 130°C after the compaction stopped. The temperature is then decreased by step up to 36°C. After irradiation at the lowest temperature, the temperature is increased monotonically except for the point at 192°C. The values of the viscosity obtained by increasing the temperature are significantly smaller. The origin of this behavior is not totally clear. One possible explanation is that during the linear regime that is necessarily measured on a limited time, there is some more additional compaction or expansion depending on whether the temperature is decreased or increased, at a rate enough low to obtain an apparent linear regime. As shown by the point at 192°C, it seems that this effect is much more important during the first increase of the temperature than during the subsequent decrease (the glass is initially very far from the equilibrium at the starting temperature). Indeed, the point obtained at 192°C by decreasing the temperature, during the increasing steps, is not significantly out of the lowest curve. So, even if the true viscosity curve is certainly slightly above the lowest curve on figure 2a, we will consider in the following that the lowest curve is the true curve.

On figure 2a, the viscosity without irradiation is also plotted [8, 9]. The slope of the viscosity at high temperature under irradiation is clearly equal to the one observed without irradiation but the temperature seems to be shifted towards low values. If we admit that the viscosity is inversely proportional to the concentration of the defects that govern the viscous flow, the irradiation must not change the viscosity at high temperature. Indeed, the number of viscous flow defects is given by a balance between a creation term that is not or only slightly temperature dependent and elimination terms (mono and/or bimolecular reaction) which are strongly temperature dependent. So, we consider that the temperature given by the pyrometer must be corrected. The superimposition of the two curves at high temperature can be obtained (figure 2b) by multiplying the increment of temperature under the electron beam by a factor of 1.23. This correction is physically justified because i) the emissivity of the glass is not equal to unity but rather around 0.9 and ii) the transmittance of the sapphire window is not unity but 0.87. The inverse of the transmittance of the window times the emissivity gives 1.27, very close to the factor used.

Let us recall that the viscosity at various temperatures is not obtained with the same electron flux ϕ. We don't know how the viscosity changes with the flux but, as discussed above, if we admit that the viscosity under electron irradiation is inversely proportional to the concentration of viscous flow defects, the viscosity must vary as ϕ^{-a} with $a \leq 1$. The exponent a is equal to 1 when the elimination term is mainly monomolecular and is equal to 1/2 when it is bimolecular. On figure 3 the viscosity is plotted versus T assuming the maximum effect (a=1) and by normalising ϕ at 4. 10^{14} e⁻cm⁻²s⁻¹. With or without the flux correction, the behavior remains qualitatively the same :
- There is no radiation effect at high temperature (the slope of the curve is the same with and without irradiation).
- At medium and low temperatures, the viscosity is drastically reduced under electron irradiation.
- At medium temperature, the behaviour is quasi-athermal.
- At low temperature, the viscosity rises again.

Comparison with swift ion irradiations

The fiber viscosity of the same glass under 1.6 GeV Ar ion irradiation, previously studied with the same experimental method [2] is reported in figure 3. The behaviour is qualitatively very similar but there is an important difference that does not appear in this figure : under swift ion irradiation, no compaction is observed whatever the temperature. This point shows that, even if the behavior of the viscosity is similar, the mechanisms are certainly very different.

An attempt to study the effect of electron irradiation on the viscosity of a borosilicate glass (Pyrex) has been performed unsuccessfully. If we clearly observe the well known compaction effect, creep is not observed in the fibers which have a diameter of the range between 1 and 1.5 mm. This is probably because at the highest flux attainable with our accelerator (5 10^{14} e-cm^{-2}s^{-1}), the values of the viscosity at the plateau level (if such a plateau exists) is higher than the highest value measurable with our simple and direct but not very sensitive experimental device (10^{14} Ns/m^2).

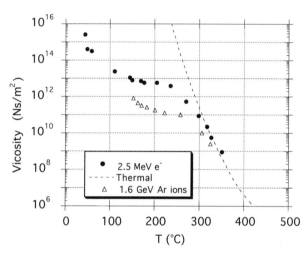

FIGURE 3: Comparison of the effect of irradiations with low flux 2.5 MeV e$^-$ and high energy ions on the viscosity of B_2O_3.

Discussion
In the discussion concerning the flux and temperature correction we explicitly admit that under electron irradiations, individual flow defects are created and that their concentrations can be described by rate equations. The nature of these defects is not known. They can be vacancies and interstitials created either by elastic collision or by some radiolytic mechanism. Then can also be broken and dandling bonds induced by electronic excitations. Within the framework of this model, the various regimes in viscosity versus temperature curve can be explained at least qualitatively : as for the diffusion in crystal under irradiation [10], the athermal plateau would correspond to a temperature range where a first order (monmolecular) elimination term is dominant and the behaviour at low temperature to a temperature range where the dominant elimination term is second order (bimolecular).
Furthermore, it is reasonable to assume that defects are also at the origin of the compaction phenomenon.
For swift ion irradiations, we envisaged a model based on point defect as for electron irradiations but we claimed that an another model totally different is more attractive

[2]. It is based on the fact that in the wake of the ions the energy deposition is so high that the material is extremely fluid for a very short time (thermal spike). Under the load, this very fluid cylinder results in an increment of deformation of the fiber. The solid in the track after the passage of the ions is more or less identical to the fresh material obtained from the melt. In this mechanism there is obviously no room for an important compaction. Trinkaus and Ryazanov [11-13] improved this model and showed that it can also explain the growth phenomenon observed in glasses under swift ion irradiation. In its present form, this model assumes that during the transient increment of temperature ΔT in the tracks, the matter in the tracks is perfectly fluid (its viscosity is so small that it can be taken equal to zero). A consequence is that it is unable to explained the increase of viscosity at low temperature. However, as the temperature in the tracks is $T+ \Delta T$ where T is the temperature of the experiment, the hypothesis that the viscosity is null is clearly no more valid at low T. By considering the variation of the viscosity with the temperature in the thermal spike it would certainly be possible to reproduce the increase of viscosity at low temperature.

ACKNOWLEDGMENTS

We gratefully acknowledge the stimulating interest of Dr D. Lesueur and Dr. Ch de Novion in this work and thank P. Laplace and C. Laumonier for technical support.

REFERENCES:

[1] G.Mayer and L Leconte J. Phys. Radium 21 (1960) 246
[2] A. Barbu, M. Bibolé, R. Le Hazif, S. Bouffard and J. C. Ramillon J. N. M. 165 (1989) 217
[3] I. Biron and A. Barbu Diffusion and Defect Data 53-54 (1987) 477
[4] I.Biron and A. Barbu Nucl. Inst. and Meth. B32 (1988) 279
[5] W. Primak, L. H. Fuchs and P. Day Phys. Rev. 92 (1953) 1064
[6] W. Primak and R. Kampwirth J. Appl. Phys. 39, 12 (1968) 5651
[7] W. Primak J. Appl. Phys. 53 (1982) 7331
[8] J. Zarzycki and F. Naudin, Phys. Chem. Glasses 8 (1967) 11
[9] P. B. Macedo and A. Napolitano, J. Chem. Phys 49 (1968) 1887
[10] R.Sizman, J. Nucl. Mat , 69 & 70 (1978) 386
[11] A. I. Ryazanov, A. E. Volkov and S. Klaumünzer, Phys. Rev. B 51, 18 (1995) 12107
[12] H. Trinkaus and A. I. Ryazanov, Phys. Rev. Let, 75, 25 (1995) 5072
[13] H. Trinkaus, Nucl. Inst. and Meth. B07 (1988) 155

SWELLING OF SIO$_2$ QUARTZ INDUCED BY ENERGETIC HEAVY IONS

C. TRAUTMANN[1], J.M. COSTANTINI[2], A. MEFTAH[3], K. SCHWARTZ[1]
J. P. STOQUERT[4], AND M. TOULEMONDE[5]

[1] Gesellschaft für Schwerionenforschung, Planckstr. 1, 64291 Darmstadt, Germany, C.Trautmann@gsi.de
[2] DPTA/PMC, BP 12, 91680 Bruyères-Le-Châtel, France
[3] ENSET, BP 26 Merj-eddib, 21000 Skikda, Algeria
[4] Laboratoire PHASE, 67037 Strasbourg - cedex 2, France
[5] CIRIL, Laboratoire commun CEA/CNRS, BP 5133, 14070 Caen-cedex 5, France

ABSTRACT

A pronounced swelling effect occurs when irradiating SiO$_2$ quartz with heavy ions (F, S, Cu, Kr, Xe, Ta, and Pb) in the electronic energy loss regime. Using a profilometer, the out-of-plane swelling was measured by scanning over the border line between an irradiated and a virgin area of the sample surface. The step height varied between 20 and 300 nm depending on the fluence, the electronic energy loss and the total range of the ions. From complementary Rutherford backscattering experiments under channelling condition (RBS-C), the damage fraction and corresponding track radii were extracted. Normalising the step height per incoming ion and by the projected range, a critical energy loss of 1.8 ± 0.5 keV/nm was found which is in good agreement with the threshold observed by RBS-C. Swelling can be explained by the amorphisation induced along the ion trajectories. The experimental results in quartz are compared to swelling data obtained under similar irradiation conditions in LiNbO$_3$.

INTRODUCTION

Radiation induced volume expansion is a general effect which has been discovered in various solids in the late fifties. The swelling under irradiation can result from point defects, but also from defect aggregates or from phase transitions. In the case of point defects, the swelling mechanism is well understood [1,2], while for more complex defects, i.e. under irradiation at higher doses, it is described only phenomenologically. In this regime, the relative volume changes are larger and, depending on the material, can reach several percent (e.g. 15% for quartz) [3]. Swelling occurs in a wide range of irradiation conditions (e.g. temperature, dose, and dose rate) and under various particle beams such as electrons, neutrons or ions [4-8]. More recently, swelling has been observed when irradiating Al$_2$O$_3$ [9] and LiNbO$_3$ [10-12] with energetic heavy ions. The macroscopic volume increase of the bulk material was clearly attributed to the damage produced by electronic excitations. The goal of our study was to extend the swelling tests to SiO$_2$ crystals. It should be mentioned that in quartz, the damage under heavy ion irradiation has been studied by various techniques [7,13-15]. Evidence was given that along the ion trajectory a cylindrical zone of amorphous silica is formed having a diameter of a few nanometers. The experimental data are in good agreement with the thermal spike model which describes the track as a cylindrical zone resulting from rapid quenching of a liquefied phase along the ion path [15].

In order to perform systematic measurements of the out-of-plane swelling in SiO_2, we irradiated quartz with various ion species in the energy regime between 15 and 830 MeV. The swelling tests were complemented by RBS-C analysis from which the damage fraction at different ion fluences was deduced. By combining the two techniques, we followed the data analysis as presented by Canut et al in ref. [10]. Extending this analysis, it was possible to extract an energy loss threshold for swelling. Finally, we will compare ion induced swelling and damage effects in quartz and in $LiNbO_3$.

EXPERIMENTAL CONDITIONS AND PHYSICAL CHARACTERISATIONS.

Irradiation

We used single crystals of synthetic quartz of a thickness of about 1.5 mm. The polished surface (optical quality) of the samples was covered with a 50 nm thin carbon layer in order to avoid electrostatic charging during irradiation or analysis. The irradiations were performed using lighter ions (^{19}F, [32]S, and [63]Cu) at the 7 MV tandem Van de Graaff in Bruyères-le-Châtel and heavier ions ([84]Kr, [129]Xe, [181]Ta, and [208]Pb) at the medium energy line of the GANIL accelerator in Caen. All irradiations were performed at room temperature under normal incidence. In some cases, thin aluminium foils were placed in front of the samples in order to vary the energy and energy loss of the ions when impinging on the sample surface. By limiting the maximum energy of the ion beam to 4 MeV/u, significant velocity effects were avoided [16]. A complete list of the irradiation parameters is given in Table I, where the values for the energy loss dE/dx and the ion range were calculated with the TRIM 91 code [17]. The flux of the ion beam was of the order of 10^9 ions $s^{-1}cm^{-2}$ at the tandem Van de Graaff accelerator and 3×10^8 ions $s^{-1}cm^{-2}$ at the GANIL accelerator. The maximum applied ion fluence ϕ depended on the ion species and was, e.g., 4×10^{11} and 2×10^{14} ions/cm^2 for Pb and F ions, respectively. During the irradiation, all crystals were partially masked in order to analyse the damage of an irradiated area in direct comparison with a virgin area of the same sample.

Table I. Parameters of ion irradiation, the damage cross section A_d and track radius R

ion	energy (MeV/u)	dE/dx (keV/nm)	range (μm)	$A_d = \pi R^2$ (10^{-13} cm^2)	radius R (nm)	reference
F	0.79	2.4	7.1	0.11 ± 0.03	0.6 ± 0.1	[14]
S	1.56	4.7	11.7	1.2 ± 0.3	2.0 ± 0.3	[14]
Cu	0.79	9.1	8.7	2.6 ± 0.5	2.9 ± 0.03	[14]
Kr	3.3	12.0	27.7	3.3 ± 1.0	3.2 ± 0.5	[14]
Xe	1.5	17.0	14.7	5.0 ± 1.0	4.0 ± 0.4	[14]
Ta	0.8	17.6	13.2	11.2 ± 1.5	6.0 ± 0.8	present work
	1.5	21.1	19.6	8.9 ± 1.0	5.3 ± 0.6	
	4.6	24.5	48.0	8.5 ± 1.0	5.2 ± 0.6	
Pb	1.2	21.9	15.7	11.2± 1.5	5.5 ± 0.5	present work
	4.0	27.6	38.8	8.9 ± 1.0	5.2 ± 0.5	

Step height measurements:

The swelling effect was studied using a profilometer (Dektak 8000) where a high precision stage moves the sample beneath a diamond-tipped stylus. The out-of-plane swelling l was measured by scanning over the border line between an irradiated and a virgin area of the sample surface (Fig. 1). Depending on the ion fluence, the electronic energy loss and the total range of the ions, l varied between 20 and 300 nm. The surface roughness of the crystals was in the order of 5 nm. For each sample, a mean step height was extracted from several measurements.

The evolution of the step height as a function of the ion fluence is presented in Figure 2. After an initial linear increase, the swelling saturates at high ion fluences where the ion tracks begin to overlap. This saturation effect was observed only when a sufficiently high fluence was obtained such as in the case of Ta ions (2×10^{12} cm^{-2}) and F ions (1×10^{14} cm^{-2}). For all other ion species, the swelling effect was restricted to the linear regime.

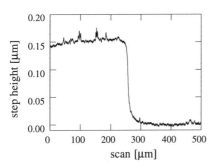

Fig. 1. Profilometer scan from the irradiated (4.6×10^{13} F-ions/cm^2) (left) to the virgin (right) area of the crystal.

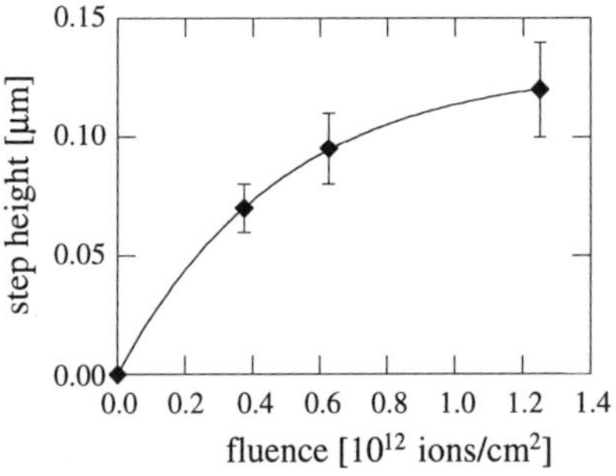

Fig. 2. Step height as a function of the fluence of the Ta ions (0.8 MeV/u).

Damage analysis by Channelling Rutherford Backscattering

The radiation damage of the irradiated crystals was analysed by means of Rutherford backscattering under channelling condition (RBS-C). The experiments were performed at the 4 MV Van de Graaff accelerator (Strasbourg) using 2 MeV He$^+$ ions and a backscattering angle of 170°. Figure 3 shows a typical RBS-C spectrum of a non-irradiated (curve a) and an irradiated area (curve b) of a SiO$_2$ quartz sample in channelling condition along the (0001) direction. Due to the damage induced by the ions, the backscattering yield increases, but is still below the yield of a randomly oriented crystal (curve c).

Using the surface approximation, the backscattering yield χ was measured by extrapolating the energy evolution of the yield over the first 500 nm up to the mean energy of the random edge. The damage fraction F_d of the material is given by

$$F_d = (\chi_i - \chi_v) / (\chi_r - \chi_v), \qquad (1)$$

where χ_i and χ_v are the backscattering yield under channelling condition of the irradiated and of the virgin sample, respectively, and χ_r corresponds to the yield of the randomly-oriented crystal.

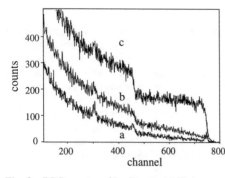

Fig. 3. RBS spectra of backscattered He ions on SiO$_2$ quartz in channelling conditions (0001): (a) virgin sample (b) sample irradiated with Pb ions (2.3 10^{11} cm^{-2} at 1.2 MeV/u) (c) virgin sample in random orientation.

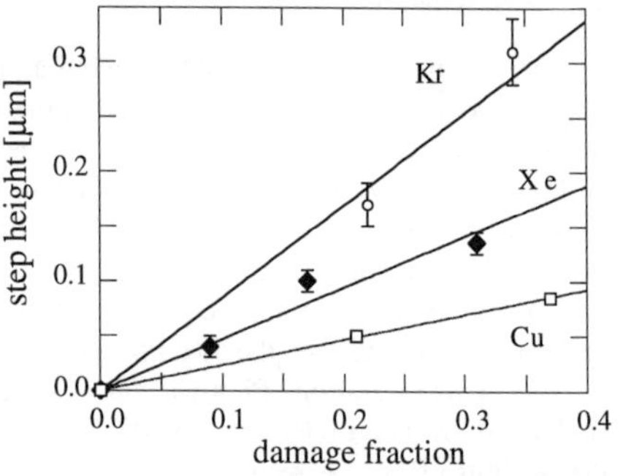

Fig. 4. Swelling versus damage fraction for Cu, Xe, and Kr ions with a total range of 8.7, 14.7 and 27.7 µm, respectively.

Assuming that at higher ion fluences ϕ, the overlapping of tracks leads to a damage distribution which can be described by a Poisson law

$$F_d = 1 - \exp(- A_d \times \phi) \qquad (2)$$

we can deduce the mean cross section A_d of the damage. Since the damage zone along each ion trajectory has a cylindrical geometry, the track radius R can be determined from $A_d = \pi R^2$. RBS-C data analysed accordingly are listed in Table 1. The track radii of Ta and Pb ions at different energies exhibit a maximum variation of the radius of around 15% for the studied energy regime from 0.8 to 4.6 MeV/u and 1.25 to 4.0 MeV/u, respectively. This observation supports the assumption that, in our situation, the track radius can be regarded as almost constant along the full length of the ion path.

DATA ANALYSIS AND DISCUSSION:

Combining the data from the swelling and the RBS-C measurements, it becomes evident that the step height has a linear dependence on the damage fraction F_d. Figure 4 demonstrates this relation for the irradiation with Cu, Xe and Kr ions. Moreover, the slope of the curve becomes steeper for larger range of the ions. The same observation has been made in LiNbO$_3$ [9,10]. In order to compare quantitatively our results to the effects observed in LiNbO$_3$, the step height l was normalised by the damage fraction F_d and then plotted versus the total range L of the ions (Fig. 5). Although the data exhibit some scattering, in particular for the lighter ions in quartz, both materials follow a linear dependence. The line fit intercepts the range axis at $L_0 \approx 5$ µm (Fig. 5) indicating that the damage responsible for swelling is not produced along the full range of the ion path. The slope of the curve has a value of about 0.04 and corresponds to the relative dimensional change l/L. Using this presentation, it is surprising to note that SiO$_2$ and LiNbO$_3$ show qualitatively and quantitatively the same dimensional change.

In our experimental set-up, the free expansion of the irradiated volume is limited by the constraint of the undamaged substrate. Therefore, the induced swelling occurs mainly normal to the sample surface and corresponds to a relative decrease of the mass density. The density decrease can be understood on the basis of the finding that tracks in quartz consist of amorphised regions [14]. As a consequence of the transition from the crystalline to the amorphous phase, each individual track undergoes a volume expansion finally leading to a macroscopic out-of-plane swelling.

In order to test the correlation between swelling and the energy loss of the ions, we determined the relative contribution of each single ion per unit damage length. This was done by dividing the initial rate of swelling ($\Delta l / \Delta\phi$) by the projected ion range L and plotting it versus the mean energy loss. Since the ions were stopped in the crystals, we used the energy loss averaged along the ion path obtained from dividing the initial energy by the total range of the ions. If we fit the experimental data by a linear curve, a threshold for the energy loss of $(dE/dx)_c = 1.8\pm0.5$ keV/nm was found. The given error includes considerations that the critical energy loss is not surpassed along the full length of the ion path. In the case of the light ions (F, S, and Cu), we took this part of the trajectory into account as possibly not contributing to the swelling. It should be noted that the threshold of swelling is in good agreement with the value determined from the RBS-C analysis of the damage induced with low-velocity ions [14].

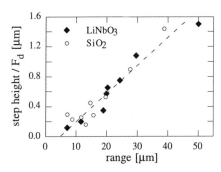

Fig. 5. Step height normalised by the damage fraction F_d as a function of the ion range for SiO_2 in quartz and $LiNbO_3$ [11].

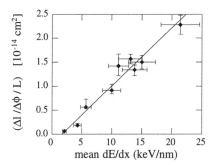

Fig. 6. The initial swelling ($\Delta l/\Delta\phi$) normalised by the ion range L versus the mean energy loss dE/dx in SiO_2 quartz.

CONCLUSIONS

Swelling effects have been found when irradiating SiO_2 quartz with various heavy ions in the electronic energy loss regime. For high fluences when track overlapping becomes significant, the step heights show saturation effects. The out-of-plane swelling depends both on the induced damage fraction and on the range of the ions. From a detailed analysis of the swelling data, a critical energy loss of 1.8 ± 0.5 keV/nm was extracted. Below this value, the damage along the ion path does not contribute to the expansion of the sample dimension. The observed macroscopic swelling can be interpreted in terms of a decrease of the mass density of around 4%. This is supported by the finding that energetic ions induce amorphisation along their trajectories [14]. The quantitative analysis of swelling and damage creation gives evidence that both oxides, SiO_2 quartz and $LiNbO_3$ undergo similar structural modifications under heavy ion irradiation.

REFERENCES :

1. R. Balzer, H. Peisl and W. Waidelich, Phys. Stat. Sol. **15**, p. 495 (1966).

2. F. Agullo-Lopez, C.A.R. Catlow and P. D. Townsend, in Point defects in Materials, Academic Press , pp.149 (1988).

3. L. Douillard and J. P. Duraud, J. Phys. III France **6**, p. 1677 (1996).

4. S.L. Chan, L.F. Gladden and S.R. Elliott, in The Physics and Technology of Amorphous SiO$_2$, edited by R.A.B. Devine, Plenum Press, pp. 83 (1987).

5. F.W. Clinard (Guest Editor) MRS Bulletin **22**, N°4 (1997).

6. D.S. Bilington and J.H. Crawford, in Radiation Damage in Solids, Princeton University Press, Princeton (1961).

7. L. Douillard, F. Jollet, J.P. Duraud, R.A.B. Devine and E. Dooryhee, Rad. Eff. Def. Sol. **124**, p. 351 (1992).

8. W. Primak, Phys. Rev. **B 14**, p.4679 (1976).

9. R.Brenier, B. Canut, S.M.M.Ramos and P. Thévenard, Nucl. Instr. Meth. Phys. Res. **B 90**, p. 339 (1994).

10. B. Canut, R. Brenier, A. Meftah, P. Moretti, S. Ould Salem, S.M.M. Ramos, P. Thévenard and M. Toulemonde, Nucl. Instr. Meth. Phys. Res. **B 91**, p. 312 (1994)

11. B. Canut, S.M.M. Ramos, R. Brenier, P. Thévenard, J.L. Loubet and M. Toulemonde, Nucl. Instr. Meth. Phys. Res. **B 107**, p. 194 (1996).

12. S.M.M. Ramos, B. Canut, M. Ambri, C. Clement, E. Dooryhee, M. Pitival, P. Thévenard and M. Toulemonde, Nucl. Instr. Meth. Phys. Res. **B 107**, p. 254 (1996).

13. R.L. Fleischer, P.B. Price and R.M. Walker, Nuclear Tracks in Solids: Principles and Applications, University of California, Berkeley, (1975).

14. A. Meftah, F. Brisard, J.M. Costantini, E. Dooryhee, M. Hage-Ali, M. Hervieu, J.P. Stoquert, F. Studer and M. Toulemonde, Phys. Rev. **B 49**, p. 12457 (1994).

15. M. Toulemonde, J.M. Costantini, Ch. Dufour, A. Meftah, E. Paumier, and F. Studer, Nucl. Instr. Meth. Phys. Res. **B 116**, p. 37 (1996).

16. A. Meftah, F. Brisard, J.M. Costantini, M. Hage-Ali, J.P. Stoquert, F. Studer and M. Toulemonde, Phys. Rev. **B 48**, p. 920 (1993).

17. J.P. Biersack and L.G. Haggmark, Nucl. Instr. Meth. **174**, p. 257 (1980).

EFFECTS OF HEAVY-ION CASCADE VOLUME ON THE SUPPRESSION OF RADIATION-INDUCED SEGREGATION IN ALLOYS

M.J. Giacobbe [1,2], L.E. Rehn [1], N.Q. Lam [1], P.R. Okamoto [1], L. Funk [1], P. Baldo [1], A. McCormick [1], and J.F. Stubbins [2]

[1] Argonne National Laboratory, Materials Science Division, Argonne, IL 60439
[2] University of Illinois at Urbana-Champaign, Dept. of Nuclear Engineering, Urbana, IL 61801

ABSTRACT

The interactions of cascade remnants with freely migrating defects (FMDs) during dual light and heavy-ion irradiations in Cu-1at.%Au at 400°C were investigated using Rutherford backscattering spectrometry. Near-surface Au depletion driven by 1.5-MeV He ion irradiation was suppressed by concurrent bombardment with 1.2-MeV Ag ions. The dual irradiation effect suggests that short-lived cascade remnants created by heavy ions act as recombination centers for FMDs, reducing radiation-induced segregation (RIS). The effects of the total cascade volume generated by heavy-ion beams on the suppression of RIS were examined. The investigation revealed when 800-keV Cu and 1.2-MeV Ag ion beams produce nearly the same total cascade volume per second, their suppression effects on 1.5-MeV He-induced Au transport are also nearly equal even though the total cascade volume produced per ion for each are different. This result indicates that the suppression effect of cascade remnants produced by heavy ions depends on the total cascade volume induced per unit time and not on the total cascade volume per ion generated by individual ions of different mass and energy.

INTRODUCTION

Radiation-induced segregation (RIS) occurs when nonuniform defect production or annihilation produces persistent fluxes of freely migrating defects (FMDs) which couple preferentially to particular alloy components and drive atom transport [1-3]. FMDs are defects that survive an initial cascade event and are free to migrate over distances which are large relative to cascade sizes. Results from ion irradiations in Cu-1at.%Au [4], Ni-12.7at.%Si [5], and Mo-Re [6] alloys demonstrate that the production rate of FMDs depends on the primary recoil energy spectrum [7]. As recoil energy increases above 2 keV, the FMD production efficiency decreases rapidly until the total number of FMDs is only a few percent of the value given by the modified Kinchin-Pease relationship [4]. Rutherford backscattering spectrometry (RBS) measurements by Iwase et al. on Cu-1at.%Au [8,9] and Ni-12.7at.%Si [10] suggest that interactions between cascade remnants produced by heavy ions (Cu or Ne) and FMDs produced by 1.5-MeV He ions are responsible for reducing steady-state defect populations during dual heavy and light-ion irradiations. The reduction in FMD concentration caused by the heavy ion results in the suppression of RIS. The suppression effect increases as the heavy-ion beam flux increases until a critical value is reached at which it is believed all FMDs produced by light ions are annihilated [8,9]. In this work using 1.5-MeV He and either 800-keV Cu or 1.2-MeV Ag ions, we show that during dual irradiations of Cu-1at.%Au at 400°C the suppression of He-induced RIS caused by heavy-ion bombardment depends on the total cascade volume per unit time produced by heavy ions.

EXPERIMENTAL PROCEDURE

Single 1.5-MeV He, 800-keV Cu [8,9], and 1.2-MeV Ag ion and dual 1.5-MeV He and heavy-ion irradiations of Cu-1at.%Au alloys were performed at the Argonne National Laboratory Tandem accelerator facility. The irradiations were conducted at 400°C, and near-surface Au depletion caused by RIS was measured in situ with RBS. Sample preparation and the experimental geometry of the ion beams, the sample, and the detector are described elsewhere [9]. 1.5-MeV He ions were chosen to produce FMDs because of both their high efficiency for producing mobile defects and their ability to simultaneously provide an easily measurable RBS signal. 800-keV Cu and 1.2-MeV Ag ions were chosen as heavy ions because of both their capability to produce energetic displacement cascades and their different primary recoil energy spectra.

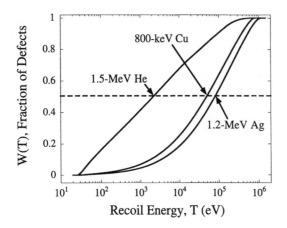

Figure 1. Weighted primary recoil energy spectra for 1.5-MeV He, 800-keV Cu, and 1.2-MeV Ag ions in Cu-1at.%Au as calculated by TRIM. The ordinate gives the fraction of defects produced by primary recoil collisions with energy less than T. $T_{1/2}$ values are indicated by the arrows.

Weighted primary recoil energy spectra for all ions used in this investigation were calculated using TRIM and are shown in Fig. 1. These curves reveal the fraction of defects produced by primary recoil collisions with energies less than T. For quantitative comparison of weighted primary recoil spectra, the recoil energy above which 50% of all displacements are produced, $T_{1/2}$, is used. $T_{1/2}$ values are 2.2, 50, and 81 keV for 1.5-MeV He, 800-keV Cu, and 1.2-MeV Ag ions in Cu-1at.%Au, respectively. Frenkel pair production cross sections, σ_d, used to calculate ion doses, and the projected ranges of all ions were also computed using TRIM. Since Au-depleted zones are ≤60 nm deep, a 100 nm depth was chosen to determine all σ_d. An average threshold displacement energy of 29 eV [11] was used to calculate the weighted primary recoil energy spectra, σ_d values, and projected ranges for all ions. Information concerning $T_{1/2}$, σ_d, and ion ranges is summarized in Table 1.

Table 1. Ion irradiation parameters.

Ion/Energy	Current (nA)	σ_d (cm²)	Range (nm)	$T_{1/2}$ (keV)	Total cascade volume per ion (Å³/ion)	Total cascade volume per s (Å³/s)
1.5-MeV He	100	1.46x10⁻¹⁸	2380	2.2	----	----
800-keV Cu	5	2.99x10⁻¹⁵	225	50	9.6x10⁵	3.0x10¹⁶
1.2-MeV Ag	2	5.81x10⁻¹⁵	200	81	2.1x10⁶	2.6x10¹⁶

The total volume of the first 100 nm of Cu-1at.%Au filled with cascades per ion (Å³/ion) and per second (Å³/s) during heavy-ion irradiation is also given in Table 1. These values were calculated by first using TRIM to determine the primary recoil energy, E, of each collision for 1000 incident ions. Then, the peak number of displacements, N, in each cascade was determined by using the relationship, $N = E(keV) * 500 - 250$, which was extracted from molecular dynamics results in Cu-1at.%Au [12]. In calculating N for each cascade, no sub-cascade formation was considered. The total cascade volume produced per ion was computed by summing all N for 1000 ions, multiplying the sum by the atomic volume of Cu-1at.%Au, and dividing by the number of ions (1000). Figures 2a and b give a schematic example of the typical cascade distribution produced in the first 100 nm of Cu-1at.%Au by one 800-keV Cu and one 1.2-MeV Ag ion. All cascades were assumed be spherical in shape with the primary collision event lying at the center. It can be seen in Figs. 2a and b that each ion generates a different total cascade volume.

During 1.5-MeV He single ion and simultaneous He and heavy-ion irradiations, RBS spectra were recorded at 20 min intervals. For single heavy-ion irradiations, RBS spectra were acquired with 1.5-MeV He ions following various heavy-ion dose intervals. Apertures for 1.5-MeV He and heavy-ion beams were 1 and 3 mm in diameter, respectively, for dual irradiations, which ensured that the heavy-ion beam overlapped the light ion beam. The heavy-ion beam aperture remained at 3 mm for single heavy-ion experiments while the 1.5-MeV He beam aperture size was increased to 1.5 mm in order to make its contribution to RIS during RBS measurements negligible [9].

RESULTS

During irradiation of Cu-1at.%Au, persistent point-defect fluxes to extended sinks such as the surface lead to a net flow of Au atoms away from the surface and, hence, Au-depletion in the near-surface region [3,4].

As previously mentioned, RBS spectra were taken at 20 min intervals for dual irradiations and at various heavy-ion dose intervals for single heavy-ion irradiations. In all cases, the initial spectrum was compared to all subsequent spectra from the same experiment in order to determine the segregation of Au away from the surface. An example of this can be seen in Figs. 3a and b. Figure 3a compares RBS spectra acquired before and after 1.1 dpa of 1.5-MeV He ion irradiation. Figure 3b shows the difference spectrum computed by subtracting the pre-irradiation spectrum from the post-irradiation spectrum. The magnitude of Au depletion can subsequently be determined by calculating the negative area (approximately from channels 860-940 in Fig. 3b) near the surface in the difference spectrum. All spectra in each experiment were normalized to a fifty channel region (channels 701-750) behind the Cu edge where RIS in not expected to make measurable changes in local sample composition. The normalization allows for all depletion values to be presented on a unified Au-depletion count scale.

The Au-depletion results from all single and dual ion irradiations are summarized in Figure 4, where the depletion is plotted as a function of total ion dose. The current for 1.2-MeV Ag ion irradiations, 2 nA, was chosen so that the Ag ions had the same total production of cascade volume per second (see Table 1) as 800-keV Cu ion irradiations at 5 nA used by Iwase et al. [8,9]. Errors in the depletion values are calculated from Poisson statistics as discussed in [9]. Using a standard weighted least-squares analysis, the depletion rate, A, and its error, ΔA, are calculated and presented in Table 2 for all ion irradiations. It should be noted that the depletion data presented here for the experiments involving 800-keV Cu ions was calculated from the raw data obtained by Iwase et al. [8], but was subject to the same analysis as all other data in this investigation.

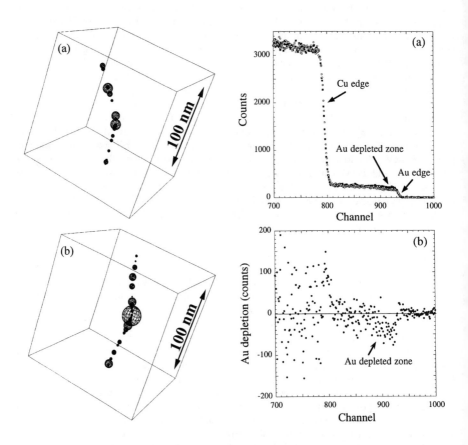

Figure 2. Typical cascade damage profiles for (a) one 800-keV Cu ion and (b) one 1.2-MeV Ag ion incident on the first 100 nm of Cu-1at.%Au. The bottom surface of the cube is the sample surface, while the top surface denotes a depth of 100 nm. The volume of each sphere represents the cascade volume induced by each primary collision, which, in turn, can be summed over many ions to determine the total cascade volume per ion.

Figure 3. (a) RBS spectra for before (solid squares) and after (open circles) 1.1 dpa of 1.5-MeV He irradiation in Cu-1at.%Au. (b) Difference spectrum of the two spectra shown in (a). The difference spectrum is calculated by subtracting the 1.1 dpa spectrum from the pre-irrradiation spectrum. The negative area denoted as the Au depleted zone in (b) can subsequently be measured to determine the total Au depletion at 1.1 dpa.

DISCUSSION

The values of A are the same within the uncertainty for single heavy-ion irradiations and their respective dual irradiations, i.e., single 5 nA Cu and dual 100 nA He and 5 nA Cu. This result shows that the addition of 5 nA of 800-keV Cu (previously reported in [8,9]) or 2 nA of

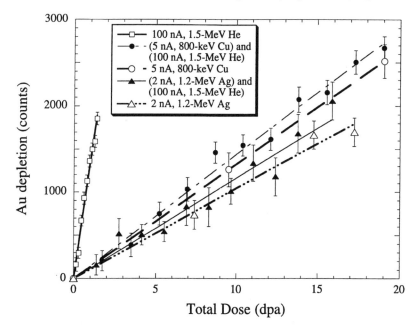

Figure 4. Au depletion as a function of total dose in dpa for single 1.5-MeV He (open squares), single 800-keV Cu (open circles), single 1.2-MeV Ag (open triangles), dual 800-keV Cu and 1.5-MeV He (solid circles), and 1.2-MeV Ag and 1.5-MeV He (solid triangles) ion irradiations. For dual irradiation experiments, the total dose represents the sum of dpa from both ion beams. Ion beam currents are given in the legend.

1.2-MeV Ag to 100 nA of 1.5-MeV He removes virtually all of the RIS driven by the He beam, and hence the RIS driven by those defects is suppressed. The reduction in FMD population is believed to occur because of the existence of short-lived cascade remnants produced by energetic heavy ions which act as FMD annihilation centers [8-10].

The Au depletion behaviors observed in both dual irradiation experiments are nearly equal within ΔA. Since A for both single heavy-ion irradiations is the same within ΔA and heavy-ion currents for the dual irradiations were chosen to produce approximately the same total cascade volume per second, the similarity in A for both dual irradiations suggests that the cascade remnant suppression effect is independent of the different total cascade volumes per ion produced by 800-keV Cu and 1.2-MeV Ag ions. It has been demonstrated in computer simulations by Wiedersich [13,14] that the cascade remnant density is independent of the cascade size, hence remnant production depends only on the total cascade volume. This finding [13,14] supports the present experimental results, and together they provide evidence that the interactions between 1.5-MeV He-generated FMDs and heavy-ion cascade remnants, which reduce RIS, only depend on the total volume of heavy-ion cascades created during irradiation. Currently, additional RBS investigations and computer modeling are underway to further test this dependence.

Table 2. Depletion rates during various irradiations.

Current/Energy/Ion(s)	Depletion rate, A (counts/dpa)	Error in depletion rate, ΔA (counts/dpa)
100 nA, 1.5-MeV He	1360	59
(5 nA, 800-keV Cu) and (100 nA, 1.5-MeV He)	143	7
5 nA, 800-keV Cu	131	29
(2 nA, 1.2-MeV Ag) and (100 nA, 1.5-MeV He)	114	12
2 nA, 1.2-MeV Ag	104	23

CONCLUSION

As RBS measurements demonstrate, the addition of 800-keV Cu and 1.2-MeV Ag ion beams to 1.5-MeV He ion irradiations of Cu-1at.%Au at 400°C resulted in the suppression of He irradiation-induced RIS of Au away from the surface. The reduction in RIS is attributed to the recombination of FMDs produced by 1.5-MeV He with cascade remnants produced by heavy ions. The RIS suppression effect of these remnants seems to be independent of the total cascade volume per ion created by individual 800-keV Cu and 1.2-MeV Ag ions and dependent on the total cascade volume generated per unit time.

ACKNOWLEDGMENTS

The authors would like to thank A. Iwase and D. Alexander for helpful discussions and T.L. Daulton for help with the three-dimensional plotting. They are also grateful to B. Kestel for assistance in sample preparation. This work was supported by the U.S. DOE under BES contract W-31-109-Eng-38.

REFERENCES

[1]   P.R. Okamoto and L.E. Rehn, J. Nucl. Mater. **83**, 2 (1979).
[2]   H. Wiedersich and N.Q. Lam, in: *Phase Transformations during Irradiation*, ed. F.V. Nolfi, Jr. (Applied Science Publishers, London, 1983) p. 1.
[3]   R.A. Johnson and N.Q. Lam, Phys. Rev. B **13**, 4364 (1976).
[4]   T. Hashimoto, L.E. Rehn, and P.R. Okamoto, Phys. Rev. B **38**, 12868 (1988).
[5]   L.E. Rehn, P.R. Okamoto, and R.S. Averback, Phys. Rev. B **30**, 3073 (1984).
[6]   R.A. Erck and L.E. Rehn, J. Nucl. Mater. **168**, 208 (1989).
[7]   R.S. Averback, J. Nucl. Mater. **216**, 49 (1994).
[8]   A. Iwase, L.E. Rehn, P.M. Baldo, and L. Funk, Appl. Phys. Lett. **67**, 229 (1995).
[9]   A. Iwase, L.E. Rehn, P.M. Baldo, and L. Funk, J. Nucl. Mater. **238**, 224 (1996).
[10] A. Iwase, L.E. Rehn, P.M. Baldo, and L. Funk, J. Nucl. Mater. **244**, 147 (1997).
[11] P. Lucasson, in: Proceedings of the International Conference on *Fundamental Aspects of Radiation Damage in Metals* (GPO, Washington DC, 1976) Vol. 1, p. 42.
[12] H.F. Deng and D.J. Bacon, Phys. Rev. B **53**, 11 376 (1996).
[13] H. Wiedersich, Mater. Sci. Forum **97-99**, 59 (1992).
[14] H. Wiedersich, J. Nucl. Mater. 205, **40** (1993).

CHARACTERIZATION OF VACANCY-TYPE DEFECTS IN ION IMPLANTED AND ANNEALED SiC BY POSITRON ANNIHILATION SPECTROSCOPY

W. ANWAND*, G. BRAUER*, P.G. COLEMAN**, W. SKORUPA*
* Institut für Ionenstrahlphysik und Materialforschung, Forschungszentrum Rossendorf, Postfach 510119, 01314 Dresden, Germany, brauer@fz-rossendorf.de
** School of Physics, University of East Anglia, Norwich NR4 7TJ, UK

ABSTRACT

New examples of characterization of vacancy-type defects in ion implanted and annealed SiC by the established technique of slow positron implantation spectroscopy are presented. In particular, the estimation of the depths of damaged regions and their change (a) after post-irradiation annealing, or (b) due to variation of substrate temperature during implantation, is addressed.

INTRODUCTION

Recent developments in slow positron beam methods have allowed the extension of the traditional Positron annihilation spectroscopy (PAS) technique to investigations of thin films, layered structures and surfaces.[1] SiC is a promising semiconductor material for high-power/high-frequency and high-temperature applications, and for selective doping of SiC ion implantation it is the only possible process.[2] However, relatively little is known about ion implantation, radiation damage caused thereby, and annealing effects. This paper outlines briefly how beam-PAS is used to extract information about vacancy-type defects in SiC due to ion implantation and presents new results on defect identification and the influence of the variation of substrate temperature during ion implantation.

EXPERIMENT

Single crystalline, 25 mm diameter n-type 6H-SiC wafers (Si surface)[3] were used as substrate material, being implanted by 200 keV Ge^+ ions to a fluence of 10^{15} cm^{-2}. As Ge^+ ions have the same valency as Si and Ge they create damage but no doping effects.

After implantation, heat treatment of three specimens was performed at predetermined temperatures of 500, 950 and 1500 °C. Treatment at the two lower temperatures was performed after encapsulation in a glass ampoule containing an Ar atmosphere. For heat treatment at 1500 °C the specimen was put into a closed graphite container evacuated to 10^{-4} Torr and filled with Ar at 760 Torr. The time of heat treatment at each temperature was 600 s.

In a second study, epilayers of 6H-SiC of a few μm thickness were implanted by Al^+ ions as follows (energy in keV first / fluence in cm^{-2} second): $500/1.1x10^{15}$, then $360/6.5x10^{14}$, then $260/4.8x10^{14}$, then $180/4.4x10^{14}$. The implantations were performed with the substrate at room temperature, 200, 400, 600, 800, and 1000 °C. Slow positron implantation spectroscopy is above all sensitive to vacancy-type defects and has been used to investigate these defects and their depth distribution after implantation and annealing processes.

The measurements were performed for all specimens at room temperature using a computer-controlled magnetic-transport beam system at UEA Norwich.[4] Incident positron energies E varied between 0.1 and 30 keV. Energy spectra of annihilation gamma rays were measured with a Ge detector having an energy resolution (FWHM) of about 1.2 keV at 511 keV. The Doppler broadening of the annihilation line is caused by the momentum of the annihilating electron-positron pair and can be characterized by the lineshape parameter S. The value of S is defined as the integral of gamma ray intensity in the central energy region divided by the total intensity of the line.

The measured value of S at each energy E is thus

$$S(E) = S_s F_s(E) + \Sigma S_d F_d(E) + S_b F_b(E) \qquad (1)$$

where S_s, S_d and S_b are the S parameters associated with annihilation with electrons at the surface, in layers containing a typical defect structure and in the unimplanted bulk material, respectively. $F(E)$ are the fractions of positrons annihilated in each state; thus $\Sigma F_n = 1$ at each E.

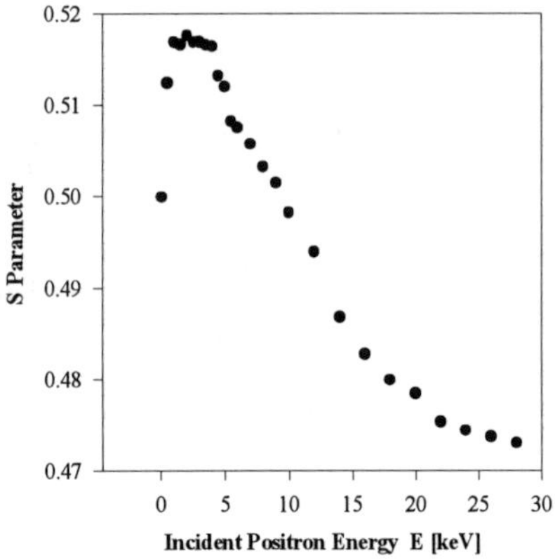

A curve of the S parameter vs. positron energy E is shown in Fig.1.

In this work the program VEPFIT[5] was used to fit possible defect distributions to the experimental results. The samples are modelled as comprising discrete homogenous layers, represented by the S-parameters according equation (1). The implantation depth profile used by VEPFIT is described by the Makhovian distribution

$$P(E, z) = (2z/z_0^2) \exp[-(z/z_0)^2] \qquad (2)$$

$$\text{with } z_0 = (A/\rho) E^n \qquad (3)$$

where ρ is the sample density. These formulae are empirical and the parameters A, m, and n are chosen to fit experimental and Monte Carlo results.

Fig. 1 S parameter vs. positron energy E of Ge^+ implanted SiC (implantation at rt)

For a given implantation energy E and in the absence of internal electrical fields, positron diffusion in the sample takes place according to

$$D_+ \frac{d^2 n(z)}{dz^2} - \left(\lambda_b + vC(z)\right)n(z) + P(E,z) = 0 \qquad (4)$$

where $n(z)$ is the thermalised positron density at depth z and D_+ the positron diffusion constant. The second term relates to the annihilation of positrons in the defect free bulk, at a rate λ_b, and at defects, at a rate $vC(z)$, where v is the specific defect trapping rate and $C(z)$ the defect concentration profile.

The VEPFIT fitting program solves the positron diffusion equation (4); positron diffusion lengths in the sample layers and the layer boundaries can be chosen as free fitting parameters.

RESULTS

Damaged layer thicknesses and defect profiles may be directly extracted from the measured $S(E)$ curves. For easier comparison of different samples the reduced parameter S/S_b is commonly used, where S_b is the bulk S value measured in a virgin reference sample. The larger S/S_b is, the larger is the dominating vacancy-type defect in a given layer.[1]

From TRIM90 Monte Carlo simulations of 200 keV Ge^+ implantation to a fluence of 10^{15} cm^{-2} vacancy formation should be expected up to a depth of only 150-200 nm. However, the experimental results shown in Fig. 2 show that vacancy-type defects were found up to a depth of

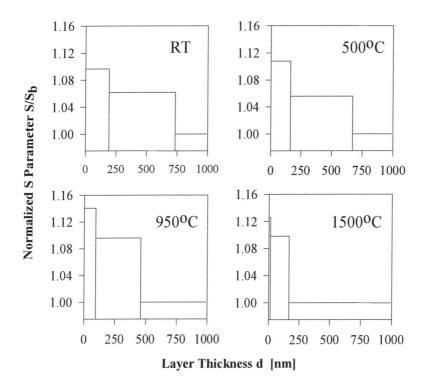

Fig. 2 Post implantation annealing of SiC implanted with Ge^+

about 750 nm. Post-irradiation annealing at temperatures up to 1500 °C does not remove this vacancy-type damage, but results in a reduction of the depth of the damaged layer and the formation of larger vacancy agglomerates close to the surface.

The prevention, or at least reduction, of the formation of vacancy-type damage by using a higher substrate temperature during ion implantation has been tested using 6H-SiC epilayer samples.[6] The multiple Al^+ implantation performed should produce a rectangular doping profile, with a plateau extending from about 200 to 600 nm. A rather complex defect structure, with several different layers, might be expected. Indeed such a structure is suggested by VEPFIT for substrate temperatures of 200, 400, 600, and 800 °C, as shown in Fig. 3. Although the general form of the defect structure is similar for all these temperatures, a reduction in the depth of the damaged region and an increase in S/S_b - indicating an increase in size of the vacancy-like defects - is observed as the temperature increases. At the highest substrate temperature (1000 °C) a less

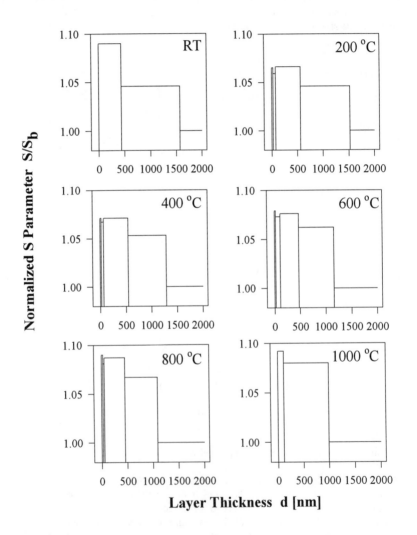

Fig. 3 Al$^+$ fourfold implantation into SiC at different substrate temperatures

complicated layer structure is seen, but with a further decrease in the depth of the damaged region and increase in size of the agglomerates. At room temperature it is impossible to fit a multilayer structure to the experimental data and the two-layer structure shown in Fig. 3 provides the best fit to the data.

200 keV Ge[+] ion implantation and post-irradiation annealing has been studied in more detail by a combination of the monoenergetic positron Doppler broadening and lifetime techniques, coupled with theoretical calculations.[7-9] In particular, it has been found that the predominant defect formed is the silicon-carbon divacancy, whose formation by irradiation has been assumed for a number of years in optical investigations.[10] Combining the results from all the cited positron studies[7-9] one arrives at a scaling curve that correlates S/S_b for positrons trapped in an agglomerate of silicon-carbon divacancies with the number of divacancies in the agglomerate (Fig.4). This may give an idea of the dominating vacancy-type defect in a certain defect layer elucidated in the present studies.

To obtain an estimate of the defect concentration C_i in defect layer i, we may use the following expression:

$$C_i = (\lambda_b/\nu) \, [(L_b/L_i)^2 - 1] \quad (5)$$

The positron diffusion lengths in a given layer (L_i) and in the bulk (L_b) are given by the VEPFIT evaluations.[5] The bulk positron lifetime λ_b is ~141ps. The specific positron trapping rate ν for a certain type of defect in SiC is still unknown. As a first approximation, the known value for monovacancies in Si, $3\times10^{14}s^{-1}$, might be used.

In addition, one might assume that in a vacancy cluster the specific positron trapping rate may be calculated simply as a multiple of the monovacancy value. However, such rough evaluations should always be taken with caution and compared to the results of other methods which may give more straightforward estimations.

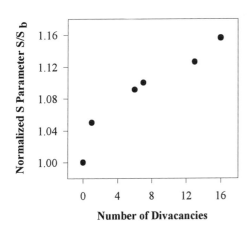

Fig. 4 Normalized S Parameter vs. number of agglomerated SiC divacancies

CONCLUSIONS

It has been clearly demonstrated that the vacancy-type damage caused by ion implantation into 6H-SiC can be monitored and well characterized by positron annihilation. Deep damage, reaching well beyond the predictions of TRIM90 calculations, is found.

Post-irradiation annealing at temperatures up to 1500 °C is found to result not in removal of vacancy-type damage but in a reduction of damage depth and the formation of larger vacancy agglomerates.

An increase in substrate temperature during ion implantation is found to result not in elimination or even reduction of vacancy-type damage, but instead has consequences similar to those for post-irradiation annealing.

The characterization of vacancy-type damage by positron annihilation is complementary to results provided by ion-based methods and thus may contribute to a complete picture and understanding of the behaviour of materials following ion implantation.

REFERENCES

1. P.J. Schultz and K.G. Lynn, Rev. Mod. Phys. **60**, 701 (1988)

2. V. Heera and W. Skorupa in Materials modification and synthesis by ion beam processes, edited by Dale E. Alexander, NathanW. Cheung, Byungwoo Park, and W. Skorupa (Mater. Res. Soc. Proc. **438**, Pittsburgh, PA, 1997) pp.241-252

3. Cree Research Inc., Durham, NC (USA)

4. N.B. Chilton and P.G. Coleman, Meas. Sci. Technol. **6**, 53 (1995)

5. A. van Veen, H. Schut, J. de Vries, R.A. Hakvoort and M.R. Ijpma, in Positron Beams for Solids and Surfaces, edited by P.J. Schultz, G.R. Massoumi and P.J. Simpson, AIP Conf. Proc. No. **218** (American Inst. of Physics, New York, 1990), p. 171

6. H. Wirth, W. Anwand, G. Brauer, P.G. Coleman, M. Voelskow, D. Panknin and W. Skorupa, Proc. of the International Conference on Silicon Carbide, III-nitrides and Related Materials, Stockholm , 1997, (in press)

7. G. Brauer, W. Anwand, P.G. Coleman, A.P. Knights, F. Plazaola, Y. Pacaud, W. Skorupa, J. Störmer and P. Willutzki, Phys. Rev. B **54**, 3084 (1996)

8. G. Brauer, W. Anwand, E.-M. Nicht, J. Kuriplach, M. Sob, N. Wagner, P.G. Coleman, M.J. Puska and T. Korhonen, Phys. Rev. B **54**, 2512 (1996)

9. G. Brauer, W. Anwand, P.G. Coleman, J. Störmer, F. Plazaola, J.M. Campillo, Y. Pacaud and W. Skorupa, J. Phys.: Condens. Matter (in press)

10. W.J. Choyke, The Physics and Chemistry of Carbides, Nitrides and Borides, Vol. **185** of NATO Advanced Study Institute (Kluwer, Dordrecht, 1990), p.563

MECHANISM OF ENHANCED DIFFUSION OF ALUMINUM IN 6H-SiC IN THE PROCESS OF HIGH-TEMPERATURE ION IMPLANTATION

Igor O. USOV[*], A.A. SUVOROVA[*], V.V. SOKOLOV[*], Y.A. KUDRYAVTSEV[*] and A.V. SUVOROV[**]
[*] Ioffe Physical Technical Institute RAS, St.Petersburg, 194021, Russia
[**] Cree Research Inc., Durham, NC 27713, USA

Abstract

The diffusion of Al in 6H-SiC during high-temperature ion implantation was studied using secondary ion mass spectrometry. A 6H-SiC wafer was implanted with 50 keV Al ions to a dose of 1.4E16 cm^{-2} in the high temperature range 1300°-1800°C and at room temperature. There are two diffusion regions that can be identified in the Al profiles. At high Al concentrations the gettering related peak and profile broadening are observed. At low Al concentrations, the profiles have a sharp kink and deep penetrating diffusion tails. In the first region, the diffusion coefficient is temperature independent, while in the second it exponentially increases as a function of temperature. The Al redistribution can be explained with the substitutional-interstitial diffusion mechanism.

Introduction

Silicon carbide is one of the most important semiconductors for electronic and optoelectronic device applications. Diffusion of Al implanted into SiC has been studied by several investigators [1-5]. In the previous studies, implants were mostly performed either at room temperature (RT) or at temperatures lower than 850°C. It was established that the 800°C implanted SiC has a broader Al distribution as compared with the RT implanted sample [1]. Only a slight Al in-diffusion of the RT implantation was observed for the annealing temperature up to 1800°C [2,3]. However, in our previous paper we have shown the anomalously high diffusion of Al implanted at 1800°C [5]. Experimental data on Al diffusion in SiC are few and the diffusion mechanism has not been well understood. In this work the implant temperature dependence of the enhanced Al diffusion is investigated.

Experiment

In this study we have used 3 μm thick n-type Si-faced 6H-SiC epitaxial layers with a carrier concentration of 5.5E17 cm^{-3} grown on n-type 6H-SiC substrates from Cree Research Inc. Ion implantation was done with 50 keV Al$^+$ ions to a dose of 1.4E16 cm^{-2} and ion current density of 8.5 μA/cm^2 (implantation time about 320 sec). The implants were performed at RT and at high temperatures (HT) ranging from 1300°C to 1800°C using a holder which could be resistevely heated. During implantation the samples were tilted 7° off the normal to minimize channeling. The secondary-ion mass spectrometry (SIMS) profiles of Al were determined using a CAMECA IMS-4F spectrometer.

Results and Discussion

The concentration profiles of Al implanted at different temperatures are shown in Figure 1 and Figure 2. The Al profile of the RT implanted sample can be fit to a Gaussian distribution with a

Mat. Res. Soc. Symp. Proc. Vol. 504 © 1998 Materials Research Society

projected range (R_p) of 59.8 nm and standard deviation (ΔR_p) of 25.2 nm. As the implant temperature is increased from 1300°C to 1800°C there is a significant difference between the profiles of RT and high temperature (HT) implanted Al. Two diffusion regions can be identified in the Al profiles.

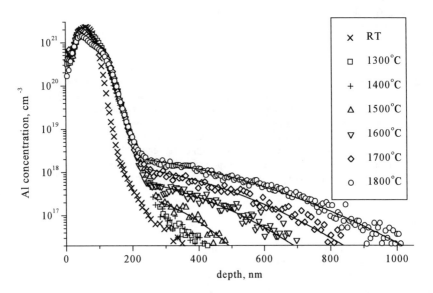

FIG. 1. The SIMS profiles of a 1.4E16 cm^{-2}, 50 keV Al implantation into 6H-SiC in the HT range from 1300 °C to 1800 °C and room temperature (RT). The solid lines are the equation (3) fitting to the diffusion tails for the implantation temperatures from 1500 °C to 1800 °C.

At high concentration region, there is a broadening from the surface side of R_p to the substrate side relative to the RT implanted profile. In this case profile broadening is independent of implantation temperature over the range between 1300°C and 1800°C. The profiles are no longer symmetrical. In order to estimate the diffusion coefficient (D), the deep-half profile was used for Gaussian curve fitting. Thus D can be calculated from the following equation [3]:

$$(\Delta R_p')^2 = \Delta R_p^2 + 2Dt , \qquad (1)$$

where ΔR_p is the standard deviation after RT implantation, $\Delta R_p'$ is the standard deviation after HT implantation and t is the duration of HT implantation. The standard deviation $\Delta R_p'$ of the deep-half profile of the sample implanted at 1300°C is 38.5 nm, yielding D = 1.3 × 10^{-14} cm^2/sec.

In addition to broadening in the high concentration region, the gettering related Al peak can be clearly seen in the linear plot of the Al depth profile in the Figure 2. With increasing implantation temperature from 1300°C to 1500°C, the implant depth profile maximum has shifted toward the surface. At the higher implantation temperatures, the gettering peak at ~0.8R_p location is observed. In our previous work [6] on a Transmission Electron Microscopy (TEM) study of HT Al implanted 6H-SiC it was established that the implanted layer contains Al precipitates close to the surface. It is believed that the process of precipitate formation is probably one of the causes that bring the Al diffusion in the high concentration region to a stop. In this region the Al concentration is well above the solid solubility limit, and the diffusion coefficient, evaluated from equation (1) satisfies the condition [7]:

$$(Dt)^{1/2} > a , \qquad (2)$$

where a corresponds to several lattice constants. The lattice constant for 6H-SiC is 1.5 nm and the diffusion length obtained from equation (2) is 20.4 nm. Therefore, the diffusion conditions in the high concentration area are favorable for Al precipitation.

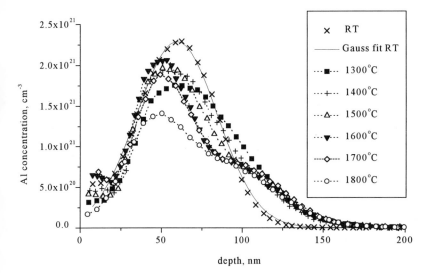

FIG. 2. The linear plots of SIMS profiles of a 1.4E16 cm^{-2}, 50 keV Al^{+} implantation into 6H-SiC in the HT range from 1300°C to 1800 °C (line and symbol curves) and room temperature (RT) (squares). The solid line is the Gaussian fit to implantation at room temperature.

The Al content of the samples implanted at the temperatures higher than 1600°C measured by SIMS is less than the implanted dose, indicating that out-diffusion has taken place. The out-diffusion of Al leads to a decrease of the gettering peak magnitude and the size of precipitates [6]. At temperatures higher than 1600°C, a shoulder-like feature develops at a depth of ~1.5R$_p$. We assume that Al atoms occupy Si lattice sites and diffuse via vacancies in this region. Thus the concentration at which the shoulder appears corresponds to the solid solubility limit. The observed solubility is constant in the temperature range 1600° - 1800°C and equals 8.5E20 cm^{-3}.

The second diffusion region is located at low Al concentrations (Fig. 1) for implantation temperatures higher than 1500°C. The concentration profiles have a sharp kink and deep penetrating diffusion tails. The penetration of the tail and the concentration at which the kink forms increase with increasing implantation temperature. The Al distribution in the low concentration region is satisfactorily described by the solution of Fick's equation for impurity diffusion from a constant source [8]:

$$N(x, t) = N_0 \left(1 - erf \frac{x}{2\sqrt{Dt}} \right),$$ (3)

where N_0 is the concentration at which the kink forms, D and t are the diffusion coefficient and duration of HT implantation, respectively. In Fig.3 the temperature dependence of the diffusion

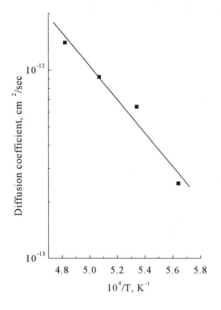

FIG. 3. The Arrhenius plot for the Al diffusion during high temperature implantation into 6H-SiC in the tail region.

coefficients calculated by equation (3) is presented. These results are well described by the Arrhenius equation:

$$D(cm^2/s) = 2.84 \times 10^{-8} \exp\left(-\frac{1.76eV}{kT} \right)$$ (4)

TEM observation revealed that the HT implanted layer contains end of range (EOR) damage below the depth containing Al precipitates [6]. With an increase in this implantation temperature, the EOR defects grow up in size and at temperatures higher than 1500°C evolve from small basal loops to elongated loops of the interstitial type. The formation of these defects may be attributed to recoiled matrix atoms created by implantation and interstitial Si atoms supersaturation induced by Al diffusion via Si substitutional sites. Consideration of TEM results in conjunction with the SIMS results shows that Al diffusion in the tail region occurs in the crystal region without extended defects [5]. The depth at which the diffusion tail starts corresponds to the depth at which EOR defects disappear. We speculate that the anomalous redistribution in the low concentration region can be attributed to the fast diffusing interstitial component of the Al.

The Al redistribution results can be explained with the substitutional-interstitial diffusion (SID) mechanism [9,10]. This mechanism implies a conversion between substitutional and interstitial dopant atoms. Since Al atoms can occupy only Si sites, we consider only the point defects of the Si sublattice: vacancies (V) and interstitials (I). There are two forms for this process:

$$Al_i + V_{Si} \leftrightarrow Al_{Si} \qquad (5)$$
$$Al_{Si} + I_{Si} \leftrightarrow Al_i , \qquad (6)$$

where Al_i and Al_{Si} are Al interstitial and substitutial atoms, respectively. Equation (5) is known as the dissociative mechanism and applies when the dominant point defects are vacancies [9]. Equation (6) is known as the kick-out mechanism and applies when the dominant point defects are matrix interstitials [10]. The dopant atoms on interstitial sites are assumed to have a low solubility and a high diffusivity whereas the dopant atoms on substitutional sites have a high solubility but a low diffusivity. The HT implanted Al profiles are of characteristic shapes, indicating that the SID mechanism is operative. Thus, in the high concentration region where the Al concentration corresponds to the solid solubility limit, the dissociative mechanism determines the diffusion, while in the low concentration region containing excess matrix interstitials the diffusion is determined by the kick-out mechanism. The interstitialy rich region can serve as a sink for vacancies and act as a barrier for vacancies in-diffusion. Thus, in addition to precipitation, the interstitial type EOR defects can stop the substitutional Al diffusion.

Conclusions

From our experiments we conclude that Al is a substitutional-interstitial impurity in SiC. The redistribution of high-temperature implanted Al differs greatly from the one obtained after the annealing of the room-temperature implantation. The diffusion can be explained by the SID mechanism. The kick-out mechanism dominates the Al diffusion in the low concentration region while the dissociative mechanism prevails in the high-concentration region.

References

1. Z. Yang, H. Du, M. Libera and I.L. Singer, J. Mater.Res., **10**(6), 1441(1995).
2. M.V. Rao, P. Griffiths, O.W. Holland, G. Kelner, J.A. Freitas, D.S. Simons, P.H. Chi and M. Chezzo, J. Appl. Phys., **77**(6), 2479(1995).
3. Y. Tajima, K. Kijima and W.D. Kingery, J.Chem.Phys., **77**(5), 2592(1982).
4. W. Lucke, J. Comas, G. Hubler and K. Dunning, J.Appl.Phys., **46**(3), 994(1975).
5. A.V. Suvorov, I.O. Usov, V.V. Sokolov and A.A. Suvorova in Ion-Solid Interaction for Materials Modification and Processing, edited by D.B. Poker, D. Ila, Y-S. Cheng, L.R. Harriot and T.W. Sigmon,(Mater. Res. Soc. Symp. Proc. **396**, Pittsburgh, PA 1996), p.239-242.
6. A.A. Suvorova, O.I. Lebedev, A.V. Suvorov and I.O. Usov, Proceedings of the 10th International Conference Microscopy of Semiconducting Materials, Oxford, UK, 1997, (Inst. of Physics Conference Ser. No. **157**, Bristol & Philadelphia, UK, 1997), p.531, and A.V. Suvorov, I.O. Lebedev, A.A. Suvorova, J. Van Landuyt, I.O. Usov, Nucl. Instr. Meth. B, **127/128**, 347 (1997), and O.I. Lebedev, G.Van Tendeloo, A.A. Suvorova, I.O. Usov, A.V. Suvorov, Journal of Electron

Microscopy **46**(4), 271 (1997).

7. S.M. Myers in Ion Implantation, edited by J.K. Hirvonen (Treatise on Materials Science and Technology Vol. 18, Academic Press, 1980), p.48.
8. D. Shaw in Atomic diffusion in semiconductors, edited by D. Shaw (Plenum Press, London and New York, 1973), p.19.
9. F.C. Frank and D. Turnbull, Phys.Rev. **104**(3), 617 (1956).
10. U. Gösele, W. Frank and A. Seeger, Appl. Phys. **23**(4), 361 (1980).

RDF ANALYSIS OF ION-AMORPHIZED SiO₂ AND SiC FROM ELECTRON DIFFRACTION USING POST-SPECIMEN SCANNING IN THE FIELD-EMISSION SCANNING TRANSMISSION ELECTRON MICROSCOPE

David C. Bell*, Anthony J. Garratt-Reed* And Linn W. Hobbs†

*Center for Materials Science and Engineering
†Department of Materials Science and Engineering
Massachusetts Institute of Technology
Cambridge, MA, 02139

ABSTRACT

Radial density functions (RDFs) provide important information about short- and intermediate-range structure of topologically-disordered materials such as glasses and irradiation-amorphized materials. We have determined RDFs for irradiation-amorphized SiO_2, $AlPO_4$ and SiC by energy-filtered electron diffraction methods in a field-emission scanning transmission electron microscope (FEG-STEM) equipped with a digital parallel-detection electron energy-loss spectrometer. Post-specimen rocking was used to minimize the effects of spherical aberration in the objective lens, which interfere with the acquisition of data collected by pre-specimen rocking. Useful energy-filtered data has been collected beyond an angular range defined by $q \equiv 2 \sin(\Theta/2)/\lambda = 25$ nm^{-1}.

BACKGROUND

Fast electrons are a particularly useful chemical and structural probe for the small sample volumes associated with ion- or fast electron-irradiation-induced amorphization, because of their much stronger interaction with matter than for X-rays or neutrons, and also because they can be readily focused to small probes. Three derivative signals are particularly rich in information: the angular distribution of scattered electrons (which is utilized in both diffraction and imaging studies); the energy loss spectrum of scattered electrons (electron energy loss spectroscopy, or EELS); and the emission spectrum of characteristic X-rays resulting from ionization energy losses (energy dispersive X-ray spectroscopy, or EDXS). We have applied the first two to the study of three amorphized compounds ($AlPO_4$, SiO_2, SiC) using MIT's Vacuum Generators HB603 field-emission (FEG) scanning transmission electron microscope (STEM), operating at 250 kV and equipped with a Gatan digital parallel-detection electron energy-loss spectrometer (digiPEELS).

Statistically significant electron diffraction (ED) patterns from topologically disordered solids can be obtained from volumes as small as $\sim 5 \times 10^{-20}$ m³, compared to $\sim 5 \times 10^{-11}$ m³ for X-rays and $\sim 5 \times 10^{-8}$ m³ for thermal neutrons. ED patterns must be energy-filtered to remove electrons scattered by inelastic processes, which predominate for low-Z elements and have a different angular distribution from those of elastically-scattered electrons. Radially-averaged real-space radial density functions (RDFs) can then be derived by Fourier transforming electron diffraction data. Neutron diffraction provides better real-space resolution, because atom scattering lengths b do not fall off with scattering angle Θ and neutron data are collectable further out in reciprocal space to $q \equiv 2 \sin(\Theta/2)/\lambda = 70$ nm⁻¹. In contrast, scattered electron intensity falls off roughly as

Mat. Res. Soc. Symp. Proc. Vol. 504 © 1998 Materials Research Society

q^{-4}, slightly faster than for X-rays at high angles. Nonetheless, energy-filtered electron diffraction (EFED) is the only practical way to investigate very small volumes, because of the much stronger interaction of the electron with matter, and can provide information about the spatial variation of atom correlations unavailable from X-ray or neutron diffraction methods, spatially averaging over much larger volumes. EFED still appears to be the *only* method by which information about changes in intermediate-range order can be assessed in the small sample volumes associated with amorphization using the focused electron or ion-beams required to effect the amorphization fluences.

EXPERIMENTAL

In the HB603 FEG-STEM instrument used for our studies, two methods of generating a selected-area diffraction pattern are possible. In the first, the incident electron beam is rocked about the scanned area to define the scattering angle with respect to the entrance aperture of the electron spectrometer. This is the mode in which our earlier EFED data were generated [1,2], but spherical aberration of the objective lens causes offsets in the effective area from which the data are collected [3]. Since the samples typically have non-uniform thickness, the result is that data collected at different angular distances originate from areas of differing thickness, effectively precluding practical data reduction for data collected at large scattering angles ($q > 16$ nm^{-1}). In the second [4], the incident beam is kept stationary and post-specimen scan coils, available in the HB603, are used to sweep the diffraction pattern across the spectrometer entrance aperture. The spherical aberration manifests itself in this mode in a defocusing of the PEELS spectrometer, but enough adjustment is available to compensate. In this way, we have recently been able to extend the collection angle to at least $q = 25$ nm^{-1} (Fig. 1), comparable to the range (and corresponding RDF peak resolution) available in X-ray diffraction [5].

Current operation of the HB603 in the EFED mode replaces the PIN-diode array of the PEELS with an optical slit, demagnified by lenses in the PEELS system to define an 8 eV energy window centered on zero energy loss into a photomultiplier detector operating in analog mode. The arrangement effectively converts the system from parallel to serial detection and is described in detail elsewhere [6]. A future implementation may employ spectrum imaging for EFED when the associated software becomes available. Energy discrimination cannot remove the low-loss but high-angle electron-phonon scattering, which comprises the principal background limiting angular collection range, so EFED at cryogenic temperatures is presently being explored.

Energy filtered diffraction-pattern line scans $I(q)$ are calibrated against a crystalline reference, which in our case is a polycrystalline gold sample, and ideally are corrected for multiple elastic scattering and scattering from any thin amorphous carbon layer deposited to obviate charging in specimens of poor electrical conductivity or present from incidental hydrocarbon contamination. The diffracted intensity distribution is then rendered into the form of a reduced intensity function $S(q)$ normalized by a sharpening function representing the average intensity scattered from a composition unit. Radial density functions

$$g_e(r) = 4\pi r^2 [\rho(r)-\rho_a] = 8\pi r \int qS(q)\ sin(2\pi qr)\ M(q)\ dq \qquad (1)$$

are derived from the Zernicke-Prins interference function $qS(q)$ by Fourier transform algorithms [1,7,8] using the Lorch modification function $M(q)$[7], where $\rho(r)$ is a number density and ρ_a the average number density in the sample.

RESULTS

Applications to electron-amorphized berlinite and quartz

Crystalline berlinite ($AlPO_4$) is nominally isostructural with quartz (SiO_2), exhibiting ordering of [AlO_4] and [PO_4] fully-shared tetrahedra. It is found to amorphize radiolytically by 200-kV electron irradiation after a fluence of 2×10^{23} e/m^2, some 50 times faster than for radiolytic amorphization of quartz [2]. By contrast, irradiation with 1.5 MeV Kr$^+$ ions results in amorphization after an ion fluence corresponding to an energy deposition density of 10 eV/atom [2], very little different from quartz and in keeping with the available structural freedom [9]. Fig. 1 illustrates a radial density function for electron-amorphized berlinite derived from EFED data taken with the HB603 operating at 250 kV, with comparison data for electron-amorphized quartz (SiO_2) and the positions of correlation peaks from an X-ray diffraction study [10] of bulk hydrothermally-grown a-AlPO$_4$ superimposed. The resolution is increased over that for previous data acquired with incident beam rocking, in particular for the M-M correlations (M = Si,Al,P). The RDF for the radiolytically-amorphized material is a better match to the RDF for amorphized quartz than to those for amorphized crystobalite or tridymite [1], which also radiolytically amorphize more efficiently than quartz.

Application to Ion-Amorphized SiC

SiC, equally in its α- or β-polymorphic forms, amorphizes readily in both cascade and single displacement irradiation modes below a threshold temperature as high as 390 K [11,12]. It was deemed of considerable interest to explore the structure of the irradiation-amorphized product and any difference in that structure as a function of irradiation temperature below the threshold. A suite of four α-SiC samples has been amorphized [11] as TEM foils in the Argonne National Laboratory (ANL) HVEM-IVEM/tandem ion accelerator facility [13] at four different temperatures (100 K, 200 K, 250 K and 383 K) with 800 keV Ne$^+$ or 1.5 MeV Xe$^+$ ions to fluences (8.9×10^{19} - 3.1×10^{20} ions/m^2) just past complete amorphization for the respective amorphization temperatures. EFED profiles were obtained from approximately 1 μm-diameter areas of these thin foil samples in the HB603 STEM, using a polycrystalline gold reference as a calibration.

The profiles, which are reproduced in Fig. 2, were Fourier transformed using (1) to RDFs which are compared in Fig. 3. The first major set of peaks at 0.16-0.18 nm corresponds principally to the Si-C correlation distance in a [SiC_4] (or [CSi_4]) tetrahedron (nominally 0.188 nm in β-SiC) and the second set at 0.27-0.29 nm to the C-C (or Si-Si) correlation distance across a tetrahedron edge. Within each set, the peaks differ slightly, in position, breadth, and magnitude, depending on irradiation temperature. The peaks for the lowest temperature (100 K) irradiation had consistently the largest correlation distance and the largest breadth. The small shifts evident in the position of the first sharp diffraction peaks at $q \approx 4$ nm^{-1} in Fig. 2 suggests that there are corresponding differences in intermediate range order as well.

149

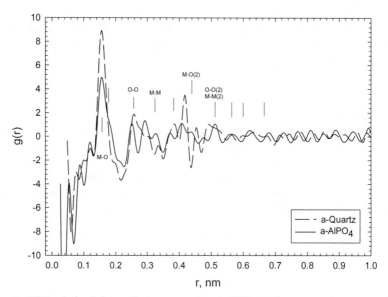

Figure 1: RDFs derived from electron-amorphized $AlPO_4$ and quartz using energy-filtered electron diffraction in the HB 603 STEM. The markers and ascriptions indicate peak positions from an X-ray study of hydrothermally grown $AlPO_4$.

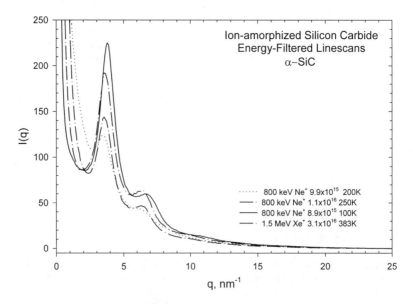

Figure 2: Energy filtered electron diffraction linescans collected in HB603 STEM from ion-amorphized SiC thin foils, yielding *I(q)* for four ion-amorphized α-silicon carbide thin foils.

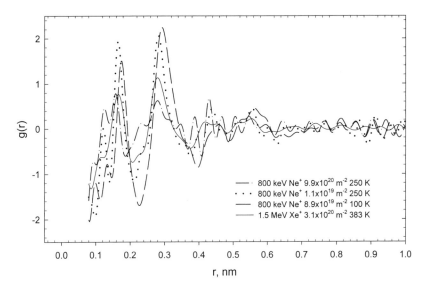

Figure 3: RDFs derived from HB603 STEM EFED data for four ion-amorphized α-SiC thin foils.

While it is too early to more than speculate about the origins of these differences from such a small sampling, the fact that there are differences suggests that the structures of ion-amorphized SiC depend at least on the temperature at which the amorphization has occurred. The result also confirms the utility of EFED techniques in obtaining useful structural information from extremely small amorphized volumes.

CONCLUSION

We have acquired energy-filtered electron diffraction patterns from three amorphized compounds ($AlPO_4$, quartz-SiO_2, α-SiC) using field-emission scanning transmission electron microscope, operating at 250 kV equipped with PEELS. Diffracted intensity has been recorded out to $q \equiv 2\sin(\Theta/2)/\lambda = 25$ nm^{-1}, using post-specimen scanning of diffracted electrons across the spectrometer entrance aperture to minimize spherical aberration in the collection. RDFs derived from the electron-amorphized quartz and $AlPO_4$ diffraction data exhibit better resolution than earlier data acquired by incident beam rocking. RDFs derived for ion-amorphized α-SiC show variations in short- and intermediate-range correlations for different irradiation temperatures between 100 K and 383 K. EFED appears to remain the only method by which information about changes in intermediate-range order can be assessed in the small sample volumes associated with amorphization using the focused electron or ion-beams required to effect the amorphization fluences.

ACKNOWLEDGMENTS

This work was supported by the U.S. Department of Energy, Office of Basic Energy Sciences, under grant DE-FG02-89ER45396 and carried out in the Center for Materials Science and Engineering, Massachusetts Institute of Technology, which is funded by National Science

Foundation MRSEC grant DMR-9400334. The authors are grateful to Drs. L.-M. Wang, University of Michigan, and W. J. Weber, Pacific Northwest National Laboratory, for kindly supplying the suite of ion-amorphized SiC specimens. We are also indebted to Drs. D.J.H. Cockayne and W. McBride of the University of Sydney for the use of computer code for Fourier transforming the I(q) data.

REFERENCES

1. L. C. Qin & L. W. Hobbs, J. Non-Cryst. Solids **192&193** (1995) p. 456-462.

2. A. N. Sreeram, L. W. Hobbs, N. Bordes and R. C. Ewing, Nucl. Instrum. Meth. **B 116** (1996) p 126-130 .

3. J.M. Cowley, Diffraction Physics, 2nd ed. (North Holland Press, New York, 1990) p. 279-284

4. D. C. Bell. A. J. Garratt-Reed and L. W. Hobbs, *Proc. 55th Ann. Mtg. MSA*, ed. G. W. Bailey p. 1019.

5. R. L. Mozzi and B. E. Warren, *J. Appl. Cryst.* **2** (1969) 164; A. C. Wright, "Neutron and X-Ray Amorphography," in: *Experimental Techniques of Glass Science*, ed. C. J. Simmons and O. H. El-Bayoumi (Ceramic Transactions, American Ceramic Society,1992), Chapter 8.

6. A. N. Sreeram, L.-C. Qin, A. J. Garratt-Reed and L. W. Hobbs, *Proc. 54th Ann. Mtg. MSA*, ed. N. Zaluzec and G. W. Bailey (San Francisco Press, 1996) p. 702.

7. E. Lorch, *J. Phys* **C 2** (1969) 229.

8. D. J. H. Cockayne and D. R. MacKenzie, *Acta Cryst.* **A44** (1988) 870; Z. Q. Liu, D. R. MacKenzie, D. J. H. Cockayne and D. M. Dwarte, *Phil. Mag.* **B57** (1988) 753; D. A. MacKenzie, C. . Davis, D. J. H. Cocakayne, D. A. Muller and A. M. Vassallo, *Nature* **355** (1992) 622.

9. L. W. Hobbs, A. N. Sreeram, C. E. Jesurum and B. A. Berger, *Nucl. Instrum. Meth.* **B 116** (1996) 18.

10. G. D. Wignall, R. N. Rohon, G. W. Longman and G. R. Woodward, *J. Mater. Sci.* **24** (1977) 1039.

11. W. J. Weber and L.-M. Wang, *Nucl. Instrum. Meth.* **B 106** (1995) 298; W. J. Weber, L.-M. Wang and N. Yu, *Nucl. Instrum. Meth.* **B 116** (1996) 322.

12. L. L. Snead and S. J. Zinkle, *Mater. Res. Soc. Symp. Proc.* **439** (1997) 595.

13. C. W. Allen, L. L. Funk, E.A. Ryan and S. T. Ockers, *Nucl. Instrum. Meth.* **B 40& 41** (1989) 553.

OXIDE AND NITRIDE FORMATION AND SEGREGATION OF METALS DURING OXYGEN AND NITROGEN BOMBARDMENT OF SILICON

J.S. WILLIAMS,* M. PETRAVIC,* M. CONWAY,* L. Fu* and D.J. CHIVERS**
* Department of Electronic Materials Engineering, Research School of Physical Sciences and Engineering, Australian National University, Canberra, ACT, 0200, Australia
** AEA Technology, Harwell, Didcot, Berks., U.K.

ABSTRACT

This study compares oxide and nitride formation during oxygen and nitrogen bombardment of Si. Ion bombardment is carried out both in a SIMS machine and in a conventional implanter at various temperatures. Stoichiometric SiO_2 and slightly N-rich Si_3N_4 are formed during bombardment even at cryogenic temperatures. Implanted metals were found to have a strong tendency to be segregated at a moving Si-SiO_2 interface during oxygen bombardment but little segregation is observed at a Si-Si_3N_4 interface.

INTRODUCTION

High dose bombardment of Si with keV oxygen or nitrogen ions forms oxides [1] and nitrides [2], respectively. Although SiO_2 formation has been extensively studied as a function of irradiation temperature [1, 3-5], including the segregation of many metals at Si-SiO_2 interfaces during oxygen bombardment [3-7], there is little similar data available for nitrogen bombardment [2,8,9]. Silicon nitride formation is becoming important in Si devices as oxidation barriers and gate insulators, primarily due to low diffusivity of many species in nitrides [10]. In addition, nitride formation and possible segregation of metal species during nitrogen bombardment is important for sputter profiling using secondary ion mass spectrometry (SIMS). In this paper we compare oxide and nitride formation during irradiation of Si with oxygen and nitrogen beams and examine segregation of metals.

EXPERIMENTAL

Cz-Si wafers of (100) orientation and with a resistivity of 1-10 Ω cm, n-type, were used. The Si was irradiated either in a keV ion implanter or a SIMS apparatus. In the case of SIMS, an 8-10 keV O_2^+ beam or a 10-12 keV N_2^+ beam was used in a Riber MIQ 256 instrument. For implantation, 15 or 25 keV O⁻ bombardment was carried out at ANU over a substrate temperature range from liquid nitrogen to 250°C. Also 30 keV N_2^+ bombardment was undertaken on a Lintott mk-4 implanter at Harwell with substrate temperatures varying from about 150-450°C. Oxygen and nitrogen doses covered the range from 10^{17} to 5×10^{18}cm^{-2}. Prior to oxygen or nitrogen bombardment, some Si samples were implanted with either Cu or Au in an energy range 70-100 keV (projected range around 300-500Å) to study possible segregation effects during oxide and nitride formation. Samples were analysed by high depth resolution Rutherford backscattering and channeling (RBS-C) using 2 MeV He$^+$ ions or in SIMS using both Cs$^+$ and O$_2^+$ beams.

RESULTS AND DISCUSSION

Fig.1 illustrates stoichiometric SiO$_2$ formation for 15 keV O$^-$ bombardment of Si to a dose of 1 x 10^{18} cm^{-2} at room temperature. The solid curve is a RUMP simulated RBS-C spectrum for the structure indicated in the inset. SiO$_2$ forms even at liquid nitrogen temperature [1] indicating the strong chemical driving force for SiO$_2$ formation and the mobility of O in SiO$_2$ under ion irradiation.

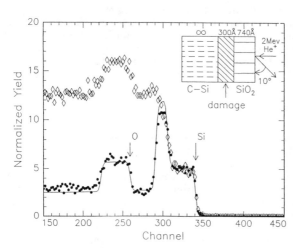

Fig.1

Fig.1

RBS-C spectra illustrating SiO$_2$ formation during 15 keV O$^-$ bombardment to a dose of 1 x 10^{18}cm^{-2} at room temperature into Si (100). The random spectrum ($\Diamond$) and the channeled spectrum ($\bullet$) confirm the structure in the inset. The solid curve is a RUMP simulation of the inset structure: the crystalline, c-Si, part of the spectrum is fitted with a straight line.

Fig.2 shows RBS-C spectra illustrating typical behaviour for 10 keV N$_2^+$ irradiation of Si in the SIMS apparatus at room temperature. At a beam incidence of 55° to the normal, sputtering does not allow the nitrogen concentration to reach Si$_3$N$_4$ even at very high doses. At normal incidence, however, a slightly N-rich Si$_3$N$_4$ layer is formed. With increasing dose the N-content remains unchanged. The solid curve is a RUMP simulation for 170Å of Si$_3$N$_4$ on an amorphous Si layer of ~ 150Å [2].

Fig.3 illustrates the angular dependence of oxygen and nitrogen content for irradiation in the SIMS. For angles less than about 30°, stoichiometric SiO$_2$ and slightly N-rich Si$_3$N$_4$ layers are formed whereas at higher angles, sputtering limits the composition. This similar behaviour for 0- and N- irradiation is not surprising and reflects similar sputtering yields for these ions at the same energy.

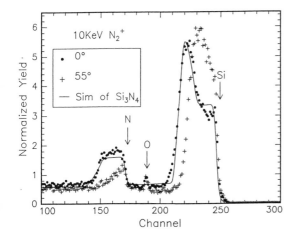

Fig.2

RBS-C spectra (scattering angle 94°) for 10 keV N_2^+ bombardment of Si in a SIMS apparatus at room temperature. The crosses are for a N-beam at 55° incidence with respect to the surface normal and the solid circles are for normal incidence. The solid curve is a RUMP simulation for ~ 170Å Si_3N_4 with ~ 150Å of underlying amorphous Si.

It is well known that metals are segregated at the Si-SiO$_2$ interface during oxygen bombardment [3-7] and this is reflected in the anomalously broad Cu SIMS profile shown in Fig.4 for 10 keV O_2^+ bombardment of Si containing a shallow Cu concentration. However, for N-bombardment at the same angle (20°), where Si_3N_4 forms, a normal Cu profile is obtained exhibiting no anomalous broadening. Indeed, irradiation at 40°, where an Si_3N_4 layer does not form gives an identical Cu profile. This result suggests little segregation of Cu occurs during

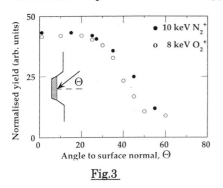

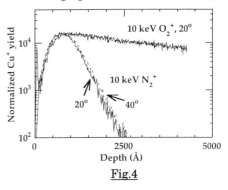

Fig.3

Normalised oxygen and nitrogen content as a function of angle of incidence in a SIMS apparatus. Data obtained from RBS-C data.

Fig.4

Comparison of measured SIMS profiles for Cu (implanted into Si) measured for O_2^+ and N_2^+ bombardment at differeng angles of incidence.

N-bombardment at room temperature regardless of whether Si_3N_4 forms. For segregation at Si-SiO_2 interfaces, a recent model [6] proposes that the low solubility of Cu in SiO_2 and high solubility in disordered Si result in segregation if Cu is mobile during oxygen bombardment. For N-bombardment, Cu would appear to be immobile under irradiation in Si_3N_4 and is sputtered away without any tendency to segregate during room temperature irradiation.

In Fig.5, the RBS-C spectrum illustrates Cu segregation in SiO_2 during 25 keV oxygen bombardment at room temperature. At a dose of $8 \times 10^{17} O^- cm^{-2}$, a buried SiO_2 layer has formed (see inset) and Cu is observed to segregate at both Si-SiO_2 interfaces. At higher doses, the near-surface Cu peak is sputtered away when the oxide becomes continuous to the surface. Fig.6 shows RBS-C spectra

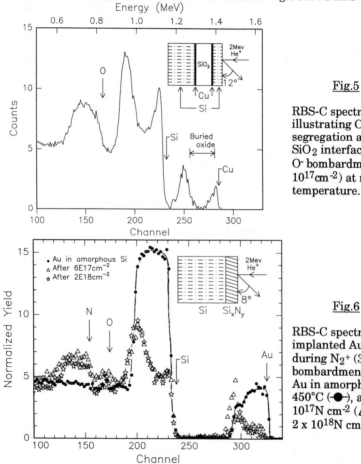

Fig.5

RBS-C spectrum illustrating Cu segregation at both Si-SiO_2 interfaces for 25 keV O^- bombardment of Si ($8 \times 10^{17} cm^{-2}$) at room temperature.

Fig.6

RBS-C spectra showing implanted Au profiles during N_2^+ (30 keV) bombardment at 450°C. Au in amorphous Si at 450°C (—●—), after $6 \times 10^{17} N\ cm^{-2}$ (△) and after $2 \times 10^{18} N\ cm^{-2}$ (☆).

which illustrate changes in an implanted Au profile with increasing dose of 30 keV N_2^+ bombardment at a temperature of ~ 450°C. At this temperature, the Au is initially uniformly distributed within an amorphous layer (formed when the Au was introduced by ion implantation into Si). As the N-dose is increased, an N-rich Si_3N_4 layer is again formed and the Au is progressively consumed into an Si_3N_4 layer. At 450°C there is a very slight tendency for Au to segregate but, compared with the total segregation which would occur for O-bombardment at this temperature [6], the effect is quite small.

CONCLUSIONS

Near-stoichiometric SiO_2 forms at all temperatures between liquid nitrogen and 250°C for oxygen bombardment of Si in cases where sputtering does not limit the oxygen concentration. Slightly N-rich Si_3N_4 is observed to form at temperatures between room temperature and 450°C for nitrogen bombardment of Si. Whereas metals (Cu and Au) strongly segregate at Si-SiO_2 interfaces during oxygen bombardment, there is little tendency for segregation to occur for nitrogen bombardment even when Si_3N_4 forms and high substrate temperatures are used (~ 450°C). As a result, whereas there are strong thermodynamic driving forces (solubility differences in SiO_2/disordered Si) for mobile metals to segregate at Si-SiO_2 interfaces, metals are readily consumed into Si_3N_4 layers during continued N-bombardment.

REFERENCES

1. J.S. Williams, M. Petravic, B.G. Svensson and M. Conway, *J.Appl.Phys.* **76** (1995) 1840.

2. M. Petravic, J.S. Williams, B.G. Svensson and M. Conway, SIMS XI, Orlando, Florida (1997) to be published in proceedings.

3. N. Menzel and K. Wittmaack, *Mat.Sci.Eng.* **B12** (1992) 91.

4. W. Vandervorst, J. Alay, B. Brijs, W. DeCoster and K. Elst, Proc. SIMS IX, ed. A. Benninghoven et al. (Wiley, N.Y., 1994) p.599.

5. J.A. Kilner, D.S. McPhail and S.D. Littlewood, *Nucl.Instrum.Meth.* **B64** (1992) 632.

6. J.S. Williams, K.T. Short and A.E. White, *Appl.Phys.Lett.* **70** (1997) 426.

7. M. Petravic, B.G. Svensson, J.S. Williams and J.M. Glasko, *Nucl.Instrum. Meth.* **B118** (1996) 151.

8. J.A. Kilner etal. *Nucl.Instrum.Meth* **B15** (1986) 214.

9. W. DeCoster, B. Brijs, H. Bender, J. Alay and W. Vandervorst, *Vacuum* **45** (1994) 389.

10. See for example, N. Herbots etal. in Low Energy Ion-Surface Interactions, ed. J.W. Rabalais (John Wiley, Chichester, 1994) p.389.

RADIATION DAMAGE EFFECTS IN FERROELECTRIC LiTaO₃ SINGLE CRYSTALS

C. J. WETTELAND, K. E. SICKAFUS, V. GOPALAN, J. N. MITCHELL, T. HARTMANN, M. NASTASI, C. J. MAGGIORE, J. R. TESMER, T. E. MITCHELL
Los Alamos National Laboratory, Materials Science and Technology, Los Alamos, NM 87545, kurt@lanl.gov.

ABSTRACT

Z-cut lithium tantalate ($LiTaO_3$) ferroelectric single crystals were irradiated with 200 keV Ar^{++} ions. $LiTaO_3$ possesses a structure that is a derivative of the corundum (Al_2O_3) crystal structure. A systematic study of the radiation damage accumulation rate as a function of ion dose was performed using ion-beam channeling experiments. An ion fluence of $2.5 \bullet 10^{18}$ Ar^{2+} ions/m² was sufficient to amorphize the irradiated volume of a $LiTaO_3$ crystal at an irradiation temperature of ~120K. This represents a rather exceptional susceptibility to ion-induced amorphization, which may be related to a highly disparate rate of knock-on of constituent lattice ions, due to the large mass difference between the Li and Ta cations. We also observed that the c⁻ end of the ferroelectric polarization exhibits slightly higher ion dechanneling along with an apparent greater susceptibility to radiation damage, as compared to the c⁺ end of the polarization.

INTRODUCTION

Lithium tantalate ($LiTaO_3$) is a ferroelectric oxide possessing large nonlinear optical coefficients [1] as well as large electro-optic coefficients [2]. As such, $LiTaO_3$ crystals are of interest for nonlinear optical applications as well as for surface acoustic wave and piezoelectric devices [3]. $LiTaO_3$ is a derivative structure of corundum (α-Al_2O_3) with space group R3c. Recently, there has been considerable interest in the radiation damage tolerance of corundum structural derivative rhombohedral oxides [4-7]. The purpose of the study presented here is to assess the ion beam radiation damage resistance of lithium tantalate, especially in the context of ion beam irradiations under cryogenic conditions.

EXPERIMENT

Synthetic Z-cut $LiTaO_3$ crystals were obtained from Yamaju Ceramics Co., Ltd., (Crystal Technology Division, No. 1123 Sango-Cho, Owariasashi-shi Aichi-ken, Japan). Crystal wafers were 51 mm diameter and 0.5 mm thick, polished to a mirror finish on both sides. Samples were irradiated with 200 keV Ar^{++} ions at an inclination of a few (~10) degrees to the c <0001> crystallographic axis, in order to minimize channeling effects. The ferroelectric polarization of these crystals, P_s, is oriented along the c axis, i.e. normal to the front and back surfaces of each $LiTaO_3$ wafer. The c⁺ axis is defined as the surface normal parallel to P_s, while the c⁻ axis is antiparallel to P_s. Samples were irradiated at cryogenic temperature using a sample stage cooled to ~120K by liquid nitrogen conduction cooling.

TRIM simulations [8] were performed to determine the projected range of 200 keV Ar ions in $LiTaO_3$ and to determine the instantaneous displacement damage induced by this irradiating

Mat. Res. Soc. Symp. Proc. Vol. 504 © 1998 Materials Research Society

species. Simulations indicate that the longitudinal range of Ar ions in LiTaO$_3$ is ~120 nm, while the longitudinal straggling is ~51 nm. The peak Ar concentration is about 0.8 at.% per 10^{19} Ar/m^2, while the peak number of displacements per atom (dpa) is approximately 0.8 dpa / 10^{19} Ar/m^2.

Unirradiated and irradiated lithium tantalate crystals were analyzed by Rutherford backscattering spectroscopy and ion channeling (RBS/C), using 2 MeV He$^+$ ions for the analyses. Minimum backscattering yield, χ_{min}, as defined by the ratio between the RBS/C ion yield in an aligned spectrum versus a random spectrum, was measured in both unirradiated and irradiated samples to determine radiation damage accumulation rates in LiTaO$_3$.

RESULTS

Figure 1 shows ion channeling spectra from both surfaces of an LiTaO$_3$ unirradiated single crystal wafer. A difference in He$^+$-ion backscattered yield is apparent in Fig. 1, wherein more dechanneling is observed along c$^-$ compared to c$^+$. This difference may be due either to: (1) dechanneling induced by the polarization $\mathbf{P}_s$; or (2) to the existence of an internal field, $\mathbf{E}_{int}$, whose presence is attributed to lattice defects such as OH- ions and nonstoichiometric, charge-compensating defects [3, 9].

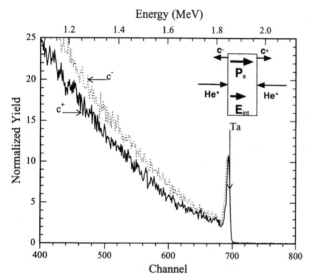

Figure 1. He$^+$ ion channeling from an unirradiated LiTaO$_3$ single crystal, with the He$^+$ beam aligned along the c-axis and with the ferroelectric polarization, $\mathbf{P}_s$, aligned either parallel (c$^+$) or antiparallel (c$^-$) to the c-axis.

Figure 2 shows He+ ion RBS/C spectra obtained from the same crystal as in Fig. 1. In this plot, more dechanneling is apparent along c$^+$ compared to c$^-$. The difference compared to Fig. 1 is that the ferroelectric polarization of the crystal has been reversed at room temperature using an electric field of ~21 kV/mm. At room-temperature, this applied field is able to reverse the polarity, $\mathbf{P}_s$, but is not of sufficient strength to reverse the sense of the internal field, $\mathbf{E}_{int}$ [3].

160

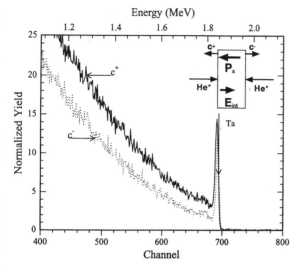

Figure 2. He⁺ ion channeling spectra from an unirradiated LiTaO₃ single crystal, with the He⁺ beam aligned along c⁺ and c⁻, respectively.

Referring to the schematic insets in Figs. 1 and 2 (representing edge-on views of a thin, single crystal LiTaO₃ wafer), the maximum dechanneling is observed on the right-hand surface, irrespective of the orientation of $\mathbf{P_s}$.

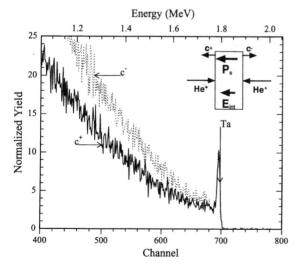

Figure 3. He⁺ ion channeling spectra from an unirradiated LiTaO₃ single crystal, with the He⁺ beam aligned along c⁺ and c⁻, respectively.

161

Figure 3 show RBS/C spectra from the same LiTaO₃ crystal as in Figs. 1 & 2. More dechanneling is apparent along c⁻ compared to c⁺. The difference here compared to Fig. 2 is that the crystal was annealed at 520K for one hour in air. This treatment is sufficient to reverse the sense of the internal field, E_{int}, without affecting the polarization, P_s [3, 9]. Clearly, the results in Figs. 1-3 indicate that c-axis dechanneling differences are dependent on the orientation of E_{int}, not P_s.

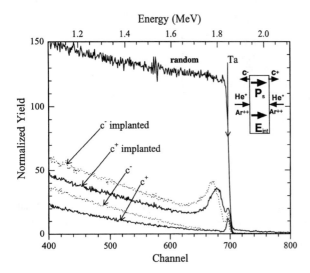

Figure 4. Ion channeling spectra from: (1) an unirradiated LiTaO₃ single crystal (spectra labeled "c⁺" and "c⁻"); and (2) an LiTaO₃ crystal irradiated with 200 keV Ar⁺⁺ ions to a fluence of 1•10¹⁸ Ar⁺⁺/m² at a substrate temperature of ~120K (spectra labeled "c⁺ implanted" and "c⁻ implanted"). The He⁺ beam is aligned along either the c⁺ or c⁻ direction in each spectrum, except for the spectrum with highest ion yield (labeled "random").

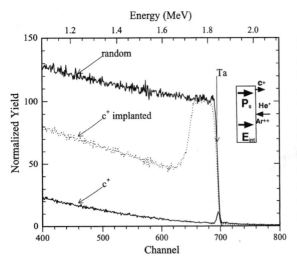

Figure 5. 2 MeV He⁺ RBS/C spectra from: (1) an unirradiated LiTaO₃ single crystal (channeling spectrum labeled "c⁺" and random spectrum labeled "random"); and (2) an LiTaO₃ crystal irradiated with 200 keV Ar⁺⁺ ions to a fluence of 2.5•10¹⁸ Ar⁺⁺/m² at a substrate temperature of ~120K (spectrum labeled "c⁺ implanted").

Figure 4 shows 2 MeV He+ RBS/C spectra from both an unirradiated LiTaO$_3$ wafer (both c- and c+ surface orientations, and a LiTaO$_3$ crystal irradiated to a fluence of $1\bullet10^{18}$ Ar^{++}/m^2 at a substrate temperature of ~120K. In this plot, more dechanneling is observed following Ar^{++} ion implantation along both c$^+$ and c$^-$ orientations and dechanneling along c$^-$ is greater than c$^+$, in either the unirradiated or irradiated condition. As indicated in the schematic inset, the enhanced dechanneling along c$^-$ is due to the antiparallel sense of the internal field, E_{int}, with respect to the c$^-$ axis used in these experiments.

Figure 5 shows RBS/C spectra from an unirradiated LiTaO$_3$ single crystal an LiTaO$_3$ crystal irradiated with 200 keV Ar^{++} ions to a fluence of $2.5\bullet10^{18}$ Ar^{++}/m^2 at a substrate temperature of ~120K. In this plot, the dechanneling yield in the near surface region (i.e., for energy losses in the vicinity of the energy for surface tantalum scattering (labeled "Ta")) reaches the random level. This is indicative of an amorphization transformation in the implanted volume. Amorphization of LiTaO$_3$ due to ion irradiation has been confirmed in a related study [4].

Figure 6 summarizes the radiation damage accumulation results of several Ar^{++} ion irradiations of c$^+$ oriented LiTaO$_3$ crystals (with E_{int} parallel to c$^+$). These results indicate that LiTaO$_3$ is rather easily amorphized, upon a peak damage level of about 0.2 dpa. This represents a rather exceptional susceptibility to ion-induced amorphization, which may be related to a highly disparate rate of knock-on of constituent lattice ions, due to the large mass difference between the Li and Ta cations.

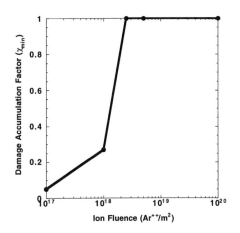

Figure 6. Summary of ion channeling measurements of Ar^{++} ion irradiated LiTaO$_3$ single crystals. This plot shows the damage accumulation fraction, χ_{min}, as a function of Ar^{++} ion fluence. χ_{min} is the ratio of He$^+$ ion backscattered yield in a c-axis channel orientation to the He$^+$ ion yield in a 'random' orientation, for ion backscattering over an energy range equivalent to 100 RBS/C multi-channel analyzer channels.

CONCLUSIONS

• c-axis ion channeling in $LiTaO_3$ single crystals is dependent on the orientation of the internal field, E_{int}, a field associated with lattice point defects.

• At cryogenic temperatures, ion irradiation damage accumulation in $LiTaO_3$ is evident at low doses and amorphization is observed at a peak damage level of about 0.2 dpa.

ACKNOWLEDGMENTS

Support for this research was provided by the Department of Energy, Office of Basic Energy Sciences, Division of Materials Science.

REFERENCES

[1] R. C. Miller and W. A. Nordland, Phys. Rev. B **2**, p.4896 (1970).

[2] K. Onuki, N. Uchida and T. Saku, J. Opt. Soc. Am. **62**, p. 1030 (1972).

[3] V. Gopalan and M. C. Gupta, J. Appl. Phys. **80**, pp. 6099-6106 (1996).

[4] J. N. Mitchell, R. Devanathan, N. Yu, K. E. Sickafus, C. Wetteland, J., V. Gopalan, M. Nastasi and K. J. McClellan, Nucl. Instr. and Meth. B, submitted for publication (1997).

[5] J. N. Mitchell, N. Yu, R. Devanathan, K. E. Sickafus, M. Nastasi and G. L. Nord, Jr., Nucl. Instr. and Meth. B **127/128**, pp. 629-633 (1997).

[6] R. Devanathan, K. E. Sickafus, W. J. Weber, J. N. Mitchell and M. Nastasi, Mater. Sci. & Eng. A, submitted for publication (1997).

[7] J. N. Mitchell, N. Yu, K. E. Sickafus, M. Nastasi, T. N. Taylor, K. J. McClellan and G. L. Nord, Jr., Mater. Res. Soc. Symp. Proc. **396**, pp. 173-178 (1996).

[8] J. F. Ziegler, J. P. Biersack and U. Littmark, The Stopping and Range of Ions in Solids (Pergamon Press, New York, 1985).

[9] V. Gopalan and M. C. Gupta, Appl. Phys. Lett. **68**, pp. 888-890 (1996).

A MODEL FOR IRRADIATION-INDUCED AMORPHIZATION

S. X. Wang, L. M. Wang and R. C. Ewing, Department of Nuclear Engineering and Radiological Sciences, University of Michigan, Ann Arbor, MI 48109

ABSTRACT

A model based on cascade melting and recrystallization is derived to describe ion irradiation-induced amorphization. The accumulation of amorphous volume fraction during irradiation is represented in a single equation. Depending on the extent of recrystallization of a subcascade, the amorphous volume accumulation can be described by a set of curves that change from exponential to sigmoidal functions. The parameters (including temperature, cascade size, crystallization rate, glass transition temperature, dose rate) that affect the extent of recrystallization are included in the model. The model also describes the temperature dependence of critical dose for amorphization.

1. INTRODUCTION

There are two general approaches used to describe radiation-induced amorphization: the direct impact model [1,2] and the defect accumulation model [1,3,4]. The direct impact model assumes that the amorphous domain forms directly in the core of a displacement cascade in a manner similar to liquid quenching [1,5-8]. The defect accumulation model assumes that the incoming particle produces defects, and the defects accumulate with continued irradiation until an amorphous phase forms when the (local) defect density reaches a threshold level [1,3,9,10].

Both the direct impact and the defect accumulation models were developed by Gibbons [1]. According to Gibbons, the change in amorphous fraction with ion dose is simply an exponential function for the direct impact model; while for the defect accumulation model (or overlap model), the function is sigmoidal. The amorphization in-growth curve given by the model is generally adopted to distinguish between direct impact or defect accumulation as amorphization mechanisms [4]. For heavy ion irradiations, especially at low temperatures, the amorphization process is assumed to be caused by direct amorphization within the cascade. While for light ions (or electrons and neutrons) the dominant mechanism is assumed to be defect accumulation [1,4,6]. Other models have been developed to interpret and to model amorphization induced by ion-bombardment [9,11-16]. Most of the recent models are based on defect accumulation or are combined with direct impact model [3,13,14].

In this paper, we present a new model. We assume that amorphization is only due to the direct impact of energetic particles (including heavy and light ions). Although direct impact amorphization is difficult to visualize in the case of light ions, they are included in the same mathematical model.

2. THE MODEL

Each incident ion is assumed to create one or several subcascades. Inside each subcascade, all atoms are mobile. Initially, the subcascade size increases (high-density energy redistribution) and shrinks (energy dissipation) after reaching its maximum size. The damage evolution is as follows: (1) When a subcascade forms, all of the structural "memory" of that volume is lost. The subcascade is similar to a melt because of the high mobility of the atoms; (2) When the subcascade cools and shrinks, epitaxial recrystallization begins and progresses inward. The epitaxial crystallization depends on the remaining crystalline fraction in the surrounding matrix. (3) Depending on the competition between the cooling process and crystal growth, the subcascade may be partially or fully recrystallized. This process is expressed by the following equation (with the assumptions that recrystallization is only epitaxial and all the subcascades are the same size) [7]:

$$\frac{dV_c}{d\widehat{N}} = -V_0 \cdot \frac{V_c}{V_T} + A \cdot V_0 \cdot \frac{V_c}{V_T} + A(1-A) \cdot V_0 \left(\frac{V_c}{V_T} \right)^3 , \tag{1}$$

where $\hat{N}$ is the number of subcascades; V_T is the total volume irradiated; V_c is the crystalline volume within V_T; V_0 is the volume of the subcascade; V_{cryst} is the volume recrystallized from V_0; A is the epitaxial crystallization efficiency, which is the recrystallized volume fraction of a subcascade ($A = V_{cryst}/V_0$) when the matrix is fully crystalline. The value of A can be between 0 (cooling rate = ∞ or crystallization rate = 0) and 1 (cooling rate = 0 or crystallization rate = ∞).

The left side of Eq. (1) represents the net change of crystalline volume after formation of a subcascade. The first term to the right in Eq. (1) is the crystalline volume that has been lost because the damage removes all structural "memory" from that subcascade volume. The second and third terms to the right in Eq. (1) represent the crystalline volume that has been recovered by recrystallization. Without the third term in Eq. (1), our model would be the same as the direct impact model of Gibbons [1] (except that we calculate the crystalline volume change, while Gibbons calculated the amorphous volume/area change). Without the third term, Eq. (1) results from the assumption that the recrystallized volume is linear with V_{cryst}/V_0. As discussed in Ref. [7], the linear assumption underestimates the amount of recrystallization. The third term in Eq. (1) is a non-linear term added to describe the recrystallization of subcascade.

Assuming one incident ion creates m subcascades, or $\hat{N} = m \cdot N$. Eq. (1) can be rewritten to represent the change in crystalline volume with ion number:

$$\frac{dV_c}{dN} = -mV_0 \cdot \frac{V_c}{V_T} + A \cdot mV_0 \cdot \frac{V_c}{V_T} + A(1-A) \cdot mV_0 \left(\frac{V_c}{V_T}\right)^3 . \tag{2}$$

Integrating the equation from zero to N ions, the crystalline fraction is:

$$\frac{V_c}{V_T} = \frac{1}{\sqrt{A + (1-A) \cdot \exp\left(\frac{mV_0}{V_T} \cdot 2(1-A)N\right)}} . \tag{3}$$

Using D for ion dose (ions/cm^2), the amorphous volume fraction, $f_a = 1 - V_c/V_T$, is:

$$f_a = 1 - \frac{1}{\sqrt{A + (1-A) \cdot \exp\left(\frac{mV_0}{h} \cdot 2(1-A) \cdot D\right)}} \tag{4}$$

where h is the sample thickness (assuming $h <$ ion range).

3. DISCUSSION

The amorphous fraction accumulation is plotted in Fig. 1. The shapes of the curves depend on the crystallization efficiency, A. When $A = 0$, Eq. (4) becomes exponential. At larger A, the curve becomes sigmoidal. As A approaches 1, the sigmoidal shape becomes more prominent. When $A = 1$, $f_a = 0$ at any dose.

Comparing Fig. 1 with the overlap model of Gibbons [1], Eq. (4) covers all of the discrete models of direct impact, single overlap, double overlap and n-tuple overlap models [1]. It is the crystallization efficiency, A, that makes the amorphization in-growth curves different. In our model, for smaller A, the amorphized volume increases with a functionality suggested by the direct impact model. As A become larger, the amorphous accumulation curve is similar to that described by the single-overlap model, double-overlap model, and n-tuple overlap model [1].

3.1 The expressions of crystallization efficiency

Currently we have two representations of crystallization efficiency, A [7]. The first is called the temperature ratio form and is based on cascade quenching [7,8]. The general form is:

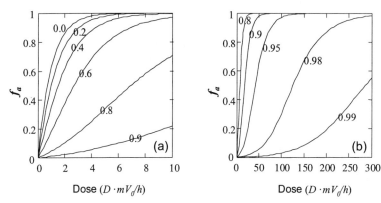

Fig. 1. Amorphous fraction accumulation as a function of ion dose according to Eq. (4). The numbers used to label the curves are values of A. (a) and (b) show curves for different ranges of A.

$$A = \frac{V_{cryst}}{V_0} = 1 - \left(\frac{T_c - T}{T_m - T}\right)^b , \qquad (5)$$

where b is a subcascade shape-parameter which has a value between 2 to 3, depending on cascade shape ($b = 2$ for a cylinder, $b = 3$ for a sphere, $2 < b < 3$ for an ellipsoid); T_m is the melting temperature; T_c is the critical temperature of the sample above which full recrystallization always occurs. We have used $b = 3$ in the following discussion.

Another form of A is exponential. As suggested by many authors [16,], we have assumed that the recrystallization rate, R, is controlled by an activation energy: $R = R_0 \exp(-E_a/kT)$. The exponential form of A is:

$$A = \exp\left[\frac{E_a}{k}\left(\frac{1}{T_c} - \frac{1}{T}\right)\right]. \qquad (6)$$

3.2 Factors affecting the crystallization efficiency

Eqs. (5) and (6) give the temperature dependence of the crystallization efficiency. We anticipate that A should also be a function of other factors such as material properties, ion mass, ion energy and dose rate. The relation of T_c to other parameters was derived from a subcascade quenching process [7]:

$$T_c \cong T_m - (T_m - T_g) \cdot R_{cryst}/(r_0 \cdot B) . \qquad (7)$$

where T_g is the glass-transition temperature; R_{cryst} is the crystallization rate; r_0 is the subcascade radius; B is a constant.

Eqs. (5), (6) and (7) show the relation between the crystallization efficiency, A, and other factors. Changes in parameters,

Table 1. The relationship between the crystallization efficiency, A, and other parameters (referring to Eqs. 4-7).

		small A	large A
parameter change	T	low	high
	T_c/T_m	large	small
	T_g/T_m	large (generally good glass formers)	small (generally less good glass formers)
	r_0	large (heavy ion, large E_d)	small (light ion, electron, large E_d)
	R_{cryst}	small (glass formers)	large (non or less good glass formers)
	E_a	large	small
	dose rate	high	low
likely shapes of amorphous accumulation curves		(curve)	(curve)

167

such as temperature, the glass-transition temperature, subcascade size, crystallization rate and dose rate affect the value of A. The value of A will determine the functionality of the amorphous fraction accumulation curve. The qualitative dependence of A on different parameters is summarized in Table 1.

The dose rate effect cannot be evaluated by Eqs. (5), (6) or (7) because the derivation does not include a time factor. Dose rate is important in two situations: (1) When the dose rate is high enough that there is significant subcascade overlap. In this situation, the higher dose rate has an effect similar to that of increasing subcascade size, thus reducing A (Eqs. 5 and 7); (2) When time scale is long enough to show significant thermal annealing, as is the case over geological time scales or at high temperatures where thermal annealing is rapid. In this situation, the lower dose rate has an effect to increase A.

3.3 Temperature dependence of amorphization dose

The temperature dependence of the amorphization dose can be easily deduced by solving D from Eq. (4). We define $\hat{f}_c$ as the detection limit of the crystalline fraction below which the sample is assumed to be "fully amorphous". Replacing f_c ($= 1 - f_a$) with $\hat{f}_c$ in Eq. (4) and because $1/\hat{f}_c^{\,2} \gg A$, the critical amorphization dose is:

$$D_c = \frac{h}{2mV_0} \cdot \frac{\ln\left(1/\hat{f}_c^{\,2}\right) - \ln(1-A)}{1-A} \qquad (8)$$

The explicit temperature dependence can be obtained by inserting either one of the expressions for A (Eqs. (5) and (6)) into Eq. 8. Thus we have two expressions of the temperature dependence of amorphization:

$$D_c = \frac{h}{2mV_0} \cdot \frac{\ln\left(1/\hat{f}_c^{\,2}\right) - 3\ln\left(\dfrac{T_c - T}{T_m - T}\right)}{\left(\dfrac{T_c - T}{T_m - T}\right)^3}, \qquad (9)$$

or,

$$D_c = \frac{h}{2mV_0} \cdot \frac{\ln\left(1/\hat{f}_c^{\,2}\right) - \ln\left(1 - \exp\left[\dfrac{E_a}{k}\left(\dfrac{1}{T_c} - \dfrac{1}{T}\right)\right]\right)}{1 - \exp\left[\dfrac{E_a}{k}\left(\dfrac{1}{T_c} - \dfrac{1}{T}\right)\right]}. \qquad (10)$$

The above equations can be further simplified by ignoring the less important terms [7]. The simplified forms of the temperature dependence of amorphization dose are:

$$D_c = \frac{D_0 \cdot (T_c / T_m)^3}{\left(\dfrac{T_c - T}{T_m - T}\right)^3}, \qquad (11)$$

or,

$$D_c = \frac{D_0}{1 - \exp\left[\dfrac{E_a}{k}\left(\dfrac{1}{T_c} - \dfrac{1}{T}\right)\right]}. \qquad (12)$$

where D_0 is the amorphization dose extrapolated to $T = 0$ K. D_0 incorporates all the temperature independent

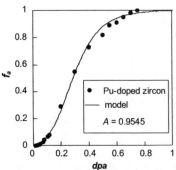

Fig. 2. Comparison of experimental data [16] for Pu-doped zircon with the model (Eq. (4). $A = 0.9545$, which has equivalent $T_c = 1050$ K.

terms (h, m, V_0 and $\hat{f}_c$). Eq. (12) is identical to that derived by Weber *et al.* [16].

3.4 Application to experimental data

Fig. 2 shows the amorphous volume fraction, f_a, as a function of dose for experimental data for Pu-doped zircon [16]. The solid line is drawn according to Eq. (4). By fitting the data with Eq. (4), The A value was determined to be 0.9545, which corresponds to $T_c = 1050$ K. This T_c value is close to that obtained in a 1.5 MeV Kr$^+$ irradiation of zircon ($T_c = 1101$ K [16]). Fig. 3 shows the amorphization volume accumulations resulting from Kr and Ni irradiation of Ni₃B at liquid nitrogen temperature and 300 K [17]. High temperature data ($A = 0.94$ for Kr, $A = 0.95$ for Ni irradiation of Ni₃B) show stronger sigmoidal features than the lower temperature data ($A = 0.9$). The ion mass effect is shown in Fig. 4 for which the data of NiAl irradiated by Xe^{++} and D$_2^+$ at 15 K are shown [18]. Because Xe^{++} is more

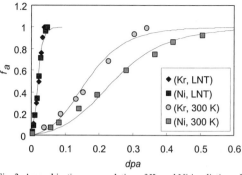

Fig. 3. Amorphization accumulation of Kr and Ni irradiation of Ni₃B at liquid nitrogen temperature and 300 K [17].

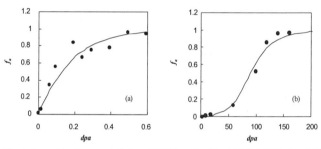

Fig. 4. Amorphization accumulation of NiAl irradiated by (a) 360 keV Xe^{++} and (b) 30 keV D$_2^+$ at 15 K [18].

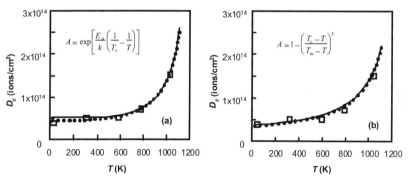

$$A = \exp\left[\frac{E_a}{k}\left(\frac{1}{T_c} - \frac{1}{T}\right)\right]$$

$$A = 1 - \left(\frac{T_c - T}{T_m - T}\right)^3$$

Fig. 5. Fit of the models of temperature dependence of amorphization dose with the experimental data (1.5 MeV Xe$^+$ radiation of quartz [8]). (a) and (b) are based on different expressions of A. In both (a) and (b) two curves are drawn according to the full equation (solid lines) of temperature dependence of amorphization dose and according to the simplified equation (dotted lines).

likely to create larger subcascades than D_2^+, the A value of Xe^+ is expected to be smaller than that of D_2^+. This is consistent with irradiation experiments on NiAl ($A = 0.1$ for Xe^{++} irradiation, $A = 0.99$ for D_2^+ irradiation).

The temperature dependence of the amorphization dose is shown in Fig. 5, for which the fits of the model to experimental data for the 1.5 MeV Xe^+ irradiation of quartz are shown. Fig. 5a is the fit to data using the exponential form of A; Fig. 5b is the fit to data using the temperature-ratio form of A. In both (a) and (b) of Fig. 5, there are two curves: the dotted lines are for the simplified temperature dependence (eqs. 11 and 12); the solid lines are for the full equations (eqs. 9 and 10). The difference between the full equation and simplified equation is small.

4. CONCLUSIONS

A model for irradiation-induced amorphization is presented. This model describes the functionality of the amorphous fraction accumulation during irradiation and the temperature dependence of the critical dose for amorphization. Depending on the crystallization efficiency, the accumulation of amorphous volume can be exponential or sigmoidal. The factors that affect the amorphization process, such as temperature, subcascade size, crystallization rate of the material, dose rate, and ion mass are discussed in the model.

ACKNOWLEDGMENT

This work was supported by the Office of Basic Energy Sciences, U. S. Department of Energy under grant DE-FG03-93ER45498 (RCE).

Reference:
[1] J. G. Gibbons, Proc. IEEE 60 (1972) 1062.
[2] F. F. Morehead, Jr. and B. L. Crowder, Radiat. Eff. 6 (1970) 27.
[3] A. I. Titov and G. Carter, Nucl. Instrum. Meth. B 119 (1996) 491.
[4] P. Ziemann, W. Miehle and A. Plewnia, Nucl. Instrum. Meth. B 80/81 (1993) 370.
[5] H. M. Naguib and R. Kelly, Radiat. Eff. 25 (1975) 1.
[6] P. Ziemann, Mater. Sci. Eng. 69 (1985) 95.
[7] S. X. Wang, Ph. D. Thesis (The University of New Mexico, 1997).
[8] S. X. Wang, L. M. Wang and R. C. Ewing, Mat. Res. Soc. Symp. Proc. 439, 619 (1997) 619.
[9] J. R. Dennis and E. B. Hale, J. Appl. Phys. 49 (1978) 1119.
[10] A. T. Motta, L. M. Howe, and P. R. Okamoto, J. Nucl. Mater. 205 (1993) 258.
[11] D. A. Thompson, A. Golanski, K. H. Haugen, D. V. Stevanovic, G. Carter and C. E. Christodoulides, Radiat. Eff. 52 (1980) 69.
[12] G. Carter, I. V. Katardjiev and M. J. Nobes, Radiat. Eff. 105 (1988) 211.
[13] A. Benyagoub and L. Thomé, Phys. Rev. B 38 (1988) 10205.
[14] G. Carter, J. Appl. Phys. 79 (1996) 8285.
[15] A. T. Motta and D. R. Olander, Acta Metall. Mater. 38 (1990) 2175.
[16] W. J. Weber, R. C. Ewing and L. M. Wang, J. Mater. Res. 9 (1994) 688.
[17] A. V. Drigo, M. Berti, A. Benyagoub, H. Bernas, J. C. Pivin, F. Pons, L. Thomé and C. Cohen, Nucl. Instrum. Meth. B19/20 (1987) 533.
[18] C. Jaouen, J. P. Riviere, and J. Delafond, L. Thomé, F. Pons, R. Danielou, J. Fontenille and E. Ligeon, J. Appl. Phys. 65 (1989) 1499.

DEFECT PRODUCTION AND ANNEALING
IN HIGH-T$_C$ SUPERCONDUCTOR EuBa$_2$Cu$_3$O$_y$
IRRADIATED WITH ENERGETIC IONS AT LOW TEMPERATURE

N. ISHIKAWA*, Y. CHIMI*, A. IWASE*, K. TSURU**, O. MICHIKAMI***
*Japan Atomic Energy Research Institute, Advanced Science Research Center, Tokai-mura, Ibaraki-ken 319-11, JAPAN, ishikawa@popsvr.tokai.jaeri.go.jp
**NTT Integrated Information & Energy Systems Laboratories, Information Hardware Systems Laboratory, Musashino-shi, Tokyo, 180, JAPAN
***Iwate University., Faculty of Engineering, Morioka-shi, Iwate-ken 020, JAPAN

ABSTRACT

The *in-situ* measurement of fluence dependence of electrical resistivity at 100K has been performed for EuBa$_2$Cu$_3$O$_y$ irradiated at 100K with various energetic ions (Cl, Ni, Br, and I) at energy of 90-200McV. Decreasing slope of resistivity-fluence curves has been observed for irradiations with 120MeV Cl, 90MeV Ni, and 185MeV Ni, while increasing slope of the curves has been observed for irradiations with 120MeV Br, 125MeV Br, and 200MeV I. It is assumed that the damaged region has a cylindrical shape along ion path and a higher resistivity than the undamaged matrix region. The calculated resistivity-fluence curve fitted well with the experimental data when using the diameter and the resistivity of the damaged region as fitting parameters. The obtained diameter and the resistivity of the damaged region have increased with increasing the electronic stopping power, S$_e$. Successive annealing of the specimens up to 300K after irradiation has resulted in 50-70% recovery of irradiation-induced resistivity change at 100K. The diameter of the damaged region has been larger than that of amorphous tracks observed by transmission electron microscope. This result is discussed in relation to the result of annealing experiment.

INTRODUCTION

There are two processes of defect production in high-T$_C$ superconductors irradiated with energetic ions (~100MeV); one is the defect production due to the elastic displacement and the other is the defect production due to the electronic excitation. When Y(Eu)Ba$_2$Cu$_3$O$_y$ (1-2-3) high-T$_C$ superconductors are irradiated with ~1MeV ions, elastic displacement dominates the defect production[1,2]. If the density of energy transfer from incident ions to the electrons of target is high enough, electronic excitation dominates defect production. The resistivity as a function of fluence in ion-irradiated 1-2-3 high-T$_C$ superconductors can be correlated to the defect production through inelastic collision when electronic stopping power, S$_e$, is above 8keV/nm[3] . The expansion of c-axis lattice parameter of EuBa$_2$Cu$_3$O$_y$ (EBCO) due to electronic excitation has been observed when the electronic stopping power is high [2,4]. In this work, EBCO have been irradiated at 100K with ions which have the electronic stopping power in the range of 7-27keV/nm. The diameter and the resistivity of the damaged region along the ion path have been determined from the resistivity-fluence curves and is discussed in terms of S$_e$ value. The result of successive annealing after the irradiation is also discussed.

EXPERIMENT

The c-axis oriented films of EBCO high-T$_C$ superconductor were prepared by rf magnetron sputtering[5]. The thickness of the films was about 300nm. EBCO films were irradiated with the following energetic ions; 120 MeV ^{35}Cl, 90 MeV and 200 MeV ^{58}Ni, 120MeV and 125 MeV ^{79}Br, 200 MeV ^{127}I from the tandem accelerator at JAERI-Tokai. The films were irradiated parallel to the c-axis under vacuum. All ion-irradiations were performed keeping the temperature at 100K. The values of $\Delta\rho/\rho_0$ were measured *in-situ* as a function of fluence, Φ, at 100K by the standard four probe method, where $\Delta\rho/\rho_0$ is the change in resistivity, $\Delta\rho$, normalized by the resistivity before irradiation, ρ_0. The nuclear stopping power, the electronic stopping power, and the ion range were calculated by TRIM-92[6]. The calculated ranges for all irradiations ($\geq$9μm) are much larger than the sample thickness, and the effect of implantation can be excluded. The incoming ions lose their energy only less than 5% after passing through the sample. After the irradiations, the samples were annealed up to 300K. The samples were then cooled again, and the electrical resistivity at 100K before annealing was compared with that after annealing.

RESULTS AND DISCUSSIONS

Fig.1 shows $\Delta\rho/\rho_0$ plotted as a function of Φ in EBCO irradiated at 100K. The increase in normal-state resistivity due to irradiation is observed. The slope of the curves decreases with increasing fluence when EBCO is irradiated with 120MeV Cl, 90MeV Ni, and 185MeV Ni. On the other hand, the slope of the curve increases with increasing fluence when EBCO is irradiated with 120MeV Br, 125MeV Br, and 200MeV I. Why are the shapes of these $\Delta\rho/\rho_0$-Φ curves different? The effect of defect production due to elastic displacement can not be the origin of this difference.

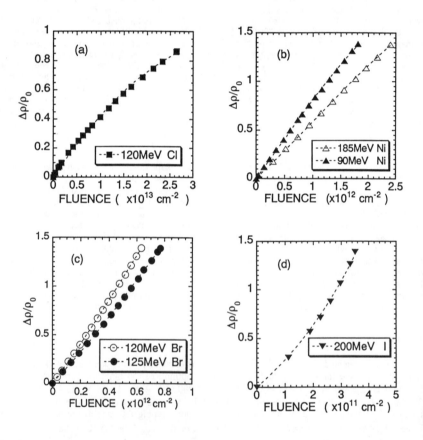

Fig.1

The fluence dependence of $\Delta\rho$ normalized by ρ_0 in EBCO irradiated with (a)120MeV Cl, (b) 90MeV and 185MeV Ni, (c)120MeV and 125MeV Br, (d)200MeV I. The dotted lines are the result of fitting the data with the equation (2).

The reason is as follows. In the previous experiment[7] we have shown that when EBCO is irradiated with ~1MeV ions, the value of $[(\Delta\rho/\rho_0)/\Phi]_{\Phi=0}$ increases linearly with increasing nuclear stopping power, S_n, and the defect production is dominated by elastic displacement. We have found that the effect of defect production due to elastic displacement is expressed by the following equation.

$$[(\Delta\rho/\rho_0)/\Phi]_{\Phi=0} \sim 3.4\times10^{-15}\, S_n, \qquad (1)$$

where $[(\Delta\rho/\rho_0)/\Phi]_{\Phi=0}$ is in the units of cm^2 and S_n in the units of $(MeV/(mg/cm^2))$. When the value of S_n is known, the effect of S_n on $[(\Delta\rho/\rho_0)/\Phi]_{\Phi=0}$ can be estimated. By using S_n values calculated by TRIM, the contribution of elastic displacement to $[(\Delta\rho/\rho_0)/\Phi]_{\Phi=0}$ was calculated by equation (1) to be 3.0×10^{-15}, 1.6×10^{-14}, 8.5×10^{-15}, 2.4×10^{-14}, 2.3×10^{-14}, and 5.1×10^{-14} for irradiations with 120 MeV Cl, 90 MeV Ni, 200 MeV Ni, 120MeV Br, 125 MeV Br, and 200 MeV I, respectively. The values of $[(\Delta\rho/\rho_0)/\Phi]_{\Phi=0}$ obtained for high energy (~100MeV) ion irradiations are more than 25 times larger than those calculated above. This means that the irradiation-induced resistivity increase for high energy ion irradiations is mainly attributed to defect production through electronic excitation. When the value of S_e is high enough, production of a cylindrical defect region along ion path is expected. According to the calculation by Bruggeman[8], when the cylindrically damaged region is introduced, the resistivity as a function of fluence is expressed by the following equation[9] .

$$(\rho(\Phi)-\rho')\,(\rho_0/\rho(\Phi))^{0.5} = (1-\delta)\,(\rho_0-\rho'). \qquad (2)$$

In this equation, $\rho(\Phi)$ is the resistivity of irradiated sample measured perpendicular to the irradiation direction. The resistivity inside the damaged region is defined to be $\rho'=k\rho_0$, where $k(>1)$ is a constant which depends on the irradiating ions and energies, δ is the volume fraction of the damaged region and is assumed to follow $\delta=1-e^{-A\Phi}$, where A is a cross section of the damaged region. The value of ρ_0 is the resistivity of the undamaged region. By varying the values of k and A, we have found that the different shapes in $\Delta\rho/\rho_0$-Φ curves are reproduced. The dotted lines in Fig.1 are the lines fitted according to the equation (2) by taking A and k as fitting parameters. The diameter of the damaged region, $D=(4A/\pi)^{1/2}$, and $k=\rho'/\rho_0$ are plotted as a function of S_e in Fig.2 and Fig.3, respectively. Both the diameter and k increase with increasing S_e. It should be noted that the observation of amorphous tracks along ion path has been reported, when S_e value is high, in ion-irradiated $YBa_2Cu_3O_y$ high-T_c superconductors using transmission electron microscope (TEM) [10]. The diameter of the damaged region obtained from our work is larger than that of amorphous region observed by TEM. For example, when S_e value is 24keV/nm, the diameter of amorphous region observed at room temperature by TEM is about 20-50nm for ion-irradiated $YBa_2Cu_3O_y$ ($6.9\leq y\leq7$)[10], while, from Fig.2, the diameter of the damaged region measured at 100K is about 120-130nm for ion-irradiated EBCO. One of the reasons for this difference in diameter can be explained as due to the difference in the measurement temperature. The existence of defect recovery during irradiation at room temperature is confirmed by the annealing experiment after the samples are irradiated at 100K. One of the results for annealing experiments is shown in Fig.4. As the figure shows, a part of the irradiation-induced resistivity change at 100K is recovered after the annealing. The irradiated samples in this experiment were all annealed up to room temperature, and it was found that there was 50-70% recovery of the irradiation-induced resistivity change at 100K.

The experimental result shown in Fig.3 indicates that not only the diameter of the damaged region but also the degree of damage is important for the understanding of columnar defect structures in high-T_c superconductors. TEM observations have usually determined the diameter of the amorphized region, while the present method suggests that the whole damaged region including amorphized region can be detected. This is another reason why the diameter of the damaged region determined in the present experiment is larger than that of the amorphized region observed by TEM.

The previous work[4] shows that the change rate of the c-axis lattice parameter as a function of

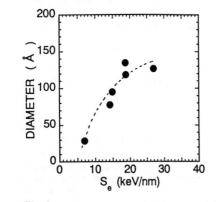

Fig.2
Diameter of the damaged region plotted as a function of electronic stopping power, S_e.

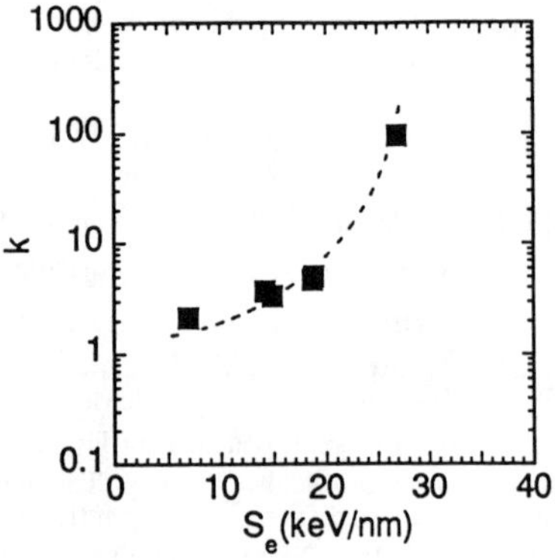

Fig.3
The value of k $(=\rho'/\rho_0)$ plotted as a function of electronic stopping power, S_e.

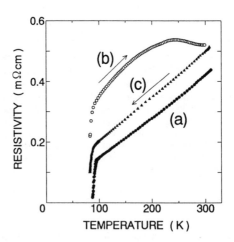

Fig.4
Resistivity-temperature curves for EBCO irradiated with 185MeV Ni ions measured before irradiation(a), during warming after ion-irradiation at 100K(b), and during successive cooling after annealing up to room temperature(c).

fluence in ion-irradiated EBCO varies, when the values of S_e are same but the velocities of the incident ions are different. In this work, however, such a difference in defect production, i.e. "the velocity effect", was not observed within the experimental error.

CONCLUSIONS

Different shapes of resistivity-fluence curves have appeared in $EuBa_2Cu_3O_y$ irradiated with energetic ions (90-200MeV, Cl, Ni, Br, and I). For Cl- and Ni-irradiations the slope of resistivity-fluence curves decreases with increasing fluence, and for Br- and I-irradiations the slope of resistivity-fluence curves increases with increasing fluence. By assuming the damaged region which has a cylindrical shape and has a resistivity higher than that of the undamaged matrix region, the diameter and the relative resistivity of the damaged region can be estimated. The diameter and the relative resistivity increase with increasing S_e value, indicating that electronic excitation play an important role in defect production. The successive annealing of the samples irradiated at 100K resulted in 50-70% recovery of irradiation-induced resistivity change. This can partly explain the difference in diameter between the damaged region observed in the present experiment at 100K and the amorphous region observed in the previous studies by TEM at room temperature.

ACKNOWLEDGEMENTS

We would like to thank the technical staffs of the Accelerators Division at JAERI-Tokai for their great help.

REFERENCES

1. G. P. Summers, E. A.. Burke, D. B. Chrisey, M. Nastasi, and J. R. Tesmer, Appl. Phys. Lett. **55**,1469 (1989).
2. N.Ishikawa, A. Iwase, Y. Chimi, H.Maeta, K. Tsuru, and O. Michikami, Physica C **259**,54 (1996).
3. V. Hardy, D. Groult, M. Hervieu, J. Provost, and B. Raveau, Nucl. Instr. Meth. in Phys. Res. B**54**, 472(1991).
4. N.Ishikawa, Y. Chimi, A. Iwase, H.Maeta, K. Tsuru, O. Michikami, T. Kambara, T. Mitamura, Y. Awaya, and M. Terasawa, to be published in Nucl. Instr. and Meth. in Phys. Res., B.
5. O.Michikami, M.Asahi, and H.Asano, Jpn. J. Appl. Phys. **29**, L298 (1990).
6. J.P.Biersack and L.G.Haggmark, Nucl. Instr. and Meth. **174**, 257 (1980).
7. N.Ishikawa, Y. Chimi, A. Iwase, H.Maeta, K. Tsuru, O. Michikami, to be published in J. of Nucl. Mat..
8. D. A. G. Bruggeman, Annalen der Physik [5] **24**, 636 (1935).
9. S.Klaumunzer, Rad. Eff. Defects Solids, **126**, 141 (1993).
10. Y. Zhu, Z. X. Cai, R. C. Budhani, M. Suenaga, and D. O. Welch, Phys. Rev. B **48**, 6436 (1993).

CORRELATION OF INTERMEDIATE ION ENERGY INDUCED EXTENDED DEFECT CONTINUITY TO ENHANCED PINNING POTENTIAL IN TL-2212 FILMS

P. Newcomer Provencio*, E. L. Venturini*, B. L. Doyle*, D.K. Brice*, H. Schöne**

*Sandia National Laboratories, Albuquerque, NM 87185
**Phillips Laboratory/VTMC, Kirtland AFB, NM 87117

ABSTRACT

Lattice defects are introduced into the structure to suppress the motion of magnetic vortices and enhance the critical current density in high temperature superconductors. Point defects are not very effective pinning sites for the cuprate superconductors; however, extended defects, such as linear tracks, have been shown to be strong pinning sites. We study the superconducting cuprate Tl-2212 (the numbers designate Tl-Ba-Ca-Cu stoichiometry). Large enhancements of vortex pinning potential were observed in Tl-2212 after high-intermediate energy heavy-ion irradiations where non-continuous extended defects were induced at dE/dx of 9 to 15.2 keV/nm (60 MeV Au, 60 MeV Cu, and 30 MeV Au) and continuous linear defects were induced at 19.5keV/nm (88MeV Au). Our research addresses the question of pinning in highly anisotropic materials like Tl-2212 where the vortices are "pancakes" rather than "rods" and suitable defect structures may be discontinuous extended damage domains. The defect microstructure and the effectiveness of the pinning potential in Tl-2212 after irradiation by intermediate energy Au at lower dE/dx of 5-15 keV/nm, where recoils are more significant, is studied using high resolution transmission electron microscopy digital imaging and a SQUID magnetometer. The nature of the ion irradiation damage at these intermediate dE/dx will be correlated to the average vortex pinning potential and the TRIMRC calculations for recoils.

INTRODUCTION

Microstructural defects are an important aspect in the performance of many high-temperature superconductors, HTS. Many applications for HTS require high critical current densities, J_c, in high magnetic fields, B, where magnetic vortices penetrate the materials and strong Lorentz forces cause dissipation [1]. One of the goals of this study is to achieve suitable defect formation in the HTS's at ion velocities that are obtainable at most accelerator facilities. The radial and longitudinal damage profile along a heavy ion track depends strongly on the rate of energy loss (electronic and nuclear stopping power), dE/dx, in the sample. The rate of energy loss is proportional to the square of the atomic number, Z^2. The velocity dependence of the incident ions on dE/dx can be summarized as follows. For very small ion velocities, v, of a few keV/amu (or more accurately for $v<<v_0$; v_0=Bohr velocity) nuclear collisions dominate dE/dx, creating a non-contiguous path of randomly displaced target ions. At a higher velocity of $0.1v_0$ to $Z^{2/3}v_0$ nuclear stopping power becomes insignificant and decreases with $1/v^2$ while the electronic stopping power increases proportional to v until it reaches a maximum. At ion velocities much larger than $Z^{2/3}v_0$, the electronic stopping power decreases again at an approximate rate of $\ln(v^2)/v^2$. The regime where extended defects and tracks can be created begins near the maximum electronic stopping power.

Early work has shown some improvement in vortex pinning at localized structural defects induced by neutron or ion irradiation [2, 3]. Significant enhancement in J_c and vortex pinning was shown in a $Tl_2Ba_2Ca_2Cu_3O_{10-\delta}$, Tl-2223, single crystal after irradiation with 4.5 MeV protons. The superconducting transition temperature, T_c, decreased, but the J_c increased by an order of magnitude [4]. Irradiation that generates point defects can enhance J_c at low fluences; however, as fluence is increased T_c decreases and an increase in normal state resistivity results [5, 6, 7]. A microstructural study with HRTEM showed damage by 1.5 MeV Kr ions in Tl-2212 single crystals consisted of clustered point defects in localized amorphized regions [7]. Strong vortex pinning was observed in $YBa_2Cu_3O_{7-x}$, YBCO, single crystals after high energy heavy ion irradiation [8, 9]. HRTEM has shown that pinning sites are amorphous, linear damage tracks [10]. Tl-based HTS ceramics [11, 12] and thin films [13, 14] showed strongly enhanced pinning when linear damage tracks were incorporated. Recent work has shown that an intermediate energy ion irradiation, 60 MeV Cu [15, 16] and a slightly higher energy 88 MeV Au [16], in Tl-2212 thin films resulted in extended defects which greatly enhanced the vortex pinning and J_c.

The goal of this study is to induce the formation of suitable defects in the HTS phase, $Tl_2Ba_2CaCu_2O_{8-\delta}$, Tl-2212, thin films at ion velocities which are widely available at accelerator facilities; and correlate the resulting microstructural defects to the HTS properties. Loss rates in this range have been shown to generate extended defects in other complex oxides [17, 18, 19]. In a recent study of Tl-2212, for dE/dx below ~19 keV/nm the damage consisted of continuous or discontinuous elongated regions along the incident ion path; above this threshold continuous, linear amorphous regions were generated [16]. It is possible that suitable defects for efficient enhancement of Tl-2212 may consist of short c-axis extended defects in Tl-2212. Ion irradiations, produced by a Tandem Ion Accelerator at 88, 60, and 30 MeV Au on Tl-2212 films are compared.

EXPERIMENTAL METHODS

Thin films of Tl-2212, 600 nm thick, were grown by Dupont [20]. These films typically exhibit strong c-axis normal orientation and are highly oriented in the plane as well. The films have superconducting onsets greater than 100K. The films have low microwave surface resistance, and high J_c's of $2X10^6$ A/cm^2 at 77K [21].

The Tl-2212 films were cleaved into ~2x2 mm^2 samples and irradiated at ambient temperature to a fluence of $1.0x10^{11}$ Au ions/cm^2 with incident energies of 30, 60 and 88 MeV at the Sandia National Laboratory Tandem Accelerator. The low dose rate of 10^9 ions/sec/cm^2 caused negligible heating during the irradiation. The incident ion direction was slightly off the normal to the film surface (the crystallographic c-axis). The energy loss rate decreases by ~5% through the film thickness. The projected range for these ions is more than an order of magnitude greater than the 600 nm film thickness.

The dE/dx in Tl-2212 was calculated using Monte-Carlo based simulation codes, TRIM [22, 23] and TRIMRC [23]. These codes do not model track formation, but are used to correlate recoil histories with the observed extent of recoil damage in the microstructure. The estimated dE/dx for these incident ions in Tl-2212 was compared to detailed damage morphology and dE/dx studies of irradiated yttrium iron garnet, YIG, [17, 18, 19]. TRIMRC calculations show dE/dx in Tl-2212 increasing from 9.0 keV/nm for 30 MeV Au to 19.5 keV/nm for 88 MeV Au.

HRTEM was done on the JEOL 2010 equipped with a Gatan 694 slow-scan camera in the Earth and Planetary Sciences Department at University of New Mexico using phase contrast imaging with the (100) diffracting beams. The data were analyzed quantitatively using the Gatan's Digital Micrograph software. Superconducting properties were measured with a Quantum Design MPMS SQUID magnetometer.

RESULTS

Using the YIG studies [17, 18, 19] as a guide, we estimated a dE/dx of more than 4.5 keV/nm is needed to form elongated defects, and approximately 10 keV/nm is needed to produce continuous cylindrical linear damage tracks in a complex oxide. TRIMRC calculations show dE/dx of 19.5 keV/nm for 88 MeV Au, 15.2 keV/nm for 60 MeV Au, 9.0 keV/nm for 30 MeV Au in Tl-2212 using crystallographically determined density and default displacement energies. Electronic and nuclear interactions between ion and target atoms are taken into account. In this study, cross-sectional HRTEM observation of Tl-2212 film shows that continuous elongated, cylindrical defects were created at 19. keV/nm by 88 MeV Au; and discontinuous elongated damage was created at 9 keV/nm by 30 MeV Au. The contrast to the YIG study may be due to the asymmetric layering of the Tl-2212 material. Th

178

damage morphology for the 60 MeV Au (calculated dE/dx of 15.2 keV/nm) in this study was compared to the 60 MeV Cu in the previous study (calculated dE/dx of 12.5 keV/nm) [16]. Cross-section HRTEM shows the continuous to discontinuous cylindrical shape of the extended amorphous domains along the ion beam direction are similar in the 60 MeV Au to the 60 MeV Cu irradiated samples with the exception there is strain surrounding the defects generated by the Au irradiation. The calculated dE/dx value of 19.5 keV/nm for 88 MeV Au is well above the 10 keV/nm threshold for forming linear tracks in YIG. Cross-section HRTEM of the resulting damage morphologies show continuous cylindrical extended damage.

HRTEM observation and comparison of the 88, 60, 30 MeV Au, and 60 MeV Cu irradiation damaged microstructures shows that more recoil damage occurred after the Au irradiations, as shown by strain surrounded defects, than the 60 MeV Cu, where only point defects surround the damage. Figure 1 shows the comparison of the damaged microstructures after the 3 Au irradiations. The 30 MeV Au irradiation generated more recoil damage than at higher energy resulting in strain surrounding the short extended defects; however the total strain volume/volume of film is much lower than that generated by 88 MeV Au. HRTEM of the region surrounding the 30 Mev Au induced defects shows stacking faults with damage along 2 to 3 planes. Dark-contrast strain lobes were also observed around the 88 MeV Au induced tracks. This strain could be caused by stacking faults in many layers surrounding the extended defects. The tracks lengths are nearly half the thickness of the film; therefore, the strain volume/volume of film is quite large resulting in broadening of the superconducting transition. Strain is also observed surrounding the 60 MeV Au induced extended amorphous defects, but was not observed after 60 MeV Cu. The extended defects are short compared to the thickness of the film and the strain volume/volume of film is much less in the 60 MeV Au than in the 88 MeV Au.

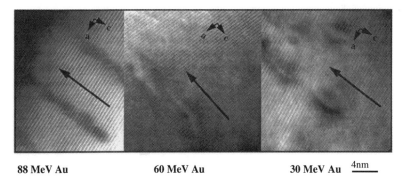

| 88 MeV Au | 60 MeV Au | 30 MeV Au | 4nm |

Figure 1 HRTEM observation and comparison of the 88, 60, 30 MeV Au, and 60 MeV Cu irradiation damaged microstructures shows that more recoil damage occurred after the Au irradiations than the Cu. This show s the comparison of the damaged microstructures.

The onset of superconductivity (T_C) was determined from the Meissner transition (field cooling) in a 0.2 mT field applied normal to the film (along the crystallographic c-axis). Figure 2 compares the high-temperature portion of the transition in an as-grown Tl-2212 film and after irradiation at three incident energies to a common fluence of 1.0×10^{11} Au ions/cm^2. T_C is decreased by ~2 K (from 101 to 99 K) by the 30 MeV Au, by ~3 K for 60 MeV Au and by 7-8 K for 88 MeV Au. The transition is increasingly broadened for higher irradiation energies. Isothermal hysteresis loops were measured to determine the magnetic field dependence of Jc. The semilog plot in Figure 3 compares Jc at 20 K versus magnetic field applied normal to the Tl-2212 film surface for an as-grown film and films irradiated with 30, 60 and 88 MeV Au ions. Jc was extracted from the hysteresis data using the Bean critical state model and the dimensions of the entire sample with the appropriate geometric correction [24].

At 80 K (approaching T_c) the damage at all three irradiation energies reduces Jc compared to the as-grown values, and the degradation increases with increasing incident ion energy.

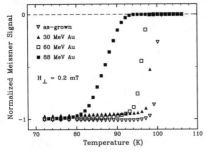

Figure 2 Comparison of Meissner transition in a Tl-2212 as-grown film and after irradiation at three incident Au ion energies at 1.0×10^{11} ions/cm^2.

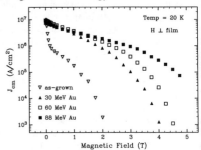

Figure 3 This semilog plot compares Jc at 20 K versus magnetic field applied normal to the Tl-2212 film surface.

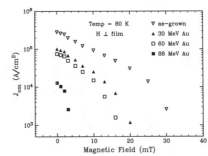

Figure 4 Comparison of Jc versus field at 80 K for the as-grown and irradiated films.

DISCUSSION

The decrease in T_C by ~2 K (from 101 to 99 K) by the 30 MeV Au, by ~3 K for 60 MeV Au and by 7-8 K for 88 MeV Au at a fluence of 1.0×10^{11} Au ions/cm^2, shown in figure 2, correlates to strain surrounding the defects. The increasingly broadened transition for 88 MeV irradiated sample correlates to the larger volume of strained microstructure surrounding the tracks. The screening supercurrents in the Meissner state flow through the superconducting regions between the amorphized extended defects, so the suppression of T_C and the broadened transition suggest damage throughout the film from secondary ions generated by collisions with the incident Au ions. This secondary damage is presumably nonuniform, resulting in the observed broadening of the Meissner transition and dark contrast regions observed by TEM.

The relatively weak vortex pinning in the as-grown film, as shown by the semilog plot in figure 3, is reflected in the decrease in J_c with increasing field. At low fields Jc is similar in the as-grown and all three irradiated films, suggesting that vortex pinning is not the dominant factor that limits Jc. As the magnetic field increases, the relatively weak vortex pinning in the as-grown film is reflected in the rapid decrease in Jc which drops by four orders of magnitude in a field of 2 tesla. In contrast, the extended defects resulting from the Au irradiations provide strong pinning sites for the vortices; Jc decreases by only an order of magnitude in the same field for all three irradiated films. At this relatively low temperature, the continuous, amorphous tracks produced by the 88 MeV Au ions offer significantly stronger vortex pinning sites as shown by the data in fields above 3 tesla. Despite the broadened Meissner transition and greater suppression of T_c for the higher energy Au ions, the continuity of the extended defects dominates the pinning force.

This is not true at higher temperatures, however. At 40 K all three Au irradiations result in a dramatic, but similar, improvement in vortex pinning with the 60 MeV Au producing the greatest improvement at higher magnetic fields. The comparison in figure 4 of Jc versus field at 80 K for the as-grown and irradiated films shows that at this relatively high temperature (approaching T_c) the damage at all three irradiation energies reduces Jc compared to the as-grown values, and the degradation increases with increasing incident ion energy. At 80 K Jc is not limited by vortex pinning and the secondary ion damage between the extended defects is very detrimental to supercurrent flow. The greater suppression of T_c and the much broader Meissner transition at higher irradiation energies, reflecting a nonuniform damage profile, are clearly dominant in reducing Jc. Hence, although the continuous amorphous tracks from the 88 MeV Au ions provide the strongest vortex pinning sites, the additional damage outside these tracks dominates the supercurrent flow. Strain free, perhaps short, extended damage with very few point defects throughout would be more efficient.

TRIM and TRIMRC alone can not predict the morphology of the damage, as this would require the calculation of time dependent electro-mechanical forces of more than 10^6 ionized target atoms for each incident ion. Using the YIG damage morphology results from the different dE/dx regimes as a guide, we determined that we need a dE/dx of approximately 20 keV/nm to produce continuous tracks, a dE/dx of approximately 10 keV/nm for extended continuous to noncontinuous. Although there is some agreement with the HRTEM observed data, the highly anisotropic layered nature of the Tl-2212 results in slightly different thresholds and damage morphologies for continuous track formation.

Since the radial and longitudinal damage profile along a heavy ion track depends strongly on the rate of energy loss dE/dx in the sample, intermediate energies can be used to obtain continuous tracks as long as the dE/dx is sufficient. There is a low and a high velocity regime were dE/dx can have the same value but differing attributes. Obviously, at low velocity the range of the ions is greatly reduced; however, this does not pose any problems for the 600 nm thick films used in this study. Secondly, and less obvious, is an increase in the track radius over the high velocity track at the same dE/dx. This can be easily explained by the increase in the impact parameter dependent ionization cross section of the target atoms as predicted in the Plain-Wave-Born approximation. At the same time, the average energy transferred to the electrons is reduced. This effect has been seen in previous studies [19].

The capability of creating continuous columnar amorphous cores in Tl-2212 films with 88 MeV Au on a Tandem Accelerator at lower energies is significant because previous track formation studies [10, 11, 12, 13, 14] were done at much higher energies with heavy ions at less accessible facilities. The fact that the continuous tracks created by the 88 MeV Au caused a broadening of the Meissner transition and breakdown at higher temperature is in part due to the strain surrounding the length of the track. The volume of the long strained cylinders is great compared to the volume of the film, while the volume of the strained damage regions from the 30 MeV Au irradiation is much less because it only involves a few planes. The 30 MeV Au induced elongate damage is odd shaped and short while the 88 MeV Au induced track lengths approach the thickness of the film and strain surrounds them. The damage volume for the 60 MeV Cu irradiated samples is much less in comparison to the 60 MeV Au irradiated samples because there is no lattice mismatch due to strain surrounding the tracks.

SUMMARY

One of the goals of this study was to induce suitable defects in Tl-2212 at ion velocities that are obtainable at most accelerator facilities. Another was to tailor the shape of defects. We have shown that extended defects are created with 30 MeV Au, 60 MeV Au, and 88 MeV Au irradiations in Tl-2212. It appears that the greater strain resulting in larger damage regions accompanying the Au irradiation severely limits the improvement in J_c. The strain is a result of the significant recoil damage, which can be correlated to the TRIMRC calculations. Suitable defects for enhancement of magnetic pinning potential may not necessarily be continuous linear defects; however shorter extended defects with insignificant surrounding damage are desirable.

ACKNOWLEDGMENT

This work was supported by Division of Material Sciences, Office of Basic Energy Sciences, US Department of Energy under Contract DE-AC04-94AL85000. Sandia is a multiprogram laboratory operated by Sandia Corporation, a Lockheed Martin Company, for the US Department of Energy.

REFERENCES

1. Y. Yeshurun, A.P. Malozemoff, Phys. Rev. Lett. **60** (1988) 2202.
2. A. Umezawa, G. W. Crabtree, J.Z. Liu, H. W. Weber, W.K. Kwok, L.H. Nunez, T.J. Moran, C.H. Sowers, H. Claus, Phys. Rev. **B36** (1987) 7151.
3. R.B. van Dover, E.M. Gyorgy, L.F. Schneemeyer, J.W. Mitchell, K.V. Rao, R. Puzniak, J.V. Waszczak, Nature **342** (1989) 55.
4. J.C. Barbour, E.L. Venturini, D.S. Ginley, Nucl. Instr. & Meth **B59/60**, 1395 (1991).
5. J.C. Barbour, E.L. Venturini, D.S. Ginley, and J.F. Kwak, Nucl. Instr. & Meth. **B65**, 531 (1992).
6. E.L. Venturini, J.C. Barbour, D.S. Ginley, R.S. Baughman, B. Morosin, Appl. Phys. Lett. **56** (1990) 2456.
7. P.P. Newcomer, J.C. Barbour, L.M. Wang, E.L. Venturini, J.F. Kwak, R.C. Ewing, M.L. Miller, B. Morosin, Physica C **267** (1996) 243.
8. L. Civale, A.D. Marwick, T.K. Worthington, M.A. Kirk, J.R., Thompson, L. Krusin-Elbaum, J. Sun, J.R. Clem, F. Holtzberg, Phys. Rev. Lett. **67** (1991)648.
9. M. Konczykowski, F. Rullier-Albenque, E.R. Jacoby, A. Shaulov, J. Yeshurun, P. Lejay, Phys. Rev. **B44** (1991) 7167.
10. Y. Zhu, Z.X. Cai, R.C. Budhani, M. Suenaga, D.O. Welch, Phys. Rev. **B48** (1993) 6436.
11. V. Hardy, D. Groult, J. Provost, M. Hervieu, B. Raveau, S. Bouffard, Phycia C **178** (1991) 255.
12. A. Wahl, M. Hervieu, G. Van Tendeloo, V. Hardy, J. Provost, D. Groult, Ch. Simon, B.Raveau, Radiation Effects and Defects in Solids **133** (1995) 293.
13. R.C. Budhani, W.L. Holstein, M. Suenaga, Phys. Rev. Lett. **72** (1994) 566.
14. R.C. Budhani, M. Suenaga, S.H. Liou, Phys. Rev. Lett. **69(26)** (1992) 3816.
15. E.L. Venturini, P.P. Newcomer, H. Schone, B.L. Doyle, K.E. Myers, Nucl. Instr. and Meth. B **127/128** (1997) 587.
16. P.P. Newcomer, E. L. Venturini, H. Schöne, B. L. Doyle, K. E. Myers, Mater. Res. Soc. Symp. Proc.**439** (1996) 639.
17. C. Houpert, F. Studer, D. Groult, M. Toulemonde, Nucl. Instrum. Methods **B39** (1989) 720.
18. F. Studer and M. Toulemonde, Nucl. Instr. and Meth. **B65** (1992) 560.
19. A. Meftah, F. Brisard, J.M. Constntini, M. Hage-Ali, J.P. Stouqert, F. Studer, M. Toulemonde, Phys. Rev. **B48** (1993) 920.
20. W.L. Holstein, C. Wilker, D.B. Laubacher, D.W. Face, P. Pang, M.S. Warrington, C.F. Carter, L.A. Parisi, J. Appl. Phys. **74** (1993) 1426.
21. A. Lauder, C. Wilker, D.J. Kountz, W.L. Holstein, D.W. Face, IEEE Trans. Appl. Supercond. **3** (1993) 1683.
22. J. Biersack, L. Haggmark, Nucl. Instr. and Meth. B **174** (1980) and J. Ziegler, Vol.2-6 Pergamon Press, 1977-1980.
23. D.K. Brice, Nucl. Instr. and Meth. Vol. B **44** (1990) 302.
24. E.M. Gyorgy, R.B. van Dover, K.A. Jackson, L.F. Schneemeyer and J.V. Waszczak, Appl. Phys. Lett. **55** (1989) 283.

HRTEM AND EELS STUDIES OF REACTED MATERIALS FROM CAF$_2$ BY ELECTRON BEAM IRRADIATION

T. KOGURE, K. SAIKI, M. KONNO*, AND T. KAMINO*

Graduate School of Science, the University of Tokyo, Tokyo 113, JAPAN
*Hitachi Instruments Engineering Co., Ltd., Ibaraki 312, JAPAN

ABSTRACT

The change of calcium fluoride (CaF$_2$) by electron beam irradiation has been investigated in TEMs operated at 200 kV. By irradiation fluorine is desorbed rapidly from CaF$_2$ crystal. Although plasmon peaks in EELS spectra suggest the formation of Ca metal, no evidence for the existence of Ca metal is found in electron diffraction pattern or HRTEM images. This is due to perfect topotactic formation of Ca metal from CaF$_2$ with a similar crystal structure and closely similar lattice parameters. Irradiation also forms amorphous material near the edge of Ca/CaF$_2$ and randomly oriented CaO crystallites grow from the amorphous material. This amorphous material is regarded as hydroxide formed through the reaction of Ca and water, considering the residual gas composition and the formation rate of the material.

INTRODUCTION

CaF$_2$ has been investigated as a potential insulating layer on Si or GaAs substrates for future semiconductor technology[1-3]. This material is also expected to be used as an inorganic resist for high-resolution electron beam lithography [4, 5], although CaF$_2$ is not the only candidate material since other inorganic halides can be used [6, 7]. Such inorganic halides are prone to lose halogen ions by intense electron radiation, which is known as electron-stimulated desorption (ESD) [8]. Some electron beam lithography experiments with inorganic halides utilize this phenomenon, changing halides to metals and forming holes by the displacement of the metals [4, 6]. On the other hand, Mankiewich et. al. [5] formed narrow lines of CaO in CaF$_2$ by electron irradiation and successive oxidation of Ca metal. Water was used as a developer to remove CaO because CaO dissolves easily in water compared to CaF$_2$.

To understand the atomistic mechanism of these processes using CaF$_2$, we have investigated the structural change of CaF$_2$ by electron beam irradiation in TEMs using high-resolution (HR) TEM and other techniques.

EXPERIMENTAL

The specimen investigated was natural CaF_2 (fluorite). No impurity was detected using energy dispersive X-ray (EDX) analysis. Synthetic MgF_2 with 99.99% (Rare Metallic Co., Ltd., Japan) was also used to investigate the difference of irradiation effects between these two materials (see the next section). They were crushed and dispersed on holy carbon films using ethanol for TEM experiments.

Two TEMs were used for the experiments. For HRTEM and EDX analyses, a JEOL JEM-2010 (nominal Cs value is 0.5 mm) with a LaB_6 filament and an ultra-thin-window (UTW) EDX detector, was used, operating at 200 kV. Electron energy loss spectroscopy (EELS) analyses were performed at 200 kV using another TEM, a Hitachi HF-2000 with a cold field emission source and a Gatan 666 PEELS system.

RESULTS AND DISCUSSION

Fig.1 shows the change of EDX spectra as a function of irradiation time. The experimental condition is described in the figure caption. The X-ray intensity of fluorine decreases rapidly with irradiation whereas that of calcium is almost constant. The left shoulder of the fluorine peak, which corresponds to oxygen, rises with beam irradiation as

Fig. 1. Change of EDX spectra by beam irradiation. The beam diameter is 200 nm. Each spectrum was acquired with a beam current of 0.6 A/cm^2 on the specimen for 1 minute after irradiation with 6 A/cm^2 for 3 minutes.

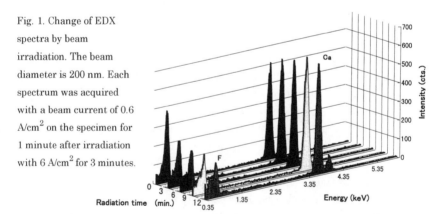

shown in the figure. Fig. 2 indicates the low-loss region of EELS spectra with similar irradiation conditions as in Fig.1. Two peaks at about 3.5 eV and 9 eV, which appear with irradiation, are regarded as surface and volume plasmons of Ca metal respectively [9]. The peak at about 17 eV, which corresponds to a plasmon of the valence electrons in CaF_2, disappears. Other peaks in the spectra, which correspond to the excitation of Ca3p state electrons, do not show any change. These changes of EELS spectra clearly indicate CaF_2 is decomposed and Ca metal is formed, as described in previous works [3, 7].

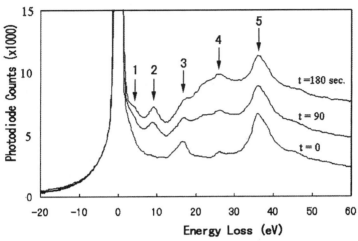

Fig. 2. EELS spectra of CaF_2 as a function of irradiation time. Peak 1 and 2 correspond to surface and volume plasmon of Ca metal respectively and peak 3 to volume plasmon of CaF_2. Peak 4 and 5 correspond to transitions of Ca3p electrons to 3d and 4s state respectively. The increase of the base line by irradiation is due to contamination with amorphous carbon.

However, no change is observed in the electron diffraction pattern (Fig. 3) or in the lattice image during beam irradiation. This is probably because CaF_2 and Ca metal belong to the same space group (Fm3m) and they have very close cell parameters (a_{CaF2}=0.546 nm and a_{Ca}=0.559 nm, see Fig. 4). It is suggested that fluorine is desorbed from CaF_2 without breaking the Ca lattice, leaving Ca metal behind. To confirm this idea, the change in MgF_2 by electron beam irradiation has been observed. MgF_2 adopts the rutile (tetragonal) structure, which is different from the hcp structure of Mg metal. There is also a volume decrease of about 30 % for metallization of MgF_2, compared to a volume increase of about 6 % for that of CaF_2. In this case, we can identify the formation of polycrystalline Mg metal using electron diffraction, as shown in Fig. 5. Salisbury et. al. reported electron beam threshold current densities to generate holes in CaF_2 [4]. This is probably due to the iso-volume Ca metal formation below threshold intensities as described above. Such iso-volume transformation may have a potential to draw nano-scale electrical circuits by Ca metal in CaF_2 insulator using electron beams.

Another phenomenon caused by the beam irradiation on CaF_2 is the formation of CaO near edges of a CaF$_2$ crystal. HRTEM images indicate that the thin area of Ca / CaF$_2$ changes to amorphous material (Fig. 6a). This amorphous material expands to the inside of

Fig. 3. (a) Bright field image of irradiated area (around the center of the edge) in CaF$_2$ and diffraction patterns from (b) irradiated and (c) fresh area. Crystal orientations are different between the image and diffraction.

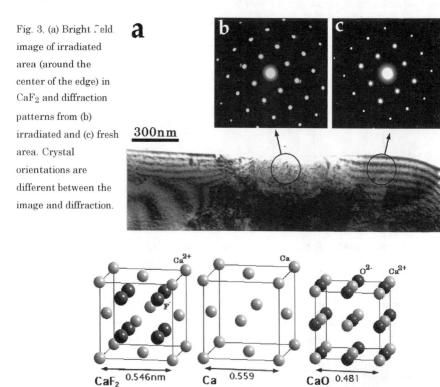

Fig. 4. Crystal structures of CaF$_2$, Ca metal and CaO. They all have a space group with Fm3m.

Fig. 5. (a) Bright field image of irradiated area (around the center of the edge) in MgF$_2$ and diffraction patterns from (b) irradiated and (c) fresh area. Crystal orientations are different between the image and diffraction.

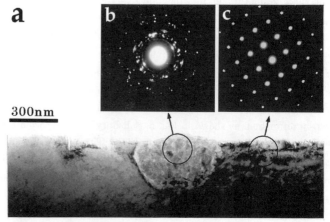

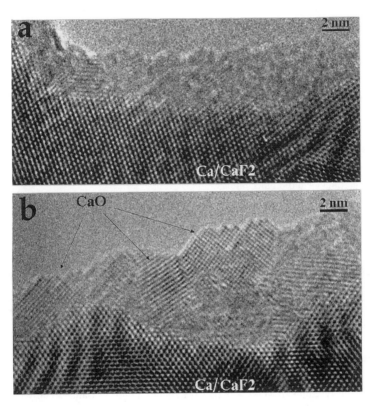

Fig. 6. HRTEM images of CaF_2 (a) at an initial stage of irradiation and (b) after several minutes. The areas are different between (a) and (b). The beam direction is parallel to [110] of CaF_2.

the crystal by further irradiation and CaO crystallites, which are identified by electron diffraction, lattice imaging and EDX analysis, start to grow from the amorphous material (Fig.6b). These CaO crystallites do not have the specific orientation relationship with Ca / CaF_2 crystal. The amorphous material can not be identified because the area is so small and it changes to CaO easily. However, it may be calcium hydroxide for the following reasons; Firstly its formation rate is very fast. The total pressure in the specimen chamber of the TEM (JEM-2010), although it was not measured directly, is estimated to be a low value in the 10^{-5} Pa range and the partial pressure of oxygen (O_2) is considered to be about two orders of magnitude smaller than the total pressure [10]. If the total pressure at the specimen is assumed to be 2.0×10^{-5} Pa, the partial pressure of O_2 is estimated to be roughly 2.0×10^{-7} Pa. A simple calculation indicates that it takes about 25 minutes to form oxygen mono-layer on the crystal surface with this O_2 partial pressure, assuming all arriving O_2 molecules are adsorbed. It is too long by several orders of magnitude compared to our observation results (for example, the image in Fig.6 was recorded after about 10 minutes irradiation). A possible explanation is that another oxygen-containing residual gas, perhaps water which is one of major residual gas molecules [10], is adsorbed and reacted with Ca. Secondly, CaO has the simple rock-salt structure and it is unlikely that this material forms an amorphous structure. Perhaps further beam irradiation dehydrates the amorphous hydroxide to form polycrystalline CaO. A similar feature was observed in the case of the electron beam radiation of ZnTe, in which surface amorphous material is formed and ZnO grows from the material [11]. Perhaps experiments using an ultra-high vacuum (UHV) TEM are useful to confirm this CaO formation mechanism.

CONCLUSIONS

Electron beam irradiation of CaF_2 causes electron-stimulated desorption of fluorine and Ca metal is formed topotactically from CaF_2. Immediately the surface of Ca reacts with residual gas in the TEM to form amorphous material which we speculated is calcium hydroxide. CaO nano-crystallites with random orientation are formed from this amorphous material by further beam radiation.

REFERENCES

1. Farrow, P.W. Sullivan, G.M. Williams, G.R. Jones, and D.C. Cameron, J. Vac. Sci. Technol. 19, 415 (1981).
2. Choi, R. Hull, H. Ishikawa, and R.J. Nemanich (editors) , Heteroepitaxy on Silicon: Fundamentals, Structure, and Devices (Mater. Res. Soc. Proc. 116, Pittsburgh, PA, 1988) pp. 401-430.
3. Saiki, Y. Sato, and A. Koma, Phys. Scrip. T41, 255 (1992).
4. Salisbury, R.S. Timsit, S.D. Berger, and C.J. Humphreys, Appl. Phys. Lett. 45, 1289 (1984).
5. Mankiewich, H.G. Craighead, T.R. Harrison, and A. Dayem, Appl. Phys. Lett. 44, 468 (1984).
6. Muray, M. Scheinfein, M. Isaacson, and I. Adesida, J. Vac. Sci. Technol. B 3, 367 (1985).
7. Egerton, P.A. Crozier, and P. Rice, Ultramicroscopy 23, 305 (1987).
8. D. Menzel, Surf. Sci. 47, 370 (1975).
9. Saiki, Y. Sato, K. Ando, and A. Koma, Surf. Sci. 192, 1(1987).
10. JEOL Ltd. (private communication).
11. Lu and D.J. Smith, Phys. Stat. Sol. (a) 107, 681 (1988).

TRANSMISSION ELECTRON MICROSCOPY STUDY OF CASCADE COLLAPSE IN COPPER DURING IN-SITU ION-IRRADIATION AT ELEVATED TEMPERATURES

T. L. Daulton, M. A. Kirk, and L. E. Rehn
Materials Science Division, Argonne National Laboratory, Argonne IL, 60439.

ABSTRACT

The basic mechanisms driving the collapse of point defects produced in collision cascades are investigated by transmission electron microscope (TEM) characterization of defect microstructures produced in fcc-Cu irradiated with low-fluences of heavy (100 keV Kr) ions at elevated temperature (23 - 600°C). Areal defect yields are determined from direct TEM observation of the total defect production integrated over the duration of the *in-situ* ion-irradiation. They are unequivocally demonstrated to decrease with increasing lattice temperature. This decrease in defect yield indicates a proportional decrease in the probability of collapse of cascade regions into defects of size where visible contrast is produced in a TEM.

INTRODUCTION

Displacement collision cascades represent the primary mechanism of lattice displacement for high-energy ions and neutrons and consequently play an important role in the modification of physical properties. Collision cascades generate localized volumes within the solid containing high concentrations of interstitial and vacancy point-defects. At sufficiently large point-defect concentrations, these highly disordered regions are unstable; recombination of interstitial and vacancy point-defects can occur along with the aggregation of point-defects into clusters. These clusters can collapse into various types of dislocation loops and stacking fault tetrahedra (SFT) which are large enough to produce lattice strain fields which are visible under diffraction-contrast imaging in a TEM.

The basic mechanisms driving the formation of cascade regions and their structural evolution into collapsed point-defect clusters are not well understood. These mechanisms are strongly driven by temperature and governed by kinetics. As temperature is varied, changes in product microstructures should reflect the manner in which cascade processes depend upon temperature. These changes include the spatial distribution of point-defects within cascade cores and their probabilities of collapse into various types of dislocation loops and SFT. In addition, changes in the distribution of defect types produced as well as changes in their respective size distributions can occur. To develop a better understanding of the kinetics governing cascade formation and subsequent collapse into point-defect clusters, the defect yield produced during 100 keV Kr ion-irradiation of fcc-Cu is measured *in situ* as a function of lattice temperature using TEM.

In a previous paper [1], we reported preliminary measurements of post-irradiation defect yields at elevated temperatures. The post-irradiation yield is defined as areal defect density per incident ion present immediately after completion of the ion-irradiation. In that work, defect loss resulting from isothermal anneal following the ion-irradiation was measured and then fitted to a simple model describing defect loss rates. Post-irradiation yields were then determined by eliminating the effects of isothermal loss. The resultant post-irradiation yields exhibited a clear dependence on lattice temperature (Fig. 1), remaining relatively constant from 23°C to 300°C then abruptly decreasing above 300°C. This suggests a lattice temperature dependence for the probability of collapse of cascade regions into defects of size which produce visible contrast. However, before this can be established definitively the question of defect loss during ion-irradiation needs to be

Post-Irradiation Defect Yield

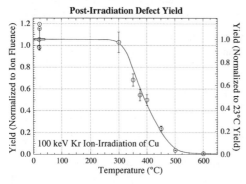

Figure 1: Post-irradiation defect yields measured from extrapolations of isothermal anneal data for 100 keV Kr ion-irradiations of fcc-Cu. The bar represents the weighted average for measurements near 23°C. The curve through the data displays a trend and is not fitted to a model.

fully addressed.

During ion-irradiation an increase in defect loss rate over those rates measured under similar conditions without ion-irradiation has been observed [2, 3]; this increase was reported to scale with ion-fluence. The immediate question arises as to whether a temperature dependence for the isothermal loss rate during ion-irradiation can account for the change in post-irradiation yield. In our previous work [1], only limited data addressing this concern was available to support a temperature dependence for the yield.

To adequately resolve this point, it is necessary to directly observe the production and loss of defects during ion-irradiation at elevated temperature. This can be accomplished through the analysis of video recorded TEM images. In this paper we discuss some of the preliminary results from our *in-situ* TEM observations of defect formation and loss during ion-irradiation at elevated temperatures.

EXPERIMENT

TEM samples were prepared from a rolled polycrystalline copper sheet of 99.99% nominal purity and a single crystal copper rod of 99.999% nominal purity cut into 3 mm diameter disks. The single crystal was sliced with the [100] zone axis near normal to the plane of the disk. The copper disks were sealed in quartz tubes, evacuated to 2×10^{-6} torr, then backfilled with high purity argon, and annealed for 2 hours (polycrystalline) or 4 hours (single crystal) at 1093°K (0.8 T_{melt}). Annealed disks were electro-polished to perforation using an electrolytic solution of 500 ml methanol, 30 ml nitric acid, and 5 ml 2-butylethanol at 238°K and stored under 10^{-8} torr prior to insertion into the TEM for examination.

The microscopy was performed in the HVEM-Tandem facility at Argonne National Laboratory using a Hitachi H-9000NAR TEM interfaced with a 2 MV tandem ion accelerator and a 0.65 MV ion implanter. The ion implanter was used to produce 100 keV Kr ion beams with typical fluences between $(9.4 \leq \Phi \leq 14.1) \times 10^{10}$ ions-cm^{-2}. *In-situ* ion-irradiations were conducted in the TEM with the ion-beam at near normal incidence to the foil surface. Further details describing the H-9000NAR TEM / accelerator arrangement are given by Allen [4].

The microscope was equipped with a Gatan model 622SC image intensified video rate camera which was connected to a VCR, enabling the *in situ* recording of images during ion-irradiations. A Gatan model 652 double-tilt, side-entry, elevated temperature goniometer stage was used. The microscope was operated at low-voltage (100 keV) to reduce the influence of electron irradiation on the production and subsequent evolution of defects produced by ion-irradiation. All TEM imaging was performed under strong dynamical 2-beam dark-field (DF) conditions utilizing the (200) operating reflection. The **g** = (200) reflection can produce strong visible contrast for the maximum number of possible defect clusters in fcc-Cu: all Frank partial (sessile) dislocation loops (**b** = a/3<111>), all partially dissociated Frank partial (sessile) dislocation loops, four of the six perfect (glissile) dislocation loop Burger's vector variants ($\mathbf{b} = a/2<110>$), and all (sessile) SFT.

RESULTS

The formation of defect microstructure during ion-irradiation is a dynamic process. Once a defect is formed, its position in the foil as well as its structure (Burger's vector, habit plane, or size) can be altered [5] or even destroyed upon further ion-irradiation. The integrated defect yield is defined as the total number of TEM visible defects per unit-area per incident ion formed over the duration of the ion-irradiation. The post-irradiation yield is obtained from the integrated yield by subtracting the number of defects per unit-area per incident ion lost during the ion irradiation.

Integrated yields can only be obtained from measurements of concurrent observations of defect formation during ion-irradiation; this was accomplished through the use of video recordings of 2-beam DF images in the TEM. The temporal resolution of VCR recordings is $\approx$ 1/30 frames/second, and consequently, any defects which are formed and lost within a shorter time interval will not be imaged. Although VCR format tape offers substantially increased time resolution over a time sequence of micrographs, its spatial resolution is far inferior to that of conventional film. Small defects and closely spaced defects arising from subcascades are more difficult to resolve.

One major difficulty encountered in these experiments was the inability to consistently establish optimum imaging conditions prior to ion-irradiation. Furthermore, it was not possible to adjust the instrumental imaging parameters during the short durations ($\leq$ 26 seconds) of the ion-irradiations limiting the clarity of the images. This was of particular concern if the ion-irradiation caused local buckling of the foil, which dramatically altered the 2-beam DF imaging condition.

Although the optimum TEM imaging condition is achieved when defects are $\leq$ 10 nm from the ion-entry surface, the influence of the near surface on the formation and stability of defects should be considered when the results of thin foil TEM studies are applied to bulk materials. To minimize the influence of the near-surface on the formation and loss of defects, near normal ion-incidence was use to distribute the defects the greatest distance from the near-surface. To allow *in-situ* observation, it is necessary to have a crystal oriented with the 2-beam DF $\mathbf{g}$ = (200) imaging condition coincident with the specimen orientation for near normal ion-incidence. Specimens satisfying this criterion were prepared from single-crystal Cu rods. However, specimens prepared from polycrystalline disks were also used when a grain possessing thin area had the required crystallographic orientation.

Once the specimen was brought to temperature, a 2-beam DF imaging condition within 12° of normal ion-incidence was determined. Residual deformation dislocations, when present, were used to adjust the 2-beam DF imaging condition, objective lens focus, and objective astigmatism prior to ion-irradiation. Images were then recorded on video tape during the ion-irradiation and later analyzed by overlaying a coordinate grid on the video monitor and examining each grid square in turn, marking the position of any defect which formed during the course of the ion-irradiation and those remaining upon completion of the ion-irradiation. The calibration of the horizontal and vertical length scales of the video images were estimated by comparing images of distinct thin-edge features recorded on both VCR tape at the magnification used during the ion-irradiation and TEM micrographs at 100 kx magnification.

The degree to which the lower resolution of VCR tape affected the identification of defects was evaluated at 23°C where the kinetics are slow enough to allow the comparison of the same post-irradiation microstructure recorded on both VCR tape and TEM micrographs. A comparison of

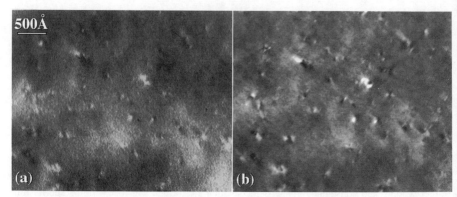

Figure 2: A comparison of the spatial resolution of the same post-irradiation defect microstructure (produced by 100 keV Kr ion-irradiations of Cu at 23°C) recorded on: a) VCR frame immediately following ion-irradiation and b) TEM micrograph. Images were formed under 2-beam DF using the (200) reflection.

these images (Fig. 2) shows that all the defects visible on the TEM micrograph are also visible on the VCR frame. However, it is evident that the closely spaced defects resulting from subcascades are not well resolved on the VCR frame. This is not an important concern for ion-irradiations performed at higher lattice temperatures where fewer subcascades are observed.

From the analysis of the VCR frames, both the post-irradiation and integrated defect yields were determined (Fig. 3 and Table I). The error estimations for the yields are based only on the combined effects of the uncertainty in the ion-fluence and area; they represent $\pm\sigma$ for the distribution of yields which are generated from all possible combinations of experimentally measured parameters falling within their respective inclusive error ranges. The loss fraction incurred by the defect population during ion-irradiation is shown by the inset in Figure 3. Fewer defects survive during ion-irradiation as the temperature is increased. There are several mechanisms which could account for the loss of defects during ion-irradiation: isothermal anneal losses through thermally activated point-defect emission, ion-irradiation induced recombination of point-defects or thermal destabilization of clusters, destruction by cascade overlap, or an increased production of glissile perfect loops through either direct formation or ion-irradiation induced conversion of nascent sessile Frank loops into perfect loops which are then lost to the surface.

The maximum possible loss fraction (Table I) predicted by the isothermal anneal loss rates measured in [1] are deficient by upwards of an order of magnitude with respect to the loss fraction observed upon completion of the ion-irradiation. Furthermore, Table I demonstrates that the

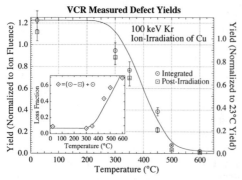

Figure 3: Post-irradiation (squares) and integrated defect yields (circles) measured by analysis of VCR data for 100 keV Kr ion-irradiations of fcc-Cu between 23°C to 600°C. The curves through the data display trends and are not fitted to a model.

Table I
100 keV Kr Ion-Irradiations of fcc-Cu

Ion-Irradiation Conditions				VCR Analysis				
Temp. (°C)	Fluence Φ ($\times 10^{10}$ Kr/cm²)	Irradiation Time (s)	Ion-Incidence Angle (±15°)	Area Analyzed ($\times 10^{-10}$cm²)	Post-Irradiation Yield (Normalized to Fluence)	Integrated Yield (Normalized to Fluence)	Loss Fraction	Predicted Isothermal Anneal Loss
23	9.4 ± 1.1	11.8	78°	7.8 ± 2.1	1.11 ± 0.08	1.12 ± 0.09	0.089	0.000
300	9.6 ± 0.8	14.4	85°	17.1 ± 4.1	0.88 ± 0.07	0.94 ± 0.07	0.065	0.004
350	9.5 ± 0.9	7.6	82°	19.1 ± 6.2	0.69 ± 0.07	0.77 ± 0.08	0.094	0.003
450	9.5 ± 1.0	15.0	80°	52.1 ± 16.2	0.22 ± 0.02	0.39 ± 0.04	0.437	0.117
500	10.8 ± 0.7	26.0	87°	138.2 ± 35.7	0.032 ± 0.003	0.075 ± 0.006	0.571	0.048
600	14.1 ± 1.4	17.9	78°	202.7 ± 59.2	0.0056 ± 0.0004	0.018 ± 0.001	0.692	0.049

fraction of defects lost during ion-irradiation does not correlate with the ion-irradiation times, contrary to that anticipated for losses by isothermal annealing. Inspection of the VCR frames reveal that over half of the defects which are lost during the ion-irradiation have short life-times (t < 1 second) and many of these are lost within only several VCR frames following formation.

In comparison, English *et al.* [6] calculated dislocation loop life-times as a function of isothermal anneal temperature and defect size to explain the temperature dependence of the yield in terms of losses from point-defect emission during ion-irradiation. Defect life-times on the order of several seconds for all but possibly the smallest defects (radius < 12 Å) were predicted. This is in excess of the observed life-times of over half of the defects lost during ion-irradiation, suggesting that the mechanism of their loss was not by simple evaporation through point-defect emission.

The short life-times of these defects further excludes their destruction by cascade overlap. The probability that a defect will be formed, directly engulfed and annihilated by a second subsequent cascade all within < 1 second is negligible for the low ion dose-rates used in these ion-irradiations. However, it is possible that the flux of energy deposited by nearby incident ions could thermally destabilize a defect quickly after its formation. The flux of point-defects generated could also recombine with these defect clusters. Alternatively, these short life-time defects could be explained as near-surface glissile perfect loops which were lost to the surface through glide. In support of this interpretation, a small fraction of defects observed on the VCR recordings were identified as glissile loops, exhibiting motion that oscillated between two fixed points before eventually disappearing from the foil. Perhaps these defects were not immediately lost after formation because they were pinned between obstructions in their glide cylinders.

Figure 3 and Table I demonstrate that defect losses through the course of the ion-irradiation are insufficient to account for the observed variation in post-irradiation yield,

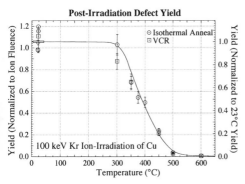

Figure 4: A comparison of the post-irradiation yields measured from extrapolations of isothermal anneal (circles) and analysis of VCR data (squares). The curve through the data displays a trend and is not fitted to a model.

the mechanism English *et al.* [6] attribute to an apparent yield decline. Defect loss during ion-irradiation is likely attributed to surface effects. Long-range strain fields are dissipated at the free surface and this establishes a decreasing gradient in the total energy of lattice displacements around dislocation cores with decreasing core distance from the surface. The forces derived from these gradients can attract glissile dislocation loops whose glide cylinders intercept the surface. The free surface is an efficient sink for mobile interstitials and glissile dislocation loops which are annihilated upon reaching the surface. The variation in the observed loss fraction with temperature may indicate a redistribution of defect types toward greater numbers of perfect loops. However, other mechanisms of defect loss such as irradiation-induced destabilization of defect clusters resulting from energy deposition or impinging point defect fluxes cannot be ruled out.

From Figure 3 it is clear that the integrated and post-irradiation yields depend upon the lattice temperature during ion-irradiation. Furthermore, as evident in Figure 4, the agreement between the post-irradiation yields estimated using two very different methods of extrapolation of isothermal anneal data [1] and direct measurement of VCR recorded images instills confidence in their determination.

CONCLUSIONS

Both the post-irradiation and the integrated defect yields were determined by direct observation through analysis of video recordings of TEM 2-beam DF images. Post-irradiation yields measured using this technique are in close agreement with values previously determined from extrapolations of isothermal anneal loss data. The defect loss fraction incurred during ion-irradiation increases with increasing temperature, suggesting a redistribution of the defect types produced. However, defect losses through the course of the ion-irradiation are insufficient to account for the observed variation in yields. Areal defect yields are unequivocally demonstrated to decrease with increasing lattice temperature in fcc-Cu. This variation with temperature can be plausibly correlated to a lattice temperature dependence for the probability of point defect collapse into defect clusters.

ACKNOWLEDGMENTS

The *in-situ* ion-irradiations and electron microscopy were performed in the HVEM-Tandem facility at Argonne National Laboratory. This research was supported by U.S. DOE under BES contract W-31-109-ENG-38. The authors thank L. L. Funk of the HVEM-Tandem facility for the operation of the ion accelerator. The authors thank, also of the HVEM-Tandem facility, E. A. Ryan and S. Ockers for valuable assistance. In addition, the authors thank B. J. Kestel for valuable assistance with sample preparation and M. L. Jenkins for providing single-crystal Cu.

REFERENCES

1. T. L. Daulton, M. A. Kirk, and L. E. Rehn, Mat. Res. Soc. Symp. Proc., **439**, 313-318 (1996).
2. N. Sekimura, Y. Yamanaka, S. Ishino, 14th International Symposium on Effects of Radiation on Materials, ASTM STP 1046, **1**, 596-608 (1990).
3. S. Ishino, N. Sekimura, K. Hirooka, and T. Muroga, J. Nucl. Mater., **141-143**, 776-780 (1986).
4. C. W. Allen, L. L. Funk, and E. A. Ryan, Mat. Res. Soc. Symp. Proc., **396**, 641-646 (1996).
5. J. S. Vetrano, I. M. Robertson, and M. A. Kirk, Scripta Met., **24**, 157-162 (1990).
6. C. A. English, B. L. Eyre, and J. Summers, Phil. Mag., **34**, 603-614 (1976).

Part II

Irradiation Effects in Metals and Alloys

ION BEAM MIXING AND THERMAL DEMIXING OF Co/Cu MULTILAYERS

M. Cai*, T. Veres*, R.W. Cochrane*, S. Roorda*, R. Abdouche** and M. Sutton**,
*Département de physique et Groupe de recherche en physique et technologie des couches minces, Université de Montréal, C.P. 6128, succ. Centre-Ville, Montréal, Canada H3C 3J7,
**Department of Physics and Center for Physics of Materials, McGill University, Montréal, Canada, H3A 2T8.

ABSTRACT

X-ray reflectivity and magnetotransport studies have been used to probe the effects of ion-beam irradiation and subsequent thermal annealing on the structure and giant magnetoresistance (GMR) in Co/Cu multilayers. Low-dose ion bombardment produces interfacial mixing which is accompanied by a systematic suppression of the antiferromagnetic (AF) coupling and the GMR. For ion doses not exceeding 5×10^{14} ions/cm^2, subsequent thermal annealing restores the abrupt interlayer structure as well as the GMR. The combination of low-dose ion bombardment and thermal annealing provides an *ex situ* technique to modify interface structure *reversibly* over a gnificant range.

'TRODUCTION

Artificially layered materials of alternating ferromagnetic and non-magnetic materials have generated considerable interest due to their unique magnetic and transport properties such as GMR and interlayer AF coupling [1]. In this context, interfacial structure plays a crucial role so that it is necessary to establish a technique capable of modifying interface structure in a systematic way. Using a co-deposition technique, Susuki *et al.* [2] reported that the MR in Co/Cu was weakened by artificial intermixing and thus concluded that the scattering centers causing the GMR were in the Co layers. However, co-deposition, like other *in-situ* techniques aimed at modifying interfaces, also affects the growth of subsequent layers and therefore the crystallography of a film. Moreover, direct comparison between different samples is often inconclusive. Post-growth thermal annealing has proven very successful in clarifying the role of intermixing in Fe/Cr [3] but has little measureable effect on the interface structure in Co/Cu [4], in part, because of the very limited equilibrium solubility of Co in Cu.

Recent studies [5,6] have demonstrated that ion-beam irradiation is a promising technique for modifying interfaces. In this paper, we report x-ray reflectivity and magnetotransport measurements on ion bombarded and annealed Co/Cu multilayers. By controlling ion dose, the interlayer magnetic coupling and the GMR can be varied systematically in a *single* sample. As a result of the strong immiscibility between Cu and Co, it is found that the metastable ion mixing can be largely reversed by thermal annealing following the irradiation. Taken together, the structural and transport data demonstrate that the ion bombardment is capable of generating a systematic intermixing at the interface and that the GMR in Co/Cu is very sensitive to such a process.

EXPERIMENTAL DETAILS

Multilayers with the configuration Cu(50 Å)/[Co(17 Å)/Cu(t Å)]$_{30}$ with t = 22 and 34 Å were prepared by rf triode sputtering onto glass (Corning 7059) substrates at deposition rates of 2 Å/s for Cu and 1 Å/s for Co [7]. With these Cu thicknesses, the multilayers are situated at the second and third peaks of the GMR oscillation. Nominal thicknesses were confirmed by surface profilometry and low-angle x-ray reflectivity measurements. Sample magnetization was measured at room temperature using a vibrating sample magnetometer and transport measurements were made, also at room temperature, with a high-resolution ac bridge. The sample structure was characterized by low- and high-angle x-ray scattering techniques using Cu K_{α} radiation.

Ion-beam irradiation experiments were performed in a vacuum of 10^{-7} Torr with 1 MeV Si^+ ions at currents below 50 nA/cm². To limit heating effects during irradiation, the samples were placed in thermal contact with a copper block kept at 77 K. Irradiation doses ranged from 10^{12} to 5×10^{14} ions/cm² which resulted in about 0.001 to 0.5 displacements per atom as estimated by TRIM simulations [8]. The energy loss of the 1 MeV Si^+ ions in the 1000 Å thick sample is less than 200 keV so that only a small fraction ($< 0.1\%$) of the implanted ions come to rest in the multilayer. Some of the irradiated multilayers were also annealed in vacuum up to 325 °C.

RESULTS

Structural Properties

Fig. 1 presents low-angle x-ray spectra of a [Co(17 Å)/Cu(34 Å)]₃₀ multilayer, (a) as-deposited, (b) after irradiation with a dose of 2×10^{14} ions/cm² and (c) after subsequent anneal for four hours at 250 °C. The as-deposited multilayer shows clear first- and second-order superlattice peaks, which confirm that the Co/Cu interfaces of the multilayer are well-defined. After irradiation, two superlattice peaks are also observed although their intensities are reduced. Given the fact that the reduction in specular superlattice peak intensity is an indication of increased interface roughness in a multilayer structure, the principal effect of the low dose irradiation has been to disorder the interfacial region.

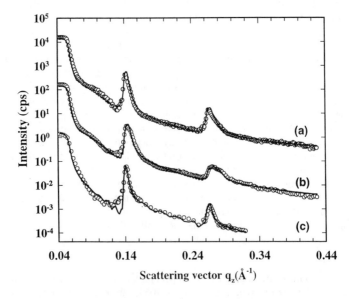

Figure 1.   Low-angle x-ray reflectivity spectra for a [Co(17 Å)/ Cu(34 Å)]₃₀ multilayer: (a) as-deposited, (b) after irradiation at 2×10^{14} ions/cm² and (c) after annealing at 250°C for 4 hours. The solid lines are fitted curves as described in the text. Curves (b) and (c) have been displaced for clarity.

To quantify the effect of the ion bombardment, the x-ray spectra have been fitted with a standard optical model [7] in which the x-ray reflectivity is calculated using a matrix method. A global interface roughness factor has been incorporated into the calculation by assuming a Gaussian form (with a Debye-Waller factor $\exp(-\sigma_r^2 q_z^2)$ where, σ_r is the root-mean-square roughness). In addition, layer thickness fluctuations and film surface oxidization are included to calculate the background intensity profile. The resulting fitted curves for all the spectra have been superimposed on the experimental spectra in Fig. 1. For the as-deposited multilayer, an interface roughness of 5.7 ± 0.5 Å has been calculated.

For the irradiated multilayers, the intermixing induced by ion irradiation is simulated in the calculation by introducing two extra layers at each interface, which correspond to a Co-rich region ($Co_{1-x}Cu_x$, where x<0.5) and a Cu-rich region ($Cu_{1-x}Co_x$, where x<0.5) respectively. Taking advantage of the fact that the x-ray reflectivity is measured on a single multilayer sample before and after irradiation, other parameters, such as the layer-thickness and roughness are kept fixed. Fig.1(b) shows the fitted curve for the multilayer after irradiation at a dose of 2×10^{14} ions/cm^2. The best fit to the data yields x=0.1 and a mixing width of about 11 Å but these parameters are highly correlated. For fixed x in the range from 0.05 to 0.3, (which is comparable to the mean number of displacements at this ion fluence), the intermixing width varies from 6 to 12 Å, all of which are small compared with the wave-length of the multilayer. This analysis leads us to conclude that the principal structural effect of the irradiation has been to blur the Cu/Co interface over a range of approximately 10 Å without significantly altering the periodic structure of the multilayer.

This point is reinforced by considering the cascade mixing width Ω of the beam [9]

$$\Omega^2 = \frac{1}{3} \Gamma_o \frac{F_D}{N} \xi_{21} \frac{R_c^2}{E_c} \Phi \quad , \tag{1}$$

where $\Gamma_o = 0.608$, N is the atomic density, $\xi_{21} = [4M_1M_2/(M_1+M_2)^2]^{1/2}$ (M_1 and M_2 are the masses of the atoms involved in the collision), F_D, the energy deposited per unit length due to nuclear collisions, E_c, a threshold displacement energy , R_c^2, the mean-square range associated with E_c, and Φ, the ion dose. Taking typical values of $E_c = 25$ eV, $R_c = 10$ Å, and $F_D = 35$ eV/Å [8], eq(1) predicts a dose of order 10^{16} to 10^{17} ions/cm^2 to reach a mixing width comparable to the wave-length of our multilayers. Our dose levels are 2 orders of magnitude smaller than these, so that eq(1) predicts a cascade mixing width of about 5 Å in line with the value of the mixing width from the x-ray analysis.

High-angle x-ray diffraction after each sample treatment reveals little change in crystallographic structure. Multilayers are textured principally in the fcc (111) direction with a relatively weak fcc (200) component. Using the Scherrer formula, we estimate a length scale normal to the surface of about 130 Å which is much larger than the individual layer thicknesses, and suggests good structural coherence across the interfaces. For ion doses less than 5×10^{14} ions/cm^2, it is found that the linewidth of the (111) Bragg peak is nearly unchanged upon irradiation, indicating the structural coherence length is not strongly influenced. As well, a small increase in the relative intensity of the (200) peak is observed.

Annealing effects on the irradiated multilayers are of particular interest. Fig. 1(c) shows that the intensities and the linewidths of the superlattice peaks have fully recovered after annealing. In contrast, x-ray reflectivity spectra of virgin Cu/Co multilayers are little affected by annealing at the same temperature [4]. Due to the equilibrium immiscibility of Cu and Co, annealing at moderate temperatures provoques a back-diffusion from the metastably mixed regions and the reformation of relatively abrupt interfaces. This spectrum can be fitted without the interfacial layers introduced for spectrum 1(b), with a slight increase of the interfacial roughness to 6.5 ± 0.3 Å.

Magnetotransport Measurements

Ion-beam mixing and demixing suggested by the x-ray analysis can be further evidenced from the resistivity and magnetoresistivity measurements. For ion-beam doses up to 10^{13} ions/cm^2, no change in the electrical resistivity of the multilayers is observed. Above this fluence, the saturation resistivity (ρ_s) of a multilayer increases progressively as a function of total ion dose. A dose of 5×10^{14} ions/cm^2 results in a 60% increase in ρ_s [5]; in contrast, the resistivities of 1000 Å Cu and Co films are nearly unchanged for this dose. Since in these multilayers the electron mean free path is comparable to the layer thickness, the increase in resistivities can be directly connected with enhanced electron scattering as a result of ion-beam mixing across interfaces.

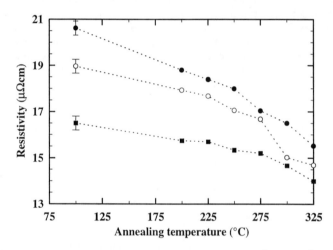

Figure 2.   Resistivity versus annealing temperature for three [Co(17 Å)/Cu(22 Å)]$_{30}$ multilayers: as-deposited (■), irradiated at 1.3×10^{14} ion/cm^2 (O) and at 2.6×10^{14} ions/cm^2 (●).

Fig. 2 shows the variations in resistivity upon thermal annealing for three nearly-identical [Co(17 Å)/Cu(22 Å)]$_{30}$ multilayers. Two of the multilayers were subjected to irradiations at 1.3×10^{14} and 2.6×10^{14} ions/cm^2 before annealing for periods of four hours at temperatures below 250 °C and two hours at higher temperatures. As shown in the figure, the resistivities of the irradiated multilayers decrease dramatically upon subsequent annealing. The resistivity of the non-irradiated multilayer also decreased but to a much smaller extent; this decrease is probably related to minor grain growth during annealing. Nevertheless, Fig. 2 demonstrates that the resistivities of the multilayers irradiated at various doses are getting very close to that of the non-irradiated sample, suggesting significant demixing occurs on heating.

Fig 3(a) shows that the GMR falls monotonically with ion dose for the multilayers [Co(17 Å)/Cu(22 Å)]$_{30}$ (at the second GMR peak) and [Co(17 Å)/Cu(34 Å)]$_{30}$ (at the third GMR peak). Given the fact that the GMR is interface related, it is a clear indication that these ion doses produce significant interface modification. It is interesting to note that in Fe/Cr multilayers, Kelly et al [6] found that ion irradiation led to an increase in the GMR in spite of a significantly reduced AF coupling. Such behaviour was explained in terms of the enhanced spin-dependent electron scattering at the Fe/Cr interfaces. In contrast, we observe no increase in the MR (either in $\Delta\rho$ or $\Delta\rho/\rho_s$) upon irradiation in any of our Co/Cu multilayers at any dosage level. This behaviour suggests that the role of interface scattering in the GMR might be quite different for Co/Cu and Fe/Cr multilayers.

In the process of interfacial demixing by annealing, the GMR increases sharply, as shown in Fig. 3(b); also shown is the minor increase in MR for the as-deposited sample. The slight increase in the MR ratio for this sample is the result of the decrease in the resistivity related to grain growth. The much more significant rise in the MR of the irradiated multilayers is, however, due mostly to the increase in $\Delta\rho$ (the field induced change in the resistivity) and is more likely to be associated with the demixing process. The MR of the multilayer irradiated at the higher dose increases nearly a factor of three from about 4% to about 12% upon annealing at 200 °C for four hours. Also, the MR of the multilayer irradiated at the lower dose has fully recovered to the as-deposited value after annealing at 250 °C. When annealing temperatures are increased above 300 °C, the MR starts to decrease for all the multilayers, as the multilayer structure begins to break down. However, at each annealing step below this temperature, the MR of all multilayers irradiated at doses $< 5 \times 10^{14}$ ions/cm² systematically increase. This increase in GMR is accompanied by an improvement in the AF coupling between Co layers.

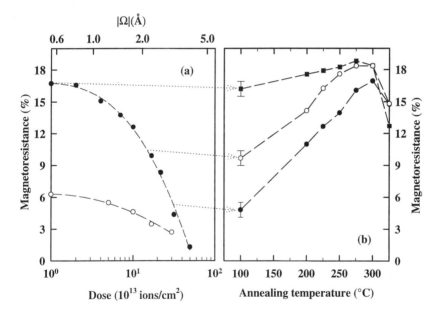

Figure 3 (a) Magnetoresistance as a function of ion dose for [Co(17 Å)/Cu(22 Å)]₃₀ (●) and [Co(17 Å)/Cu(34 Å)]₃₀ (○) multilayers. The Ω scale at the top of the figure has been calculated using eq.(1). (b) Variations in the GMR with annealing temperature for three [Co(17 Å)/Cu(22 Å)]₃₀ multilayers subjected to ion doses of 0 (■), 1.3×10^{14} (○) and 2.6×10^{14} (●) ions/cm², respectively.

There are two principal ways in which the GMR can be modified by interfacial mixing: the introduction of additional electron scattering centers with scattering asymmetry different from other mechanisms, a direct effect on the GMR; or the modification of the interlayer magnetic coupling, an indirect effect. For the latter, a first-principles calculation [10] demonstrates that a small amount of interdiffusion at the interface can dramatically suppress the AF coupling due to strong disorder between Cu and down-spin Co states. This situation is expected to be applicable to the Co-Cu interface after ion-beam mixing. An analysis of magnetization data indicates that the AF coupling is systematically reduced by irradiation, a result which is qualitatively consistent with the calculation [10]. In contrast, annealing of the irradiated multilayers always leads to an improvement of the AF coupling. This variation can be explained by the back-diffusion upon annealing which produces a sharper and more atomically-ordered interface region.

CONCLUSIONS

Ion-beam irradiation at low doses provides a means of intermixing and roughening interfaces between metallic layers of multilayer samples. In the case of the Co/Cu films studied here, the AF interlayer coupling and the GMR can be reversibly altered *ex situ* over a wide range in a single Co/Cu multilayer. Ion-irradiation monotonically decreases the GMR while subsequent annealing increases it. Compared with other techniques, this method has little effect on the crystallographic texture of the multilayer. Taken together, these results demonstrate that MeV ion bombardment can be a sensitive tool for interface modification in metallic multilayers and that, for Co/Cu multilayers, the GMR is indeed strongly dependent on the structure of the interface region.

ACKNOWLEGMENTS

This research has been supported by the NSERC of Canada and the Fonds FCAR of Quebec. It is a pleasure to acknowledge the assistance of P. Bérichon and R. Gosselin with the operation of the accelerator.

REFERENCES

1. M.N. Baibich, J.M. Broto, A. Fert, F. Nguyen Van Dau, F. Petroff, P. Étienne, G. Creuzet, A. Friederich and J. Chazelas, Phys. Rev. Lett. **61**, 2472 (1988).

2. M. Susuki and Y. Taga, Phys. Rev. **B52**, 361 (1995); J. Appl. Phys. **74**, 4661 (1993).

3. Eric E. Fullerton, David M. Kelly, J. Guimpel, Ivan K. Schuller, Phys. Rev. Lett. **68**, 859 (1992).

4. H. Zhang, R.W. Cochrane, Y. Huai, M. Mao, X. Bian and W.B. Muir, J. Appl. Phys. **75**, 6535 (1994); H. Lailer and B.J. Hickey, ibid **79**, 6250 (1996).

5. M. Cai, T. Veres, S. Roorda, R.W. Cochrane, R. Abdouche and M. Sutton, J. Appl. Phys. **81**, 5200 (1997).

6. D.M. Kelly, I.K. Schuller, K. Korenivski, K.V. Rao, K.K. Larsen, J. Bottinger, E.M. Gyorgy and R.B. van Dover, Phys. Rev. **B50** 3481 (1994); K. Temst, G. Verbanck, R. Schad, G. Gladyszewski and M. Hennion, Physica **B234-236**, 467 (1997).

7. Y. Huai, R.W. Cochrane and M. Sutton, Phys. Rev. **B48**, 2568 (1993).

8. J.F. Ziegler and J.P. Biersack, *The Stopping and Range of Ions in Solids*, Pergamon Press, New York, 1985.

9. P. Sigmund and A. Gras-Marti, Nucl. Instrum. Methods **182/183**, 25 (1993).

10. J. Kudrnovsky, V. Drchal, I. Turek, M. Sob, P. Weinberger, Phys. Rev. **B53**, 5125 (1996).

GIANT MAGNETORESISTANCE IN IRON AND COBALT IMPLANTED SILVER THIN FILMS

C.M. DE JESUS *, J.G. MARQUES *, J.C. SOARES *, L.M. REDONDO †,
M.F. DA SILVA †, M.M. PEREIRA DE AZEVEDO ‡, J.B. SOUSA ‡
* CFN, Universidade de Lisboa, Av. Prof. Gama Pinto, 2, P-1699 Lisboa Codex, Portugal
† Instituto Tecnológico e Nuclear, Estrada Nacional 10, P-2685 Sacavém, Portugal
‡ IFIMUP, Universidade do Porto, P-4150 Porto, Portugal

ABSTRACT

The magnetoresistive behavior of granular thin films prepared by Fe and Co implantation in Ag thin films is reported. Ag thin films (~2000Å) were implanted with Fe or Co at fluences up to 8×10^{16} at./cm^2. The magnetoresistive response obtained after implantation was found to increase with the implanted fluence. A further increase by a factor of 3-4 can be achieved annealing the films in a conventional furnace at 620 K under vacuum. The best value of the magnetoresistance obtained so far is 9% at 10 K for a film implanted with Co at a fluence of 8×10^{16} at./cm^2.

INTRODUCTION

Giant magnetoresistance (GMR) is exhibited by a variety of inhomogeneous metallic magnetic systems comprising magnetic layers or particles separated by non-ferromagnetic material. Since its discovery in magnetic multilayers [1] and granular materials [2,3] this effect was found to depend strongly on the materials' parameters and processing conditions. Granular alloys are normally prepared using non-equilibrium techniques such as sputtering co-deposition, melt-spinning or splat cooling [2-4]. The magnitude of the GMR effect in these materials depends on the size, concentration and distribution of the magnetic clusters in the non-magnetic matrix [5]. However, the optimal composition and annealing conditions are not easily predictable. Therefore, alternative preparation techniques are of great interest.

Recently we reported on the behavior of diluted granular Fe-Ag and Fe-Cu films prepared by ^{57}Fe implantation at a fluence of 1×10^{16} at./cm^2, and we have shown that the implanted Fe forms both small clusters and large α-Fe particles [6]. In this work we present new results on Fe-Ag granular films prepared by Fe implantation at much higher fluences, up to 8×10^{16} at/cm^2, we study the influence of thermal treatments in the optimization of GMR, and we extend this work to the Co-Ag system.

EXPERIMENTAL DETAILS

Ag thin films with 1800-2600 Å thickness were deposited by evaporation or pulsed laser ablation from 99.99 at.% pure Ag onto Si and SiO$_2$ substrates. The films were implanted with Fe$^+$ and Co$^+$ ions at 1-8×10^{16} at./cm^2 fluences and 150 keV energy. The TRIM code [7] yields a projected range of 480 Å for 150 keV Fe/Co ions in Ag, with a standard deviation of 240 Å, and a sputter coefficient of 8 for Fe/Co in Ag. The implantations were performed using a DanFysik 1090 High Current Ion Implanter at Sacavém. Fe$^+$ and Co$^+$ ions were produced in a CHORDIS ion source (Cold HOt Reflex Discharge Ion Source), model 920, using Fe or Co sputter targets. Surfaces of 30x30 cm^2 can be homogeneously implanted with our setup.

The thicknesses of the Ag films and the retained Fe and Co fluences were determined by Rutherford Backscattering Spectroscopy (RBS), using a 2.5 MeV He$^+$ collimated beam of the 3.1

Table 1 - Implantation parameters and Ag film thickness before and after implantation.

Table 1 - Implantation parameters and Ag film thickness before and after implantation.

Sample	Initial Ag thickness [Å]	Implanted isotope	Implanted fluence [10^{16}at./cm^2]	Energy [keV]	Final Ag thickness [Å]	Retained fluence [10^{16}at./cm^2]
FeAg1	2600	Fe	1.0	180	2450	a)
FeAg2	2200	Fe	6.0	150	800	5.0
FeAg3	2600	Fe	6.0	150	1400	5.0
FeAg4	2000	Fe	7.5	150	650	7.0
CoAg5	2400	Co	3.0	150	1700	2.2
CoAg6	1800	Co	4.0	150	850	2.9
CoAg7	2400	Co	8.0	150	900	7.5

a) not measured

MV Van de Graaff accelerator at Sacavém. The backscattered particles were detected at 140° and 180° with silicon surface barrier detectors with 13 and 18 keV energy resolution, respectively. Table 1 shows the thickness of the films before and after implantation, as well as the nominal and retained doses for all the investigated films. The magnetic properties of the samples were measured with a Vibrating Sample Magnetometer (VSM) in the field range of 0 to 0.2 T at room temperature. Magnetoresistance measurements (MR) were performed with a standard electromagnet (0 to 1 T), under DC current using a four-point probe technique. Annealing of the samples was performed for 20 min, under vacuum, at temperatures between 520 and 850 K.

RESULTS

Fig. 1 shows the RBS spectra taken in the FeAg4 film after implantation and after annealing at 620 K. The Fe yield was magnified by a factor of 5. After implantation Fe is distributed mainly in the near-surface region of the film and its profile deviates significantly from the one predicted by TRIM. The broadening of the Fe peak after annealing shows that there was a change in the Fe profile in the Ag film with the thermal treatment. The films implanted with Co show the same behavior (data not shown).

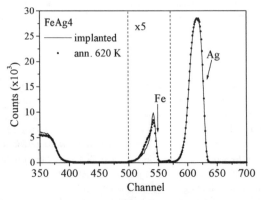

Fig. 1 - RBS spectra taken in the film implanted with 7.5x10^{16} Fe/cm^2 (FeAg4) after implantation and after annealing at 620 K. The Fe yield was magnified by a factor of 5.

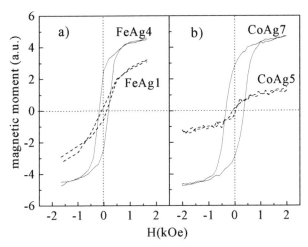

Fig. 2 - VSM results obtained at room temperature for the films implanted with a) Fe at fluences of 1x10^{16} (FeAg1) and 7.5x10^{16} Fe/cm^2 (FeAg4); b) Co at fluences of 3x10^{16} at./cm^2 (CoAg5) and 8x10^{16} Co/cm^2 (CoAg7).

Fig. 2a) shows the VSM results obtained at room temperature for the Fe implanted films FeAg1 (implanted with 1x10^{16} at./cm^2) and FeAg4 (7.5x10^{16} at./cm^2) in the as-implanted state. In the case of the FeAg4 film there is already a ferromagnetic-like behavior in contrast with the dominant superparamagnetic-like characteristic of the FeAg1 film. Fig. 2b) shows corresponding results for the Co implanted films CoAg5 (implanted with 3x10^{16} at./cm^2) and CoAg7 (8x10^{16} at./cm^2). A ferromagnetic-like behavior at room temperature is again seen in the case of the film implanted with the higher fluence.

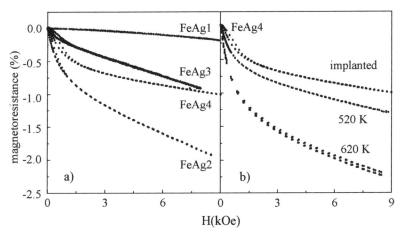

Fig. 3 - a) Magnetoresistance of the Fe-Ag films at T=10 K and fields up to 0.9 T; b) magnetoresistance of the FeAg4 film after annealing at 520 and 620 K.

205

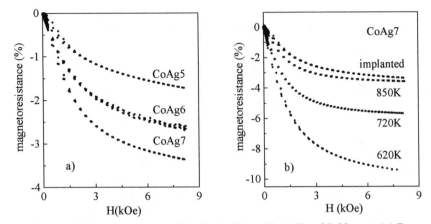

Fig. 4 - a) Magnetoresistance of the Co-Ag films at T=10 K and fields up to 0.9 T;
b) magnetoresistance of the CoAg7 film after annealing at 620, 720 and 850 K.

Fig. 3a) shows the MR at T=10 K and fields up to 0.9 T for all the Fe films in the as-implanted state. The increase of the MR response for the samples FeAg2, FeAg3 and FeAg4, relatively to the sample FeAg1, is attributed to the increase in the ferromagnetic behavior due to the higher implanted dose. The best result obtained for the MR is 2% at 10 K and 0.9 T field. Fig. 3b) shows the effect of annealing at 520 and 620 K, for 20 min on the MR curve for the sample FeAg4.

Fig. 4a) shows the MR for all Co films at T=10 K under fields up to 0.9 T. We observe an increase of the MR response which, as in the case of Fe, can be mainly attributed to the enhancement of the implanted fluence. Fig. 4b) shows the effect of thermal treatments at 620, 720 and 850 K, for 20 minutes, for the sample implanted with a fluence of $8x10^{16}$ at./cm^2. The MR amplitude increases with the annealing at 620 K, but then decreases for higher temperatures.

Fig. 5a) shows the maximum MR amplitude, $\Delta\rho/\rho$, as function of the annealing temperature, measured at 10 K and 290 K. Fig. 5b) shows the corresponding variation of the residual resistivity $\rho_0 = \rho$ (T=10 K) and the resistivity change with the magnetic field, $\Delta\rho(T) = \rho(T,H=0) - \rho(T, H=9$ kOe).

DISCUSSION

The RBS spectra taken in the films implanted with Fe and Co show that both elements are located mainly in the surface region, mostly due to the surface erosion caused by the sputtering of Ag. Due to this sputtering the thickness of the films decreased ~200 Å for each $1x10^{16}$ at./cm^2 implanted. The estimated error in the thickness of the implanted films due to the deviation from the bulk density of silver is about 70 Å for all films. The sputter coefficient for Fe and Co in Ag derived from the differences in the thicknesses before and after implantation is 12(1), much higher than the value predicted by TRIM. This can be due to the surface enrichment on Fe and Co due to the implantations. The retained fluence was in all cases about 80% of the nominal implanted one.

The VSM results show a ferromagnetic-like behavior at room temperature for the films implanted with fluences of about $8x10^{16}$ at/cm^2 of Fe and Co (films FeAg4 and CoAg7). The MR increases with the increase of the implanted fluence of Fe or Co. The MR curves at 10 K for

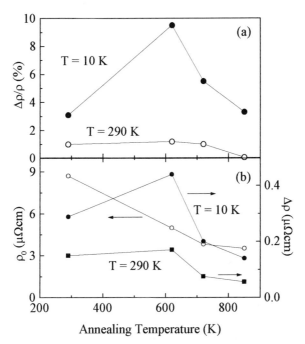

Fig. 5 - a) Magnetoresistance amplitude, $\Delta\rho/\rho$, as function of the annealing temperature, measured at 10 K and 300 K; b) corresponding variation of the residual resistivity $\rho_0 = \rho$ (10K) and the resistivity change with magnetic field, $\Delta\rho(T) = \rho(T, H = 0) - \rho(T, H = 9 \, kOe)$.

similar films show a MR of about ~1% for the Fe-Ag sample and ~3% for Co-Ag. Both MR curves exhibit a sharp decrease at low fields followed by a long tail at high fields. The initial resistivity decrease has the same physical origin as the GMR in magnetic multilayers and is associated with the rotation of the magnetization in the different grains towards the direction of the magnetic field. In constrast, the long MR tail at high fields can be associated with the slow ordering process of the magnetic moments of paramagnetic or superparamagnetic clusters towards saturation.

A clear improvement of the MR response was obtained after annealing. An increase was obtained after annealing at 520 K and 620 K, but then was followed by a decrease with further annealings at higher temperatures. This means that 620 K is close to the optimal annealing temperature for our implanted Co-Ag and Fe-Ag systems, i.e., it optimizes the Co (or Fe) grain size distribution and inter-cluster distances to enhance the GMR effect.

The MR amplitude measured at 10 K and 300 K as function of the annealing temperature shows a more pronounced variation at low temperatures than at room temperature, where it is rather insensitive to annealing. As could be expected the residual resistivity $\rho_0 = \rho(10\ K)$ decreases monotonically with the annealing temperature, due to the reduction of the number of defects during the thermal treatments. The measured residual resistivity values are quite low and reveal the presence of a relatively low content of magnetic material in the Ag matrix. The diference between the resistivity at room and low temperature, $\Delta\rho = \rho(300K) - \rho(10K)$, increased significantly with the annealing temperature as shown in Fig. 5b). Thus the temperature dependence of the MR amplitude seems to be mainly controlled by the magnetic contribution $\Delta\rho(T)$. Since this contribution depends on the scattering of conduction electrons by the ferromagnetic particles, it is sensitive to the size and number of the magnetic particles (Co, Fe) within the mean free path. We take these results together with the changes in the Fe profile shown in Fig. 1 as starting points for further work on the preparation by ion implantation of granular films for GMR applications. More data is also required in order to optimize the thickness of the films. In fact, one can expect that the true GMR amplitudes of our films are better than the values measured, since only part of the Ag film has Fe or Co grains and contributes to GMR.

CONCLUSIONS

Granular thin films prepared by high fluence Fe and Co implantation into Ag thin films (~2000Å) show a significant magnetoresistive response already after implantation. Annealing at 620 K optimizes the MR amplitude and increases its value by a factor of 3-4. The best value of the magnetoresistance obtained so far is 9% at 10 K and 1.5% at room temperature for a film implanted with Co at a fluence of 8×10^{16} at./cm^2 and annealed at 620 K. Further studies are underway to optimize the GMR effect.

ACKNOWLEDGMENTS

This work was supported by the PRAXIS XXI Programme, Portugal, through project PRAXIS/3/3.1/FIS/21/94 and individual grants (C.M.J., J.G.M. and M.M.P.A.). We gratefully acknowledge P.P. Freitas (INESC, Lisboa) for the use of the VSM setup and for valuable discussions.

REFERENCES

1. M.N. Baibich, J.M. Broto, A. Fert, F. Nguyen Van Dau, F. Petroff, P. Etienne, G. Creuzet, A. Friedrich, and J. Chazelas, Phys. Rev. Lett. **61**, 2472 (1988).
2. A.E. Berkowitz, J.R. Mitchell, M.J. Carey, A.P. Young, S. Zhang, F.E. Spada, F.T. Parker, A. Hutten, and G. Thomas, Phys. Rev. Lett. **68**, 3745 (1992).
3. J.X. Xiao, J.S. Jiang, C.L. Chien, Phys. Rev. Lett. **68**, 3749 (1992).
4. M.S. Rogalski, M.M. Pereira de Azevedo, and J.B. Sousa, J. Magn. Magn. Mater. **163**, L256 (1996).
5. B. Dieny, S.R. Teixeira, B. Rodmacq, C. Cowache, S. Aufret, O. Redon, and J. Pierre, J. Magn. Magn. Mater. **130**, 197 (1994).
6. M.M. Pereira de Azevedo, J.A. Mendes, M.S. Rogalski, J.B. Sousa, L.M. Redondo, C.M. de Jesus, J.G. Marques, M.F. da Silva, and J.C. Soares, J. Magn. Magn. Mater. **173**, 230 (1997).
7. J.F. Ziegler, J.P. Biersack, and U. Littmark, The Stopping and Range of Ions in Solids, (Pergamon Press, New York, 1985).

MAGNETOSTRICTION RELATED EFFECTS IN PULSED LASER IRRADIATION OF AMORPHOUS MAGNETS

MONICA SORESCU*, S.A. SCHAFER**
*Duquesne University, Physics Department, Pittsburgh, Pennsylvania 15282-1503
**Oklahoma State University, Physics Department, Stillwater, Oklahoma 74078-0444

ABSTRACT

Samples of $Fe_{66}Co_{18}B_{15}Si_1$, $Fe_{40}Ni_{38}Mo_4B_{18}$ and $Fe_{72.6}Cr_{22}Al_{4.8}Si_{0.3}Y_{0.3}$ metallic glasses were irradiated with a pulsed alexandrite laser (λ=750 nm, τ=60 μs) using different laser fluences. Irradiation-induced structural and property modifications were characterized using Mössbauer spectroscopy. Complementary information was obtained from hysteresis loop measurements and scanning electron microscopy (SEM). The experimental data obtained demonstrate the key role played by the magnetostriction constant in explaining the mechanism of irradiation-induced phase transformations in amorphous magnets.

INTRODUCTION

We have recently proposed a complex methodological approach for the purpose of understanding the evolution of phases and microstructure in amorphous and nanocrystalline magnetic materials. The method relies on the combined use of pulsed laser irradiation and Mössbauer spectroscopy to study unconventional aspects of materials responses by selectively inducing localized changes in the system under investigation. Through this unexpected marriage of pulsed laser processing and Mössbauer spectroscopic techniques, fundamental effects underlying the interaction of excimer laser radiation (λ=308 nm, τ=10 ns) with Fe-based and FeNi-based amorphous metals, such as excimer laser induced magnetic anisotropy,[1,2] excimer laser induced crystallization,[3] and excimer laser induced amorphization of thermally annealed systems,[4,5] were successfully demonstrated.

In the present work we report new investigations on the fundamentals of pulsed laser interaction with glassy ferromagnets by considering laser sources with longer wavelengths and pulse widths. The kinetics of the amorphous-to-crystalline transformation in iron-based glassy ferromagnets is studied as a function of laser processing parameters and the composition of ferromagnetic alloy systems.

EXPERIMENTAL

Amorphous alloys $Fe_{66}Co_{18}B_{15}Si_1$, $Fe_{40}Ni_{38}Mo_4B_{18}$ and $Fe_{72.6}Cr_{22}Al_{4.8}Si_{0.3}Y_{0.3}$ were supplied by Allied Signal Inc. and Goodfellow Corporation in the form of 20 μm thick ribbons. The magnetostriction constants had the values 35, 12 and 0 ppm, respectively. Square samples (2x2 cm) were cut from the foils and exposed on the shiny side to the λ=750 nm radiation generated by a pulsed alexandrite laser (model PAL 101), with the pulse width τ=60 μs. The samples were irradiated for 15 s/spot at a repetition rate of 20 Hz, using different laser fluences (Φ=2.8; 4.1; 4.9; and 5.2 J/cm^2). An acceptable degree of homogeneity was obtained by laser-beam scanning of the sample surface, which was placed on an x-y-z micrometer translation stage.

Room-temperature transmission Mössbauer spectra were recorded with the γ rays perpendicular to the ribbon plane using a constant acceleration spectrometer (Ranger Scientific). Hysteresis loops were recorded with a susceptometer-magnetometer system at 4.2 K in a magnetic field of 1.5 kOe, applied parallel to the sample surface. SEM investigations were carried out without further surface preparation using a JEOL electron microscope at 25 keV, operating in the secondary electron emission mode.

Mat. Res. Soc. Symp. Proc. Vol. 504 © 1998 Materials Research Society

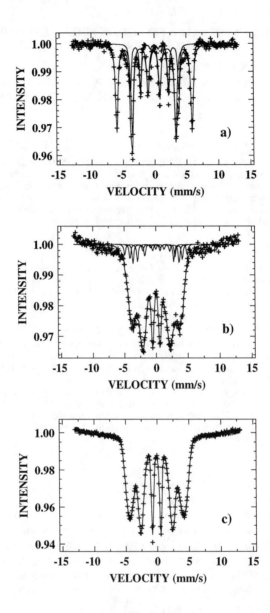

Fig. 1. Room temperature transmission Mössbauer spectra of (a) $Fe_{66}Co_{18}B_{15}Si_1$, (b) $Fe_{40}Ni_{38}Mo_4B_{18}$ and (c) $Fe_{72.6}Cr_{22}Al_{4.8}Si_{0.3}Y_{0.3}$ alloys, after pulsed alexandrite laser irradiation at a fluence of 5.2 J/cm².

RESULTS AND DISCUSSION

Room temperature transmission Mössbauer spectra of the $Fe_{66}Co_{18}B_{15}Si_1$, $Fe_{40}Ni_{38}Mo_4B_{18}$ and $Fe_{72.6}Cr_{22}Al_{4.8}Si_{0.3}Y_{0.3}$ samples after pulsed alexandrite laser irradiation at a fluence of 5.2 J/cm² are shown in Fig. 1 (a)-(c), respectively. The fitted values of the hyperfine parameters are listed in Table I.

TABLE I. Hyperfine magnetic field H_{hf}, isomer shift δ (relative to α-Fe at 300 K), and relative areas corresponding to the component patterns in the Mössbauer spectra of iron-based metallic glasses exposed to pulsed alexandrite laser processing.

Sample	λ_s (ppm)	H_{hf} (kOe)	δ (mm/s)	Rel. areas (%)	Assignment of sites
$Fe_{66}Co_{18}B_{15}Si_1$	35	366.7	0.05	75.3	α-(FeCo)
		232.5	0.09	24.7	$(FeCo)_3(BSi)$
$Fe_{40}Ni_{38}Mo_4B_{18}$	12	305.0	0.04	4.0	t-$(FeNi)_3B$
		255.0	0.55	4.0	t-$(FeNi)_3B$
		225.0	0.11	4.0	t-$(FeNi)_3B$
		203.9	0.02	88.0	amorphous
$Fe_{72.6}Cr_{22}Al_{4.8}Si_{0.3}Y_{0.3}$	0	248.1	0.10	100.0	amorphous

On the grounds of the values obtained for room-temperature Mössbauer parameters, the first two sextets in Table I can be assigned to α-(FeCo) and $(FeCo)_3(BSi)$ crystalline phases, respectively. This identification is in agreement with results of thermal annealing studies of Fe-Co alloys. Consequently, pulsed alexandrite laser irradiation of the $Fe_{66}Co_{18}B_{15}Si_1$ amorphous alloy at the highest fluence employed resulted in complete crystallization of the laser-exposed, highly magnetostrictive specimen.

The next three sextets in Table I can be assigned to the three inequivalent sites of t-$(FeNi)_3B$. This identification is in agreement with results on the phase transformation of $Fe_{62}Ni_{16}B_{14}Si_8$ amorphous system by isothermal annealing.[6] The previously reported study indicates a metastable equilibrium of the amorphous phase with (FeNi), (FeNi)Si and t-$(FeNi)_3B$, followed by the decomposition of t-$(FeNi)_3B$ into t-$(FeNi)_2B$ and (FeNi). In the present pulsed laser irradiation study of $Fe_{40}Ni_{38}Mo_4B_{18}$ metallic glass, the t-$(FeNi)_3B$ crystalline phase is in equilibrium with a dominant amorphous component. Consequently, pulsed alexandrite laser processing of the $Fe_{40}Ni_{38}Mo_4B_{18}$ amorphous alloy resulted in the onset of the amorphous-to-crystalline phase transformation in the irradiated system.

It can be seen in Fig. 1(c) and Table I, however, that pulsed alexandrite laser irradiation of the $Fe_{72.6}Cr_{22}Al_{4.8}Si_{0.3}Y_{0.3}$ alloy preserved the amorphous nature of the laser-exposed specimen. Thus, the phenomenon of laser-induced crystallization is not observed in the zero-magnetostrictive system. It is concluded that, for the same values of the irradiation parameters, the onset of irradiation-driven phase transformation is determined by the magnetostriction constant of the material.

Figure 2 shows the hysteresis loop measurements performed at 4.2 K in a parallel applied magnetic field of 1.5 kOe for the $Fe_{72.6}Cr_{22}Al_{4.8}Si_{0.3}Y_{0.3}$ system, in the amorphous as-quenched state and after pulsed alexandrite laser irradiation at fluences of 2.8, 4.1, 4.9 and 5.2 J/cm², respectively. It can be seen that the saturation magnetic moment and coercive field are controlled by the values of the irradiation parameters. Similar measurements recorded on the $Fe_{66}Co_{18}B_{15}Si_1$ alloy system showed that the hysteresis phenomenon increased in the laser-exposed samples, as compared to the amorphous specimen.

$Fe_{72.6}Cr_{22}Al_{4.8}Si_{0.3}Y_{0.3}$

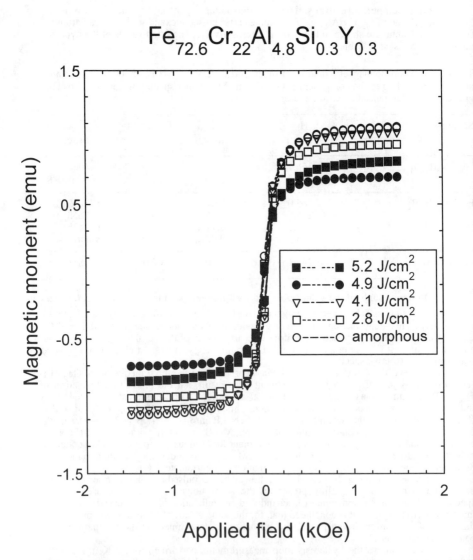

Fig. 2. Hysteresis loops recorded at 4.2 K in an applied magnetic field of 1.5 kOe for the $Fe_{72.6}Cr_{22}Al_{4.8}Si_{0.3}Y_{0.3}$ alloy, in the amorphous as-quenched state and after pulsed alexandrite laser processing.

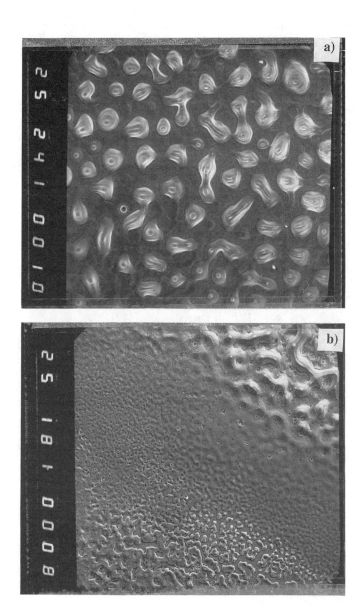

Fig. 3. SEM micrographs of the $Fe_{72.6}Cr_{22}Al_{4.8}Si_{0.3}Y_{0.3}$ system, after pulsed alexandrite laser irradiation at 5.2 J/cm^2. The magnification employed had the values: (a) x240; (b) x180.

Figure 3 (a)-(b) shows the SEM micrographs of the $Fe_{72.6}Cr_{22}Al_{4.8}Si_{0.3}Y_{0.3}$ alloy, after pulsed alexandrite laser irradiation at the highest fluence employed. The surface morphological modifications consist of molten zones, which subsequently resolidified preserving the amorphous nature of the specimen.

CONCLUSIONS

This work reports on the kinetics of laser-induced phase transformations in several Fe-, FeNi-, and FeCo-based amorphous ferromagnetic alloys with magnetostriction constants λ_s ranging from zero to 35 ppm. For the same values of the irradiation parameters, the onset of bulk crystallization is determined by the magnetostriction constant of the material. Laser-induced crystallization was not observed in the zero-magnetostrictive sample. The experimental data obtained in this work demonstrates the key role played by λ_s in explaining the mechanism of irradiation-induced phase transformations in amorphous magnets.

ACKNOWLEDGMENT

This work was supported by a Cottrell College Science Award of Research Corporation.

REFERENCES

1. D. Sorescu, M. Sorescu, I.N. Mihailescu, D. Barb, and A. Hening, Solid State Commun. **85**, 717 (1993).
2. M. Sorescu and E.T. Knobbe, J. Mater. Res. **8**, 3078 (1993).
3. M. Sorescu and E.T. Knobbe, Phys. Rev. B**49**, 3253 (1994).
4. M. Sorescu, E.T. Knobbe, and D. Barb, Phys. Rev. B**51**, 840 (1995).
5. M. Sorescu and E.T. Knobbe, Phys. Rev. B**52**, 16086 (1995).
6. D. Barb, I. Bibicu, M.S. Rogalski, and M. Sorescu, Hyperfine Interact. **94**, 2187 (1994).

INFLUENCE OF ION IRRADIATION ON THE MICROSTRUCTURE OF Fe/Ti MULTILAYERS

M. KOPCEWICZ, A. GRABIAS, J. JAGIELSKI, T. STOBIECKI *
Institute of Electronic Materials Technology, Wólczyńska 133, 01-919 Warsaw, Poland
*Dept. of Electronics, University of Mining and Metallurgy, Al. Mickiewicza 30, Cracow, Poland

ABSTRACT

The amorphization of the Fe/Ti multilayers due to ion-beam mixing induced by Ar and Kr ions is studied by conversion electron Mössbauer spectroscopy. Formation of the bcc-FeTi solid solution and the amorphous FeTi phase is studied as a function of ion dose for samples with the Fe to Ti thickness ratio of 1 and the modulation wavelengths of 20 and 60 nm. After reaching maximum abundance the amorphous fraction decreases at high ion doses.

INTRODUCTION

The Fe/Ti multilayer system (ML) is an interesting composition-modulated material whose microstructure and magnetic properties are governed by the thickness of the individual elemental layers. In equilibrium conditions the bulk FeTi system exists either as crystalline intermetallic compounds (FeTi and Fe_2Ti) and bcc-FeTi alloy or as an amorphous phase. The amorphous FeTi phase is fairly difficult to obtain. The most common method of rapid quenching from the melt is unable to provide the amorphous FeTi state. Amorphous FeTi phase can be formed by vapour quenching [1] and sputtering techniques [2,3]. Amorphization of the crystalline Fe/Ti multilayers by ion-beam mixing with Xe ions of the energy of 800 keV was also demonstrated [4]. However, the possibility of the formation of the amorphous phase by the latter method depends strongly on the purity of the Fe and Ti layers and may be prohibited by carbon and oxygen contamination [5]. Recently the structural and magnetic properties of Fe/Ti ML prepared by rf sputtering deposition were studied for Ti to Fe thickness ratios $\beta=d_{Ti}/d_{Fe}=1$ and 1.5 and with modulation wavelength $\Lambda=d_{Ti}+d_{Fe}$ ranging from 5 to 80 nm [3]. It was shown that for $\Lambda=5$ nm the as-deposited Fe/Ti ML were almost entirely amorphous.

It is relatively easy to distinguish by Mössbauer spectroscopy various crystalline FeTi phases from the amorphous phase thanks to the clearly different hyperfine parameters, such as the quadrupole splitting (QS), isomer shift (IS) and the magnetic hyperfine field (H_{hf}). The bcc-FeTi alloy is ferromagnetic at room temperature unlike the intermetallic compounds (FeTi and Fe_2Ti) and the amorphous phase which are paramagnetic [1]. However, the QS and IS values characteristic for the amorphous phase are clearly different from those of FeTi and Fe_2Ti [1,3].

In the present study we analyzed the structural transformations induced by Ar- and Kr-ion irradiation in Fe/Ti ML with $\Lambda=20$ and 60 nm and $\beta=1$. In particular we followed by conversion electron Mössbauer spectroscopy (CEMS) the formation of the bcc-FeTi solid solution and amorphous FeTi phase due to ion-beam mixing. The results obtained for Fe/Ti ML differ considerably from those obtained recently for the corresponding Fe/Zr ML in which ion-beam mixing induced complete amorphization [6].

EXPERIMENT

Multilayer Fe/Ti films were prepared by the rf sputtering deposition by using Fe and Ti targets at a base pressure of 1×10^{-6} Pa. Samples with the Ti to Fe thickness ratio $\beta=d_{Ti}/d_{Fe}=1$ (nominal composition $Fe_{0.6}Ti_{0.4}$) and the elemental thicknesses of d_{Fe} and d_{Ti} of 10 and 30 nm were prepared. The modulation wavelengths Λ were 20 and 60 nm, respectively. The thickness of elemental layers was measured during deposition with a quartz sensor. Total thickness of the films deposited on Corning glass was about 240 nm.

215

In order to induce ion-beam mixing the multilayers were irradiated with 200 keV Ar^{++} or 350 keV Kr^{++} ions with doses (Φ) ranging from 5×10^{13} to 7×10^{16} Ar/cm^2 and $\Phi = 5\times10^{13} \div 1.6\times10^{16}$ Kr/cm^2. The ranges of Ar and Kr ions, calculated by TRIM [7], were about 105 and 95 nm, respectively. The beam current density was less than 1 $\mu A/cm^2$ to avoid extensive heating and keep the target temperature to less than about 50°C.

After each irradiation step the samples were analyzed by conversion electron Mössbauer spectroscopy (CEMS). The CEMS measurements were performed at room temperature by using a He-6%CH_4 gas-flow electron counter. A conventional Mössbauer spectrometer with a ^{57}Co-in-Rh source was used. The isomer shift data are given with respect to the α-Fe standard.

RESULTS

As-deposited Fe/Ti multilayers

It was shown recently [3], that the as-deposited Fe/Ti ML with Λ=20 and 60 nm prepared by rf sputtering consist of predominantly crystalline close packed bcc (110) planes of Fe and (100) hcp Ti. The plane spacings were about 0.2028 nm and 0.2558 nm for Fe and Ti, respectively. The CEMS spectrum of the as-deposited film with Λ=20 nm consists of 3 spectral components (Fig. 1a): (1) a sextet with $H_{hf}\cong$32.9T and IS=0.00 mm/s corresponding to the α-Fe layers, (2) a broad sextet with $H_{hf}\cong$29T and IS=0.00 mm/s corresponding to the interfacial crystalline region, most probably bcc-FeTi, and (3) a quadrupole doublet with QS$\cong$0.40 mm/s and IS$\cong$ -0.09 mm/s corresponding to the amorphous iron-poor FeTi phase [3]. The Λ=60 nm sample is almost completely crystalline; the CEMS spectrum consists of the dominating α-Fe sextet and the QS doublet does not exceed 2% of the total spectral area (Fig. 1g).

Irradiations with Ar and Kr ions

The dose dependence of the ion-beam mixing was studied for MLs with Λ=20 and 60 nm irradiated with Ar^{++} ions with doses up to 7×10^{16} Ar/cm^2 (Fig. 1). The relative fractions of various phases detected in CEMS spectra vs. $\Phi^{1/2}$ are shown in Fig. 2. The CEMS results for Λ=20 nm (Figs. 1b-1f) reveal three distinct ranges of ion dose in which different phases dominate: (i) at low ion doses (up to about 1.3×10^{15} Ar/cm^2) the sextet corresponding to α-Fe, which dominated in the as-deposited ML (Fig. 1a), decreases markedly with increasing ion dose and the phase formed due to ion-beam mixing is the bcc-FeTi solid solution identified by its characteristic broad sextet ($H_{hf}\cong$29T, IS$\cong$0.00 mm/s, Fig. 1b). (ii) At intermediate ion doses ($1.5\div8\times10^{15}$ Ar/cm^2) the bcc-FeTi component dominates strongly in the spectra (Figs. 1c, 1d) which were fitted with the combination of the distribution of the hyperfine fields P(H) and the α-Fe sextet. The P(H) distributions were calculated by using a constrained Hesse-Rübartsch method [8,9] with the NORMOS program [10]. The contribution of the α-Fe sextet decreases further with increasing ion dose and for ML irradiated with 8×10^{15} Ar/cm^2 (Fig. 1d) it contributes to only 3% of the total spectral area (Fig. 2a). In this dose range no other phases were detected. (iii) At high ion doses ($1.5\div6\times10^{16}$ Ar/cm^2) a new phase was formed as evidenced in the CEMS spectra by the appearance of the QS doublet. The parameters of the doublet (QS$\cong$0.31 mm/s, IS$\cong$-0.20 mm/s) allowed the identification of this phase as the amorphous FeTi (Figs. 1e, 1f). The spectral contribution of the QS doublet increases rapidly to 40% of the total spectral area for 2.5×10^{16} Ar/cm^2, at the expense of the sextet due to the bcc-FeTi solid solution, and then starts to decrease at doses exceeding $4\times10^{16}Ar/cm^2$ to about 26% at $6.5\times10^{16}Ar/cm^2$ leading to a partial restoration of the bcc-FeTi sextet (Fig. 1f, Fig. 2a).

A similar behaviour was observed for Fe/Ti ML with Λ=60 nm (Figs. 1h - 1l). Considerably higher ion doses are required for a substantial transformation of the 30 nm Fe layers into the

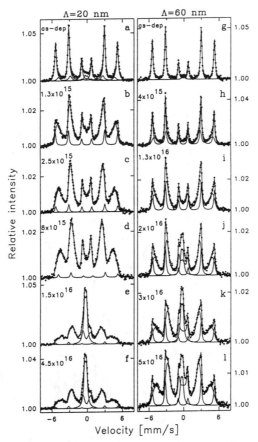

Fig. 1. CEMS spectra measured as a function of Ar-ion dose for Fe/Ti multilayers with Λ=20 nm (a-f) and 60 nm (g-l).

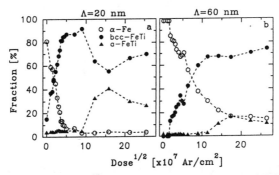

Fig. 2. Phase abundance vs. (dose)$^{1/2}$ determined for Ar-ion-beam mixed Fe/Ti multilayers with Λ=20 nm and 60 nm.

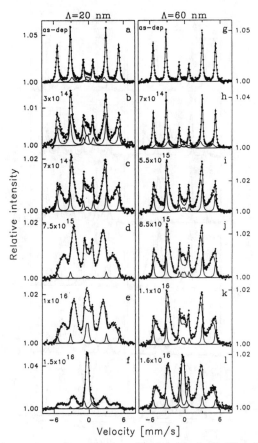

Fig. 3. CEMS spectra measured as a function of Kr-ion dose for Fe/Ti multilayers with Λ=20 nm (a-f) and 60 nm (g-l).

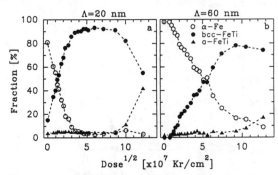

Fig. 4. Phase abundance vs. (dose)$^{1/2}$ determined for Kr-ion-beam mixed Fe/Ti multilayers with Λ=20 and 60 nm.

bcc-FeTi solid solution. As can be seen from Fig. 1i, even after irradiation with 1.3×10^{16} Ar/cm^2 the α-Fe sextet still contributes to about 30% of the total spectral area. The QS doublet characteristic for the amorphous FeTi phase appears when the relative fraction of the bcc-FeTi phase reaches saturation (Figs. 1j, 1k, Fig. 2b). The difference between the ML with Λ=20 nm (Fig. 2a) and Λ=60 nm (Fig. 2b) is that the amorphous phase appears in ML with Λ=20 nm only after the entire Fe-layer was consumed for the formation of the bcc-FeTi phase (Figs. 1d, 1e, Fig. 2a) whereas in ML with Λ=60 nm the Fe-layer is still substantial. Even at high Ar-ion doses (3×10^{16} Ar/cm^2) the QS doublet due to the amorphous phase contributes only to 17% of the total spectral area and with further increase of the ion dose it starts to decrease (to 13% at 5×10^{16} Ar/cm^2, Fig. 1l, Fig. 2b) similarly to the case of the ML with Λ=20 nm.

The increase of the relative fraction of the bcc-FeTi phase in the ML with Λ=20 nm is almost linear vs. $\Phi^{1/2}$ in the dose range of 1×10^{14} to 5×10^{15} Ar/cm^2 (Fig. 2a). Only after saturation of the bcc-FeTi fraction is reached at about 97%, the amorphous FeTi phase is rapidly formed at the expense of the bcc-FeTi phase. In the ML with Λ=60 nm (Fig. 2b) the bcc-FeTi fraction also increases linearly vs. $\Phi^{1/2}$ in the dose range up to 8×10^{15} Ar/cm^2 and then almost saturates at about 70%. A significant fraction of the amorphous phase appears at 2×10^{16} Ar/cm^2 and reaches its maximum of 17% at 3×10^{16} Ar/cm^2 (Fig. 2b).

The dose dependence of the ion-beam mixing induced by Kr ions is shown in Fig. 3 for the Fe/Ti ML with Λ=20 nm and Λ=60 nm and the relative fractions of various phases vs. $\Phi^{1/2}$ are presented in Fig. 4. As can be seen, the results obtained for Kr-irradiated ML are qualitatively very similar to those obtained by Ar-irradiations (Figs. 1 and 2). At low Kr-doses the spectral contribution due to the original α-Fe sextet decreases with increasing ion dose (Figs. 3b, 3c) and the dominating mixed phase is the bcc-FeTi solid solution. When the α-Fe layer in the ML with Λ=20 nm is almost completely converted into the bcc-FeTi phase (Fig. 3d) saturation of the relative fraction of the bcc-FeTi phase is reached at the dose of about 1.2×10^{15} Kr/cm^2 (Fig. 4a). The shape of the CEMS spectra remains virtually unchanged for $\Phi \leq 7.5 \times 10^{15}$ Kr/cm^2 (Fig. 3d) - the spectra consist almost exclusively of the broad sextet due to a broad P(H) distribution characteristic for the bcc-FeTi solid solution. The residual α-Fe component does not exceed 4% of the total spectral area (Fig. 4a). The amorphous phase appears clearly at $\Phi \approx 1 \times 10^{16}$ Kr/cm^2 (Fig. 3e). Its spectral contribution increases to about 40% at 1.5×10^{16} Kr/cm^2 (Fig. 3f, Fig. 4a). In the ML with Λ=60 nm considerably higher doses are required for the formation of the substantial fraction of the bcc-FeTi phase. E.g., in order to obtain 50% relative fraction of this phase in the Λ=60 nm ML the dose of about 4×10^{15} Kr/cm^2 is required whereas in the ML with Λ=20 nm the same bcc-FeTi fraction is obtained at already 2×10^{14} Kr/cm^2.

Similarly to the case of Ar-irradiations the relative fraction of the mixed bcc-FeTi phase is linear vs. $\Phi^{1/2}$ at doses up to 1.2×10^{15} Kr/cm^2 and 5.5×10^{15} Kr/cm^2 for ML with Λ=20 nm and Λ=60 nm, respectively (Fig. 4).

DISCUSSION

The structural transformations observed in the Fe/Ti ML due to Ar- and Kr-ion-beam mixing and the changes of the relative fractions of the relevant phases can be associated with the changes of the iron content in the mixed phases. As shown in ref. [4], the enthalpy curves for the amorphous FeTi and bcc-FeTi solid solution show three ranges for the stability of these phases depending on the Fe content, x: (1) a single glassy state for 0.36<x<0.72, (2) two phases bcc-FeTi and amorphous FeTi, can exist for 0.10<x<0.36, and (3) a single bcc-FeTi phase exists for x<0.10 and x>0.81. We observed in our study the formation of the bcc-FeTi phase in Λ=20 nm ML at low ion doses and its increase up to saturation at 8×10^{15} Ar/cm^2 is related to a complete consumption of the starting α-Fe layers by the formation of the bcc-FeTi solid solution. At ion doses exceeding 2.5×10^{15} Ar/cm^2 (saturation range, Fig. 2a) the iron content in the solid solution remains unchanged but further increase of the ion dose causes an increase of the lattice strain in the bcc structure. The amorphous phase appears rapidly when the strain becomes sufficiently large [11]. The formation of this phase releases the strain. However, further increase of ion dose causes a decrease of the amorphous fraction and partial restoration of the bcc-FeTi fraction (Fig. 2a). This effect is not fully understood at the moment. It may be related to a partial segregation of Ti from the bcc phase (e.g. due to the formation of TiC with carbon contamination) that would cause an increase of iron content in the solid solution sufficient for the transition from the iron range (2) in the direction of range (3).

From the relative fraction of the mixed phase it is possible to estimate the average thickness of the iron layer converted into the bcc-FeTi phase. The mixed layer thickness is $\Delta x = nd\Delta F/2$, where n is the number of monolayers of Fe per Fe-Ti bilayer, d=0.2028 nm (d spacing in the Fe layer), ΔF is the spectral area fraction of a given phase and the factor 2 accounts for two Fe-Ti interfaces per Fe layer [3,12]. The thickness of the Fe layer converted into the bcc-FeTi phase when it reaches its maximum abundance estimated in this way for ML with Λ=60 nm irradiated with 1.3×10^{16} Ar/cm^2 is about 10 nm. Similar estimation for Λ=20 nm shows that entire Fe layers are consumed for the formation of the bcc-FeTi phase. The 3% spectral contribution of the α-Fe component results most probably from deep Fe layers in ML film, beyond the range of Ar ions. Similar estimation for Λ=60 nm ML irradiated with 8.5×10^{15} Kr/cm^2 gives $\Delta x \cong 11.7$ nm.

A linear dependence of the relative fraction of the mixed bcc-FeTi phase vs. $\Phi^{1/2}$ at low Ar- and Kr-ion doses (Figs. 2 and 4) strongly suggests that the mass transport during mixing occurs due to "random walk" mechanism and that the chemical driving force plays a minor role.

It is worth noting that the relative fractions of the mixed phases (bcc-FeTi and amorphous FeTi) are considerably larger in ML with small Λ (Figs. 2 and 4) despite the fact that the nominal compositions of all our MLs are the same. Therefore not only the nominal composition of the ML structure determines the formation of a given phase, but the thickness of the elemental layers in the starting system is also important.

ACKNOWLEDGEMENT
The financial support from the Grant No. 2P 03B 098 08 from the Polish Committee of Scientific research is gratefully acknowledged. Sample preparation was supported from UMM 11.120.68.

REFERENCES
1. C.L. Chien, S.H. Liou, Phys. Rev. B **31**, 8238 (1985).
2. B. Rodmacq, J. Hillairet, J. Laugier, A. Chamberod, J. Phys.:Condens. Matter **2**, 95 (1990).
3. M. Kopcewicz, T. Stobiecki, M. Czapkiewicz, A. Grabias, J. Phys.:Condens. Matter **9**, 103 (1997).
4. R. Brenier, T. Capra, P. Thevenard, A. Perez, M. Treilleux, J. Rivory, J. Dupuy, G. Guiraud, Phys. Rev. B **41**,11784 (1990).
5. R. Brenier, A. Perez, P. Thevenard, T. Capra, Mater. Sci. Eng. **69**, 83 (1985).
6. M. Kopcewicz, J. Jagielski, T. Stobiecki, F. Stobiecki, G. Gawlik, J. Appl. Phys. **76**, 5232 (1994)
7. J.P. Biersack, L.G. Haggmark, Nucl. Instr. Meth. **174**, 257 (1980).
8. J. Hesse, A. Rübartsch, J. Phys. E **7**, 526 (1974).
9. G. LeCaer, J.M. Dubois, J. Phys. E **12**, 1083 (1979).
10. R.A. Brand, J. Lauer, D.M. Herlach, J. Phys. F **31**, 675 (1983).
11. G. Linker, Nucl. Instr. Meth. B **19/20**, 526 (1987).
12. D.L. Williamson, B. Clemens, Hyperfine Inter. **42**, 967 (1988).

ELECTRONIC EXCITATION EFFECTS ON DEFECT PRODUCTION AND RADIATION ANNEALING IN Fe IRRADIATED AT ~80K WITH ENERGETIC PARTICLES

Y. CHIMI, A. IWASE and N. ISHIKAWA
Advanced Science Research Center, Japan Atomic Energy Research Institute, Tokai-mura, Naka-gun, Ibaraki-ken 319-11, Japan, chimi@popsvr.tokai.jaeri.go.jp

ABSTRACT

Defect accumulation behavior in Fe irradiated with high energy (~100MeV) ions, low energy (~1MeV) ions and 2MeV electrons was studied by measuring the electrical resistivity change of the specimen at ~80K as a function of particle fluence. From the experimental results, the defect production cross-section, the defect annihilation cross-section and the damage efficiency (the ratio of the experimental defect production cross-section to the calculated one) were derived for each irradiation. By comparing the results for high energy ion-irradiations with those for low energy ion- and electron-irradiations, the dependence of defect production and radiation annealing on the electronic excitation is discussed.

INTRODUCTION

In the last decade, atomic displacements due to the high density electronic excitation have been found in several FCC metals irradiated with high energy (~100MeV) heavy ions [1]. Also, in the case of Fe which is one of the typical BCC metals, it has been observed that damage recombination and damage production are induced by the electronic excitation due to GeV-ion irradiations [2]. In the present work, we performed ~100MeV heavy ion irradiations in Fe at low temperature, and measured the electrical resistivity change as a function of ion fluence. By comparing the results with those for low energy (~1MeV) ion- and electron-irradiations, the effects of electronic excitation on radiation annealing and defect production in Fe could be observed clearly. The present results qualitatively agree with those for GeV-ion irradiations [2].

EXPERIMENTAL PROCEDURE

The specimens were polycrystalline Fe thin films about 190~370nm thick deposited on α-Al_2O_3 single crystal substrates by rf magnetron sputtering with an Fe target (99.99%) using Ar gas. The electrical resistivity of the specimen at room temperature was typically $11\mu\Omega \cdot cm$. These films were irradiated with the following particles; 0.5MeV ^1H, 1.0MeV ^4He, 1.0MeV ^{12}C, 1.0MeV ^{20}Ne, 2.0MeV ^{40}Ar, 120MeV ^{35}Cl, 150MeV ^{58}Ni, 125MeV ^{79}Br, 185MeV ^{127}I, 200MeV ^{127}I and 200MeV ^{197}Au ions and 2.0MeV electrons. As the projected ranges of these particles in Fe are much larger than the film thickness, most of the incident particles can pass through the specimen without remaining as impurities and the irradiation-produced defects are distributed uniformly in the specimen. In order to suppress the thermal motion of irradiation-produced defects in Fe, the irradiations were performed at liquid-N_2 temperature (~80K), because most of interstitials in Fe cannot move up to ~80K. During irradiation, the electrical resistivity of the specimen was measured at appropriate particle fluence intervals.

221

RESULTS

Figure 1(a) shows the resistivity change rate, $d(\Delta\rho)/d\Phi$, as a function of ion fluence, Φ, for 0.5MeV ^{1}H ion-irradiation. The logarithm of $d(\Delta\rho)/d\Phi$ decreases linearly with increasing Φ. This means that the irradiation-produced defects are annihilated during irradiation due to the interaction between defects and incident ions. The ion fluence dependence of $d(\Delta\rho)/d\Phi$, for the other low energy ion-irradiations is quite similar to that for 0.5MeV ^{1}H ion-irradiation. Therefore, the resistivity change rate for low energy ion-irradiations can be analyzed by using the following equation,

$$\ln\left\{\frac{d(\Delta\rho)}{d\Phi}\right\} = \ln\left\{\rho_F\frac{dC}{d\Phi}\right\} = \ln(\rho_F\sigma_d) - \sigma_r\Phi, \tag{1}$$

where $\Delta\rho$ is the resistivity change by irradiation, ρ_F is the resistivity increase per unit concentration of Frenkel pairs, σ_d is the defect production cross-section, σ_r is the defect annihilation cross-section, and C is the concentration of the irradiation-produced defects. In the present work, the value of ρ_F=1250$\mu\Omega\cdot$cm was used [3]. From the value at Φ=0 and the slope of $d(\Delta\rho)/d\Phi$ curve, σ_d and σ_r were determined, respectively. In the case of electron-irradiation, because of the small electron fluence, we could not observe the saturation of the resistivity change, and only the value of σ_d was obtained.

Figure 1(b) shows $d(\Delta\rho)/d\Phi$ as a function of Φ for 200MeV ^{197}Au ion-irradiation. For high energy (~100MeV) ion-irradiations, in contrast with low energy (~1MeV) ion-irradiations, the resistivity change rate cannot be described by eq. (1), but by the following equation,

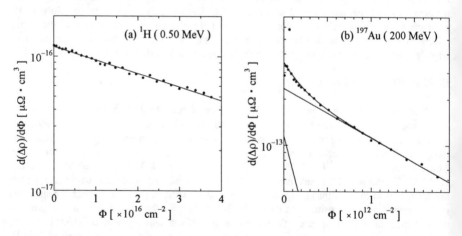

Fig. 1. Electrical resistivity change rate, $d(\Delta\rho)/d\Phi$, as a function of ion fluence, Φ, for (a) 0.5MeV ^{1}H and (b) 200MeV ^{197}Au ion-irradiations.

$$\frac{d(\Delta\rho)}{d\Phi} = \rho_F \sum_i \left\{ \sigma_{di} \exp\left(-\sigma_{ri}\Phi\right) \right\},\tag{2}$$

where it is assumed that the subscript i denotes the sort of defect stability against radiation annealing. The present analysis was performed in the region of Φ in which the defect production rate consists of two component ones (i=1 and 2). Assuming $\sigma_{r1} \gg \sigma_{r2}$, the values of σ_{d1}, σ_{d2}, σ_{r1} and σ_{r2} were derived by fitting the experimental $d(\Delta\rho)/d\Phi$ curve to eq. (2).

DISCUSSION

Damage Efficiency

The defect production cross-section, σ_d, means the concentration of defects produced by the unit fluence of incident particles. The damage efficiency, ξ, is defined as $\xi = \sigma_d^{exp}/\sigma_d^{cal}$, where σ_d^{exp} is the experimental value of σ_d, and σ_d^{cal} is the value calculated by assuming that the defect production occurs only by the elastic interaction between incident particles and target atoms. The values of σ_d^{exp} for high energy ion-irradiations were obtained by summing all component of σ_{di} for each irradiation ($\sigma_d^{exp} = \sigma_{d1} + \sigma_{d2}$). The values of σ_d^{cal} for all ion-irradiations were calculated by using TRIM-92 [4]. For electron-irradiation, where the velocity was relativistic, the differential scattering cross-section analytically evaluated by McKinley and Feshbach [5] was used for the calculation of σ_d^{cal}.

In Fig. 2, the normalized damage efficiency, ξ/ξ_e, for each irradiation is plotted as a function of PKA median energy, $T_{1/2}$, where ξ_e is the damage efficiency for electron-irradiation and $T_{1/2}$ is characteristic of the PKA energy spectrum [6,7]. The value of ξ_e should be close to 1, because the incident electrons produce mainly simple Frenkel pairs and not cascade damage. The normalized damage efficiencies, ξ/ξ_es, for low energy ion-irradiations show the same tendency as previous works in several FCC metals [1,6,7]. It indicates that the elastic interactions are dominant for the defect production during low energy ion-irradiations. In the case of 125MeV ^{79}Br, 185MeV ^{127}I, 200MeV ^{127}I and 200MeV ^{197}Au ion-irradiations, the value of ξ/ξ_e is much larger than that for low energy ion-irradiation at the same $T_{1/2}$. It implies that the electronic excitation by these irradiations, which have comparatively large S_e above ~30MeV/(mg/cm^2), contributes a great deal to the defect production. A similar behavior was observed in GeV-ion irradiated Fe [2], where the damage efficiency increases abruptly above S_e~50MeV/(mg/cm^2).

Defect Annihilation Cross-sections

Figure 3 shows the defect annihilation cross-section, σ_r, for each irradiation as a function of nuclear stopping power, S_n, which corresponds to the energy transferred from an incident particle to target atoms through the elastic interactions per unit length of the ion path. For low energy ion-irradiation, σ_r is proportional to S_n. In addition, the recombination volumes obtained in a conventional manner [8] are the same order of magnitude as that for electron-irradiation [9], so that the defect annihilation for low energy ion-irradiation occurs mainly by the elastic interaction. Therefore, we can define the elastic part of the defect annihilation cross-section as $\sigma_r^{elastic} = a \cdot S_n$ (a=9.58×10^{-14}). In the case of high energy ion-irradiation, each value of σ_{ri} is one or two orders of

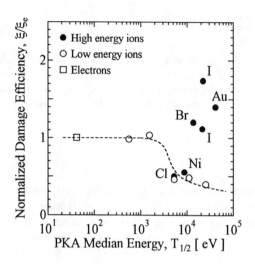

Fig. 2. Normalized damage efficiency, ξ/ξ_e, plotted against the PKA median energy, $T_{1/2}$, for each irradiation.

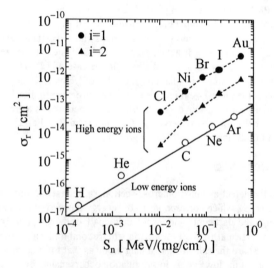

Fig. 3. Defect annihilation cross-section, σ_r, as a function of nuclear stopping power, S_n, for each ion-irradiation.

magnitude larger than that for low energy ion-irradiation at the same S_n. The value of $\left(\sigma_{ri} - \sigma_r^{elastic}\right)$ at the same S_n for each high energy ion-irradiation is plotted in Fig. 4 as a function of electronic stopping power, S_e, which corresponds to the transferred energy through the electronic excitation per unit length of the ion path. As can be seen in Fig. 4, $\left(\sigma_{ri} - \sigma_r^{elastic}\right)$ is strongly correlated with S_e, and this seems to be non-linear relation as $\left(\sigma_{ri} - \sigma_r^{elastic}\right) \sim S_e^n$, where the parameter n is larger than 1; ~2.9 for i=1 and ~3.6 for i=2. It suggests that the defect annihilation during high energy ion-irradiations, *i.e.*, the radiation annealing occurs mainly due to the electronic excitation. On the other hand, Dunlop et. al have obtained the recombination cross-section for heavy GeV-ion irradiated Fe in the region of S_e=60~85MeV/(mg/cm^2) [2]. The recombination cross-section increases with increasing S_e, and the values are 6×10^{-14}~5×10^{-13}cm^2.

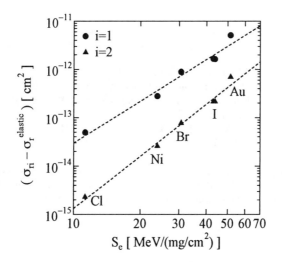

Fig. 4. Value of $\left(\sigma_{ri} - \sigma_r^{elastic}\right)$ as a function of electronic stopping power, S_e. The dotted lines show the relation of $\left(\sigma_{ri} - \sigma_r^{elastic}\right) \sim S_e^n$.

SUMMARY

Iron thin films were irradiated with high energy (~100MeV) ions, low energy (~1MeV) ions and 2MeV electrons. The defect accumulation behavior was obtained by measuring the resistivity change of the specimen at ~80K. The defect production cross-section, the damage efficiency and the defect annihilation cross-section were evaluated from the experimental results. By analyzing the resistivity change rate, it seems that defects which have two sorts of stability for the radiation annealing were produced by high energy ion-irradiation. The damage efficiencies for some high energy ion-irradiations which have comparatively large S_e are much larger than those for low energy ion-irradiations at the same $T_{1/2}$. It implies that the electronic excitation contributes to the defect production. Also, the defect annihilation cross-section for each high energy ion-irradiation is larger by one or two orders of magnitude than that for low energy ion-irradiation at the same S_n. The value of $\left(\sigma_{ri} - \sigma_r^{elastic}\right)$ is strongly correlated with S_e and, therefore, it suggests that the radiation annealing occurs mainly due to the electronic excitation. As regards both the defect production and the radiation annealing, the present results qualitatively agree with those for heavy GeV-ion irradiations [2].

ACKNOWLEDGMENTS

We are grateful to the technical staff of the Accelerators Division at JAERI(Japan Atomic Energy Research Institute)-Tokai and JAERI-Takasaki for their help.

REFERENCES

1. A.Iwase and T.Iwata, Nucl. Instr. and Meth. B **90**, 322 (1994); and references therein.
2. A.Dunlop, D.Lesueur, P.Legrand, H.Dammak and J.Dural, Nucl. Instr. and Meth. B **90**, 330 (1994).
3. P.G.Lucasson and R.M.Walker, Phys. Rev. **127**, 485 (1962).
4. J.P.Biersack and L.G.Haggmark, Nucl. Instr. and Meth. **174**, 257 (1980).
5. F.Seitz and J.S.Koehler, in Solid State Physics, edited by F.Seitz and D.Turnbull, Vol.2 (Academic Press Inc., Publishers, New York, 1956), p. 305.
6. R.S.Averback, R.Benedek and K.L.Merkle, Phys. Rev. B **18**, 4156 (1978).
7. R.S.Averback, R.Benedek, K.L.Merkle, J.Sprinkle and L.J.Thompson, J. Nucl. Mater. **113**, 211 (1983).
8. H.J.Wollenberger, in Vacancies and Interstitials in Metals, edited by A.Seeger, D.Schumacher, W.Schilling and J.Diehl (North-Holland Publishing Company, Amsterdam, 1970), p. 215.
9. J.Dural, J.Ardonceau and J.C.Jousset, J. Phys.(Paris) **38**, 1007 (1977).

IMPROVEMENT OF HYDRIDING PROPERTY IN ZrNi ALLOY BY MEANS OF IMPLANTATION

T. Suda*, T. Yonezawa*, H. Arashima*, S. Ohnuki*, T.Kabutomori**
*Hokkaido Univ, Dept. of Engineering, Sapporo 060 Japan
** Japan Steel Works, Muroran 050 Japan

ABSTRACT

Ion-implantation was applied to the surface modification of ZrNi for improving the surface activation of hydrogen absorbing alloys. At the condition of low temperature, only implanted areas showed micro-cracks and pulverization. At high temperature, the reaction was very rapid and both implanted and shielded areas showed the pulverization. The results indicate hydrogen ion-implantation can enhance the activation of this alloy, and the roles of radiation-induced implanted hydrogen and surface film are discussed.

INTRODUCTION

Hydrogen absorbing alloys have several material issues concerning the reaction properties[1]; reducing the activation temperature and pressure in the initiation of using, increasing the hydrogen storage capacity and preventing the pulverization. The initial activation properties are especially dependent on the materials. Fig. 1 shows typical pressure-composition-diagrams for the initial activation of the most common alloys, LaNi5,TiMn1.5 and ZrNi[2]. In these, ZrNi has several unique properties; the hydrogenation pressure is lower by two orders of magnitude than others, however, the initial reaction temperature is higher by 170 K than others. For improving the property of ZrNi, the most important issue is reducing the activation temperature. Unfortunately, no clear evidence has been reported for the difficulty of the activation, and the possible interpretation have been assumed to be a stable surface oxide, impurity absorption and so on. In this study, hydrogen ion-implantation has been applied for improving of activation process in ZrNi.

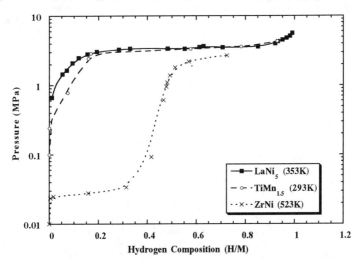

Fig. 1. Pressure-composition-diagram for initial activation of LaNi5,TiMn1.5 and ZrNi.

EXPERIMENT

The intermetallic compound ZrNi was melted in an arc furnace under an argon atmosphere[3][4]. The specimen of button shape was homogenized for 24 h at 1373 K under an argon atmosphere and then sectioned carefully to the sheets of about 500 μ m in thickness. Afterward, the sheet was polished to 100~200 μ m in thickness. A half area of the sheet was shielded with Al foil, followed by the implantation of 100 keV H$^+$ ions up to 3.6 x 10^{16} ions/cm^2 at room temperature. The specimen was encapsulated in a pressurized capsule and reacted with high pressure hydrogen gas at 353 K and 475 K. The pressurized capsule was connected with a miniature pressure sensor for checking hydrogenation. SEM observation was carried out for checking the pulverization. The process of hydrogenation and observation was repeated until completing the pulverization to the whole specimen.

Thin film specimens for TEM was carefully sectioned and polished to about 80 μ m in thickness, and then pasted to a supporting ring. Thinning was carried out in Gatan ion-milling machine at about 5 kV. Moreover, the specimen was covered with a Cu-grid, and implanted by 100keV H$^+$ ions up to 3.6 x 10^{16} ions/cm^2 at room temperature. The implanted area and shielded areas were observed by 200 kV TEM.

RESULTS AND DISCUSSION

Fig. 2 shows a typical result after the hydrogenation at 353 K for the specimen containing two areas; shielded and implanted to 3.37 x 10^{15}ions/cm^2. The shielded area did not change until 3.25 MPa, however, the implanted area turn to powder. In those cases, the critical pressure for the pulverization which is the proof of the surface activation, started from 2.2 - 2.7 MPa, where the enhancement of pulverization was not depend on the ion-dose which means the effect was probably saturated in this experimental condition. These simple data indicate that the implantation can enhance clearly the hydrogenation of this alloy.

Fig. 3 shows the critical pressure for completing pulverization at two different temperatures, as the function of ion dose. Each of the critical pressure data contained relatively large scattering, since those specimens had so small mass, typically 100 mg for a specimen. At lower temperature (353 K) , data points can be categorized into two levels and the ion-implantation enabled the pulverization at lower hydrogen pressure. The enhancement for the pulverization due to implantation was confirmed with a lower ion dose (3 x 10^{15} ions/cm^2) , however, the dose dependence of the critical pressure did not show clear decrease. At higher temperature (473 K) , both of the implanted and shielded areas showed the same level, and the pulverization was very rapid for both areas. The results mean that the temperature was enough high to activate the surface of the entire area. ZrNi alloy is generally hard to activate compared to other absorbing alloys, where the pulverization starts from 673 K for bulk specimens at common experimental conditions[5]. There was a difference in the specimen mass and shape, however. The result of this simple experiment succeeded to reduce the activation temperature by about 120 K, which means that H$^+$ ion-implantation can enhance the activation of this hydrogen absorbing alloy through surface modification.

In general, the ion-implantation can produce several structural changes not only on the surface but in sub-surface areas, where the maximum ranges calculated from TRIM code were 500 and 450 nm for hydrogen and defects. We can consider the structural changes could include the removing of the surface passive film, producing large amount of radiation-induced defects, amorphous structure and accumulation of hydrogen. All of those factors are possibly enable to activate surface reaction of ZrNi against hydrogen.

Fig. 4 shows typical dark field images and diffraction patterns from implanted and shielded areas in the same TEM specimen implanted to 3.6 x 10^{16} H$^+$/cm^2 at room temperature. In the shielded area, only small amounts of damage were detected and the surface showed a thin limited amorphous layer about 80 nm in thickness, which is probably caused during the Ar ion-milling. In the H$^+$ ion-implanted area, large amounts of defect clusters and intensive

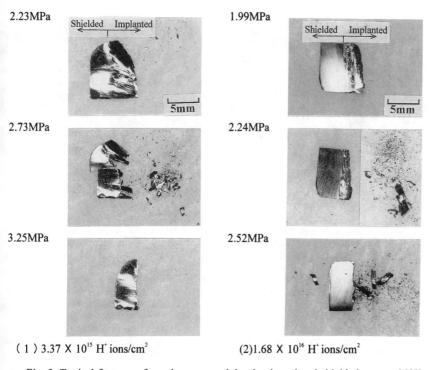

2.23MPa

Shielded   Implanted

5mm

2.73MPa

3.25MPa

1.99MPa

Shielded   Implanted

5mm

2.24MPa

2.52MPa

(1) 3.37 X 10^{15} H$^+$ ions/cm^2 (2)1.68 X 10^{16} H$^+$ ions/cm^2

Fig. 2. Typical features of specimens containing implanted and shielded area at 353K.

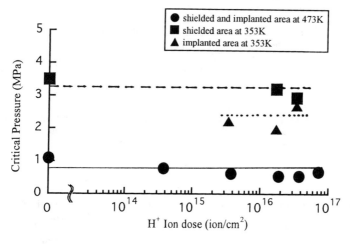

- ● shielded and implanted area at 473K
- ■ shielded area at 353K
- ▲ implanted area at 353K

Critical Pressure (MPa)

H$^+$ Ion dose (ion/cm^2)

Fig. 3. Critical pressure for completing pulverization in implanted and shielded specimens at two different temperature.

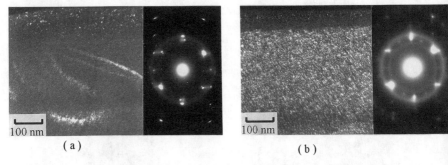

Fig. 4. Typical dark field images from { 111 } diffraction and diffraction patterns.
(a) shielded area, and (b) implanted area .

amorphous halo structure was observed. The observed area in the TEM specimen was not located in the maximum damaged depth, because the thickness of the observed area was less than 100 nm, however, the implantation can produce the mixture of damaged and amorphous zones in the observed area.

We can consider several reasons for enhancement of the pulverization of this specimen. The ion-sputtering could remove surface passive film and enhance the surface reaction directly. The large amount of radiation-induced defects and the amorphous structure can enhance the hydrogen diffusion from surface to inside. The accumulation of hydrogen at sub-surface area can be helpful for hydride formation. In this experiment, the most notable change of the implanted structure was clear damage and amorphous structures, which should have the important role for the enhancement of the pulverization of ZrNi.

CONCLUSION

The hydrogen ion-implantation has been applied to improve the initial activation of the hydrogen absorbing alloy, ZrNi.
(1) After the hydrogenation at 353 K, the implanted area showed a typical pulverization at lower critical hydrogen pressure comparing to shielded area. The result indicates that the implantation can enhance the surface reaction at low temperature.
(2) After the hydrogenation at 473 K, no difference can be clearly detected in both the implanted and shielded areas, which means that surface activation is predominate at this temperature.
(3) From those results, it can be proved that H^+ ion-implantation can reduce the activation temperature by about 120 K in limited experimental conditions.
(4) From TEM observations, radiation-induced defect clusters and amorphous structures were observed intensively at sub-surface areas. Such damaged structures can be the predominate factors for enhancing hydrogen absorbing activity.

REFERENCES

1. O. Bernauer, Int. J. Hydrogen Energy. **14**, 727-735 (1978) .
2. T. Suda, H. Arashima, T. Ynozewa, S. Ohnuki, T. Kabutomori, to be published.
3. D.G. Westlake, H. Shaked, P.R. Mason, B.R. Mcart and M.H. Mueller, J. Less-Common Met. **88**, 17-23 (1982) .
4. K. Watanabe, K. Tanaka, M. Matsuyama and K. Hasegawa, Fusion Eng. and Design. **18**, 27-3 (1991) .
5. T. Kabutomori, Doctoral Thesis, Hokkaidou Univ., (1997) .

THERMAL BEHAVIOUR OF IMPLANTED NITROGEN AND ACCUMULATED HYDROGEN IN TITANIUM

M. SOLTANI-FARSHI[a], H. BAUMANN[a], D. RÜCK[b], G. WALTER[c] AND K. BETHGE[a]

[a] Institute of Nuclear Physics, University of Frankfurt, August-Euler-Str. 6, 60486 Frankfurt, Germany, M.Soltani@gsi.de

[b] Center for Heavy Ion Research, Planckstr. 1, 64291 Darmstadt, Germany

[c] Institute of Nuclear Physics, University of Technology, Schloßgartenstr. 9, 64289 Darmstadt, Germany

ABSTRACT

The influence of nitrogen ion implantation on the hydrogen accumulation in titanium was investigated as function of sample temperature and ion fluence. 150 keV nitrogen (^{15}N) ions were implanted at different sample temperatures up to 700°C with fluences ranging from 1×10^{17} to 1×10^{18} ions/cm^2. The amount of accumulated hydrogen and its depth distribution was measured quantitatively with the ^{15}N depth profiling method. The implanted ^{15}N depth profiles were measured by the reverse reaction ^{15}N(p, $\alpha\gamma$)^{12}C at 429 keV. The binary phases of the implanted nitrogen with titanium are detected by grazing incidence x-ray diffraction. The results are compared with those obtained for samples implanted at RT and subsequently thermally treated.

INTRODUCTION

Titanium nitride is rather widely used as a coating material due to its physical properties, such as hardness, high melting point, good dielectric and heat conductivity. To prepare TiN coatings by nitrogen (N) implantation it is important to know both the absolute N concentration and the accurate N depth profile and its alteration by thermal treatment. Ti has a strong chemical affinity to hydrogen (H) and can absorb and store large amounts of it. It is well-established that H may be trapped at particular defect sites such as precipitates, grain boundaries or dislocations[1-4]. The H atoms diffuse to dislocations produced by the implanted N and is accumulated in the implanted region[5-8]. Furthermore N implantation affects the solubility and mobility of H in the surface and near surface region. The implanted N layer present an effective barrier to the H migration [7,8]. In application of the implantation technique heating up the sample caused by high ion current density (>50μA/cm^2) can be a serious problem. A sample temperature above 100°C may cause phase transformations and diffusion of N. Thus, it is desirable, on the one hand, to restrict the sample temperature during implantation to less than 100°C and, on the other hand, to know the influence of higher temperature on the depth profiles of N and accumulated H. To our knowledge few groups have focused their investigation on the influence of temperature during N implantation but only for Ti-alloys, iron, aluminium and stainless steel [9,10]. The thermal behaviour of accumulated H in Ti caused by N implantation was not investigated by theses groups. In this contribution we report the results of N implantation and H accumulation in Ti at sample temperatures between RT and 700°C during implantation. These results are compared with those obtained for samples implanted at RT and treated thermally up to 700°C afterwards.

EXPERIMENTAL

Sample preparation and implantation

Commercially available Ti (H impurity 0.3 at.%) was used as sample material. The surface of the samples (thickness 1 mm) was mechanically polished using 3 μm diamond paste in the final polishing step and treated for 10 sec with a mixture of HF and HNO_3 -acids (1:4) to guarantee a near surface region with less H concentration [11]. ^{15}N-ions were implanted with an energy of 150 keV. Ion fluences range from 1 x 10^{17} to 1 x 10^{18} ^{15}N-ions/cm^2 and the ion flux amounts to 20μA/cm^2. The samples were mounted on a water-cooled sample holder for room temperature (RT) implantation. Post-annealing experiments (PA) were carried out under high vacuum conditions at 700 °C for one hour. Samples implanted at high temperature (HT) were heated from the back side by a heating wire. The temperature was measured with a thermocouple in contact with the sample holder. The temperature was controlled with an accuracy of ±10 °C. The HT-implantations were carried out in the range of 300 to 700°C. During RT- and HT-implantation a total pressure of 6 x 10^{-7} mbar was maintained in the implantation chamber.

H determination by NRA

H in the samples was profiled using the resonant nuclear reaction $^{1}H(^{15}N, \alpha\gamma)^{12}C$ at E_{res}=6.396 MeV [12]. H depth distribution was obtained by measuring the yield of γ-rays (E_{γ} = 3 to 5 MeV) as a function of the ^{15}N ion energy. The emitted γ-rays were measured with two 8″x4″ NaI(Tl) detectors. No alteration of the H distribution caused by the analysing ^{15}N-beam was observed. NH_4Cl-standard was used for H calibration. The uncertainty of the measured H concentration is about 5%. Stopping power data were taken from [7,13].

N determination by NRA

^{15}N depth profiles of the implanted Ti were measured using the resonant nuclear reaction $^{15}N(p, \alpha\gamma)^{12}C$ (Γ_{res}=129 eV) at a proton energy of 429 keV [14]. The proton beam hit the target surface at normal incidence. The 4.432 MeV γ-rays emitted by the excited ^{12}C nucleus were detected with a 6 x 4 inch NaI(Tl) scintillation detector. A ^{15}N-implanted glassy carbon sample with a known ^{15}N-dose served as calibration standard. The N depth profiles were measured by varying the proton energy in steps of 1 keV. The uncertainty of the N concentration values in the samples was estimated to be less than 3%.

X-ray diffractometry

Phases formed by N implantation in Ti were characterised by GIXD with an equipment using a Θ-2Θ goniometer with an option for grazing incident angle and scintillation counter for peak detection. The radiation used was Cu Kα. The dispersion of the X-ray beam is limited to 0.3° by a diaphragm system and Soller slits. To obtain well-detected interference signals the measurements with a focused Bragg-Brentano configuration and fixed incident angle Θ=2° were done with a step scan of 0.02° and a step time of 5s. The penetration depth of the radiation (1/e intensity) ranged up to 380 nm. The peaks were identified with the peak-fitting program [15] and the powder diffraction file [16].

RESULTS

As reported in [5-8] N-implantation into Ti at RT leads to an enhancement of H concentration (up to 8 at.% for 3 x 10^{17} ions/cm^2) in the implanted layer with increasing N content. For N concentrations up to 25 at.% both the H and N profiles are nearly Gaussian with peak maxima lying in the same depth with respect to the sample surface. At N concentration above 25 at.% (fluence >3 x 10^{17} ions/cm^2) the H concentration decreases and a saddle-shaped H profile occurs

The two peaks are separated by a plateau with a H concentration less than 1 at.% at N concentration above 45 at.%. Fig. 1 shows the N and H depth profiles after N implantation at RT, 300, 400, 500, 600 and 700°C with a fluence of 6×10^{17} ions/cm². The H and N concentration values are drawn on the left and right Y-axis, respectively. Up to 500°C the N depth profiles show the same distribution with a maximum concentration of about 37.5 at.% at a depth of 300 nm. For sample temperatures over 600°C the implanted N starts to diffuse and the maximum concentration decreases (30 at.%). This behaviour is forced up to a temperature of 700°C and the N concentration is reduced to 20 at.%. The N atoms diffuse during implantation up to a depth of 1.4 μm. The dose of the implanted N ions is not altered at different temperatures. Contrary to that the amount of accumulated H decreases with increasing implantation temperature. The observed saddle-shaped H profile for RT implanted sample is almost vanished at 600°C and entirely disappears at 700°C. This behaviour of H was found for fluences of N ranging from 1×10^{17} to 1×10^{18} ions/cm².

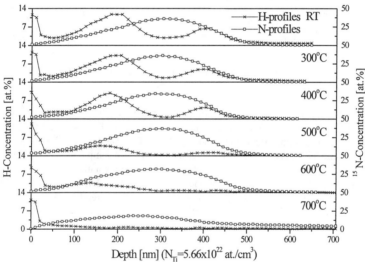

Fig.1 Depth profiles of N (6×10^{17} ions/cm²) and accumulated H in Ti for RT- and HT-implantation

Figure 2 shows the measured diffractograms of the Ti samples implanted at different sample temperatures with a N fluence of 6×10^{17} ions/cm². The 2Θ-region between 34 to 54° was chosen for phase identification because high Bragg angles lead to very low peak intensities for thin films [17].

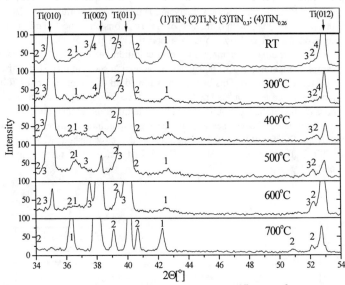

Fig. 2 Diffractograms of Ti samples implanted with N (6×10^{17} ions/cm^2) at different sample temperatures.

The observed Ti-N$_x$ phases are summarised in table I. The broadening of the Ti peaks at RT, 300, 400°C can be explained both by free N atoms in the Ti crystal and the formed phases which leads to a non-uniform lattice constant. For RT implanted samples the phases TiN$_{0.26}$, TiN$_{0.3}$, Ti$_2$N and TiN were found. The higher the fluence and sample temperature are the greater the fraction of Ti nitrides compounds. Increasing of the implantation temperature up to 700°C results in the greatest match with Ti$_2$N and TiN (over seven defined matches). This suggest some TiN$_{0.26}$ and TiN$_{0.3}$ decomposition. This result confirms the trend that HT implantation above 400°C exhibits a reduced percentages of TiN$_{0.26}$ and TiN$_{0.3}$ compounds. The phases of the 700°C implanted sample are shifted to lower 2Θ angle caused by oxygen diffusion at higher temperatures from the surface into the bulk.

Table I. TiN$_x$-Compounds formed by N implantation as shown in fig. 2

Temperature [°C]	Fluence [10^{17} ions/cm^2]	Phases indicated by GIXD
RT	<4	None
	>5	TiN$_{0.26}$(4), TiN$_{0.3}$(3), Ti$_2$N(2), TiN(1
300	>5	TiN$_{0.26}$, TiN$_{0.3}$, Ti$_2$N, TiN
400	>4.5	TiN$_{0.3}$, Ti$_2$N, TiN
500	>4	TiN$_{0.3}$, Ti$_2$N, TiN
600	>3.5	TiN$_{0.3}$, Ti$_2$N, TiN
700	>3	TiN$_{0.3}$, Ti$_2$N, TiN

N depth profiles of samples implanted with various fluences (1, 6 and 10 x 10^{17} ions/cm^2) at RT, post-annealed at 700°C and implanted at 700°C are compared in fig. 3. A strong N diffusion into the bulk is observed for post-annealed and HT implanted samples at a fluence of 1 x 10^{17} ions/cm^2. For HT implanted samples the maximum concentration is enhanced (up to 30 at.% for 1 x 10^{18} ions/cm^2) with increasing N fluence and this leads to a broadening of the N distribution in a depth range around the projected ion range.

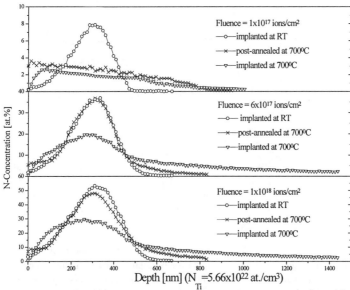

Fig. 3 N depth profiles of Ti samples implanted with various N fluences (1, 6 and 10 x 10^{17} ions/cm^2) at RT, post-annealed at 700°C and implanted at 700°C.

PA samples (6 and 10 x 10^{17} ions/cm^2) show a diffusion behaviour at the lower and upper tail of the N depth distribution. The maximum N concentration does not decreases for 6 x 10^{17} ions/cm^2 whereas for the fluence of 1 x 10^{18} ions/cm^2 the maximum N concentration decreases from 50 at.% (RT-implantation) to 47 at.% by diffusion. The accumulated H at RT implanted samples vanishes during post-annealing. When the samples are cooled down the H diffuses back to the N containing region. For PA samples the H content is three atomic percent higher than for HT-implanted ones.

GIXD measurements of samples implanted at RT, post-annealed at 700°C and implanted at 700°C carried out for a N fluence of 1 x 10^{18} ions/cm^2 are shown in figure 4. The observed TiN$_x$ phases formed for post-annealed and HT implanted samples do not exhibit marked differences. With higher N fluence the phases TiN$_{0.26}$, TiN$_{0.3}$, Ti$_2$N, TiN are formed for RT and PA specimens. Whereas at 700°C implanted samples well defined Ti$_2$N and TiN peaks appear.

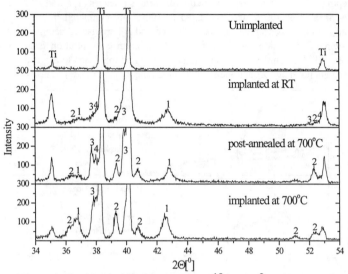

Fig. 4 GIXD of Ti samples implanted with a fluence of 1 x 10^{18} ions/cm^2) at RT, post-annealed at 700°C and implanted at 700°C.

DISCUSSION

H atoms diffuse to the dislocations or defects that were produced by N-implantation at RT. The content of accumulated H depends on the N fluence. The formation of a saddle-shaped H distribution with a concentration in the minimum of less than 3 at.% is attributed to the nitride phase formation [7,8]. Two effects are responsible for the lower H content in samples implanted at HT (between 300-700°C): Firstly, the defects produced by N implantation can be annealed during the HT implantation so that the defect production rate is reduced and as a consequence the accumulated H continuously decreases corresponding to increasing sample temperature. Secondly, the size diameter of the nitride precipitation influences the amount of the accumulated H. At RT implantation (fluence=3.5 x 10^{17} ions/cm^2) small f.c.c TiN precipitates with size diameter of 5-10 nm were observed in α-Ti [18-20]. The N implantation at higher temperatures produce enhanced growth of TiN and Ti$_2$N by diffusion of dissolved N into the precipitations (up to 20 nm) [18,21]. The GIXD results (fig. 2 and 4) of HT implanted and post-annealed samples show that with increasing implantation temperature the formation of TiN and Ti$_2$N phases is intensified. Although the maximum N concentration of the samples implanted at 700°C is 20 at.% lower than that of the PA ones, the diffractograms exhibit a greater percentages of TiN and Ti$_2$N compounds. In this study we found minor evidence for TiN$_{0.26}$, TiN$_{0.3}$ and Ti$_2$N formation by GIXD for low fluences and low temperatures (RT-400°C). It was observed that an increase in fluence up to 1 x 10^{18} ions/cm^2 at higher temperatures is followed by an increase of Ti$_2$N and TiN formation. On the other hand increasing the implantation temperature results in TiN$_{0.26}$ and TiN$_{0.3}$ dissolution and deeper N penetrations (fig.3, up to 1.4µm). These results qualitatively support the N mobility in Ti for temperatures above 600°C [19,22]. As shown in fig.3 the N atoms diffuse in PA samples implanted with 1 x 10^{17} ions/cm^2, but no diffusion in the N maximum concentration is observed at fluence of 6 x 10^{17} ions/cm^2. This effect can be explained by the diffusion behaviour of N in Ti and titanium nitrides: At 1×10^{17} ions/cm^2 (RT) no phases were observed by GIXD with the consequence that the implanted N is only dissolved. For 6×10^{17} ions/cm^2 (RT) the N with a maximum concentration of 37 at.% forms nitride phases which prevents the diffusion of N atoms. But at fluences above 9×10^{17} ions/cm^2 with a maximum concentration around 50 at.% (RT) a less N diffusion is observed once more. This behaviour is not yet understood; it might be related to the formation of N_2 bubbles.

CONCLUSIONS

H accumulation was observed in the implanted layer at RT. At N-concentration above 20 at.% the H content decreases with increasing formation of nitride precipitations depending on the ion fluence and the sample temperature during implantation and post-annealing. Titanium nitrides were observed for high temperature implantation in the range between 300°C to 700°C depending on the N fluence. High temperature implantation (>600°C) of N into Ti produces enhanced penetration depths (>1μm) of N atoms.

ACKNOWLEDGMENTS

The authors would like to thank M. Ghafari, University of Technology Darmstadt for assistance at the GIXD measurements. This work was financially supported by GSI.

REFERENCES

1. E. Abramov and D. Eliezer, Hydrogen effect on material behaviour; Publication of The Minerals, Metals & Materials Society (1989).
2. T. Asaoka et.al., Corrosion, 34 (1978) 39-47.
3. M. Aucouturier et. Al., Metallography, 11 (1978) 5-21.
4. G.M. Pressouyre and I.M. Bernstein, Metall. Trans., 9A (1978) 1571-1580.
5. K. Neu, H. Baumann, N. Angert, D. Rück and K. Bethge, NIM B89 (1994) 379.
6. K. Neu, PhD. thesis, Institute for Nuclear Physics, Frankfurt am Main, Germany (1995).
7. M. Soltani-Farshi, H. Baumann, D. Rück and K. Bethge, NIM B127/128 (1997) 787.
8. M. Soltani-Farshi, H. Baumann, D. Rück, E. Richter, U. Kreissig and K. Bethge, submitted for publication in Surface Coatings and Technology (presented on SMMIB-97).
9. S. Lucas, G. Terwagne, M. Piette and F. Bodart, NIM B59/60 (1991) 952.
10. Th. Barnavon, H. Jaffrezic, G. Marest, and N. Moncoffre, Mat. Sci. Eng., 69 (1985) 31.
11. B. Hoffmann, Ph-D thesis, Inst of Nucl. Phys., Frankfurt/Main, Germany, 1987.
12. Lanford, Nucl. Instr. and Meth. 149 (1978) 1.
13. J.F. Ziegler et.al., The Stopping and Range of Ions in Solids, 1 (Pergamon, New York, 1985).
14. C. Rolfs and W.S. Rodney, Nucl. Phys. A 235 (1974) 450.
15. E.J. Sonneveld and J.W. Visser, J. Appl. Cryst. 8 (1975) 1.
16. Joint Committee on powder diffraction standards, 1601 Park Lane, Swarthmore, Pa. 1081.
17. H. Jehn, G. Reiners and N. Siegel, DIN Fachbericht 39, Charakterisierung dünner Schichten, Berlin, 1993.
18. J.C. Pivin, P. Zheng and M. O. Ruault, Mater. Sci. Eng. A115 (1989) 83.
19. F.M. Kustas and M.S. Misra, Surf. Coat. and Tech., 51 (1992) 100.
20. R. Hutchings, Mater. Lett. 1 (1983) 137.
21. R. Martinella, S. Giovanardi, and M. Villani, Mat. Sci. Eng. 69 (1985)247.
22. G.K. Hubler, Mat. Res. Soc. Symp. 7 (1982) 345.

Part III

Energetic Particle Synthesis and Mechanical Properties

ENERGETIC PARTICLE SYNTHESIS OF METASTABLE LAYERS
FOR SUPERIOR MECHANCIAL PROPERTIES

D.M. Follstaedt,* J.A. Knapp,* S.M. Myers,* M.T. Dugger,* T.A. Friedmann,* J.P. Sullivan,*
O.R. Monteiro,** J.W. Ager III,** I.G. Brown** and T. Christenson*

*Sandia National Laboratories, Albuquerque, NM 87185-1056
**Lawrence Berkeley National Laboratory, Berkeley, CA 94720

ABSTRACT

Energetic particle methods have been used to synthesize two metastable layers with superior mechanical properties: amorphous Ni implanted with overlapping Ti and C, and amorphous diamond-like carbon (DLC) formed by vacuum-arc deposition or pulsed laser deposition. Elastic modulus, yield stress and hardness were reliably determined for both materials by fitting finite-element simulations to the observed layer/substrate responses during nanoindentation. Both materials show exceptional properties, i.e., the yield stress of amorphous Ni(Ti,C) exceeds that of hardened steels and other metallic glasses, and the hardness of DLC (up to 88 GPa) approaches that of crystalline diamond (~100 GPa). Tribological performance of the layers during unlubricated sliding contact appears favorable for treating Ni-based micro-electromechanical systems: stick-slip adhesion to Ni is eliminated, giving a low coefficient of friction (~0.3-0.2) and greatly reduced wear. We discuss how energetic particle synthesis is critical to forming these phases and manipulating their properties for optimum performance.

INTRODUCTION

Energetic particle methods are widely used to synthesize materials with special properties. The energies of impinging atoms vary greatly, from 100's of keV for ion implantation down to a few eV for plasma depositions [1]. We have used such methods to explore synthesis of two types of metastable layers with very high hardness and strength. First, ion implantation of Ti plus C into Ni with overlapping concentrations of ~20 at.% forms an amorphous phase [2]. Second, deposition of C at ~100 eV produces a hard, covalently bonded amorphous material known as diamond-like carbon (DLC) [3]. We have developed methods to quantify accurately the yield strength, elastic modulus and intrinsic hardness of these sub-micrometer layers on soft substrates for the first time: finite-element modeling is used to simulate and fit the response of the layer-substrate combinations to ultra-low load indentation, or "nanoindentation". We show that amorphous Ni(Ti,C) has exceptional strength for a metal, exceeding that of hard bearing steels, and that DLC can have hardness approaching that of crystalline diamond. This method is detailed elsewhere [4,5] but considerations specific to these two materials are given here.

Energetic particles are essential for producing these materials because the kinetic energy of the atoms forces them into the layer being formed, as opposed to being deposited on the surface with only thermal energy, and the additional energy can modify the atomic structure. The modifications range from atomic displacements and rearrangements at high energy to breaking and re-forming bonds at low energy. We show how mechanical properties can be manipulated by varying particle energies and other methods to optimize the structure and performance of each material. One motivation for our work is improved tribological performance of Ni components in micro-electromechanical systems (MEMS) [6,7]. Both layers give sizable reductions in friction and wear of Ni, and the two synthesis methods appear compatible with Ni-based MEMS.

AMORPHOUS Ni(Ti,C)

Previously, the ion implantation of Ti and C into steels was examined extensively for reducing friction and wear [8,9]. To the best of our knowledge, this treatment produced benefits for every steel tested, as well as a Co-based alloy [10]. A key feature of the treatment was the formation of an amorphous surface layer, whose presence was found to correlate directly with the benefits obtained [11]. In addition, Ti + C implantation was found to amorphize Ni [12]. With the recent interest in using x-ray lithography to form high-precision, miniature molds and electro-plating to invest them with Ni alloy (German process acronym LIGA [1,2]), we have examined Ti + C implantation of Ni as a treatment to improve the performance of Ni-based MEMS. Our recent work with pure Ni quantifies the intrinsic mechanical properties of the amorphous phase for the first time, and also evaluates the treatment of electroformed Ni for MEMS.

Microstructure and Composition of Ni Implanted with Ti and C

High-purity (99.995) Ni was annealed at 1000°C for 2 hours in a high vacuum of ~10^{-5} Pa (~10^{-7} Torr) to remove lattice damage. The anneal produced grain growth to sizes of several tenths of a millimeter and also grain relief. Specimens (12 mm x 12 mm x 250 μm) were polished with diamond suspensions down to 0.25 μm to obtain a mirror finish without relief, and then re-annealed. Titanium was implanted to a fluence of $2x10^{17}$ Ti/cm^2 at 180 keV which is predicted by TRIM97 [13] to give a projected range (R_P) of 62 nm. Carbon was then implanted to $2x10^{17}$ C/cm^2 at 45 keV for an expected R_P = 58 nm. These implantations produce nearly overlapping profiles of Ti and C, as seen with 6 MeV ^{4}He backscattering spectrometry in Fig. 1.

Plan-view transmission electron microscopy (TEM) showed that the implanted layer was only partially amorphous with remaining fcc Ni nanocrystals [14]. To determine the depth-dependent microstructure precisely, cross-section specimens were made by sandwiching the Ni between Si, extracting a cylindrical core centered on the implanted layer, slicing a disk from the core, and using metallographic polishing and ion milling to produce thin area including the implanted layer [15]. The microstructure of the layer is seen in Fig. 2: a sub-surface, fully amorphous layer that extends from 30 nm to 80 nm in depth (layer II), with a two-phase layer of

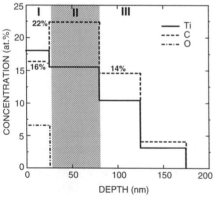

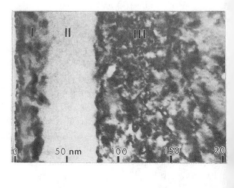

Figure 1. Depth profiles of Ti, C, and O in Ni implanted with Ti + C, obtained with 6 MeV ^{4}He$^+$ backscattering at 164°. Amorphous layer is shaded.

Figure 2. Bright-field, cross-section TEM image of Ni implanted with $2x10^{17}$ Ti/cm^2 at 180 keV and $2x10^{17}$ C/cm^2 at 45 keV. Layers I, II & III discussed in the text, are indicated.

amorphous alloy with embedded Ni crystallites above it (layer I). Careful examinations of this and other images also reveal a thin (~6 nm) oxide at the surface. Beneath layer II is a highly dislocated layer extending from 80 to 180 nm (layer III); its deep boundary is somewhat difficult to identify in Fig. 2 due to damage in the Ni substrate from ion milling, but becomes apparent when wider areas are examined. The dislocations in layer III are due to ions in the tail of the implanted distribution extending into the Ni (see Fig. 2). Dark-field imaging with fcc Ni reflections and the diffuse ring of the amorphous phase confirmed these identifications and allowed us to estimate that the fraction of fcc material in layer I is ~20 vol.% [15].

By combining the depth variations of composition (Fig. 1) and microstructure (Fig. 2), the solute concentrations for amorphous phase formation can be identified. Between layers I and II, the C concentration falls from 22 to 16 at.% for a nearly constant Ti concentration of 16-18 at.%; the minimum composition for single-phase material thus lies between these limits. In the center of layer II, 22 at.% C and 16 at.% Ti give fully amorphous material. Beneath layer II, an abrupt transition occurs from amorphous alloy to crystalline fcc metal with 14 at.% C and 10 at.% Ti. Two additional pieces of information support our interpretation that the two-phase alloy in layer I forms due to insufficient C. First, the amount of lattice damage during the C implantation is sufficient for amorphization; TRIM97 simulations [13] predict ~100 dpa, so that even the last $1/10^{th}$ of the C implantation produces sufficient displacements. Second, when a low-fluence implantation of 3×10^{16} C/cm^2 is added at the lower energy of 20 keV (R_p = 28 nm) to introduce additional C into layer I, the implanted region is fully amorphous to the surface. This second implantation was applied to layers for sliding contact testing, since the poor tribological properties of untreated Ni suggest that the fcc crystallites would impair performance. Thus the metallurgy of Ti+C -implanted Ni can be assessed by combining ion beam analysis with TEM, and manipulated by adjusting ion energies to improve mechanical properties.

Composition limits have been studied more extensively for amorphous Fe implanted with Ti and C [16], and indicate that minimum C concentrations decrease with increasing Ti content. For our Ni alloy with 16 at.% Ti, an increased C content is needed relative to Fe: ~20 at.% C is needed for Ni, whereas $\leq$ 10 at.% C is required for Fe. This finding can guide composition choices in future work; a similar trend of decreasing C with increasing Ti is likely to be found for amorphous Ni(Ti,C) also.

Mechanical Properties of Bulk Ni Implanted with Ti and C

Mechanical properties of amorphous Ni(Ti,C) were evaluated by nanoindentation of annealed, high purity Ni implanted with Ti and C as in Figs. 1 and 2. Typical indenter force versus depth data are shown in Fig. 3 for indentation to 70 nm depth, near the center of the implanted layer. The dashed lines show the standard deviation of the data obtained for 10 indentations. The methods used to produce the fitted simulation shown in Fig. 3 are discussed in detail in [5,14]. Most of our modeling of hard-layer indentations results in deduced properties with absolute uncertainties judged to be about 10% or less, but this structure requires greater attention to obtain an accurate assessment. Careful examination of the indentation response data and attempts to fit it at all depths indicate the presence of a softer surface layer and a hardened layer extending beyond 100 nm depth where the implanted Ti and C concentrations are much reduced from their peak values (Fig. 1). The cross-section image in Fig. 2 shows the same features, and layers I (softer, 2-phase), II (fully amorphous, expected to be hardest) and III (reduced hardness, extends to 180 nm) were assigned separate properties in the model. In addition, knowing the depth intervals of layers I, II and III from TEM allows more accurate

243

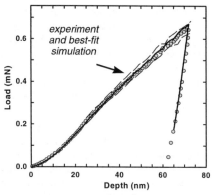

Figure 3. Nanoindentation load vs. depth data for Ni implanted with Ti + C and fitted simulation.

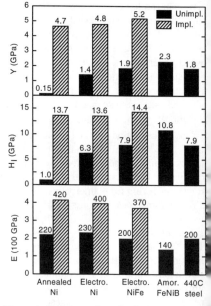

Figure 4. Summary of mechanical properties of Ni alloys, implanted and unimplanted, and reference alloys for comparison.

modeling of multiple layers with fewer variables to be fitted. The fitted simulation determined a yield stress Y.S. = 4.70+0.13 GPa and an elastic modulus E = 416+104 GPa for amorphous Ni(Ti,C). The "intrinsic hardness" (independent of substrate) of the amorphous phase was determined by an additional simulation for a hypothetical "bulk" material with these values of Y.S. and E that gave H = 13.7+2 GPa.

These values greatly exceed those of pure Ni, as seen in Fig. 4. The Y.S. is 30 times that of Ni (0.15 GPa) and the elastic modulus is doubled, producing an order of magnitude increase in hardenss. The values also exceed those of hardened 440C bearing steel and hard amorphous Fe-Ni stabilized by B additions. Such metalloid-stabilized amorphous phases typically have high hardnesses in the range 7 - 11 GPa due to the absence of crystallinity and associated plastic deformation by dislocation glide [17]. We hypothesize that the increased hardness of amorphous Ni(Ti,C) is due to a high concentration of Ti atoms that bond to C in the amorphous matrix. Evidence for this hardening is reported for melt-quenched alloys for which up to 2 at.% Ti was retained in solid solution and increased hardness by about 1 GPa [18]. Extrapolating this increment to our alloy with 16 at.% Ti predicts a hardness of ~15-19 GPa, reasonably accounting for our observed value. It is notable that amorphous alloys with such high Ti content are not attainable by melt quenching, even by very rapidly quenching with pulsed lasers [19], whereas they are readily formed by ion implantation and by pulsed-laser deposition [20].

Tribological improvements obtained with the amorphous layer under representative MEMS operating conditions are seen in Fig. 5. For unimplanted annealed Ni disks (25 mm diameter, 1.9 mm thick) the unlubricated coefficient of friction is high (>1.0) and shows erratic behavior due to stick-slip adhesion to the sliding steel ball during testing in dry N_2; see Fig. 5a. However, for Ti + C -implanted Ni, the coefficient of friction is lower (0.75) and the frictional force does not vary greatly, implying that the adhesion has been eliminated. This effect is confirmed by examining corresponding wear tracks in Fig. 5b. The unimplanted surface is very rough and worn deeply (1.82 μm) in local areas due to adhesion and removal; it is elevated in some areas,

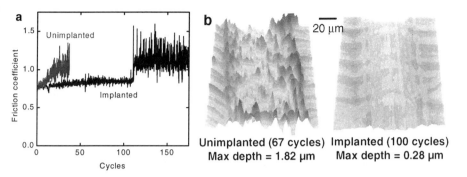

a Unimplanted

Friction coefficient

Implanted

Cycles

b 20 μm

Unimplanted (67 cycles) Implanted (100 cycles)
Max depth = 1.82 μm   Max depth = 0.28 μm

Figure 5. a) Friction coefficient obtained with bulk Ni disks, unimplanted and implanted with Ti and C, rotating in dry N_2 against a 1.6 mm radius 440C steel with 9 gf load (Hertzian stress: 43 MPa). b) Optical interferrometry images of unimplanted (left) and implanted (right) wear tracks.

indicating that material has been re-deposited on the surface. In contrast, the track for implanted material shows minimal wear with only shallow grooves. Thus implantation has changed the wear from adhesive to mild abrasive, which persists until the amorphous layer is worn through beyond 100 cycles in Fig. 5a. Stick-slip adhesion and wear are eliminated by hardening the surface, which deforms less during contact with the slider giving less area for adherence. In addition, the stronger implanted surface layer probably requires greater force to be torn away.

Properties of Electro-formed Ni

To assess Ti plus C implantation for Ni-based MEMS devices, we examined two alloys (Ni and $Ni_{80}Fe_{20}$) electro-plated during the production of LIGA components. The plating conditions and microstructure of the as-deposited metals are discussed elsewhere [15]; in both cases, fine-grained (20-600 nm) fcc material was produced. Specimens (150 μm thick) were evaluated with indentation and modeling and found to be much stronger than pure Ni (Y.S. = 1.4 and 1.9 GPa, respectively) and also harder, although the elastic modulus is not significantly changed [21,22]; see Fig. 4. The increased strength and hardness are inferred to result from the fine grain sizes.

The same implantations of Ti and C used for pure Ni produced a fully amorphous phase extending to the surface without the second, low-energy C implantation. The high fraction of amorphous phase in layer I for pure Ni (~80 vol.%) indicates that the $2x10^{17}$ C/cm^2, 45 keV implantation is almost sufficient for full amorphization, and we infer that defects in the electro-formed material and perhaps C impurities introduced during plating produce full amorphization. Nanoindentation and modeling of the implanted, electro-formed alloys gave mechanical properties for the amorphous phase [22] essentially the same as those deduced with pure Ni. This agreement with substrates having such different mechanical properties supports our evaluations since the amorphous phase should be same whether formed from large-grain, annealed Ni or fine-grain, electroplated Ni, and substituting 20 % Fe for Ni should have minimal effect.

The critical property for ion-implanted MEMS alloys is tribological performance, and the coefficient of friction is seen in Fig 6 to be quite improved over that of unimplanted material. In this test with reciprocating sliding of a steel ball on electro-deposited Ni in laboratory ambient, the unimplanted material shows a relatively high coefficient of friction with excursions to ~1.0.

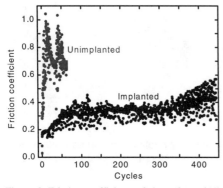

Figure 6. Friction coefficients of electroformed Ni, unimplanted and implanted with Ti + C, obtained in laboratory ambient during unidirectional, reciprocating sliding against a 1.6 mm radius 440C steel ball with 9 gf load (Hertzian stress: 43 MPa).

In contrast, that of the implanted material rises to 0.35 at 50 cycles and remains nearly constant until 350 cycles; the subsequent rise may reflect partially wearing through the amorphous layer. The large excursions in frictional force for untreated material are due to adhesion to the substrate; this behavior is again greatly reduced by implantation. It is notable that friction is lower and remains so for more cycles with implanted electroformed Ni than with implanted pure Ni (Fig. 5). This improvement is consistent with the greater hardness of electroformed Ni, although the change of atmosphere from dry N_2 to lab ambient may have also reduced the friction. The harder substrate beneath the implanted layer deforms less, further reducing the contact area for adhesion to the counterface.

DIAMOND-LIKE CARBON (DLC) DEPOSITED LAYERS

There are several types of amorphous C-based materials with varying degrees of hardness due to their composition, chemical bonding and synthesis conditions. A recent review indicates the differences between these materials [3]. Materials containing H, such as those produced by chemical vapor deposition, are generally softer and may not adhere as well due to H disrupting bonds across the interface with the substrate. For H-free materials, plasma processes produce harder layers due to higher atom energies. Hardness varies with bond type: high fractions of sp^3 bonds (found in crystalline diamond) produce higher hardness than do high sp^2 (graphite) fractions. Methods including electron energy loss spectroscopy have been used to evaluate sp^3 fraction and large values have been found (~85%) for hard, vacuum-arc deposited DLC [23]. Recent molecular dynamics simulations of the atomic structure of amorphous DLC quenched from the liquid state with a density of 3.0 g/cm^2 give a somewhat lower sp^3 fraction, ~65% [24]. An optimum energy has been identified for deposited atoms to produce maximum sp^3 content, ~100 eV/C [3]. Details of how hard DLC forms during implantation are still under discussion. Key ideas include "subplantation", subsurface implantation of atoms at low energy that results in accumulation and growth of sp^3-bonded C due to preferred displacement of sp^2-bonded C [25], and compressive stress-induced DLC formation similar to diamond at high-pressure [26]. Optimum deposition energies for high sp^3 content near ~100 eV emerge from both models. The role of "thermal spikes", melt-like zones that quench along an ion track, is not yet clear [25].

Here we evaluate hard, H-free material, known as diamond-like carbon, hard carbon, or amorphous-tetrahedral carbon (a-tC [27]). We find that vacuum-arc and pulsed-laser deposited DLC can have hardnesses approaching that of diamond. Neuville and Matthews [3] state that measuring the hardness of DLC is important to understand the material, but indentation is difficult due to substrate effects, surface effects, and deformation of the indenter tip. We show that finite-element modeling correctly accounts for these effects during nanoindentation. The dependence of sp^3 content on C^+ energy is demonstrated, and the compressive stress often found for DLC is shown to be removable by annealing without significant reduction in hardness.

Vacuum-Arc Deposited DLC

Vacuum-arc deposition was done at Lawrence Berkeley National Laboratory using methods described previously [28]. Briefly, a vacuum arc discharge is produced on a carbon cathode to generate a plasma composed largely of C^+ ions with a directed kinetic energy of 20-30 eV. The plasma passes through a 90° magnetic filter to remove macroparticles. The substrate is pulse-biased with a duty cycles of 25%. A bias of -100V has been shown to produce high hardness DLC, but with high compressive stress, 10.5 GPa. When the bias is increased to -2 kV, the stress decreases to 2.5 GPa but the material is softer. This higher bias gives deeper penetration of C^+ into the layer/substrate (R_P ~ 4 nm in Ni) and is used at the beginning of even the hard-layer depositions to "stitch" the layer to the substrate by intermixing their interface, with the bias subsequently being decreased to -100 V to obtain the higher hardness.

We evaluated a hard DLC layer 700 nm thick and a soft one 400 nm thick deposited on Si substrates (500 μm thick). The load vs. depth response data for both are given in Fig. 7; these layers require greater force on the indenter than used for Ni (Fig. 3) because the DLC layers are thicker and harder. The hard material shows a very elastic response, with data taken during indenter retraction falling only slightly below those taken during insertion. Accurate evaluation of DLC requires careful modeling with a sufficiently fine mesh around the contact area with elements as small as 2 nm x 20 nm for convergence of the numerical code. The diamond indenter must also be treated accurately by calibrating the rounded shape at the tip and modeling a sufficiently thick section to account for its elastic response correctly. The tip showed no indication of yielding between successive indentations and was modeled as a purely elastic solid. The compressive stress in the layers was included in the model. Values of yield stress and elastic modulus deduced from the best-fit simulations are given in Table I, and those for -100 V bias are noticeably higher than for -2 kV. Using these properties in additional simulations of hypothetical "bulk" materials gives intrinsic hardnesses of 68.4±2.5 GPa and 27.5±0.7 GPa, respectively. It is

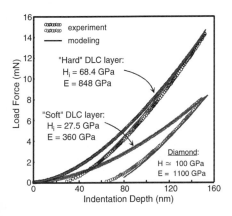

Figure 7. Nanoindentation load vs. depth data of vacuum-arc deposited DLC layers on Si produced with -100 V bias and -2 kV bias, and their best-fit finite-element simulations.

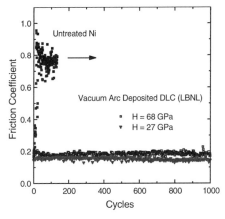

Figure 8. Friction coefficient of DLC layers on Ni, obtained in laboratory ambient during unidirectional, reciprocating sliding against a 1.6 mm-radius 440C steel ball with 9 gf load (Hertzian stress: ~500 MPa).

Table I. Mechanical Properties of Diamond-Like Carbon and Diamond

Material[+]	Elastic Modulus	Yield Stress	Hardness
Soft vacuum-arc deposited DLC	360±10 GPa	14.1±0.4 GPa	27.5±0.7 GPa
Hard vacuum-arc deposited DLC	848±10 GPa	49.1±2.9 GPa	68.4±2.5 GPa
Pulsed-laser deposited DLC	1101±18 GPa	77.6±2.4 GPa	88.1±2.2 GPa
Diamond	1141 GPa	>100 GPa*	~100 GPa

[+] Poisson's ratio, $\upsilon = 0.100$ used for DLC, 0.070 for diamond.
* Deduced from nanoindentation and finite-element modeling.

clear that the deposition energy greatly affects mechanical properties. High hardness is obtained for a total C^+ energy of ~100 eV + ~20 eV = ~120 eV, as expected [3]. The hard DLC was evaluated previously with analytical methods developed for nanoindentation [23] and a lower bound placed on its hardness: $\gtrsim$ 60 GPa, in agreement with our findings. The analytical methods [29] have limited ability to evaluate thin hard layers on softer substrates for reasons stated above.

The tribological properties of vacuum-arc deposited DLC on Si have been assessed [30], and low coefficients of friction, $\lesssim$ 0.15, were found. To evaluate the DLC for application to MEMS, the two layers were deposited on annealed Ni (1.9 mm thick) to ~200 nm thickness. The high-stress, hard layer was partially delaminated from the Ni substrate, but the low-stress, soft layer was fully adherent. The layers were tested in the reciprocating sliding tester described above for implanted Ni. Because of the high elastic modulus of DLC, the calculated Hertzian stresses are higher, ~500 GPa. As seen in Fig. 8, both layers exhibited low coefficients of friction in this test, $\lesssim$ 0.2, that persist for more than 1000 cycles. Notably, the softer layer performed quite well under these representative MEMS operating conditions; since it showed no signs of delamination before or after testing, it appears to be the better treatment for this application.

Pulsed-Laser Deposited DLC

A pulsed laser was used to deposit DLC at Sandia National Laboratories. The excimer laser (248 nm) was operated at 20 Hz repetition rate and a high energy fluence (>100 J/cm^2), giving a deposition rate of ~1 Å/sec on a (100) Si substrate place in the ablation plume [27]. This method produces DLC with high compressive stress, ~6-8 GPa. Recently, annealing methods have been identified to reduce this stress essentially to zero [31], and 2 minutes at 600°C was used for the material evaluated here. Raman and electron energy loss spectra showed only subtle changes between as-deposited and annealed DLC layers [27]. To make thick material, a layer 0.15-0.2 μm thick was deposited on Si (300 μm thick) and annealed in situ; subsequent layers were added and the structure was annealed after each to form a total thickness of 1.2 μm of stress-free DLC.

Representative nanoindentation response data for this layer indicate a very elastic response in Fig. 9a. The best-fit simulation is seen in Fig. 9a to fit the data well, and the fitted parameters are listed in Table I. The indenter was modeled as a purely elastic material [27] and the finite-element mesh for indentation to 160 nm depth is shown in Fig. 9b. This material has the highest yield stress of those tested and an elastic modulus that is ~90% of that of diamond. Using these values for bulk material gave the intrinsic hardness versus depth curve shown in the upper part of

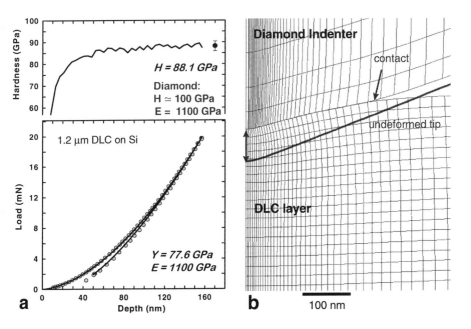

Figure 9. a) (Lower) Nanoindentation load vs. depth data for pulsed-laser deposited DLC (1.2 μm thick) on Si, and best-fit simulation. (Upper) Calculated hardness vs. depth. b) Finite element mesh of indenter and DLC layer modeled at 160 nm depth using best-fit parameters. The undeformed tip shape is shown.

Fig. 9a indicating a very high hardness, 88.1 GPa, also near that of diamond. The curve is flat beyond ~50 nm depth, indicating that the indentation is sufficiently deep for plastic deformation and maximum hardness to be assessed. Due to the thickness of this layer (1.2 μm), this curve differs little from that for the DLC/Si composite, and the analytical methods developed for nanoindentation [29] correctly assess the layer hardness. However, the high elastic modulus determined by modeling is not obtained with those methods, apparently due to larger-than-expected influence of the softer Si substrate and possibly also the substantial deformation of the indenter tip [5]. Notably, the Si substrate also plastically deforms, even with this thick hard layer.

Superimposed on the deformed mesh of elements for the indenter and DLC layer in Fig. 9b is the position and shape that the diamond tip would have if it were infinitely stiff and remained undeformed. The tip is depressed by ~65 nm at the radial center of contact (left side of Fig. 9b) and a relatively abrupt change in its surface contour occurs where it contacts the DLC. These features show the importance of modeling the indenter correctly and handling it carefully. To avoid accidentally deforming the tip during testing, the specimen surface was initially located in an area where the DLC layer was absent. To verify that plastic deformation did not occur in the tip during the indentations, the tip shape was recalibrated afterwards, with no change detected. The indentation data obtained for the DLC layers are very reproducible and overlay each other. We have considered this absence of evidence for tip deformation between individual indentations [4], and with results from this very hard specimen, a lower limit can be placed on the yield strength of diamond, Y.S. > 100 GPa in compression, since stresses inside the tip are calculated

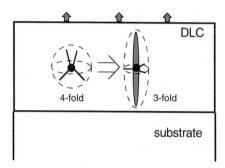

Figure 10. Schematic showing bond conversion of C in DLC from sp^3 (4-fold coordination) to sp^2 (3-fold) and the resulting outward relaxation of the layer.

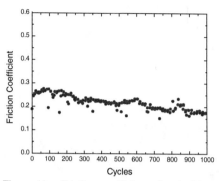

Figure 11. Friction coefficient of pulsed-laser deposited DLC on Si, obtained in lab ambient during unidirectional, reciprocating sliding against a 1.6 mm-radius 440C steel with 9 g_f load (Hertzian stress: ~600 MPa).

to have reached 106 GPa. The above discussions indicates how valid concerns about indentation of DLC [3] have been addressed in our work by using finite-element modeling to interpret the material response. This approach accounts for plastic and elastic deformations of the tip and substrate, allowing DLC mechanical properties to be assessed accurately. Preliminary work indicates that the same yield strength and modulus are obtained with layers as thin as 50 nm [32].

The elimination of compressive stress allows DLC layers to be more widely used and appears not to reduce their hardness significantly. An atomic model has been developed to explain the stress reduction based on conversion of some C atoms from sp^3 to sp^2 bonding [31], as indicated schematically in Fig. 10. This conversion is supported by several observations occurring during annealing: increased optical absorption, reduced atomic density, increased conductivity, and conversion of the layer to nanocrystalline graphite at 800°C [33]. This bonding change actually increases the atomic volume of the C atom, just as the atomic density of graphite is lower than that of diamond. However, the sp^2 atomic volume is not equiaxed, but is extended in one direction, as indicated in Fig. 10. In a compressive layer, this direction is energetically preferred to align normal to the layer, which can then relax outward for a net reduction in stress. For an initial stress of ~7 GPa, 6% of the C atoms are required to transform to completely relieve stress; using the lower value of 65% for the initial sp^3 content, the number of sp^3 atoms in the layer would then decrease by only ~1/10, consistent with the very high remaining hardness determined here. Such annealing should relieve stress in DLC layers produced by other methods also.

The viability of pulsed-laser deposited, annealed DLC layers as tribological coatings was examined for a 0.17 μm layer on Si with the above same reciprocating sliding conditions. The friction coefficient is seen in Fig. 11 to be low, ~0.2, but decreasing slightly with continuing sliding. This decrease possibly represents a change on the DLC surface from sp^3 to sp^2 bonds and a graphitic surface during sliding contact. Further testing is needed to determine if the coefficient of friction will stabilize at a low level after a greater number of cycles, like those seen in Fig. 8. To verify the applicability of this material to Ni-based MEMS, the effects of the stress-relieving anneal on the fine-grained microstructure of electroformed Ni need to be assessed. It is also possible that thin ($\lesssim 0.1$ μm) as-deposited layers give extended benefits without annealing.

DISCUSSION

The amorphous phases examined here show exceptional properties compared to their crystalline counterparts. Amorphous Ni(Ti,C) is harder than bearing steel and other hard amorphous metals of Ni or Fe. Approximately 20 at.% each of Ti and C are required for its formation, which can be achieved by adjusting ion energies and fluences to produce overlapping profiles with appropriate concentrations. The increased hardness gives a reduced contact area on the surface during sliding contact, resulting in much less adhesion and a change to mild abrasive wear. We have also examined amorphous Fe(Ti,C) produced by pulsed laser deposition and find very similar hardness[20]. It is reasonable to predict that steels implanted with Ti and C are similarly hardened, and that this hardness contributes to the reduced friction and wear observed for them [8,9,11]. Such hardening probably contributes to the improved tribological performance observed for Co alloys [10] as well.

The hardness of DLC can approach that of crystalline diamond, the hardest known substance. This is achieved by adjusting deposited ion energies to near ~100 eV. The plasma energy is not known for pulsed-laser deposition, but the very hard DLC layer examined here (88 GPa) was produced with high laser fluences, which presumably give higher deposition energies perhaps approaching the optimum value. With vacuum-arc deposition, such energies occur only when the pulsed bias is on, ~25% of the time, which could account for the lesser hardness (68 GPa) of its DLC at 100V bias. Increasing the bias to 2 kV significantly reduced the hardness.

Both materials provide low coefficients of friction for electroformed Ni used for MEMS components. Ion implantation matches key requirements for such a treatment: it is done at room temperature, produces no dimensional change, and the layer does not have an abrupt interface with the substrate thus insuring good adherence. Moreover, the resistance of the implanted layer to electropolishing observed during our TEM specimen preparation implies that the surface may be corrosion resistant. The small areas of miniature components should allow implantation at lower cost than for macroscopic components currently being commercially treated.

Deposition of DLC is also done at room temperature, and good performance is obtained with thicknesses (~0.2 μm) that perturb component tolerances only minimally. We are currently investigating the tribological performance of even thinner layers (<0.1 μm). Layer adherence is good with higher-energy ions that intermix the interface and produce less compressive stress. Thin (~20 nm) DLC layers are already applied commercially as protective coatings for computer disk drives to prevent damage when the head contacts the spinning platter. In addition, DLC layers are smooth and pin-hole free at ~10 nm thickness, and their chemical inertness is expected to inhibit corrosion. The anneal treatments identified to reduce compressive stress [30] can be expected to broaden the applicability of DLC coatings.

ACKNOWLEDGMENTS

The authors thank G. A. Petersen for performing the ion implantations and M. P. Moran for assistance with the TEM analysis. We are grateful to B. N. Lucas of Nano Instruments, Inc., for capably performing the indentations and the extra care used to obtain accurate data for hard DLC. This work was supported by Division of Materials Sciences, Office of Basic Energy Sciences and its Synthesis and Processing Center, of the U. S. Department of Energy under contract DE-AC04-94AL85000. Sandia is a multiprogram laboratory operated by Sandia Corporation, a Lockheed Martin Company, for the U. S. Department of Energy.

REFERENCES

1. S. M. Fossnagel, J. J. Cuomo and W. D. Westwood, Handbook of Plasma Processing Technology, (Noyes Publications, Park Ridge, NJ, 1990).
2. D. M. Follstaedt, Nucl. Inst. Meth. **B 7/8**, 11 (1985).
3. S. Neuville and A. Matthews, MRS Bulletin **22**, 22 (Sept. 1997).
4. J. A. Knapp, D. M. Follstaedt, J. C. Barbour, S. M. Myers, J. W. Ager III, O. R. Monteiro and I. G. Brown, Mat. Res. Soc. Symp. Proc. **438**, 617 (1997).
5. J. A. Knapp, D. M. Follstaedt, S. M. Myers, J. C. Barbour, T. A. Friedmann, J. W. Ager III, O. R. Monteiro and I. G. Brown, Surface Coatings and Technology, in press.
6. E. W. Becker, W. Ehrfeld, P. Hagmann, A. Maner and D. Münchmeyer, Microelect. Eng. **4**, 35 (1986).
7. H. Guckel, K. J. Skrobis, J. Klein and T. R. Christenson, J. Vac. Sci. Technol. **A 12**, 2559 (1994).
8. I. L. Singer, C. A. Carosella and J. R. Reed, Nucl. Instrum. Meth. **182-183**, 923 (1981).
9. L. E. Pope, F. G. Yost, D. M. Follstaedt, J. A. Knapp and S. T. Picraux, in Wear of Materials 1983, ed. K. C. Ludema (ASME, New York, 1983), pp. 280.
10. S. A. Dillich, R. N. Bolster and I. L. Singer, Mat. Res. Soc. Symp. Proc. **27**, 673 (1984).
11. D. M. Follstaedt, F. G. Yost, L. E. Pope, S. T. Picraux and J. A. Knapp, Appl. Phys. Lett. **43**, 358 (1983); also Appl. Phys. Lett. **45**, 529 (1984) and erratum Appl. Phys. Lett. **46**, 207 (1985).
12. A. W. Mullendore, L. E. Pope, A. K. Hays, G. C. Nelson, C. R. Hills and B. G. Lefevre, Thin Solid Films **186**, 215 (1990).
13. J. F. Ziegler, J. P. Biersack and U. Littmark, The Stopping and Range of Ions in Solids (Pergamon, New York, 1995).
14. S. M. Myers, J. A. Knapp, D. M. Follstaedt and M. T. Dugger, J. Applied Physics, in press.
15. D. M. Follstaedt, S. M. Myers, J. A. Knapp, M. T. Dugger and T. A. Christenson, Surface Coatings and Technology, in press.
16. J. A. Knapp, D. M. Follstaedt and B. L. Doyle, Nucl. Inst. Meth. **B 7/8**, 38 (1985).
17. J. Niebuhr, R. Gerber, A. Schaller and H.-W. Müller, Physical Data of Amorhous Metals, Part B: Data Tables (Fachinformationszentrum Karlsruhe, Freiburg, Germany, 1991) p. 239.
18. A. Inoue, T. Iwadachi, T. Minemura and T. Masumoto, Trans. Jap. Inst. Metals **22**, 197 (1981).
19. D. M. Follstaedt, J. A. Knapp and P. S. Peercy, J. Non-Crystal. Solids **61/62**, 451 (1984).
20. J. A. Knapp, D. M. Follstaedt and S. M. Myers, unpublished results.
21. S. M. Myers, D. M. Follstaedt, J. A. Knapp and T. R. Christenson, Res. Soc. Symp. Proc. <u>444</u>, 99 (1997).
22. S. M. Myers, J. A. Knapp, D. M. Follstaedt, M. T. Dugger and T. R. Christenson, Surface Coatings and Technology, in press.
23. G. M. Pharr, D. L. Callahan, S. D. McAdams, T. Y. Tsui, S. Anders, A. Anders, J. W. Ager III, I. G.Brown, C. S. Bhatia and S. R. P. Silva, Appl. Phys. Lett. **68**, 779 (1996).
24. N. A. Marks, D. R. McKenzie, B. A. Pailthorpe, M. Bernasconi and M. Parrinello, Phys. Rev.Lett. **76**, 768 (1996).
25. Y. Lifshitz, S. R. Kasi and J. W. Rabalais, Phys. Rev. Lett. **62**, 1290 (1989).
26. D. R. McKenzie, D. Muller and B. A. Pailthorpe, Phys. Rev. Lett. **67**, 773 (1991).
27. T. A. Friedmann, J. P. Sullivan, J. A. Knapp, D. R. Tallant, D. M. Follstaedt, D. L. Medlin and P. B. Mirkarimi, Applied Physics Letters, scheduled for vol. **77**, Dec. 29, 1997.
28. S. Anders, A. Anders, I. G. Brown, B. Wei, K. Komvopoulos, J. W. Ager III, and K. M. Yu, Surf. Coat. Technol. **68/69**, 388 (1994).
29. W. C. Oliver and G. M. Pharr, J. Mater. Res. **7**, 1564 (1992).
30. J. W. Ager III, S. Anders, I. G. Brown, M. Nastasi and K. C. Walter, Surf. Coat. Technol. **91**, 91 (1997).
31. J. P. Sullivan, T. A. Friedmann and A. G. Baca, J. Electron. Mat. **26**, 1021 (1997).
32. J. A. Knapp, D. M. Follstaedt, T. A. Friedmann, A. J. Magerkurth, S. W. Clarke, O. R. Monteiro, J. W. Ager III, I. G. Brown, B. N. Lucas and W. C. Oliver, presented Fall'97 MRS, Symp. NN.
33. T. A. Friedmann, K. F. McCarty, J. C. Barbour, M. P. Siegal and D. C. Dibble, Appl. Phys. Lett. **68**, 1643 (1996).

CHARACTERIZATION OF DIAMOND LIKE CARBON FILM FABRICATED BY ECR PLASMA CVD AT ROOM TEMPERATURE

K. KURAMOTO, Y. DOMOTO, H. HIRANO, H. TARUI AND S. KIYAMA
New Materials Research Center, Sanyo Electric Co., Ltd.
1-18-13 Hashiridani, Hirakata, Osaka 573, Japan

ABSTRACT

Low temperature (about 50°C) fabrication of diamond like carbon (DLC) films with a high hardness (>3000Hv) and a high electrical resistivity (>10^{11} Ωcm) has been achieved.

In order to obtain such a result, the effect of ion impingement on the growth and structural change of DLC films in an electron cyclotron resonance (ECR) plasma enhanced chemical vapor deposition (CVD) method was investigated. It was confirmed that ion impingement was fundamentally required in the growth of DLC films. Furthermore, impingement with ions energized by bias voltages between 50V and 150V had a major influence on the sp^2/sp^3 configuration in DLC films. This configuration is found to be rather sensitive to optoelectronic properties but not so sensitive to film hardness.

Additionally, this method could fabricate ultrathin DLC films that exhibited excellent wear resistance for protective applications.

INTRODUCTION

Diamond like carbon (DLC) films have a variety of useful properties such as high hardness, smooth surface morphology, wear resistance and chemical inertness[1,2]. For advanced coating with DLC films, low temperature fabrication is favorable to prevent thermal damage to the substrate and to prevent excessive graphitization which degrades the properties. Fabrication of DLC films have been widely investigated by sputtering[3,4] and plasma deposition[1,5] methods. A common feature of these techniques is exposure of the film during growth to ions energized usually up to several hundred electron volts[6]; accordingly, increased substrate temperatures as a result of excessive ion impingement cannot be discounted.

In order to decrease the substrate temperature of DLC films, two significant points should be considered. One is the separation of a plasma generating region from a deposition region to avoid direct thermal radiation from the plasma, and the other is the minimization of thermal damage caused by excessive ion impingement during deposition.

The electron cyclotron resonance (ECR) plasma enhanced chemical vapor deposition (CVD) method is expected to reduce substrate temperatures because the ECR plasma source is separated from the deposition area. Additionally, independent control of the bias voltage is expected to control the energy of ions impinging upon the substrate better.

In this paper, the effect of ion impingement on the growth and structural changes of DLC films in ECR CVD is investigated for the purposes of controlling film properties and low temperature fabrication. Furthermore, ultrathin DLC films were fabricated and their protective characteristics were studied.

EXPERIMENT

A schematic view of the experimental apparatus is shown in Fig. 1. Argon plasma was generated in the ECR cavity (875Gauss, 2.45GHz) to decompose methane gas, which was introduced near the surface of the substrate. The substrate holder was powered by an RF (13.56MHz) source to control the bias voltage to the substrate. The argon and methane partial pressures were set at 7.6×10^{-2}Pa and 1.3×10^{-1}Pa, respectively. Microwave power was maintained

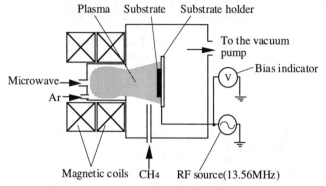

Fig. 1. Schematic view of experimental apparatus.

at 200W, and bias voltage was varied from 0 to 150V by adjusting the RF power to control the ion impinging energy to the substrate.

RESULTS

<u>Dependence of substrate temperature and film hardness on bias voltage</u>

Figure 2 shows the dependence of substrate temperature and film hardness on the bias voltage. The substrate temperature was measured with a thermocouple in contact with the substrate surface after deposition of 250nm thick films. As the bias voltage was increased, the substrate temperature gradually increased. The reason for the increased substrate temperature was considered to be ion impingement on the substrate throughout the process; however, it was nearly at the room temperature level around 50°C.

On the other hand, the hardness of the films drastically increased up to 3000Hv by applying a bias voltage of 50V and then remained nearly constant as the bias voltage was increased to 150V.

In order to investigate the structures of the films fabricated at the bias voltages of 0V and 50V, Raman spectroscopy and scanning electron microscopy were performed. The results are shown in Fig. 3. By applying a bias voltage, there clearly emerged a typical spectrum

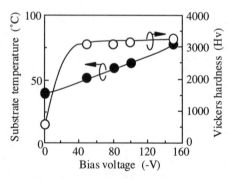

Fig. 2. Dependences of substrate temperature and film hardness on bias voltage.

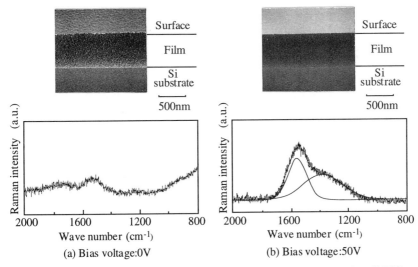

Fig. 3. Raman spectra and scanning electron micrographs of films fabricated on Si(100)
substrates at bias voltages of 0V(a) and 50V(b).

corresponding to DLC, which had a peak at around 1530cm⁻¹, called the graphite band
(G-band), with a shoulder at around 1400cm⁻¹, called the disorder band (D-band)[4].
Furthermore, the film exhibited a minute structure and a smooth surface.

Accordingly, this result suggests that impingement with ions energized by a bias voltage
of 50V is fundamentally required in the growth of DLC films and that DLC films with a
hardness of more than 3000Hv can be fabricated at about 50°C.

Effect of bias voltage on structural change of DLC films

It was impossible to study the effect of ion impingement on the properties of DLC films
fabricated above the bias voltage of 50V through the hardness test, as shown in Fig. 2.

Therefore, optical and electrical measurements were performed for DLC films fabricated
on a glass substrate. The optical gaps were calculated by the Tauc's plot[7]. Evaporated Al
electrodes were used in a gap cell configuration of 0.5mm width for the electrical resistivity
measurement. Figures 4 and 5 show the dependence of the electrical resistivity and the optical
gap on bias voltage, which was varied from 50 to 150V. In contrast to the hardness shown in
Fig. 2, both optical gap and electrical resistivity decreased as the bias voltage was increased,
and an extremely high electrical resistivity of more than 10^{11} Ωcm was obtained at the bias
voltage of 50V.

Variations in electrical resistivity and optical gap with bias voltage seemed to be
influenced by the microstructure of DLC films, such as the proportions of C-H and C-C atomic
bonds. Accordingly, Raman spectroscopy and IR absorption spectroscopy were performed on
films fabricated at bias voltage above 50V. In Fig. 6, the intensity ratio of the D-band to the G-
band (I_d/I_g ratio) and the hydrogen concentration are plotted as functions of bias voltage. The
hydrogen concentration was previously calculated from the C-H absorption area observed
around 3000cm⁻¹ in the IR absorption spectroscopy[1]. It has been reported that the increase in

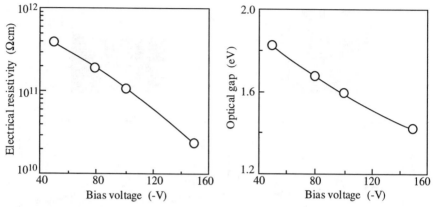

Fig. 4 Dependence of electrical resistivity on bias voltage.

Fig. 5 Dependence of optical gap on bias voltage.

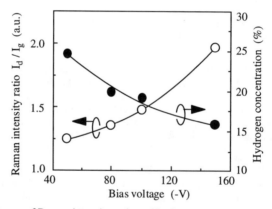

Fig. 6. Dependences of Raman intensity ratio and hydrogen concentration on bias voltage.

the I_d/I_g ratio was related to the increase in the sp^2/sp^3 ratio in DLC films[4]. As seen in Fig. 6, the I_d/I_g ratio increased as the hydrogen concentration decreased by varying the bias voltage. Thus, the results of Fig. 6 suggest that ion impingement promoted by bias voltage above 50V has the effect of decrease in carbon and hydrogen atomic bonds and increase in sp^2 configuration.

It has also been reported that the increase in sp^2 configuration with π electrons lowered the band gap in DLC films[7]. Thus, taking into account the results of both Figs. 2 and 6, it is considered that the increase in sp^2 configuration is rather sensitive to optical and electrical properties but not so sensitive to the hardness of DLC films. Moreover, the decrease in the optical gap is considered a significant reason for the decrease in electrical resistivity. Accordingly, it was confirmed that impingement with ions energized by bias voltages between 50V and 150V played an important role in changing the microstructure of DLC films.

From the above results, high quality DLC films with a hardness of more than 3000Hv and an electrical resistivity of more then 10^{11} Ωcm could be fabricated at an approximately room temperature level of around 50°C by applying the optimum bias voltage of 50V in ECR plasma CVD.

Protective characteristics of ultrathin DLC films

In this method, an improvement in surface smoothness due to ion impingement was confirmed (Fig. 3). Based on this result, fabrication of ultrathin DLC films was attempted.

The scanning electron micrograph and the atomic force micrograph of a sample fabricated by this method are shown in Figs. 7 and 8. The deposition thickness and bias voltage were set at 10nm and 50V, respectively. The results confirmed the formation of a uniform film with a smooth surface. Figure 9 shows the Raman spectrum of the film shown in Figs. 7 and 8. This film was also confirmed to be a typical DLC film having almost the same spectrum and the I_d/I_g ratio as shown in Fig. 3.

Wear resistance is an important factor for protective applications. This was estimated by a friction test in which the ultrathin DLC film was reciprocally rubbed 400 times by a loaded (50gf) alumina (Al_2O_3) ball with a diameter of 10mm. Figure 10 shows wear tracks formed on Si substrates with and without the ultrathin DLC film at 10nm thickness. An obvious wear track could be observed on the bare Si substrate, but it was quite faint on the ultrathin DLC film.

This result indicated that the DLC film was conceivably formed by a uniform growth process at the initial stage of deposition and that ultrathin DLC films fabricated by this method provide satisfactory wear resistance for protecting surfaces.

Bias voltage:-50V

Fig. 7 SEM image of ultrathin film fabricated on Si(100).

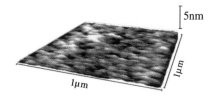

Fig. 8 AFM image of ultrathin film fabricated on Si(100).

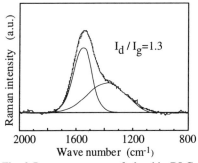

Fig. 9 Raman spectrum of ultrathin DLC film grown on Si(100).

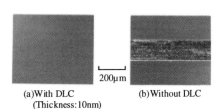

(a)With DLC (Thickness:10nm) (b)Without DLC

Fig. 10 Optical micrographs of wear tracks formed on Si(100) with(a) and without(b) ultrathin DLC film after friction test.

CONCLUSIONS

Low temperature (about 50°C) fabrication of diamond like carbon (DLC) films with a high hardness (>3000Hv) and a high resistivity (>$10^{11}\Omega$cm) was achieved.

In order to obtain such a result, the effect of ion impingement on the growth and structural change of DLC films fabricated by the ECR CVD method was investigated. It was confirmed that ion impingement promoted by a bias voltage of 50V was fundamentally required in the growth of DLC films. Furthermore, experimental results indicated that impingement with ions energized by bias voltages between 50V and 150V played an important role in changing the sp^2/sp^3 configuration of DLC films. This configuration seemed to be sensitive to optoelectronic properties but not so sensitive to film hardness.

Finally, this method was able to fabricate ultrathin DLC films with a smooth surface. In a friction test, this film exhibited excellent wear resistance for protective applications.

ACKNOWLEDGMENT

The authors would like to thank Prof. Dr. Y. Mori of Osaka University for his constant guidance and encouragement.

REFERENCES

1. A. Bubenzer, B. Dischler, G. Brandt and P. Koidl, J. Appl. Phys. 54, 4590 (1983).
2. P. Koidl, Proc. Electrochem. Soc. 89, 237 (1989).
3. X. -M. Tang, J. Weber, Y. Baer, C. Müller, W. Hänni and H. E. Hintermann, Phys. Rev. B 48, 10, 124 (1993).
4. M. Yoshikawa, G. Katagiri, H. Ishida, A. Ishitani and T. Akamatsu, J. Appl. Phys. 64, 6464 (1988).
5. E. H. A. Dekempeneer, R. Jacobs, J. Smeets, J. Meneve, L. Eersels, B. Blanpain, J. Roos and D. J. Oostra, Thin Solid Films 217, 56 (1992).
6. P. Koidl, Ch. Wild, B. Dischler, J.Wagner and M. Ramsteiner, Mater. Res. Forum. 52&53, 41 (1989).
7. S. R. P. Silva, G. A. J. Amaratunga, Thin Solid Films. 270, 194 (1995).

TRIBOLOGICAL PROPERTIES OF Si-DLC COATINGS SYNTHESIZED WITH NITROGEN, ARGON PLUS NITROGEN, AND ARGON ION BEAMS

C.G. FOUNTZOULAS[1], J.D. DEMAREE[1], L.C. SENGUPTA[1], J.K. HIRVONEN[1] and D. DIMITROV[2]

[1] U. S. Army Research Laboratory, Materials Division, APG, MD 21005-5069, cfount@arl.mil
[2] Dept. of Physics and Astronomy, University of Delaware, Newark, DE 19711, dimitar@udel.edu

ABSTRACT

Hard, adherent, and low-friction silicon-containing diamond-like carbon coatings (Si-DLC) have been synthesized at room temperature by 40 keV (N^+ plus N_2^+), 50%Ar^+/50% (N^+ plus N_2^+), and Ar^+ ion beam assisted deposition (IBAD) of a tetraphenyl-tetramethyl-trisiloxane oil on silicon and sapphire substrates. X-ray diffraction analysis indicated that all coatings were amorphous. The average coating wear rate and the average unlubricated steel ball-on-disk friction coefficient, μ, decreased with increasing fraction of nitrogen in the ion beam, along with an increase in the average coating growth rate. The Knoop microhardness and nanohardness values of the coatings synthesized by the mixed argon and nitrogen ion beam were higher than the values for the coatings synthesized with 100% nitrogen or 100% argon ion beams. These friction/wear improvements are tentatively attributed to both increased hardening due to greater penetration and ionization induced hardening by the lighter (N) ions and to the presence of SiO_2 on the surface of N-bombarded samples.

INTRODUCTION

Films of many promising tribological materials, including conventional diamond-like carbon (DLC) have been successfully deposited by ion beam assisted deposition (IBAD). The friction coefficient of unlubricated DLC films in dry gases can be as low as 0.01, but this value can reach values as high as 0.10 and 0.20 when measured in a 10% relative humidity [1-3]. However, it has been shown by various researchers [2-4] that DLC films containing elements such as Si and Ti retain low pin-on-disk friction coefficients in humid environments. DLC films containing silicon (Si-DLC) exhibit friction coefficients as low as 0.04 [2-4] at ambient humidity and temperature and so are highly promising for tribological applications. Several trial industrial applications of DLC including protective wear coatings on bearings and forming tools [5], are limited because of the poor thermal stability of DLC above 350°C. However, there are data [5,6] indicating that the presence of additional elements (F, Si and N) in the coating can increase the range of the thermal structural stability of DLC by as much as 100°C.

In this paper we report on the properties (stoichiometry, thickness, microhardness, bonding, adhesion, friction and wear) of IBAD Si-DLC coatings synthesized with: nitrogen ion beams (N/Si-DLC), argon plus nitrogen ion beams ((Ar+N)/Si-DLC)), and argon (Ar/Si-DLC) ion beams.

EXPERIMENTAL DETAILS

A ZYMET 100 non-mass analyzed ion implanter was used for the synthesis of Si-DLC coatings using energetic 40 keV ion bombardment of a vapor deposited tetraphenyl-tetramethyl-trisiloxane (Dow-Corning 704) diffusion pump oil. The nitrogen beam consisted of a mixture of roughly 40% N^+ and 60% N_2^+, yielding about 1.6 nitrogen atoms per unit charge. The diffusion pump oil precursor was evaporated from a heated [145°C] copper oil container through a 3-mm diameter, 2-mm thick aperture. The substrates, silicon and sapphire (for RAMAN analysis only), were initially cleaned in methanol and acetone and then sputter-cleaned with a 40 keV ion beam (10 $\mu A/cm^2$) of the aforementioned gaseous species for 10 minutes. The temperature of the substrate was maintained close to room temperature using heat conducting vacuum grease to hold the sample on a water-cooled sample stage. The substrate was inclined at 45° with respect to both the horizontal ion beam and the vertical flow direction of the vaporized oil. The aperture to substrate distance was 0.15 m with a shutter placed above the oil container to start and stop the oil deposition. The growing film surface was continuously bombarded by the aforementioned various ion beams at 40 keV. The base pressure was 2.66 x 10^{-4} Pa (2 x 10^{-6}Torr) and the deposition was carried out at 4 x 10^{-3} Pa (3 x 10^{-5} Torr) pressure as in previous work [4]. All coating deposition runs lasted 100 minutes resulting in coating thicknesses ranging from 420 nm (Ar/Si-DLC), 505 nm ((Ar+N)/Si-DLC) and 714 nm (N/Si-DLC).

The thicknesses of the films were measured with the aid of a profilometer. The microhardness values of the coatings were measured using a Knoop microhardness tester with a 0.15 N load. The nanohardness values of the coatings were measured with a Nano Instruments XP nanoindenter using an average load of 20 mN [7]. A ball-on disk tribometer with a 1.27 cm (1/2 in.) diameter AISI 52100 alloy steel ball under 0.5 N load was used to determine the unlubricated sliding friction coefficient μ. Rutherford Backscattering Spectrometry (RBS) was performed on the films using a 2 MeV He^+ beam and a backscattering angle of 170° to determine their near surface elemental composition using the RBS simulation program RUMP [8]. An APD 1710 (Automated Power Diffractometer) System 1 was used to study the crystallinity of the coating. Cu $K_{\alpha 1}$ and $K_{\alpha 2}$ X-ray, 0.154060 nm and 0.154439 nm wavelength respectively, were used for the X-ray analysis for diffraction angles 2θ from 15 to 65 degrees at a scan rate of 0.020 degrees/sec.

RESULTS AND DISCUSSION

Coating Appearance

Optical microscopy showed that while there were pinholes in the Ar/Si-DLC, both the N/Si-DLC and the (Ar+N)/Si-DLC coatings were practically pinhole free. The surface of the N/Si-DLC and (Ar+N)/Si-DLC coatings was stain-free and more reflective than the surface of the Ar/Si-DLC coating. The surface reflectivity was observed to increase with increasing proportions of nitrogen in the ion beam. These observations are not yet understood.

Compositional Analysis

The composition of the Si-DLC films was measured using Rutherford backscattering (2 MeV He^+, 170°) and the simulation program RUMP [8]. In all cases, the relative ratios of C, Si, and O in the IBAD coatings were found to be the approximately the same as the precursor: C:Si:O = 14:1.5:1. This strongly suggests that the siloxane backbone (Si-O-Si-O-Si) of the precursor molecule remains intact during the ion irradiation process, and only C:H and C:C bonds are broken to convert the oil to hard, diamond-like carbon.

The measured compositional differences noted between Ar/Si-DLC, (Ar+N)/Si-DLC, and N/Si-DLC coatings involved the implanted species themselves and their H content. Si-DLC formed with an argon ion beam contained 3-5 at % Ar, and coatings produced using only nitrogen ions contained 7-12 at % nitrogen. In both cases there is a zone near the surface, approximately equal to the predicted ion range, which is relatively deficient in the implanted species. The coatings produced by N-containing beams contained nearly twice as much H [circa 28 at.%] as Ar-only produced coatings [12-15 at. %].

In the case of Si-DLC produced using a mixture of argon and nitrogen ions, RBS revealed a distinct two-layer structure. In these coatings, a 250-nm layer near the surface contains 0.5 at % argon, but the underlying material contains little or no argon. It is possible that the nitrogen ions, traveling farther into the growing film than the argon ions, displaced the implanted argon either directly or by facilitating its escape through defect production. This effect would be less prominent near the surface, where nitrogen ions interact primarily via electronic stopping, leaving an argon-rich surface layer. The nitrogen content of the coatings appears to be similar to that found using only nitrogen ions (7-12 at %).

Growth Rate, Microstructure and Electrical Properties

The Si-DLC coatings produced using N^+, Ar^++N^+ and Ar^+ alone were 420 nm, 505 nm and 714 nm thick, corresponding to average growth rates of 7.14 nm/min, 5.05 nm/min and 4.2 nm/min respectively (Table I). The growth rate difference of these coatings may be attributed on the one hand to the higher surface sputtering rate (etching) caused by the heavier argon ion; and on the other hand to the higher number of ionization events created by the nitrogen ion which presumably enhanced the N/Si-DLC growth rate [9]. X-ray analysis showed that all films were amorphous, in agreement with our previous results [6]. The resistivity of all coatings was above 30 kΩcm, that was beyond the maximum measurable value of the four-point probe apparatus used. All coatings appeared to be featureless when examined under an ordinary optical microscope (200X).

Property	Ar	Ar+N	N
Thickness (nm)	420	505	714
Growth Rate (nm/min.)	4.2	5.05	7.14
Microhardness (GPa)	10±0.1	14±0.1	11±0.1
Nanohardness (GPa)	10±0.1	12.8±0.8	11.6±1.65
Modulus of Elasticity (GPa)	100±1	160±8	141±14.1
Friction Coeff. μ (average)	0.12	0.12	0.07
Wear Volume (m^3)	1.6×10^{-13}	1.54×10^{-13}	0.86×10^{-13}

Table I. Summary of measured properties of Si-DLC coatings deposited on silicon with the assistance of Ar, Ar plus N, and N ion beams.

Raman Spectra

The Raman spectra of the Ar/Si-DLC and N/Si-DLC coatings were measured [not shown] and resolved into a "D " peak" and a "G" peak associated with the sp^3 and sp^2 bond respectively. The "D" peak, occurring at 1340 cm^{-1}, results from defect-induced disorder in the microcrystalline graphite. The "G" peak, occurring at 1600 cm^{-1}, arises from the scattering sp^2 bonded carbon in graphite crystals. The relative intensity ratio of the two bands is indicative of the disorder present in the material with increased disorder accompanied by increased intensity of the "D" mode. The Si-DLC coatings produced from these two ion beams did not display any significant differences in their Raman spectra indicating no significant disorder differences between N and Ar only samples.

FTIR Spectra

Fig.1 shows the FTIR spectra of the Si-DLC coatings. Table II shows the peak position and mode assignment for the FTIR Ar/Si-DLC and N/Si-DLC spectra . FTIR spectra were taken for the N-only and Ar-only produced coatings. They show a 610 and 660 cm^{-1} IR mode which are assigned to Si-O bonds. The 809 and 820 cm^{-1} mode is assigned to the $(Si-H)^2$ vibrations. Other notable differences between the spectra of the two films are seen for the modes observed at 998, 1065 and 1250 cm^{-1}. For the N/Si-DLC coating there is a sharp mode present at 1065 cm^{-1} assigned to SiO_2, whereas for the Ar/Si-DLC coating there is a broad combined mode at 998 and 1250 cm^{-1}. The mode at 998 cm^{-1} is attributed to $C-H_3$ while the higher frequency 1250 cm^{-1} mode is indicative of $Si-CH_3$ bonds. These results indicate that Si is tetrahedrally bonded to hydrogen for both Ar/Si-DLC and N/Si-DLC coatings. However, SiO_2 bonding is present only in the N/Si-DLC coating.

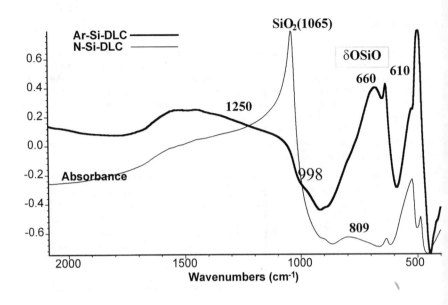

Fig. 1. FTIR spectra of N/Si-DLC and Ar/Si-DLC coatings on sapphire

Peak (cm^{-1})	Assignment
610	(äOSiO)^3
660	(äOSiO)^3
809	(Si-H)^2
820	(Si-H)^2
998	$-C\!\!\diagdown\!\!\begin{smallmatrix}H\\H\\H\end{smallmatrix}$
1065	(OSiO)^3
1250	(Si-CH)^2

Table II. Summary of FTIR modes

Microhardness, Nanohardness, Modulus of Elasticity, Adhesion, Wear

Knoop microhardness measurements were made using loads of 0.15 N which correspond to penetration depths of 1000 nm, larger than the maximum coating thickness for N/Si-DLC of 710 nm. The average Knoop microhardness of the (Ar+N)/Si-DLC coating (uncorrected for substrate effect) was 14 Gpa, 27% higher than the Knoop microhardness of the N/Si-DLC (11 GPa) and 40% higher than the Knoop microhardness of the Ar/Si-DLC coating (10 GPa). Nanoindentation measurements were made using a Nano Instruments XP nanoindenter, with a Berkovich 3-sided pyramid diamond indenter with controlled penetration depths of 300 nm. The instrument allowed a indenter penetration vs force curve to be determined allowing the determination of both (nano) hardness and the effective elastic modulus from the slope of the hysteresis curve [7]. These measurements follow the same pattern as the Knoop microhardness ones. The nanohardness of the (Ar+N)/Si-DLC coating was 12.8±0.8 Gpa, about 20% higher than the nanohardness of the N/Si-DLC (11.6±1.65 GPa) and 40% higher than the Knoop microhardness of the Ar/Si-DLC coating (10±0.1 GPa). The modulus of elasticity of the (Ar+N)/Si-DLC, N/Si-DLC and Ar/Si-DLC coatings was 160±8 GPa, 141±14.1 GPa and 100±1 GPa respectively.

No delamination was observed while testing the adhesion of the coatings to their underlying silicon substrates either by the so-called Scotch-Tape test or during ball-on-disk wear test measurements. The wear volumes of the N/Si-DLC, (Ar+N)/Si-DLC and Ar/Si-DLC coatings, determined with the aid of a stylus profilometer, were 0.86×10^{-13} m^3, 1.54×10^{-13} m^3, and 1.6×10^{-13} m^3 respectively [0.54::0.96::1.0]. The wear tracks on the surface of all coatings were barely visible to the unaided eye. Extrapolating the model developed by Rao and Lee [9] about the ion implantation on surface properties of polymers on our Si-DLC coatings we may attribute the improvements in properties to a consequence of cross-linking of the precursor material caused by the ion irradiation. The dual N-plus-Ar irradiation was better because it combined a deeper implant, in the form of N, along with Ar irradiation which resulted in a shallower but more highly crosslinked layer at the near-surface. Thus a deeper and graded crosslinked surface region is thought to have been formed. However, the wear rate may also be, at least partially, attributed to the different thickness of these Si-DLC coatings. Table I shows the microhardness and nanohardness results of all Si-DLC coatings.

Sliding Friction Coefficient

The variation of the unlubricated friction coefficient of the N/Si-DLC, (Ar+N)/Si-DLC and Ar/Si-DLC coating is shown on Fig. 2. The unlubricated friction coefficient of the N/Si-DLC coating is, for the most part of the 400 m distance traveled by the steel ball, significantly smaller than the friction

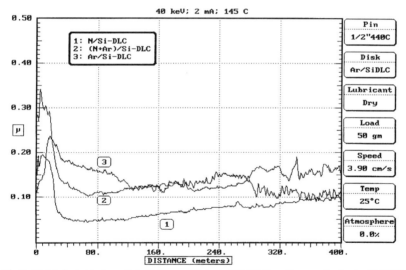

Fig 2. Unlubricated ball-on-disk friction coefficient of Si-DLC coatings synthesized with N[1], (Ar+N)[2], and Ar[3] ion beams

coefficient of the (Ar+N)/Si-DLC and Ar/Si-DLC coatings. Meletis et al. [10, 11] has attributed the smaller friction coefficient of his conventional DLC and our Si-DLC coatings to their graphitization due to frictional heating, generated by the rotating steel ball on the coating surface, during the ball-on-disk testing. The beginning of graphitization of the Si-DLC coating is indicated by the abrupt (downward) change of the slope of the curves during the initial 40 m travel distance of Fig. 2. However, the ball-on-disk friction coefficient of the N/Si-DLC and (Ar+N)/Si-DLC coatings, unlike the Ar/Si-DLC coating, for reasons not understood yet, increased monotonically during the course of the 400 m traveled distance. The latter may be attributed, as indicated by the FTIR results, to the formation of SiO_2 bonding, observed only in the coatings formed with ion beams containing nitrogen.

CONCLUSIONS AND FUTURE PLANS

Synthesis of Si-DLC coatings with N-containing ion beams, yields amorphous, non-conductive coatings with lower friction coefficients and significantly lower wear rates than the Si-DLC coatings we have produced to-date using only argon ions. This trend is tentatively attributed to the greater ion penetration of the N ions and to the observed correlation between SiO_2 content and lowered friction. XPS and Auger analysis within the wear track is planned to test this correlation. Nitrogen, and other light element ions will be used for the synthesis of the Si-DLC coating on other substrates including glass and composite materials of practical importance for Army applications. In addition the effect of the thickness of the Si-DLC coating on its wear and mechanical properties will also be examined.

REFERENCES

1. K. Enke, H. Dimingen and H. Huebsch, Appl. Phys. Lett., **36**, 291 (1980).

2. T. Hioki, S. Hibi and J. Kawamoto, Surf. Coat. Technol., **46**, 233 (1991).

3. K. Oguri and T. Arai, J. Mater. Res,. 7, 1313 (1992).

4. C. G. Fountzoulas, J. D. Demaree, W. E. Kosik, W. Franzen, W. Croft and J. K. Hirvonen in Beam-Solid Interactions (Mat. Res. Soc. Symp. Proc. **279**, Pittsburgh, PA, 1993), pp. 645-650.

5. U. Müller, R. Hauert, B. Oral, M. Tobler, Surf. Coat. Technol., **367**, 76-77 (1995).

6. C.G. Fountzoulas, J. D. Demaree, L. C. Sengupta, and J. K. Hirvonen, Materials Modification and Synthesis by Beam Processing, **438** (1997).

7. W. C. Oliver and G. M. Pharr, J. of Mater. Res. **7(6)**, 1564 (1992).

8. R. Doolittle, Nucl. Inst. Meth. B, **15**, 227 (1986).

9. P. Rao and E. H. Lee, J. of Mater. Sci., **10**, 2661 (1996).

10. E. I. Meletis, A. Erdemir and G. R. Fenske, Surface and Coatings Technol., **73**, pp. 39-45 (1995).

11. E. I. Meletis, State University of Louisiana (private communication).

PREPARATION OF HARD COATINGS ON POLYCARBONATE SUBSTRATE BY HIGH FREQUENCY ION BEAM DEPOSITION USING CH_4/H_2 GASES

S.R. KIM*, J.S. SONG*, Y.J. CHOI*, J.H. KIM**
* SamYang R&D Center, 63-2 Whaam-dong, Yusung-gu, Taejon, Korea
**Department of Materials Science, Korea University, Sungbuk-gu, Seoul, Korea

ABSTRACT

Polycarbonate is one of the most widely used engineering plastics because of its transparency and high impact strength. The poor wear and scratch properties of polycarbonate have limited its application in many fields. In order to improve the wear and scratch properties of polycarbonate we have deposited diamond like carbon (DLC) coatings. The diamond like carbon coatings were made using a high frequency ion beam gun by introducing H_2 and CH_4 gases. The coatings were characterized with scanning electron microscope, Raman spectroscopy, ellipsometer, and microscratch tester. Polymer hard coating was applied onto the polycarbonate substrate before depositing a diamond like carbon coating to see the effect of interlayer on the system's failure mechanism.

INTRODUCTION

Polycarbonate has an unusual combination of high impact strength, transparency, and good electrical insulation properties. The properties of polycarbonate have led to many successful glazing applications, such as bus shelter, telephone booths, safety goggles, and lamp housings for automobiles. However, the poor wear and scratch properties of polycarbonate have limited its application in many fields.

There have been many studies on improving the scratch and wear properties of polymer by applying coatings [1-4]. The DLC coatings have numerous useful properties such as transparency for visible and infrared light, high hardness and chemical inertness. They can be applied as a scratch-resistant coating. DLC coatings on a polycarbonate substrate maintained the high-speed impact strength [1].

Depositing DLC coating using a high frequency ion gun in this study made the polycarbonate surface wear-resistant and scratch-resistant. The effect of gas compositions and interlayer coating on deposited DLC coatings were characterized using a scanning electron microscope, Raman spectroscopy, ellipsometer, and microscratch tester.

MATERIALS

A proprietary acrylic polymer hard coating was applied to a Sam Yang TRIREX® polycarbonate substrate with thickness of 1/8". The pencil hardness of polycarbonate and polymer hard coating were EE and 4H, respectively. The polymer hard coating was applied by dipping method and the coating thickness was about 10 μm.

EXPERIMENTAL

The DLC coatings were deposited on a polymer substrate with a hard coating by a direct ion beam gun with a 60 MHz: high-frequency ion beam source operated on a mixture of hydrogen and methane gases. The beam energy was 500 eV and ion current was 25 mA. The base pressure of vacuum chamber before deposition was 3 x 10^{-5} torr and it became 2.5 x 10^{-4} torr when reaction gases were introduced. The distance between ion gun source and the substrate was 22 cm. During deposition, the substrates were put on a rotating sample holder to have coating uniformity.

Microscratch Tests

The critical load for coating failure was measured by a scratch test. A CSEM micro-scratch tester under loading rate of 100 N/min and a scratch speed of 1 cm/min was used. A diamond indenter, ground to a 120° cone with a spherical apex of 80 μm radius, was used in all measurements. The acoustic signal and friction coefficient corresponding to coating failure was recorded. The coating failures could be cracking, delamination, or the indenter plowing through the coating or substrate. The failure morphologies of coating or substrate layer were examined via scanning electron microscope.

RESULTS AND DISCUSSION

Characterization of DLC

In order to see the effect of gas composition on the deposited DLC coatings, we deposited DLC coatings with two gas compositions of CH_4: H_2 ratio of 1:1 and CH_4 to H_2 ratio of 1:2 on a silicon wafer. The deposition rate was measured with a scanning electron microscope. It was 4.8 nm/min for CH_4: H_2=1:1 and 4.0 nm/min for CH_4: H_2=1:2. The Raman spectrum and refractive index of DLC coatings with the two gas compositions were the same. Raman spectroscopy is often used to characterize the crystallinity of diamond and graphite thin films. The Raman spectrum of crystalline diamond consists of a single sharp peak at 1332 cm^{-1}. The Raman spectrum of crystalline graphite consists of two peaks: The G peak centered on 1550 cm^{-1} is the zone center E_{2g} mode of the perfect crystal and the D peak centered on 1350cm^{-1} is a zone edge A_{1g} mode activated by disorder in the graphite crystal [8]. The D mode is a common feature of disordered graphite carbon. Its intensity relative to the G mode (as measured by the ration I_D/I_G) changes with the disorder. We have deconvoluted the Raman spectra with two Gaussian peaks which correspond to G peak with frequency at 1558 cm^{-1} and D peak with frequency at 1395 cm^{-1}. The I_D/I_G was 2.2 for the two gas compositions. The refractive index of 2.0 at 632 μm was the same for the two DLC coatings.

Microscratch Results

The scratch properties of deposited DLC coatings were characterized using a microscratch tester. The microscratch results of DLC coating on polycarbonate are shown in Fig. 1(a) and its morphology is shown in Fig. 2. The scratch results and the fracture surface of DLC coating/hard coating/polycarbonate system are shown in Fig. 1(b) and Fig. 3, respectively. Figure 2(a) shows a scratched surface of DLC/polycarbonate system with increasing load to 60 N. The morphology of scratches on the DLC/polycarbonate showed good adhesion as in elsewhere [1]. Only slight scratches or indent marks were shown. DLC/polycarbonate system showed a low coefficient of fraction (approximately 0.4) at high load, while DLC/hard coating/polycarbonate system showed a higher value (approximately 0.5). The low coefficient of fraction of DLC/polycarbonate system

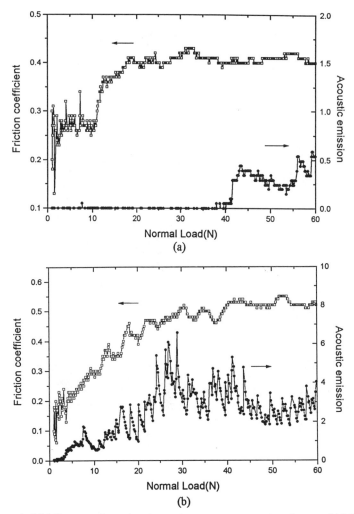

Figure 1. (a) Microscratch results of DLC (CH_4:H_2=1:1) on polycarbonate. (b) Microscratch results of DLC(CH_4:H_2=1:1) on hard coating/polycarbonate.

may be result of stylus sliding on the DLC coating layer due to good adhesion between DLC coating and polycarbonate substrate. Figure 2(b) shows a magnified view of scratch at 4 N. The cracks by indentation were occured in the direction of stylus movement without any noticeble delamination. The acoustic pattern showed a long base line that is the characteristic of smooth sliding without extensive coating failures. This coating did not show any significant acoustic signals up to 40 N or distinctive acoustic events of the coating failure. The surface morphology at 40 N of applied load is shown in Fig. 2(c). Some cracked coating layer were delaminated but most of them were well adhered to the polycarbonate substrate. The similiar surface morphology

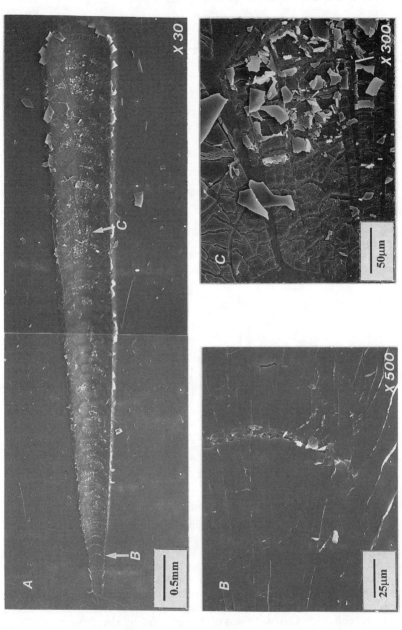

Figure 2. SEM micrographs of scratch on a DLC / polycarbonate system by microscratch test. (a) Total view of a scratch (0 N ~ 60 N), (b) Magnified view of scratched region at 5 N, (c) Magnified view of scratched region at 45 N.

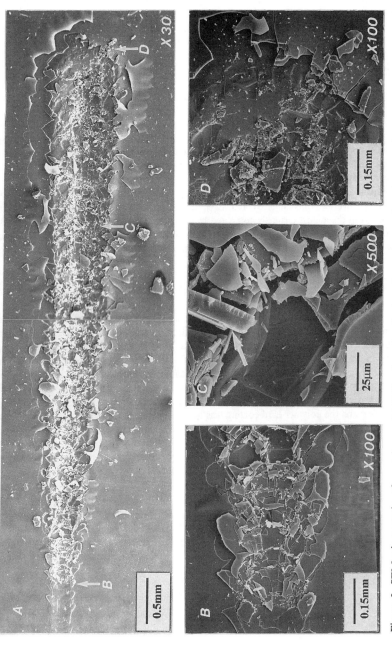

Figure 3. SEM micrographs of scratch on a DLC / hard coating / polycarbonate system by microscratch test. (a) Total view of a scratch (0 N ~ 60 N), (b) Magnifed view of a crack at 5 N, (c) Magnifed view of delaminated cracks at 40 N, (d) Magnifed view of scratched region at 60 N.

269

was retained up to 60 N. It is an example of good adhesion between DLC coating and polycarbonate. It is speculated that the DLC coating did not delaminate from the polycarbonate substrate because of its higher compliance and toughness. Hsieh et al. [1] tested DLC coated polycarbonate substrate. Delamination was not found with either the microscratch or the three-point bend tests.

A different failure mechanism was observed for a DLC/hard coating/polycarbonate system. DLC coating was deposited on a hard coated polycarbonate at a gas ratio of CH_4: H_2 =1:1. Figure 3(a) clearly shows substantial crackings and delaminations of coatings. Figures 3(b), 3(c), and 3(d) are magnified view of three different regions in Fig. 3(a). The rough surface of the scratched sample, and the debris of coating were observed. The coatings were cracked and delaminated at the initial loading of 5 N (Fig. 3(b)). An arrow mark in Fig. 3(c) shows a delaminated coating piece with DLC coating remained attached to the hard coating. It implies better adhesion between DLC coating and hard coating than between hard coating and polycarbonate substrate. A reproducible characteristic acoustic emission envelope was observed for this system.

The critical load at failure initiation on DLC coating deposited on polycarbonate or polymer hard coating interlayer was investigated. However, it was difficult to define a critical load for DLC coatings from the acoustic signals compared to that of DLC coatings on a silicon wafer. A characteristic acoustic envelope and clearly distinguishable signal for critical load for failure initiation were developed for a DLC coating on a silicon wafer. The critical load of DLC coating was 15 N on a silicon wafer.

CONCLUSIONS

DLC coatings on a polycarbonate and a hard-coated polycarbonate were studied. The interlayer effect on the failure properties of ion beam deposited DLC coatings was investigated using a scanning electron microscope and microscratch tester. Two distinctive failure modes were observed by microscratch tests. The introduction of interlayer has led to different failure mophologies. The deposited DLC coating on polycarbonate did not show any delaminations after microscratch tests. However, the DLC coating deposited on the hard-coated polycarbonate substrate showed coating crackings and delaminations. It was shown that the adhesion between the DLC coating and the polymer hard coating was good. The delamination between hard coating and the polycarbonate substrate was observed instead of delamination between DLC coating and hard coating.

REFERENCES

1. A.J. Hsieh, P. Huang, S.K. Venkataraman, and D.L. Kohlstedt, Mat. Res. Soc. Symp. Proc. **308**. (1993) pp. 653-658
2. B.J. Knapp, R.M. Kimock, R.H. Petrmchl, U.S. Patent, No. 5,508,368
3. S.J. Rzad, M.W. Devre, U.S. Patent, No. 5,431,963
4. G. Dearnaley, U.S. Patent, No.5.393.572
5. M. Yoshikawa, Materials Science Forum Vols. 52 & 53 (1989) pp.365-385
6. J.U. Oh, K.R. Lee, K.Y. Eun, Thin Solids Films **146**, (1987) pp. 173-176
7. H. Hofsass, H. Binder, T. Klumpp and E. Recknagel, Diamond and Related Materials, **3** (1993) pp.137-142
8. R.J. Nemanich and S.A. Solin, Phys Rev. B20, 392 (1979)

SYNTHESIS OF HARD NITRIDE COATINGS BY ION BEAM ASSISTED DEPOSITION

J. D. DEMAREE, C. G. FOUNTZOULAS, J. K. HIRVONEN
U. S. Army Research Laboratory, APG, MD

M. E. MONSERRAT, G. P. HALADA, C. R. CLAYTON
State University of New York at Stony Brook, Stony Brook, NY

ABSTRACT

Ion beam assisted deposition (IBAD) has been used to deposit chromium nitride coatings using 1200 eV nitrogen ions from an RF-type ion source and thermally evaporated chromium. The ion/atom arrival ratio R was varied from 0 to 8 to modify the coating composition, microstructure, growth rate and stress state in order to optimize the properties of the material for use as a possible substitute for electroplated chromium in a number of anti-corrosion and tribological applications. The coatings were examined using Rutherford backscattering spectrometry and x-ray photoelectron spectroscopy, and contained up to 44 at % nitrogen in a mixture of bcc Cr-N, Cr_2N, and CrN. The microstructure of the coatings was examined by scanning electron microscopy, and the tribological behavior of the coatings was examined using an automated scratch testing and nanoindentation. XPS examination of the coatings indicates that nitrogen near the surface was bound to the metal as CrN and Cr_2N in most of the coatings studied by XPS, which is expected to significantly affect their corrosion behavior. The high R values needed to form large amounts of the CrN phase in the bulk of the coating causes significant sputtering during deposition. This study indicates that it is not possible to form a coating consisting solely of cubic CrN by IBAD under these experimental conditions (room temperature substrate and partial pressure of nitrogen of 1.8 x 10^{-2} Pa). Nevertheless, the IBAD coatings produced were hard, in a compressive stress state, and highly adherent, all properties that make them candidates for use in selected Army applications.

INTRODUCTION

Environmental concerns regarding the disposal of toxic byproducts from the production of electroplated hard chromium (EHC), widely used as a wear- and corrosion-resistant tribological coating, have led many to consider ion beam assisted deposition (IBAD) and similar coating processes as potential replacements for electroplating technology. The ability of the IBAD process to carefully control coating properties like adhesion and stoichiometry allows us to produce coating systems that surpass the durability of EHC, with virtually no hazardous waste or chemical exposure during production. Specifically, IBAD chromium nitride (Cr_xN_y) is considered a candidate for EHC replacement because of its good corrosion resistance, high hardness, and overall similarity (in appearance and chemical compatibility) to EHC. Several studies have investigated the deposition of Cr_xN_y coatings using reactive ion plating,[1-4] reactive sputtering,[5-6] and IBAD,[7-8] and have concluded that Cr_xN_y coatings can be formed with a hardness greater than that of ordinary Cr[1-4,7] and a wear rate an order of magnitude lower.[7] To be considered a useful replacement process, however, IBAD must overcome some difficulties inherent in most low temperature PVD processes, e.g., high residual internal coating stresses[5] which can lead to coating delamination. The present study concerns the optimization of IBAD deposition parameters to maximize wear-resistance.

EXPERIMENTAL

The coatings were deposited in a vacuum chamber with a base pressure of 1.3 x 10^{-7} Pa. The operating pressure during deposition was 2.7 x 10^{-2} Pa, and consisted of approximately two parts N_2 (from the ion source) and one part Ar (from a plasma bridge neutralizer), yielding a nitrogen partial pressure of approximately 1.8 x 10^{-2} Pa. Chromium was thermally evaporated from a boron nitride crucible inside a tantalum foil resistive heater, and the deposition rate measured using a quartz crystal thickness monitor located near the substrate holder and shielded from the ion beam. The chromium deposition rate varied from 0.10 to 0.15 nm/s. The nitrogen ion beam was produced using an Ion Tech 3-cm RF ion source, operating at 1200 eV. The beam area at the target was approximately 10^{-2} m² (100 cm²), and the ion beam

271

current density was varied from 1 to 8 A/m^2 (100 to 800 $\mu A/cm^2$). The arrival rate of nitrogen ions at the target surface is not trivial to determine, since the ion beam contains both N^+ and N_2^+ ions, and measurement of the ion beam current density is not sufficient to determine the ion arrival rate. The relative amounts of N^+ and N_2^+ in the ion beam have not been measured in this study, but other studies[9] have found that such plasma sources consist of around 11% N^+ and 89 % N_2^+, yielding 1.89 nitrogen atoms per unit charge. This value has been used in calculating the ion/atom (i.e., N/Cr) arrival ratios in this study. This calculation ignores the arrival of background nitrogen gas. The substrates used included silicon and stainless steel (for microscopy, XPS, and mechanical testing), carbon foil (for Rutherford backscattering analysis), and thin microscopy cover slips made of glass (for assessment of the stress state of the deposited films).

The composition of the deposited films was measured by Rutherford backscattering spectrometry (RBS), using a 2 MeV He^+ beam and a scattering angle of 170°. The simulation program RUMP[10] was used to fit the spectra and determine the composition and thickness of the films. The bonding of the coating material near the surface was examined with X-ray photoelectron spectroscopy (XPS), using a V.G. Scientific (Fisons) CLAM2 hemispherical analyzer with an $AlK\alpha_{1,2}$ x-ray source and 20 eV pass energy. Specifics of the calibration and curve fitting procedure, including specific fitting parameters for each speci shown, are given elsewhere.[11]

The residual stress in the films was measured by depositing them onto 0.254-mm (0.01-inch) thick, 22.2-mm (0.875-inch) diameter glass cover slips. The stress in the deposited film caused the cover slips to bend, resulting in a convex or concave reflecting surface, depending on whether the film stress was tensile or compressive. The radius of curvature was estimated by finding the focal length of the concave side of t cover slip, and the average stress in the film was estimated using the stress/deflection equation for a thin film[12] using values of 70 GPa and 0.22 for the elastic modulus and Poisson's ratio of the glass slides. A CSEM-Revetest automatic scratch tester was used to evaluate the cohesive and adhesive strength of some the coatings.[13] This test used a 200 μm diameter diamond tip and a load gradient of 5 N/mm; the resultin acoustic emission signal was monitored to note the relative loads at which the coating underwent cohesive and adhesive failure.

RESULTS

Rutherford backscattering analysis showed that the films consisted entirely of Cr and N, with only 3-5 atomic percent (at. %) oxygen as a contaminant. The nitrogen content of the coatings is shown in Figure as a function of the approximate ion/atom arrival ratio. A clear dependence on the ion/atom arrival ratio i seen, with N compositions of 0 to 44 at. % achieved by varying the arrival ratio between 0 and 8. RBS als revealed that the ion/atom arrival ratio had a strong effect on the thickness of the coatings deposited. As shown in Figure 2, the highest ion/atom arrival ratios removed a significant amount of the evaporated Cr ion sputtering. Note that a N/Cr arrival ratio of nearly 10 is needed to completely remove the evaporating Cr, indicating a sputter yield on the order of 0.10. Little sputtering is seen at R values below 2. This is in qualitative agreement with predictions based on the Dynamic TRIM Monte Carlo code (Figure 2).

XPS spectra show that chromium atoms in the outer 4 nm of the coatings are bound to both oxygen ar to nitrogen, as well as to other Cr atoms in a metallic bond (Figure 3). Both Cr 2p and N 1s spectra show Cr-N bonds identical to those found in stoichiometric CrN and Cr_2N. Interestingly, the coating with the least amount of nitrogen (11 at. %), showed CrN bonding but not Cr_2N; other coatings had a constant CrN /Cr_2N ratio of roughly 4. Note, however, that we do not know the Cr 2p binding energy of Cr with N in b substitutional sites (Cr-N), and it may be similar to that of CrN (i.e., between the binding energy of metal Cr and that of Cr_2N). For this reason, the percentage of "CrN" in the Figure 3 schematic may include Cr- which would be consistent with earlier studies which found no CrN phase until N concentrations were mu higher.

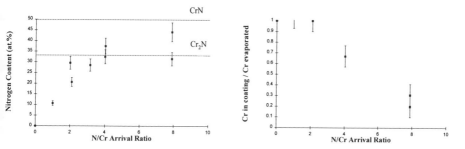

Figure 1.   Nitrogen content of deposited films and fraction of evaporated Cr remaining in the coating, as a function of the ion/atom arrival ratio.

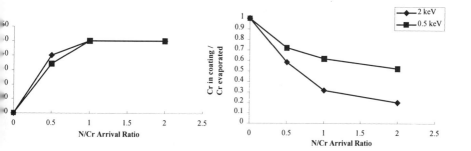

Figure 2.   Dynamic TRIM code prediction for nitrogen composition and sputtering loss, assuming all nitrogen ions are incorporated into the coating (maximum 50% nitrogen).

The average residual stress of the coatings was strongly affected by the ion/atom arrival rate (Figure 4). Cr films deposited with no ion beam bombardment were in tension (0.36 GPa), while all ion assisted films were in compression. The magnitude of the residual compressive stress in the ion assisted films appeared to reach a maximum of about 3 GPa around an ion/atom arrival ratio of 1 to 2, but then decreased with increasing ion bombardment. This behavior is typical of many IBAD coatings, and has been noted in chromium films growing under noble gas ion irradiation.[14]

The critical loads needed to fracture the films (cohesive failure) and to cause delamination (adhesive failure) in the automated scratch evaluation were influenced by the ion/atom arrival ratio (Figure 5). The increase in L_C is likely due to the high compressive stresses in that coating. The enhanced adhesion measured for the ion assisted films is likely a result of enhanced substrate cleaning and interface mixing in the early stages of film deposition.

273

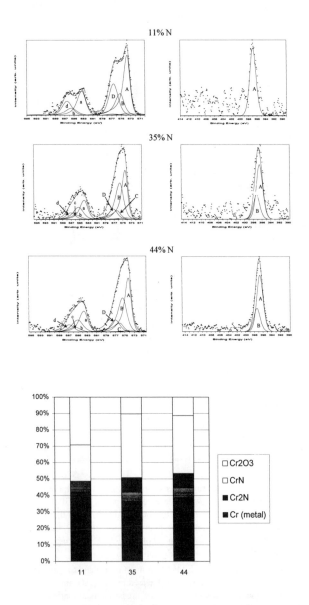

Figure 3.   Above: XPS spectra of Cr_xN_y films, showing Cr 2p (left) and N 1s (right), with peaks
              identified as Cr metal (A,a), CrN (B,b), Cr_2N (C,c), and Cr_2O_3 (D,d) for Cr 2p, and as CrN
              (A) and Cr_2N (B) for N 1s.
              Below: Schematic of Cr bonding distribution near coating surface.

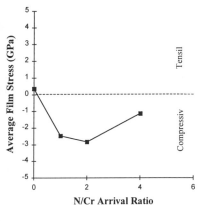
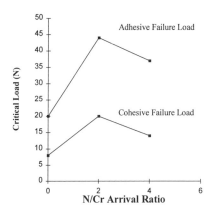

Figure 4. Average film stress, as estimated from the deflection of glass cover slips.

Figure 5. Critical loads for cohesive and adhesive failure during an automated scratch test

In a scanning electron microscope, a cross section of the Cr film deposited without ion bombardment clearly shows the columnar structure typical of low temperature PVD films (Figure 6a) and had a tendency to crack and delaminate. The $Cr_{70}N_{30}$ film produced in the same deposition run with ion beam assistance was thinner as a result of sputtering, but showed no evidence of columnar structure (Figure 6b). In addition, the $Cr_{70}N_{30}$ film remained adherent to the silicon substrate even when a scratch test was severe enough to fracture the substrate, showing no tendency to delaminate or crack.

Figure 6. Scanning electron micrographs of deposited coatings. a) 0.83 μm thick Cr deposition with no ion assist, and b) 0.42 μm thick $Cr_{70}N_{30}$ film with ion assist

DISCUSSION

The composition measurements show that the incorporation of N into the growing Cr film is a function of the ion/atom arrival ratio, and is not simply a reaction with the background nitrogen gas in the chamber. This is consistent with results from previous studies which have shown that a significantly higher pressure of nitrogen gas is needed to form nitride phases in Cr without energetic stimulation; e.g., in order to form the hard cubic CrN phase, nitrogen pressures of over 0.1 Pa are needed, and the phase does not form exclusively until pressures reach 0.5 Pa or higher.[3] In this study, the background pressure of N_2 during each run was only 0.018 Pa, so the energetic ion beam was necessary for the incorporation of N, either by direct implantation of nitrogen or by the stimulation of the film surface to react with the nitrogen gas in the chamber.

This study found that the ion/atom arrival ratios needed for significant N incorporation were much higher than unity, which is to be expected in ion irradiated thin films where defect production and the large number of grain boundaries stimulate the outward diffusion of nitrogen. The high irradiation rate results in very low deposition rates due to ion sputtering, however, which severely limits the ability of this technique to deposit thick films of the stoichiometric nitrides. In order to produce coatings consisting entirely of CrN, one might have to employ higher nitrogen partial pressures during deposition (perhaps by directed jets of nitrogen gas near the substrate) or the use of higher substrate temperatures to enhance the reactivity of the film surface with background nitrogen gas. Dynamic TRIM predictions (Figure 2) indicate that the lower sputter yields of lower energy ions may help overcome this limitation, so one may consider producing these films with End Hall or cluster beam ion sources.

The presence of the energetic ion beam has a strong effect on the stress state of the film (Figure 4), which in turn affects how the film behaves under a mechanical load (Figure 5). A tensilely stressed Cr film produced by vapor deposition without ion beam assistance is susceptible to cracking, and exhibits a relatively low critical load for cohesive failure in a scratch test. In contrast, ion assisted coatings with a compressive residual stress resist cracking, and their performance in the scratch test is enhanced. The high compressive stress state found in IBAD coatings must be carefully managed by appropriate deposition conditions in order to avoid delamination when thicker films are deposited and the integrated stress approaches the limit of coating adhesion. With the dual benefit of harder, crack-resistant films and enhanced adhesion, IBAD holds the promise of producing coatings with tribological performance comparable to or exceeding that of electroplated chromium.

CONCLUSIONS

Films of Cr_xN_y have been deposited by ion beam assisted deposition, using evaporated Cr and a nitrogen ion beam from an RF-type ion source, at a variety of ion/atom arrival ratios. The composition of the films ranged from pure Cr (no ion assist) to $Cr_{56}N_{44}$, and XPS indicated that Cr-N bonding is similar to CrN and Cr_2N. The ion/atom arrival rates needed to induce significant N incorporation were high enough to cause a large loss of coating material due to ion sputtering. For the conditions used in this study, it is unlikely that thick films of stoichiometric CrN can be produced. The presence of the energetic N ion beam induced a compressive stress in the growing films, converting the tensilely stressed Cr film into a compressively stressed Cr_xN_y film which was resistant to cracking and highly adherent to the substrates.

ACKNOWLEDGMENTS

The authors wish to thank C. Hubbard and P. Huang for assistance with the SEM examination, and W. E. Kosik for assistance with the RBS analysis of the films. In addition, we would like to acknowledge the assistance of S. Kirkpatrick of Epion Corp. in fabricating the heater used for Cr evaporation.

REFERENCES

1.   S. Komiya, S. Ono, N. Umezu and T. Narusawa, Thin Solid Films, 45 (1977) 433.

2.   T. Sato, M. Tada and Y. C. Huang, Thin Solid Films, 54 (1978) 61.

3.   D. Wang and T. Oki, Thin Solid Films, 185 (1990) 219.

4.   K. Kashiwagi, K. Kobayashi, A. Masuyama and Y. Murayama, J. Vac. Sci. Technol. A, 4(2) (1986) 210.

5.   P. M. Fabis, R. A. Cooke and S. McDonough, J. Vac. Sci. Technol. A, 8(5) (1990) 3809.

6.   K. K. Shih, D. B. Dove and J. R. Crowe, J. Vac. Sci. Technol. A, 4(3) (1986) 564.

7.   K. Sugiyama, K. Hayashi, J. Sasaki, O. Ichiko and Y. Hashiguchi, Surf. Coat. Technol., 66 (1994) 505.

8.   W. Ensinger, M. Kiuchi, Y. Horino, A. Chayahara, K. Fujii and M. Satou, Nucl. Inst. Meth. B, 59/60 (1991) 259.

9.   D. VanVechten, G. K. Hubler, E. P. Donovan and F. D. Correll, J. Vac. Sci. Technol. A, 8(2) (1990) 821.

10.   L. R. Doolittle, Nucl. Inst. Meth. B, 15 (1986) 227.

11.   G. P. Halada, D. Kim and C. R. Clayton, Corrosion, 52 (1) (1996) 36.

12.   J. A. Thornton, J. Tabock and D. W. Hoffman, Thin Solid Films, 64 (1979) 111.

13.   K. J. Bhansali and T. Z. Kattamis, Wear, 141 (1990) 59.

14.   D. W. Hoffman and M. R. Gaerttner, J. Vac. Sci. Technol. A, 17 (1980).

A DEPTH SELECTIVE MÖSSBAUER STUDY OF ION IMPLANTED STAINLESS STEEL

G. WALTER*, B. STAHL*, R. NAGEL*, R. GELLERT*, D.M. RÜCK**, M. MÜLLER*, G. KLINGELHÖFER*, E. KANKELEIT*, M. SOLTANI-FARSHI***, H. BAUMANN***
*Institute of Nuclear Physics, Darmstadt University of Technology, 64289 Darmstadt, Germany
**Heavy Ions Research Center (GSI), 64291 Darmstadt, Germany
***Institute of Nuclear Physics, Johann Wolfgang Goethe University, 60486 Frankfurt, Germany

ABSTRACT

High austenitic stainless steel of composition Fe62Ni20Cr18 was implanted with Eu ions and analyzed with Depth Selective Conversion Electron Mössbauer Spectroscopy (DCEMS). DCEMS gives information about the **depth profile of phases**, in this case about implantation induced changes in phase composition as function of depth. The samples were complementary analyzed with Rutherford Backscattering Spectrometry (RBS) to get the element profile of the implanted ions. The main experimental result is a martensitic transformed depth region that coincides with the Eu depth distribution. The potential of DCEMS for application in the field of materials modification by implantation techniques is demonstrated.

INTRODUCTION

The goal of our investigations is to study the dynamic evolution of the depth profile of implanted ions, which is determined by the nuclear and electronic energy loss of the ions in the host matrix, by sputtering of surface atoms and by ion induced or thermally driven diffusion. The basic parameters describing these processes, change dynamically with increasing ion fluence due to changes in atomic and phase composition within the implanted volume. To study these effects, originally high austenitic stainless steel of composition Fe62Ni20Cr18 (95% ^{57}Fe enrichment of iron) was implanted with Eu ions of different fluences. The samples were analyzed by Depth Selective Conversion Electron Mössbauer Spectroscopy (**DCEMS**) [1-2,4] to get information about the implantation induced phase composition as function of depth. The concentration profile of the implanted Eu ions was analyzed by Rutherford Backscattering Spectrometry (RBS).

EXPERIMENT

Sample preparation

The stainless steel samples were cold worked to a thickness of about 2 μm and implanted with 400 keV Eu ions of different fluences, ranging from $3 \cdot 10^{16}$ to $12 \cdot 10^{16}$ ions/cm^2. The implantations were done using the 300 kV implantation facility [5] at GSI, Darmstadt. The temperature during implantation was held at 77 K to minimize thermally driven diffusion processes. The residual gas pressure in the vacuum chamber ranged from $9 \cdot 10^{-8}$ to $2 \cdot 10^{-6}$ mbar, details are discussed in the next section. Neither cold working, nor cooling (77K) lead to a phase transformation, shown by Mössbauer experiments with an unimplanted sample of same composition.

DCEMS

To determine the phase composition in the implanted volume, the samples were analyzed with ^{57}Fe DCEMS at room temperature. The principle of this method can be summarized in the following way (details are described elsewhere, see e.g. refs. [2,4,6-8]): By Mössbauer Spectroscopy one can get information about the chemical and magnetic phase composition due to the hyperfine interactions of the ^{57}Fe probe nucleus with its electronic surrounding. They lead to isomer shift, quadrupole and/or magnetic splitting of the resonant signal, resulting in the phase information. Detecting the conversion electrons (e.g. K conversion electron energy 7.3 keV) and their energy loss, the method is sensitive to the topmost 100 nm of a sample, the typical depth range of implantation experiments. Using an orange type magnetic electron spectrometer [3,4], a set of CEM spectra can be measured at different electron energy settings. Caused by the energy loss of the electrons during their transport through the sample, the electron energy can be corellated with their depth of origin. This is visualized in fig. 1, where a DCEMS matrix as function of velocity and electron energy is shown for the sample implanted with $3 \cdot 10^{16}$ ions/cm^2: The measured Mössbauer spectra are plotted versus electron energy. The velocity axis describes the phase information via the hyperfine parameters, the energy axis is corellated to the depth information, via the energy loss of K and L conversion and Auger electrons.

RBS

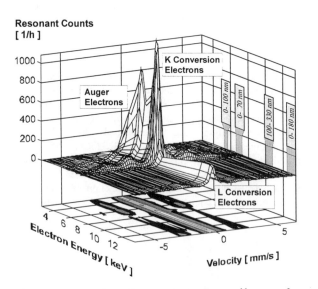

Figure 1: Measured DCEMS matrix for the $3 \cdot 10^{16}$ ions/cm^2 implanted sample (for details, see text).

Rutherford Backscattering Spectra were measured at the Institute of Nuclear Physics, University Frankfurt. Details of the method are described e.g. in [16]. Different samples were analyzed using 1.6 MeV protons or 7.6 MeV He ions as projectiles. The analysis was done using the RUMP code [9]: The sample is divided into discrete layers, RBS spectra are simulated varying the composition as function of depth. This procedure results in the element profile of the implanted Eu ions, assuming Bragg's rule to describe the mass density distribution. The energy resolution was 8 keV in the case of proton RBS and 15 keV in the case of He-RBS, resulting in depth resolutions of 16 nm and 5 nm at the sample surface, respectively.

RESULTS

The DCEMS analysis shows a martensitic transformed layer within the original high austenitic stainless steel matrix. As an example, this is demonstrated in fig.2, where a Mössbauer spectrum is shown for the $3 \cdot 10^{16}$ Eu/cm^2 implanted sample, measured at an elecron energy setting of 7.2 keV (K conversion electrons, energy loss of 0.1 keV) . The resulting Mössbauer components are

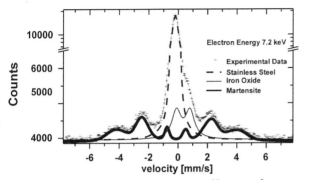

- a single line, describing the original austenitic phase
- a magnetically split component, that originates from a martensite phase, which is reported in refs. [10,11]. The average magnetic hyperfine field is distributed around 25 T, as observed for all samples.
- Furthermore, a broad doublet is observed that can be identified as

Figure 2: Mössbauer spectrum of the $3 \cdot 10^{16}$ ions/cm^2 implanted sample, measured at an electron energy setting of 7.2 keV.

iron oxide $Fe_{1-x}O$ ($x \approx 0$). It shows the oxidation of the surface near region caused by an adsorbate layer during implantation.

The sets of DCEM spectra are analyzed using extensive calculations developed in our group during the last years [2,4,6]: Monte Carlo simulations, modeling the transport of conversion electrons within the sample, are combined with a least square routine to fit the parameters of depth profile simultaneously with those of the hyperfine interactions. The resulting depth profiles of phases are shown in fig. 3 in comparison with the element profile of the implanted Eu ions. The left axis refers to the Eu concentration profiles, measured by RBS, and the right axis to the phase composition. One has to realize, that different depth scales are compared in the pictures: The DCEMS analysis results in a depth scale in nm α-iron, assuming specific electron stopping cross sections for iron, while the element profile contains the systematic error of unknown mass density distribution.

Looking at the dynamic evolution of the depth profiles as function of the ion fluence, one can see the following results: The maximum Eu concentration is increased from 8 at% at low ($3 \cdot 10^{16}$ ions/cm^2) to 17 at% at medium ($6 \cdot 10^{16}$ ions/cm^2) fluence and further decreased to 7 at% at the highest implanted fluence of $12 \cdot 10^{16}$ ions/cm^2. Furthermore, for the lowest fluence the maximum of Eu concentration is shifted to larger depths. Comparing the RBS with the DCEMS results, one can see clearly, that in all cases the element profile of the implanted Eu coincides well with the martensitic transformed region, concerning also higher moments of the distributions. In the case of highest fluence, the topmost 20 nm of the sample show a component with an average magnetic hyperfine field of 33.6 T. Furthermore, DCEMS shows an oxidation of the samples, the depth distribution of $Fe_{1-x}O$ strongly depends on the implantation conditions:

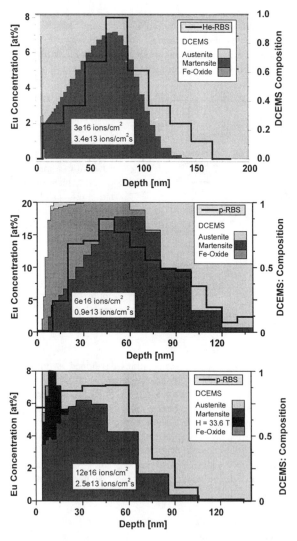

The samples implanted with high and low ion fluence were implanted at residual pressures of $<7 \cdot 10^{-7}$ mbar and $1 \cdot 10^{-7}$ mbar, respectively, and with ion fluxes of around $3 \cdot 10^{13}$ ions/(cm^{2}s). This results in a thin oxidized surface layer caused by surface adsorbates during implantation. In the case of medium ion fluence ($6 \cdot 10^{16}$ions/cm^2), the oxidation is present in the whole implanted volume. This sample was implanted with lower ion flux of $0.9 \cdot 10^{13}$ ions/(cm^{2}s) at a residual pressure of $2 \cdot 10^{-6}$ mbar. This allows the conclusion, that constituents of the adsorbate layer are continuously mixed into the implanted depth region. Fig. 3 also indicates the sensitivity of DCEMS to adsorbate covering during analysis, shown by the white bars at the surface of the depth profiles: a non-resonant adsorbate covering of the sample surface causes an additional energy loss of conversion and secondary electrons, which has to be taken into account for the quantitative analysis of the DCEM spectra.

Figure 3: Depth profiles from DCEMS and RBS analysis.

CONCLUSIONS

The dynamic evolution of the element profile during implantation is determined by the energy loss of the implanted ions and by the sputtering of surface atoms. Both depend strongly on the change of element and phase composition in the implanted depth region with increasing fluence. The evolution of the Eu profiles, as described above, can be understood qualitatively in the following way: The sucessive deposition of Eu in the implanted depth region leads to a decrease of projected range. Simultaneously the surface is sputtered, the sputtering yield varies also dynamically, if the surface composition of the sample is affected with increasing fluence [12,13], changing surface binding energy. This can explain the fluence dependence of the Eu distribution's mean depth. If the thickness of sputtered layer per fluence step increases stronger than the decrease in mean projected range, the maximum Eu concentration can decrease with higher fluences, as shown in fig. 3. In addition, in the case of medium fluence, the higher Eu concentration is also caused by the significant adsorbate covering during implantation, protecting the sample from being sputtered. A further aspect is the role of ion flux. Although the bulk of the sample is held at 77 K during implantation, it cannot be excluded, that the topmost adsorbate layer of the sample is strongly affected by thermal activation caused by the higher ion flux, resulting in an enhanced sputter yield. To quantify these effects, it is planned to complete the implantation setup with a pyrometer to detect the surface temperature in-situ during implantation.

Implantation induced martensitic transformations of austenitic stainless steels have also been reported by other authors (see refs. [10,11]). They implanted different stainless steels systematically with noble gas ions and analyzed the samples also with DCEMS. They were able to give crude information about the phase profile. The measured hyperfine parameters of martensite are in agreement with our results, but the detailed knowledge of the depth profiles of the implantation induced phases is due to the combination of high energy resolution and the high transmission of the orange spectrometer, which results in high statistical quality of the data. A rough measure of the depth resolution is indicated in fig. 3 by the varying binsize (to be published elsewhere).

In principle, the martensitic transformation can be understood as a stress induced plastic deformation of the austenitic fcc- to a bcc (bct)-phase, as shown in [10,11]. Especially the phase profile of the low fluence implanted sample shows, that ion beam induced defect production is not the primary mechanism to induce the transformation: The depth region near the surface shows a significant amount of the original austenitic phase. The agreement of the depth profiles of martensitic phase and Eu concentration shows, that deposited Eu atoms are responsible for the induced transformation. The Miedema parameters [14] show a low solubility of rare earth ions in α-iron [15], confirming the assumption, that the Eu atoms precipitate in the austenitic host matrix, causing a stress induced lattice distortion. To analyse the chemical state of Eu, further Mössbauer experiments using the [151]Eu Mössbauer isotope as probe nucleus for DCEMS are planned. Furthermore, systematic temperature dependent implantations are necessary to clarify the influence of temperature on the onset of the martensitic transformation process. E.g. the appearance of the 33.6 T component (which can be interpreted as a thin layer of α-Fe) in the sample implanted with highest fluence: It is not yet clear, if it is a boundary effect, connected with the significant amount of martensite reaching the sample surface. It is still an open question, which role plays the implantation prehistory, concerning the above mentioned dynamic profile evolution, possibly connected with a segregation of the alloy constituents.

The presented work demonstrates the high potential of DCEMS for non-destructive depth selective phase analysis, to study the dynamics of the implantation process. The detailed knowledge of the phase profile is necessary to understand the complex basic mechanisms underlying the process of ion implantation on the atomic scale. This is also of great interest for technological relevant applications. Especially our developments during the last few years, concerning the experimental setup and the analysis of DCEMS, opens up a number of perspectives to apply this method in the field of materials science. A new UHV orange spectrometer allows temperature dependent studies at low temperature, which is of importance for analysis of nanostructured materials. The development of a multi channel detection system improves the detection efficiency significantly. The drastic reduction of measuring time allows to apply the method to technological relevant systems with iron in natural isotope abundance and to other Mössbauer isotopes. Furthermore, the implantation facility at GSI is improved by a new setup, containing a CEM Spectrometer. Temperature dependent implantations combined with on-line Mössbauer analysis will be possible to minimize the influence of after-effects, e.g. caused by thermal activation.

ACKNOWLEDGEMENTS

We thank Prof. Dr. K. Bethge, Dr. K.E. Stiebing and Dr. O. Fröhlich, Institute for Nuclear Physics, University Frankfurt, for their support and the possibility to carry out the RBS measurements. This work is supported by GSI and BMBF.

REFERENCES

[1]   E. Kankeleit, Z. Phys. **164**, 442 (1961)
[2]   R. Gellert, O. Geiss, G. Klingelhöfer, H. Ladstätter, B. Stahl, G. Walter, E. Kankeleit, Nucl. Inst. Meth. **B76**, 381 (1993)
[3]   E. Moll, E. Kankeleit, Nukleonik **7**, 180 (1965)
[4]   B. Stahl, E. Kankeleit, Nucl. Inst. Meth. **B122**, 381 (1997)
[5]   D.M. Rück, N. Angert, H. Emig, K.D. Leible, P. Spädtke, D. Vogt, B.H. Wolf, Nucl. Tracks Rad. Meas, Vol. **19**, Nos 1-4, 951 (1991)
[6]   R. Gellert, PhD thesis, Institute of Nuclear Physics, Darmstadt University of Technology 1998, in preparation
[7]   G. Walter, B. Stahl, R. Gellert, R. Nagel, D.M. Rück, E. Kankeleit, Materials Science Forum, Vols. **248-249**, 189 (1997)
[8]   K. Nomura, Y. Ujihira, A. Vértes, Jounal of Radioanalytical and Nuclear Chemistry, Articles, Vol. 202, Nos 1-2, 103 (1996)
[9]   L.R. Doolittle, Nucl. Inst. Meth. **B9**, 344 (1985)
[10]   E. Johnson, A. Johansen, L. Sarholt Kristensen, L. Grabaek, N. Hayashi, I. Sakamoto, Nucl. Inst. Meth. **B19/20**, 71 (1987)
[11]   A. Johansen, E. Johnson, L. Sarholt-Kristensen, S. Steenstrup, E. Gerritsen, C.J.M. Denissen, H. Keetels, J. Politiek, N. Hayashi, I. Sakamoto, Nucl. Inst. Meth. **B50**, 119 (1990)
[12]   G. Walter, R. Heitzmann, D.M. Rück, B. Stahl, R. Gellert, O. Geiss, G. Klingelhöfer, E. Kankeleit, Nucl. Inst. Meth. **B113**, 167 (1996)
[13]   G. Walter, PhD thesis, Institute of Nuclear Physics, Darmstadt University of Technology 1997, to be published
[14]   A.R. Miedema, R. Boom, F.R. de Boer, J. Less-Common Mat. **41**, 283 (1975)
[15]   B.D. Sawicki, J.A. Sawicka, Nucl. Inst. Meth. **209/210**, 799 (1983)
[16]   J.A. Leavitt, L.C. McIntyre, M.R. Weller in <u>Handbook of Modern Ion Beam Materials Analysis</u>, edited by J.R. Tesmer, M. Nastasi, J.C. Barbour, C.J. Maggiore, J.C. Mayer, Materials Research Society, 37 (1996)

CRACK NUCLEATION IN ION BEAM IRRADIATED MAGNESIUM OXIDE AND SAPPHIRE CRYSTALS

V.N. GURARIE*, D.N. JAMIESON*, R. SZYMANSKI*, A.V. ORLOV*, J.S. WILLIAMS**
*School of Physics, MARC, University of Melbourne, Parkville VIC. 3052 Australia
**Department of Electronic Materials Engineering, Research School of Physical Sciences and Engineering, ANU, Canberra, 0200, Australia

ABSTRACT

Monocrystals of magnesium oxide and sapphire have been subjected to ion implantation with 86 keV Si⁻ ions to a dose of $5x10^{16}$ cm^{-2} and with 3 MeV H$^+$ ions with a dose of $4.8x10^{17}$ cm^{-2} prior to thermal stress testing in a pulsed plasma. Fracture and deformation characteristics of the surface layer were measured in ion implanted and unimplanted samples using optical and scanning electron microscopy. Ion implantation is shown to modify the near-surface structure of samples by introducing damage, which makes crack nucleation easier under the applied stress. The effect of ion dose on the thermal stress resistance is investigated and the critical doses which produce a noticeable change in the stress resistance is determined for sapphire crystals implanted with 86 keV Si⁻. In comparison with 86 keV Si⁻ ions the high energy implantation of sapphire and magnesium oxide crystals with 3 MeV H$^+$ ions results in the formation of large-scale defects, which produce a low density crack system and cause a considerable reduction in the resistance to damage. Fracture mechanics principles are applied to evaluate the size of the implantation-induced microcracks which are shown to be comparable with the ion range and the damage range in the crystals tested. Possible mechanisms of crack nucleation for a low and high energy ion implantation are discussed.

INTRODUCTION

Surface modification by ion implantation has been previously shown to alter a number of strength and fracture characteristics of lithium fluoride, magnesium oxide and glass samples [1,2]. In particular, thermal shock testing of LiF and MgO crystals implanted with Ar$^+$ or Si⁻ ions has revealed that the fracture threshold is lowered by ion implantation, allowing fracture to be initiated at lower surface temperatures. At the same time, ion implantation produces a higher density of cracks, but such cracks penetrate smaller distances into the material. This effectively raises the damage resistance parameter [3,4]. The observed modification of fracture behaviour is due to the formation of surface energy- absorbing layers which are known to improve the impact and thermal shock resistance of ceramic materials. Fine cracks developed in such layers limit the strength of the material, but provide an effective mechanism for absorbing strain energy during thermal shock and preventing catastrophic crack propagation [5].

Ion implantation has been shown to be effective in generating numerous crack nucleating centers in lithium fluoride, magnesium oxide and glass samples [1,2]. These centers are readily activated to develop multiple microcracking and thus to effectively absorb the strain energy under the applied stress. Ion implantation using 86 keV Si⁻ and Cr$^+$ ions is shown to be capable of producing highly efficient surface energy-absorbing layers in magnesium oxide crystals [6].

This study aims at (1) analysing the effect of implantation dose on the thermal stress resistance of sapphire crystals implanted with 86 keV Si⁻ ions, (2) evaluating the size of the implantation-induced microcracks and their relation to the ion range and the damage dimensions in the implanted crystals and (3) comparing the crack initiation effects produced by the 80 keV Si⁻ and 3 MeV H^+ irradiation in both sapphire and MgO crystals.

EXPERIMENTAL METHOD

The (0001) faces of sapphire monocrystals were implanted with 86 keV Si⁻ ions to doses ranging from 5×10^{14} cm⁻² to 5×10^{16} cm⁻² at room temperature. Sapphire and magnesium oxide crystals were also irradiated with 3 MeV H^+ ions with doses of 4.8×10^{17} cm⁻² at room temperature. Only half of the crystal surface was subjected to implantation to compare the fracture behaviour and characteristics for implanted and unimplanted regions subjected to identical thermal stress loading.

Thermal shock was produced by exposing the sample surface of ~ 1.5 x 3 cm² to a plasma jet produced by a plasma gun [1,2]. The plasma pulse duration was ~40 μs. The samples were placed at some distance from the gun with the plasma jet propagating perpendicular to the sample surface. The temperature and heat flux were continuously reduced from the central part of the plasma-affected area in a radial direction along the x axis which separates the implanted and unimplanted areas.

This arrangement allows both implanted and unimplanted regions to be tested under similar thermal conditions with the surface temperature varying from room temperature up to the melting point and above within roughly a 10-20 mm range along the x axis [1]. The temperature gradient is estimated to be ~ 200 °C/mm in the x direction. However, due to the short duration of the plasma pulse, the temperature gradient in the direction perpendicular to the surface is about three orders of magnitude higher than that across the surface. Therefore, the thermal stresses produced under these conditions are mainly due to the temperature gradient in the direction perpendicular

Fig. 1. SEM photograph of cracking in sapphire crystal.

to the surface. This produces a planar stress, with a zero stress component normal to the surface.

In these experiments the surface temperature along the x axis is calibrated by measuring the size of fragments, bounded by cracks, and the gap between them. This gap is formed as a result of the contraction of adjacent fragments on cooling from the fracture temperature. Thus, by the end of cooling the relative temperature-induced deformation of the fragment is $\Delta b/b$, where Δb is the gap between the fragments and b is the fragment size. On the other hand, the relative (temperature) deformation is known to be equal to $\Delta b/b = \alpha (T_f - T_0)$, where α is the thermal expansion coefficient, T_f is the fracture temperature, at which a crack separating adjacent fragments originates, and T_0 is the final (room) sample temperature. The fracture temperature is then determined from the expression: $T_f = T_0 + \Delta b / \alpha b$. Details of this treatment are given elsewhere [1].

For a pure elastic behaviour the fracture temperature is equal to the actual temperature T_m corresponding to cooling from a zero stress level. Ion implantation generally suppresses the plastic strain, so for the maximum dose of 5×10^{16} cm^{-2} used in the experiments we assumed the plastic strain prior to fracture to be negligible, so $T_m \cong T_f$. The gaps between fragments are often very small, particularly between small fragments and at low peak temperatures, so it is difficult to resolve them under an optical microscope and an SEM is used in these cases. Fig. 1 shows SEM micrograph of the fragmentation pattern in the sapphire crystal where the gaps between the fragments are well resolved.

RESULTS AND DISCUSSION

An optical photograph of the fracture pattern of the sapphire crystal, subjected to thermal shock in a plasma, is shown in Fig. 2. The temperature is increased along the x axis in an upward direction towards the center of the plasma affected area. The x axis separates the implanted and unimplanted regions with the left hand side of the crystal irradiated with 86 keV Si$^-$ ions to a dose of 3.0×10^{16} cm^{-2}. The low temperature edge of the fracture zone corresponds to the start of fracture where the peak surface temperature corresponds to the fracture threshold, that is, the minimum temperature variation (relative to the mean sample temperature) required to initiate fracture. The photograph illustrates that in the irradiated region the fracture extends further into a

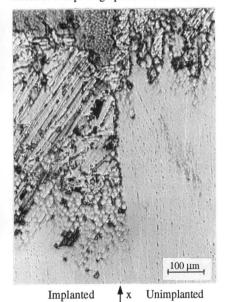

region of lower temperature (lower x values), indicating the reduction in the thermal shock resistance following implantation. Similar effects have been observed in lithium fluoride and magnesium oxide crystals irradiated with Ar$^+$ and Si$^-$ ions respectively [1,2].

The temperature along the x axis has been calibrated by analysing the fragment size and the gaps between them as described above. For the maximum dose of 5×10^{16} cm^{-2} used in the experiments the plastic strain prior to fracture was taken to be negligible and the fracture temperature is then equal to the actual peak temperature corresponding to the start of cooling. In this way the minimum temperature variation necessary to initiate fracture has been evaluated for different implantation doses for sapphire crystals irradiated with 86 keV Si ions and the data are presented in Fig. 3.

Fig. 2. Fracture zones in sapphire crystal. Implantation with 86 keV Si$^-$ ions, dose: 3×10^{16} cm^{-2}

Implanted ↑x Unimplanted

The graph shows that the minimum temperature variation necessary to initiate fracture is decreased with the dose. However, this effect is only noticeable above a certain critical dose, which corresponds to ~1.0×10^{15} cm^{-2}. Below the critical dose the implanted

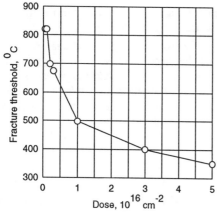

Fig. 3. Thermal stress resistance parameter
of sapphire crystals versus dose, implantation
with 86 keV Si⁻ ions.

crystals have the same thermal stress resistance parameter as the unimplanted ones. This suggests that below the critical dose ion implantation produces crack nucleating defects which are less severe than the ones preexisting in the crystal. With this in mind it is worth noting that the critical dose and the corresponding lattice damage are obviously dependent on the size and distribution of the preexisting crack nucleating defects in the crystal.

To identify the properties affected by ion implantation and analyse the effect of ion dose, the minimum temperature variation required to initiate fracture is presented in the form:

$$\Delta T_m = T_m - T_0 = (\epsilon_p/\alpha) + (1-\mu)(\sigma_f/\alpha E) \qquad (1)$$

where the terms $(1-\mu)(\sigma_f/\alpha E)$ and (ϵ_p/α) are the effective temperature variations corresponding to the elastic and plastic strain prior to fracture respectively; ϵ_p is the plastic strain, E is the elastic modulus, σ_f is the fracture stress. The fracture stress is known to depend on the length of existing crack-nucleating defects according to the expression [7]:

$$\sigma_f = K / \sqrt{\pi c} \qquad (2)$$

where K is the fracture toughness and c is the length of a single-ended crack.

Equations (1) and (2) are used to estimate the size of crack-nucleating defects in the implanted crystals, which corresponds to the experimentally observed temperature threshold of cracking, ΔT_m, and to compare it with the dimensions of the ion beam-induced damage. The temperature threshold of fracture is measured at the coordinate x corresponding to the edge of the fracture zone, where the fracture temperature is minimum and $\Delta T_m \approx T_f - T_0$. For sapphire crystals implanted to a maximum dose of 5×10^{16} cm^{-2} the temperature threshold of fracture $\Delta T_m \approx 360^0$ C. Since the dose increases the amount of damage which limits the dislocation mobility, the plastic strain prior to fracture is likely to decrease with increasing dose. For the purpose of estimation at the maximum dose used in the experiments the quantity ϵ_p is ignored. The estimation uses the material constants for unimplanted sapphire crystals as follows: $E=3.5 \times 10^{11}$ Pa [8], the surface energy $\gamma = 12.0$ Jm^{-2} [9], $\alpha = 9.0 \times 10^{-6}$ deg^{-1} [10], $\mu = 0.3$ [11]. The fracture toughness is determined using its relation to the surface energy [7]: $K = (2\gamma E)^{1/2} \approx 2.7$ MPa.m$^{1/2}$. The estimation gives the size of the crack-like defects in the implanted crystals of ~0.29 μm.

On the other hand, from TRIM calculations the range for the 86 keV Si⁻ ions in sapphire is ~0.1 μm. However, ion channelling measurements, to be reported in detail elsewhere [12], indicate that the lattice damage introduced by the implantation extends beyond the ion range to about 0.3 μm for Si ions. Thus, the microcrack size corresponding to the strength of the implanted crystals and the dimensions of ion beam induced damage are comparable. This suggests that the implantation-induced defects are developed into crack-nucleating defects under applied stress by growing with relative ease within the damaged layer. This happens because the lattice damage obviously decreases the fracture toughness K, so according to expression (2) the stress required for growing even very small implantation-induced crack-nucleating defects can be substantially reduced.

The ease with which these defects are developed into microcracks depends on the dose. The graph in Fig. 3 demonstrates that these microcracks may dominate preexisting defects in determining the strength and thermal shock resistance of the ion implanted crystals. However, microcrack growth beyond the damaged layer is determined by the fracture toughness of the unimplanted material and therefore requires higher stresses. Therefore, once the microcracks reach the damage range a further increase in the ion dose is not likely to significantly affect the fracture stress. With this in mind, the fracture stress is not expected to be the only factor that reduces the thermal stress resistance parameter as shown in Fig. 3. As pointed out earlier the plastic strain prior to fracture should decrease with dose due to the increase in the dislocation density that tends to suppress plastic deformation. Other properties involved in equation (2), in particular the thermal expansion, can also be affected by ion implantation, but such effects have not been considered in the treatment here.

IRRADIATION WITH 3 MEV PROTONS

Implantation with 3 MeV H^+ ions to a dose of 4.8×10^{17} cm^{-2} results in a number of characteristic features as compared to the 86 keV Si$^-$ implantation and unimplanted sapphire crystals. In the irradiated region the fracture pattern extends further into a region of much lower temperature. The crack density in the region is much lower, ~10 times, and the depth of crack penetration is much higher. The fracture morphology is also different. The cracks propagate along a single crystallographic system of cleavage planes, whereas in the unirradiated crystals and in the crystals implanted with the 86 keV Si$^-$ ions the cracks usually propagate along two crystallographic systems of cleavage planes. In many respects similar effects, which will be reported elsewhere, have been observed in magnesium oxide crystals implanted with 2.9 MeV H^+ ions.

A deep penetration of cracks extending into a region of low temperature is the result of a considerable reduction in the thermal stress resistance parameter following high energy proton implantation. Measurements of the fragment size and the separation between them near the edge of the fracture zone in the implanted area show that the minimum temperature variation required to initiate fracture here is ~40°C. This is about 10 times lower than in the crystals implanted with 86 keV Si$^-$ ions to a dose of 3×10^{16} cm^{-2}.

From TRIM calculations the range of the 3 MeV protons in sapphire crystals is ~50 μm, which is two orders of magnitude higher than the range of 86 keV Si$^-$ ions. Assuming the microcrack length c to be equal to the range and ignoring the plastic strain prior to fracture for ion implanted crystals, $\varepsilon_p = 0$, from equations (1) and (2) the fracture stress and the thermal stress resistance parameter for sapphire crystals irradiated with the 3 MeV protons should be about one order of magnitude less than those quantities for the crystals irradiated with the 86 keV Si$^-$ ions. This result is consistent with the experimental data suggesting that ion implantation generates microcracks of a size comparable with the ion range and the depth of the implantation-induced damage.

However, the mechanism of crack nucleation is apparently different for 86 keV Si ions and 3 MeV protons. TRIM calculations show that the 86 keV Si^- ions of a dose of $1.0x10^{15}$ cm^{-2} produce the maximum lattice damage of ~$1.6x10^{15}$ displ./$\overset{\circ}{A}$. According to the graph in Fig. 3 this dose and the corresponding damage can be considered as the critical ones below which they are insufficient to change the stress resistance of sapphire crystals. However, the damage produced by the 3 MeV protons at a dose of $4.8x10^{17}$ cm^{-2} is calculated to be ~$3.2x10^{14}$ displ./$\overset{\circ}{A}$ that is much below the critical one. This result suggests that the lattice damage is presumably not the only factor which leads to crack nucleation in sapphire crystals irradiated with 3 MeV protons. A major cause for crack nucleation in this case may be the formation of gas bubbles due to coalescence of hydrogen. Indeed, coalescence of hydrogen in gas bubbles is well known to occur in Si substrates after high dose proton irradiation [13]. And have been also observed in other materials [14,15]. The cavities are likely to be formed at the ion range where the concentration of hydrogen ions is maximum. Under applied stress or pressure developed in the cavities they can grow towards the surface forming microcracks of length equal to the range.

CONCLUSION

Ion implantation is shown to modify the near-surface structure of samples by introducing damage, which makes crack nucleation easier under the applied stress. Above a critical dose the thermal stress resistance parameter is decreased with increasing ion dose. The effect is largely associated with the reduction in the fracture stress and plastic strain prior to fracture. In comparison with 86 keV Si^- ions the high energy implantation of sapphire and magnesium oxide crystals with 3 MeV H^+ ions results in the formation of large-scale defects, which produce a low density crack system and cause a considerable reduction in the resistance to damage. The results suggest that in crystals irradiated with 86 keV Si^- ions crack nucleation is mainly due to lattice damage, while in the crystals irradiated with the 3 MeV protons it is likely to be associated with gas bubble formation.

REFERENCES.

1. V.N. Gurarie, J.S. Williams and A.J. Watt, J. Mat. Sci. and Eng., A189, p.319 (1994).
2. V. N. Gurarie and J. S. Williams, J. of Mat. Res., 5, No. 6, 1257 (1990).
3. D.P.H. Hasselman, Mat. Sci. and Eng., 71, 251 (1985).
4. D.P.H. Hasselman and J.P. Singh, Am. Cer. Soc. Bull., 58, No.9, 856 (1979).
5. D.W. Richerson, Modern Ceramic Engineering, Marcel Derrer, Inc. Ch. 4.5 (1982)
6. V. N. Gurarie, A.V. Orlov and J. S. Williams, Nucl. Instr. and Meth. in Phys. Res. B 127/128, 616-620 (1997)
7. R.A. Flinn and P.K. Trojan, Eng. Mater. and Their Appl., Houghton Miffin Co., 2nd Ed, (1981)
8. R.W. Davidge and D. Tappin, Trans. of Brit. Cer. Soc., 66, 8, 405 (1967).
9. F.J.P. Clarke, H.G. Tattersall and G. Tappin, Proc. Brit. Cer. Soc., 6, 163 (1966).
10. J.R. Hague, J.F. Lynch, A Rudnick, F.C. Holden, and W.H. Duckworth, In "Refractory Ceramics for Airospace", Am. Cer. Soc., Inc. (1964).
11. M.F. Ashby and D.R.H. Jones, Engineering Materials 2, Pergamon Press (1986)
12. V.N. Gurarie, D.N. Jamieson, J.S. Williams, J. Appl. Phys. (to be submitted).
13. J. Wong-Leung, E. Nygren and J.S. Williams, Appl. Phys. Lett. 68, 417 (1995)
14. P.B. Johnson, K.L. Reader, R.W. Thomson, J. of Nuclear Mater., 231 (1-2), 92, Jul (1996).
15. G.A. Hishmeh, L. Cartz, F. Desage, C. Templier, J.C. Desoyer, R.C. Birtcher, J. Mat. Res. 9(12), 3095-3107 (1994).

METASTABLE MATERIALS FORMATION BY ION BEAM ASSISTED DEPOSITION: APPLICATION TO METAL CLUSTERS IN CERAMIC MATRICES

* C.A. CAROSELLA, *G.K. HUBLER, *C.M. COTELL, AND **S. SCHIESTEL
*Surface Modification Branch, US Naval Research Laboratory, Washington, DC 20375
**University of Heidelberg, Germany

ABSTRACT:

The collision cascade, the fundamental event in ion-solid interactions, is responsible for the beneficial effects on thin films deposited by low energy ion beam assisted deposition (IBAD) or by energetic ion assisted deposition processes in general. However, the fundamental implications of the marriage of collision cascades and film growth processes have yet to be fully realized. The first half of this paper reviews the effects of ion bombardment on film growth and reaches some new conclusions. We propose that IBAD represents a different ion-solid interaction in a fundamental sense, and that as such, it should lead to new microstructures unattainable by other materials synthesis methods.

The second part of this paper discusses the deposition of metal nanoclusters in a dielectric matrix by means of beam assisted phase separation (BAPS), a term coined here to describe deposition of phase-separated multicomponent materials. Examples discussed are gold nanoparticles in both niobium oxide and silica matrices.

INTRODUCTION

Ion beam assisted deposition (IBAD) has achieved significant penetration into the optical thin-film industry due primarily to the fact that IBAD films are denser than evaporated or sputtered films. The higher densities increase both optical performance and production yield. The possibility for the production of unique metastable materials by IBAD has not been a focus of the research using the technique, with some exceptions (e.g., c-BN, MoN (B1 phase), Cu_5O_4) [1,2,3]. In this paper we explore the fundamental aspects of ion-solid interactions and reach the following conclusions: IBAD processing drives material synthesis parameter space into new regimes in the time-temperature-flux continuum, and, this attribute of IBAD processing leads to microstructures far from equilibrium that are unattainable by other materials synthesis methods.

Energetic deposition techniques have been studied for over 30 years [4] and intensively investigated for the last 15 years [5, 6]. The term "energetic deposition" is blanket term used to describe many different deposition techniques that possess one common feature - the application of energetic species to the surface of the growing film [7-9]. Ion beam assisted deposition is singled out in this paper, but the same principles will apply to other energetic deposition techniques as well.

We will first discuss the basic mechanisms of ion-solid interactions and film growth. This follows with an example of a process that uses IBAD to control the kinetics of phase transformations that are thermodynamically favored but kinetically inhibited. We call this process beam assisted phase separation, or BAPS.

ION BEAM ASSISTED DEPOSITION (RELATIONSHIP TO ION-SOLID INTERACTIONS)

Since the mid 1980's, ion beam assisted deposition (IBAD) has been heavily investigated for production of thin films [5, 6, 10]. The IBAD process, schematically represented at the top of Fig. 1, incorporates all of the ion-solid interaction phenomena that influence the phases which form - namely, radiation enhanced diffusion (RED) [11-14], radiation induced segregation (RIS) [15], ion implantation (II) [16, 17] and ion mixing (IM) [18 19, 20]. The question to address is, how does IBAD differ from ion irradiation that is used to investigate RED, RIS and IM, and from ion implantation that is used to alter the material composition?

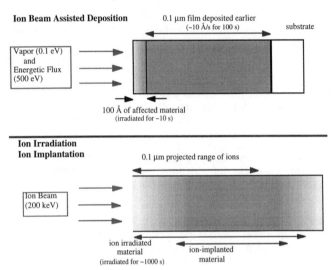

Figure 1:   Schematic representation of and ion beam assisted deposition (top) and ion implantation (bottom).

Recall that the IBAD consists of the build-up of a coating by energetic species (see Fig. 1) that impinge at high rates onto a substrate. The low energy ions used in IBAD cause displacement damage when they collide with the atoms already deposited. The typical depth affected by the low energy ions is about 100 Å. On the other hand, ion irradiation consists of passing ions though a solid and creating copious amounts of point defects through atomic collisions. The defects dynamically interact with the lattice and modify the structure, even to the extent of changing the phase. Ion irradiation occurs in a material with constant composition. Finally, ion implantation is the same as ion irradiation with one important difference - the ion species becomes one of the constituents of the material.

FUNDAMENTAL DIFFERENCES BETWEEN ION IMPLANTATION AND IBAD

Figure 1 schematically depicts the processes of ion beam assisted deposition (top) versus ion irradiation and ion implantation (bottom) in such a way as to emphasize the major differences in each process. For this comparison, assume that; a) the final average material composition represented by the solid shading, is the same for each process, b) the thickness of material in each case that is modified or produced is equal to about 0.1 μm (a very practical consideration), and c) the implanted ion is a major constituent (~20 at. %) of the final material. The latter implies high implantation dose and significant sputtering so that the implanted-ion profile is not a simple Gaussian distribution.

The differences in the processes to note are: i.) the energy required to affect the 0.1 μm of material is ~400x smaller (500 eV vs 200 keV) in the IBAD process than in the ion implantation process. (This is due to the fact treatment depth is accomplished by deposition in the case of IBAD, and ion penetration in the case of ion implantation); ii.) the total dpa damage to achieve the desired composition is much larger for ion implantation than for IBAD, because ion implantation supplies a constituent of the material; and, iii.) the chemical potential during formation of the material is different. This difference arises from two effects: 1) For ion implantation, the material is formed in a changing concentration of constituents as the implanted ion concentration builds up. On the other hand, for IBAD, the material is formed in a constant or steady-state concentration of constituents; 2) In IBAD, the irradiation flux is larger than for ion implantation. For example, 10 Å/s is a typical IBAD deposition rate and 100 Å is a typical depth of material affected by the low energy beam. The time of irradiation is only 100 Å ÷ 10 Å/s = 10 s due to the fact that in 10 s, a new 100 Å layer is deposited which shields the ion bombardment of the underlying material. In 10 seconds, the 100 Å layer typically receives several dpa of damage. In ion implantation, the time that the sample is under irradiation is on the order of 1000 s to achieve a 20% concentration of the implanted species and the typical total dpa is about 10 times higher than for IBAD. The comparative flux for ion implantation is then (IBAD flux) x (10x higher total damage)/(100x more irradiation time) = (IBAD flux) x 0.1. Therefore, the local defect density in the irradiated region during the time the irradiation is active is much larger in IBAD than in ion implantation which means that the materials formed have greatly different average defect densities.

There are several ways to compare the implications of the differences noted above. Table 1 shows approximate values of relevant parameters calculated for the condition of comparable defect flux (within a factor of 5) in IBAD and ion implantation. For IBAD, the vapor flux is Fe and the ion flux is 1 keV N or Ti at 1 mA of current. The damage and flux are given in units of displacements per atom (dpa). For ion implantation the conditions are given in the first column. The temperature rise assumes no active cooling. Note that to irradiate at similar defect flux levels, the temperature rise for ion implantation is substantial.

	Power (Watts)	Damage (dpa)	Flux (dpa/s)	ΔT (°C)	Dose (/cm²)	Time (s)
IBAD (1 keV, 1 mA)	1	12	1.2	10		10
N-->Fe (100 keV, 3 mA)	300	108	5.4	969	4×10^{17}	20
Ti-->Fe (270 keV, 0.3 mA)	87	440	4.4	607	2×10^{17}	100

Table 1. Calculated parameters for IBAD and ion implantation for assumptions in the text and for approximately equivalent defect flux. The ΔT assumes no active cooling.

Table 2 shows the parameters for irradiation where the temperature rise is small for both processes. Here the defect flux is much smaller in ion implantation than in IBAD.

	Power (Watts)	Damage (dpa)	Flux (dpa/s)	ΔT (°C)	Dose (/cm^2)	Time (s)
IBAD (1 keV, 1 mA)	1	12	1.2	10		10
N-->Fe (100 keV, 0.3 mA)	30	108	0.12	10	4×10^{17}	200
Ti-->Fe (270 keV, 0.04 mA)	9	440	0.04	10	2×10^{17}	1000

Table 2. Calculated parameters for IBAD and ion implantation for the assumptions in the text and for conditions of negligible temperature rise.

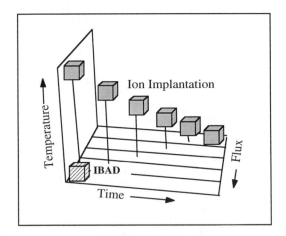

Figure 2: Three dimensional plot of temperature-time-flux-relationships between IBAD and ion implantation for the assumptions in the text.

Figure 2 illustrates some of these differences showing that IBAD and ion implantation occupy different space in the three dimensional plot of time, temperature and flux. It shows that if the time of irradiation and the defect flux are the same for IBAD and ion implantation, then the temperature of the material must be much higher for ion implantation, or, if the temperature of the material during irradiation is the same for the two processes, then the time of irradiation is much longer and the flux much less for ion implantation. This sets up different chemical potential conditions for formation of metastable materials.

Finally, Table 3 summarizes the differences in the synthesis methods. The table entries above the thick solid line (total damage dose and composition) always differ between IBAD and ion implantation, independent of the defect flux.

	Ion Implantation	IBAD
• Total damage dose • Composition	**10x** **changing with dose**	**1x** **steady state**
and either • Temperature rise • Defect flux	**substantial** 1x	**none** 1x
or		
• Temperature rise • Defect flux	none 1x	none **10x**

Table 3. Summary of the differences in material synthesis conditions between IBAD and ion implantation.

The two cases below the thick line in bold type (temperature rise, defect flux) depend on defect flux and are either/or conditions. The chemical environment during phase formation in the two processes is characterized by the conditions above the thick line plus either one of the conditions below the thick line. Tables 1-3 and Fig. 2 serve to support the premise that IBAD processing drives material synthesis parameter phase space into new regimes in the time-temperature-flux continuum, and, that this attribute of IBAD processing leads to microstructures far from equilibrium that are unattainable by other materials synthesis methods.

BEAM ASSISTED PHASE SEPARATION

The unique ion-solid interactions that occur during IBAD can control the kinetics of phase transformations that are thermodynamically favored but kinetically inhibited. We coin the term, beam assisted phase separation, BAPS to describe this process. BAPS is discussed below using as an example, metallic nanoclusters in dielectric matrices produced by IBAD. The deposition of metallic clusters by IBAD is also an example of the production of unique metastable materials by energetic deposition as compared to materials produced by thermal evaporation or ion implantation methods.

Electron beam evaporation or sputtering of two-component thin films have been used for decades to achieve desired phase separated microstructures [21]. Applications of these films include magnetic recording tapes and heads and passive optics. For phase separated films, the only parameters one can vary to control the process are substrate temperature, component arrival ratio (ion/atom), and post-deposition annealing. Such films often have undesirable characteristics such as columnar microstructure, low density, and a broad distribution of grain sizes and shapes. At the same time, deposition at low temperature frequently results in solid solution alloys of binary systems, even for systems with low mutual solubility or immiscibility, as a result of kinetic limitations to atom movement.

Process control and improved film characterisation result from application of a low energy (100-900 eV Ar^+, O_2^+, N_2^+) ion beam during deposition of these films onto low temperature substrates. Atomic collision cascades during stopping of the ions displace atoms (i.e., create vacancy and interstitial defects), causing a variety of collision induced effects that provide the control.

Figure 3 summarizes the processes occurring in BAPS. The upper portion is a schematic of the substrate/film/film surface and the collision cascade in the film surface showing the ballistic portion of the collision cascade. The "Damage Calculations" box in Fig. 3 calculates the dpa used in the IBAD process and shows how it varies from 0.3 to 50 dpa. For a deposition rate of 1.0 nm/s, the damage rate is enormous, up to 6 dpa/s and typical values are 1-2 dpa/s. In the lower box, the processes that influence the outcome of the deposition are grouped according to the trend each process has on film formation. Recall it has already been determined that IBAD or BAPS occurs in implanted material and in material with a steady-state concentration of constituents. Defects induced in the film (vacancies) are locations of probable nucleation of a second phase and radiation enhanced diffusion (RED) provides transport of atoms for growth of the phase. The growth of the second phase is favored by RED and radiation induced segregation (RIS). Disruption of phase separation occurs simultaneously by ballistic mixing and thermal spike mixing that break up the phase. The situation is even more complex because RED and thermal spike mixing are dependent on thermodynamics and can contribute to either the growth or the dissolution of phase. The factors for disruption are also responsible for the suppression of the columnar microstructure and film densification.

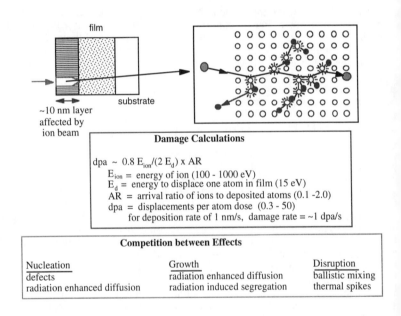

Damage Calculations

$dpa \sim 0.8\, E_{ion}/(2\, E_d) \times AR$

E_{ion} = energy of ion (100 - 1000 eV)
E_d = energy to displace one atom in film (15 eV)
AR = arrival ratio of ions to deposited atoms (0.1 -2.0)
dpa = displacements per atom dose (0.3 - 50)
    for deposition rate of 1 nm/s, damage rate = ~1 dpa/s

Competition between Effects

Nucleation	Growth	Disruption
defects	radiation enhanced diffusion	ballistic mixing
radiation enhanced diffusion	radiation induced segregation	thermal spikes

Figure 3: Characteristics of beam assisted phase separation (BAPS).

All of these processes occur simultaneously at high rate and result in a uniform homogeneous mixture of the film components and associated defects. At any given depth in the film, however, all of these processes are quenched rather abruptly when the overlying film reaches a thickness that exceeds the range of the ion damage (~10 nm). This leads to the distinct possibility that, in the case where phase separation is thermodynamically favored, the second phase will have only a brief time to ripen and therefore, be small (i.e., nm in size) and, that a uniform grain size may be achieved as a result of nucleation and growth from a homogeneous material. The system is constantly nucleating, growing, and dissolving thermodynamically favored second phase. With additional experimental variables with which to control kinetics of the process (ion energy, ion/atom arrival ratio, ion species, background gas, substrate temperature) it should be possible to change the relative rates of these physical processes and, therefore, film properties.

In the case of the two-phase systems for metal particles in a dielectric matrix, systems for study can be readily chosen from equilibrium phase diagrams and/or Gibbs free energy of compound formation. Binary systems that are expected to phase separate (e.g., Ag/Si, Au/Si, etc.), binary or ternary oxides with greatly different heats of formation for their respective oxides (e.g., SiO/Au, SiO/Ag, SiO/Pt, SiO/Cu, NbO/Pt, etc.) and ternary nitrides and carbides with greatly different heats of formation for the nitride or carbide (e.g., AlN/Ge, SiN/Au, SiC/Ag, SiN/Pt, SiN/Cu, etc.) are all appropriate.

METAL NANOCLUSTERS IN DIELECTRIC MATRIX BY BAPS

There has been much interest recently in the optical properties of metal nanocluster composites formed by ion implantation of metals such as Cu and Au into glass substrates [22-24]. These nanocomposites have picosecond nonlinear response times with effective third order susceptibilities several hundred times larger than those of colloidal melt glasses [25, 26].

Ion beam assisted deposition is an alternative processing method for such nanocomposite structures [27]. IBAD offers a number of potential advantages over ion implantation for introducing metal nanoclusters to a dielectric matrix. Simultaneous deposition of the oxide and the metal allows deposition of cluster-containing films of greater thicknesses than those achievable by ion implantation and very high concentrations of metal clusters are possible. By combining IBAD with material combinations expected to phase separate according to equilibrium thermodynamics considerations, the technique of beam assisted phase separation (BAPS) can be explored.

EXPERIMENTAL

The thin films were deposited by IBAD in a multi-hearth vacuum chamber with a base pressure of 5×10^{-9} mm Hg. The evaporants consisted of chemically active elements Si and Nb and the noble metal Au. Films were deposited simultaneously on Si and SiO_2 substrates. The substrate temperatures ranged from room temperature to 730°C. Room temperature refers to a condition in which no heat was applied to the substrate; however, during deposition the substrate surface temperature rose to ~90°C. The films were 200-500 nm thick.

Nanocluster sizes and distributions were determined from TEM photomicrographs on Si substrates. The sizes were compared to those predicted by Mie scattering theory based on the linear absorption characteristics as measured by VIS/UV spectroscopy.

Figure 4 shows an example of the linear optical absorption characteristics of Au/Nb_2O_5 films deposited by IBAD [27]. This film was deposited at room temperature with an O_2^+ ion beam at 150 eV and 200 μAcm^{-2}. The evaporation rates for Au and Nb were 0.04 and 0.12 nms^{-1}, corresponding to a 2.8 Nb/Au arrival rate. After deposition, the film was annealed in flowing oxygen at 600°C for 10 min.

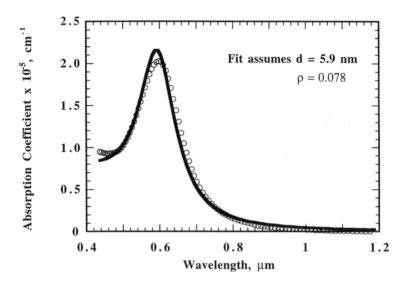

Figure 4: Optical absorption coefficient as a function of wavelength for a Au/Nb$_2$O$_5$ composite film described in text.

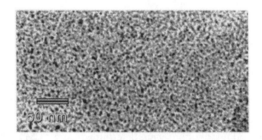

Figure 5: Transmission electron micrograph of the film whose optical absorption characteristics are plotted in Fig. 4. The average Au particle size is ~5 nm.

The hollow circle data points in Fig 4 represent the visible spectrum, while the solid curve shows a fit to the data derived from Mie theory for scattering and absorption of small Au spheres [27]. The fit to the data in Fig. 4 determined that the cluster size was 5.9 nm and the volume fraction of Au clusters was 0.078. These data are in excellent agreement with the size and density of clusters measured by TEM.

Figure 5 shows a bright field TEM image from the film in Fig. 4, showing particles with an average circular cross section of ~5 nm. Transverse section TEM samples also showed a circular shape to the particles in cross section. Circular cross sections in both transverse and plan-view sample orientations demonstrate that the particles are spherical. The silica films prepared by IBAD had a higher index of refraction (1.46 at 550 nm) than the traditional value [29]. RBS data indicates that the stoichiometry is SiO_2, and remains SiO_2, even for atomic arrival rate ratios as low as O/Si ~ 0.5. We surmise that the oxygen from the vacuum system (8×10^{-5} Torr) during IBAD completes the oxidation of the deposited silicon. The pure silica films are clear and have no measurable absorption in and near the visible wavelength.

Figure 6 shows the dependence of the linear absorption coefficient as a function of photon energy for a samples deposited by straight evaporation of SiO_2 and Au, and for ion assist with O_2 (150 eV, 135 mA/cm2). The deposition rates for both samples was 0.2 nm/s for Si and 0.04 nm/s for Au, both were given a post-deposition anneal (1000 C, 2h, in air) and the fill factor was measured as 0.09 for both. The Mie absorption peak is much sharper than for PVD alone. This indirectly indicates a much more uniform particles size is achieved by IBAD over straight PVD.

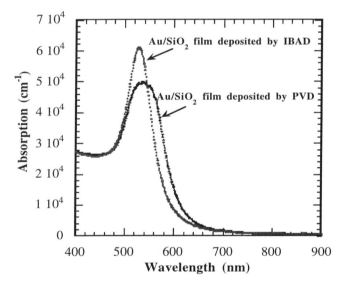

Figure 6: Optical absorption coefficient vs. wavelength for Au/SiO2 composite films described in text.

CONCLUSIONS

We propose that IBAD represents a different ion-solid interaction in a fundamental sense, and that as such, it should lead to new metastable microstructures unattainable by other materials synthesis methods. Some representative quantities in the time-temperature-flux continuum supporting this conclusion were presented.

It was demonstrated that IBAD is a very effective way to prepare noble metal nanoclusters in niobium and silicon oxide dielectric films by means of beam assisted phase separation (BAPS). Both Si- and Nb- IBAD-grown oxide dielectrics are compatible with the gold nanoclusters. The only criteria seems to be that the dielectric metal oxidizes more readily than the cluster metal.

The control exercised over the size of the Au clusters give evidence that beam assisted phase segregation is a viable experimental technique. These promising initial results suggest that the investigation of different material systems by BAPS will be a fruitful avenue for future research as a means to investigate materials far from equilibrium, and as a means to obtain fine control over the approach to equilibrium.

ACKNOWLEDGMENTS

We thank Ken Grabowski and Jim Sprague for helpful comments regarding ion-solid interactions. The work of S. Schiestel supported by Deutsche Forschungsgemeinschaft.

REFERENCES

1. N. Tanabe and M. Iwaki, Nucl. Instrm. Methods Phys. Res. **B80/81**, 1349 (1993).

2. E. P. Donovan, G. K. Hubler, M. Mudholkar, et al., Surf. Coat. Technol., **66**, 499 (1994).

3. C. R. Guarnieri, S. D. Offsey, and J. J. Cuomo, in MRS Bull. (S.M. Rossnagel and J.J. Cuomo, 1987), Vol. 12, p. 40.

4. D. M. Mattox, (Sandia Corp., 1963).

5. F. A. Smidt, Int. Mat. Rev. **35**, 61 (1990).

6. J. K. Hirvonen, Mat. Sci. Rep. **6**, 215 (1991).

7. G. K. Hubler, in Beam-Solid Interactions for Materials Synthesis and Characterization, edited by D. C. Jacobson, D. E. Luzzi, T. F. Heinz and M. Iwaki (MRS, 1995), Vol. Vol. 354, p. 45.

8. G. K. Hubler and J. A. Sprague, Surf. & Coatings Technol. **81**, 29 (1996).

9. G. K. Hubler, in American Society of Metals Annual Meeting (ASM, Cleveland OH, 1996).

10. G. K. Hubler and J. K. Hirvonen, Ion Beam Assisted Deposition (American Society of Metals, Metals Park, OH, 1994).

11. S. T. Picraux, in Site Characterization and Aggregation of Implanted Atoms in Materials, edited by A. P. a. R. Coussement (Plenum, NY, 1980), p. 325.

12. S. T. Picraux, in Site Characterization and Aggregation of Implanted Atoms in Materials, edited by A. P. a. R. Coussement (Plenum, NY, 1980), p. 307.

13. J. R. Holland, L. K. Mansur, and D. I. Potter, (AIME, New York, NY, 1981).

14. R. S. Nelson, in Radiation Processes in Materials, edited by C. H. S. Dupuy (Noordhoff, Leyden, 1975), p. 477.

15. H. Wiedersich and P. R. Okamoto, in Phase Stability During Irradiation, edited by J. R. Holland, L. K. Mansur and D. I. Potter (AIME, New York, 1981), p. 23-41.

16. W. A. Grant, Nucl. Instrum. & Meth. **182/183**, 809-825 (1981).

17. W. A. Grant, J. L. Whitton, and J. S. Williams, Rad. Eff. **49**, 65 (1980).

18. J. W. Mayer, B. Y. Tsaur, S. S. Lau, et al., Nucl. Instrum. & Meth. **182/183**, 1-13 (1981).

19. Y.-T. Cheng, M. V. Rossum, M. A. Nicolet, et al., Appl. Phys. Lett. **45**, 185 (1984).

20. Y.-T. Cheng, S. T. Simko, M. C. Militello, *et al.*, Nucl. Inst. Meth. Phys. Res. **B64**, 38 (1992).

21. C. F. Bohren and D. R. Huffman, *Absorption and Scattering of Light by Small Particles* (Wiley, New York, 1983).

22. R. F. Haglund, L. Y. Jr., I. R.H. Magruder, *et al.*, Nuc. Instr. Methods in Phys. Res. **B91**, 493-504 (1994).

23. L. Yang, K. Becker, F. M. Smith, *et al.*, J. Opt. Soc. Am. **B11(3)**, 457-461 (1994).

24. R. H. Magruder, I. , J. R.F. Haglund, *et al.*, J. Appl. Phys. **76(2)**, 708-715 (1994).

25. R. F. Haglund, L. Y. Jr., I. R.H. Magruder, *et al.*, Opt. Lett. **18**, 373 (1993).

26. C. Flytzanis, F. Hache, M. C. Klein, *et al.*, in *Progress in Optics*, edited by E. Wolf (North-Holland, Amsterdam, 1991), Vol. Vol. 29.

27. C. M. Cotell, C. A. Carosella, S. R. Flom, *et al.*, in *Ion Solid Interactions for Materials Modification and Processing*, edited by D. B. P. D. Ila, L.R. Harriott, and Y.-T. Cheng (Materials Research Society,, Boston, MA, 1996), Vol. Vol. 396.

28. U. Kreibig, J. de Physique **C-2 38**, 97-103 (1977).

29. P. J. Martin, J. Mat. Sci. **21**, 1-25 (1986).

GROWTH AND STRUCTURE OF IN-PLANE TEXTURED MgO THIN FILMS DEPOSITED ON AMORPHOUS SUBSTRATES USING ION-BEAM-ASSISTED E-BEAM EVAPORATION

Connie P. Wang, Khiem B. Do, Ann F. Marshall*, Theodore H. Geballe, Malcolm R. Beasley, and Robert H. Hammond, Ginzton Laboratory and *Center of Materials Research, Stanford University, Stanford, CA

ABSTRACT

In-plane aligned MgO thin films ($\sim$100Å) have been obtained on various amorphous substrates by Ar^+ ion-assisted electron-beam evaporation. Based on RHEED and cross-section TEM, we have shown that the MgO texture appears at a very early stage of film growth and is optimized at a thickness of around 100Å. Optimal thickness is the stage at which the surface is fully covered by MgO crystallites. The planar-view TEM of grain structure evolution in samples at different stages of growth reveals the dynamics of the texture developing process. Small, (100)-faceted MgO grains were observed both in planar-view and cross-section TEM images.

INTRODUCTION

Ion-beam assisted deposition (IBAD) has been demonstrated to be an effective way to control thin-film texture without the need of a single crystal substrate. Various groups have controlled the out-of-plane alignment in metal films (i.e., the fiber texture)[1,2], or the in-plane alignment in oxide films (YSZ [4-7], CeO_2 [8], and MgO [9,10]), in order to use these films as structural templates for subsequent epitaxial film deposition or to alter the physical properties of the films themselves. Among these various materials, MgO has shown considerable promise as a templates for subsequent epitaxial growth. We have previously reported a 7° in-plane aligned IBAD (100) MgO on flat amorphous Si_3N_4/Si substrates.[10] The alignment of MgO is found to be optimal at a thickness of around 100Å, which is different from what is suggested by Bradley's model.[3] A nucleation-controlled model was proposed to describe the MgO texturing process. The model is based on cross-sectional TEM (X-TEM) and *in situ* reflection high energy electron diffraction (RHEED) studies.[10] The RHEED result suggested that the MgO texture was developed at the initial stage of growth, but unlike in the case of IBAD (100)YSZ[4,6,7], was not an evolutionary process. This result was confirmed with high resolution X-TEM images, which clearly showed that the MgO texture starts at the substrate interface. The optimal alignment was achieved at a thickness, corresponding to the point at which the substrate surface is fully covered by MgO crystallites ($\sim$100Å). In this paper, we discuss a planar-view TEM (P-TEM) study of the film structure at different texture stages. Combining all the results, we propose a microscopic texture development process.

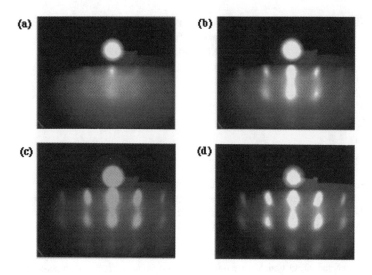

Figure 1. RHEED observation of IBAD MgO texture development: (a) the MgO diffraction
pattern just started to form, (b) at a thickness of 30Å, (c) ~70Å, and (d) ~100Å.

EXPERIMENT

The MgO was deposited in an e-beam evaporation chamber with a Kaufman-type ion
source bombarding the substrate surface during deposition. An *in situ* RHEED system was
used to monitor the film texture development in *real time*. The Ar^+ ion flux was measured by
a Faraday cup biased at -20V. The MgO was evaporated from MgO source with oxygen
pressures ranging from 0 to $1*10^{-5}$ torr. The evaporation rate of MgO molecules was
measured by a quartz crystal monitor (QCM), which was shadowed to prevent bombardment
from the ion beam. The film thickness was determined starting at the stage at which the
crystalline diffraction pattern has barely formed in RHEED. However, the actual film
thickness may differ from the value obtained in this manner both because of the ion sputtering
of the film and the detection limit of RHEED for a very thin film. Since the growth process is
a balance between ion-sputtering and deposition rate, the QCM film thickness should be
considered as a record of time of the deposition rather than an absolute number.

The evaporation rate of the MgO molecules was selected by carefully monitoring changes in
the RHEED pattern. A low evaporation rate was chosen to start the run, then the rate was
increased incrementally with a 20 second dwell time. The critical growth rate Rc, (i.e., the rate
at which MgO barely has a net growth) is around 1.3 Å/sec near room temperature with
$110\mu A/cm^2$, 700eV Ar^+ ion bombardment. The best-textured MgO film obtained at room
temperature was grown at an evaporation rate of ~1.5Å/sec with 700eV and $110\mu A/cm^2$ Ar
ion bombarding at 45°. The corresponding ion-to-molecule ratio is around 0.64.

Figure 2. Cross-section TEM images of IBAD MgO deposited on Si_3N_4/Si substrate.

For the study of P-TEM, an IBAD MgO film was deposited on a TEM-ready Si_3N_4/Si substrate. The Si_3N_4/Si substrate was pre-etched at the center, leaving a ~500Å Si_3N_4 film free-standing. Thus, the MgO film deposited on the substrate could be viewed in the microscope directly, eliminating the artifact introduced during the TEM sample preparation process. The X-TEM sample was prepared following a standard thin-film-sample preparation procedure.

DISCUSSION OF RESULTS

In Fig. 1, the RHEED patterns show the process of the texture development. The spot-type diffraction pattern indicates that the MgO forms small, well-aligned 3D (100) islands with <110> facing the ion source (the ion angle was 45°) from the beginning of growth. The diffraction becomes more intense as the thickness increases and it is optimized at around 100Å. This observation is confirmed by the X-TEM study of MgO/Si_3N_4/Si, which reveals the grain structure pictorially. As shown in Fig. 2, the IBAD MgO film grew as a pseudo-columnar type, with each column consisting of small, well-aligned, rectangular-shaped grains beginning right from the interface with Si_3N_4. A careful study of texture area vs. film thickness suggested that the optimal thickness observed in RHEED corresponds essentially to a stage at which the substrate is fully covered by MgO crystallites. [10]

The X-TEM image provides important information about the substrate/film interface and texture development process. However, it shows only the final film structure, but not the process by which the film structure developed. Two P-TEM samples were prepared to study

(a)   **(b)**

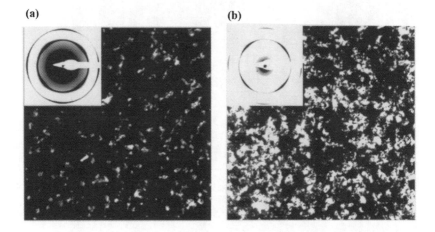

Figure 3. Planar-view TEM of IBAD MgO film deposited on TEM-ready Si3N4/Si substrate. (a) MgO at a thickness of around 30Å and (b) around 70Å.

the evolution of the texture development process. They were roughly 30Å and 70Å thick (as inferred by the QCM), both of which are less than the optimal texture stage (~100Å).

In Fig. 3, both the diffraction patterns and the dark-field images of the two samples are shown. In the dark-field TEM images, the bright region roughly corresponds to the area that is well-aligned. It is clear that the textured area in the thicker films is much larger than in the thinner one. This result can also be interpreted directly from the diffraction pattern. In addition, the diffraction pattern indicates that the un-textured area in Fig. 3(b) was reduced compared to that shown in Fig. 3(a), suggesting the ion-beam selectively removed the un-textured grains.

The high magnification images in Fig. 4 reveal the differences of the grain structure in the two samples. In the dark regions in Fig. 4(a), the MgO lattice fringes can be seen, while the rest of the image has no visible structure, corresponding to the bare amorphous Si_3N_4 substrate. However, the grain structure of the thicker sample (Fig. 4(b)) is very different. Its grain structure is similar to what was observed in X-TEM: clusters of small rectangular-shaped grains ranging in size from 50 to 100Å with (100)-oriented facets. The MgO lattice fringes are clear in 4(b). In several places, small areas of amorphous Si_3N_4 substrate can be observed between the saw-tooth shaped MgO grain boundaries.

A similar range of size, (100)-faceted, rectangular-shaped surface step structure was also observed in single crystal MgO substrate ion-milled by a 3KeV Ar^+ ion at a grazing angle (see Fig. 5). This suggests that this rectangular structure results from an ion-bombarded crystalline MgO sample and that the MgO (100) planes are the planes most resistant to ion sputtering.

The comparison of a very thin sample with the thicker sample (Fig. 3 and 4) provided information about the evolution of the film-forming process. When the film just started to develop, a thin layer of MgO clusters formed covering part of the Si_3N_4 substrates. As the

(a)   **(b)**

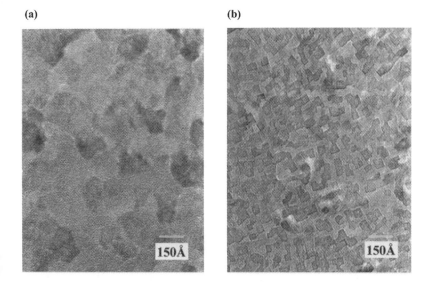

Figure 4. High magnification planar-view TEM images of the IBAD MgO
film at a thickness of (a) 30Å and (b) 70Å.

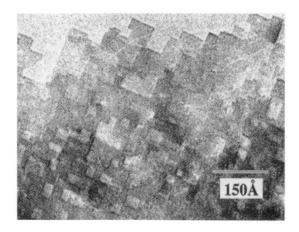

Figure 5. Planar-view TEM of an MgO single crystal sample bombarded by
3KeV Ar ion at a grazing angle.

population of clusters grew, ion etching/sputtering induced a (100)-facet formation. The early stage of the (100)-facet formation may also correspond to the onset of the observation of the 3D RHEED pattern. The texture development was a dynamical balance among ion sputtering, cluster forming, and (100)-facet forming in the clusters. As the process kept on developing, following the trends seen in Fig. 4(a) and 4(b), the film finally achieved the structure observed in X-TEM.

CONCLUSION

In this paper, we discussed an experiment of IBAD MgO deposition. The structure and dynamics of the texture development were explored based on *in situ* RHEED observation and characterization by cross-section and planar-view TEM. A microscopic picture of the IBAD MgO texture development process was proposed.

ACKNOWLEDGMENT

We thank the IBM Almaden research center for providing its TEM facility and Dr. Robert Beyers' supervision of the cross-section TEM work. We are grateful for the support of the Electrical Power Research Institute and the encouragement of Dr. Paul Grant.

REFERENCES

[1] J. M. E. Harper, J. J. Cuomo, and H. R. Kaufman, J. Vac. Sci. Technol., **21** (3), 737 (1982).

[2] L S. Yu, J. M. E. Harper, J. J. Cuomo, and D. A. Smith, Appl. Phys. Lett. **47** (9), 932 (1985).

[3] R. M. Bradley, J. M. E. Harper, and D. A. Smith, J. Appl. Phys. **60** (12), 4160 (1986).

[4] Y. Iijima, N. Tanabe, Y. Ikeno, O. Kohno, Physics C **185** (3), 1959 (1991).

[5] R. P. Reade, P. Berdahl, R. E. Russo, S. M. Garrison, Appl. Phys. Lett. 61 (18), 2231 (1992).

[6] N. Sonnenberg, A. S. Longo, M. J. Cima, B. P. Chang, K. G. Ressler, P. C. McIntyre, Y. P. Liu, J. Appl. Phys. **74** (2), 1027 (1993).

[7] X. D. Wu, S. R. Foltyn, P. Arendt, J. Townsend, C. Adams, I. H. Campbell, P. Tiwari, Y. Coulter, and D. E. Peterson, Appl. Phys. Lett. **65** (15), 1961 (1994).

[8] S. Zhu, D. H. Lowndes; J. D. Budai, and D. P. Norton, Appl. Phys. Lett. 65(16), 2012 (1994).

[9] K. B. Do, presented at the Fall Materials Research Society Conference, Boston, MA, 1995 (unpublished).

[10] C. P. Wang, K. B. Do, M. R. Beasley, T. H. Geballe, and R. H. Hammond, Appl. Phys. Lett., Nov. 17, 1997.

PROPERTIES OF C AND CN FILMS PREPARED BY MASS-SEPARATED HYPER-THERMAL CARBON AND NITROGEN IONS

N. Tsubouchi, Y. Horino, B. Enders, A. Chayahara, A. Kinomura and K. Fujii
Osaka National Research Institute, AIST, Ikeda, Osaka 563, Japan, tsubouchi@onri.go.jp

ABSTRACT

Amorphous carbon films and CN films were prepared by ion beam deposition using isotopically mass-separated, hyper-thermal (50-400eV) ion species such as $^{12}C_2^-$, $^{12}C^{14}N^-$ or simultaneous $^{12}C_2^-$ and $^{14}N^+$ under high vacuum condition. Film properties such as structures, bonding states, compositions, etc. were investigated. The optical band gaps of amorphous carbon films deposited by using $^{12}C_2^-$ ions or CN films deposited by using $^{12}C^{14}N^-$ ions were estimated from optical constants. The gaps were about 2.3eV for the amorphous carbon films and about 0.8 eV for the CN films. Energy dependence of optical constants of CN films formed by simultaneous deposition of C_2^- and N^+ was investigated. The refractive index (n) and extinction coefficient (k) at $\lambda = 675$nm increased 1.5 to 2.4 and 0.3 to 1.3, respectively, with increasing N^+ kinetic energy. The relations between bonding states and kinetic energy of hyper-thermal ions are discussed.

INTRODUCTION

Amorphous carbon (a-C) has attracted a lot of researchers for a long time because it has interesting properties such as high hardness, chemical inertness, high electric resistivity, etc. Nowadays it is used to produce a protective coating material for hard discs. There are a lot of ways to produce a-C films like filtered cathodic vacuum arc [1], laser ablation of graphite [2,3], etc. Among these methods, an ion beam deposition (IBD) method is an interesting way to produce a-C films. The films are deposited by irradiation of direct hyper-thermal (several tens to hundreds eV) carbon ions on a substrate. Using the IBD method, the sp^3/sp^2 bonding ratio of a-C is able to be varied up to the maximum value, 0.8 in accordance with kinetic energies of carbon ions [4,5]. J. L. Robertson et al. [6] reported that a diamond epitaxial film was also formed by the IBD method.

As interest of a hard material, undiscovered metastable crystalline β-C_3N_4 has been predicted by Liu and Cohen in 1989 [7]. It is expected to have hardness comparable to or more than that of diamond. However there is no clear evidence of its existence in spite of much efforts of producing it by means of various ways such as arc discharge deposition [8], chemical vapor deposition [9], sputter deposition [10], ion beam deposition [11,12,13], reactive pulsed laser ablation [14], etc. Another group discusses the stability of other polytypes of crystalline C_3N_4 like an α phase, a cubic phase, etc [15]. From the above discussions, it is difficult to say that C-N system is clearly understood.

In this study, properties such as composition, impurity, structure, bonding states of C and CN films deposited by hyper-thermal $^{12}C_2^-$ and $^{12}C^{14}N^-$ were compared to each other. Furthermore, correlation between ion's kinetic energy and ratios of bonding states in CN films by simultaneous deposition of C_2^- and N^+ was investigated.

EXPERIMENT

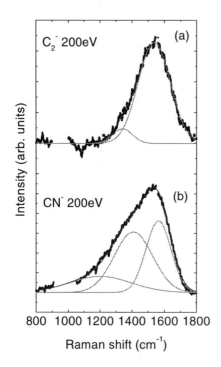

C$_2^-$ 200eV   (a)

CN$^-$ 200eV   (b)

Intensity (arb. units)

800 1000 1200 1400 1600 1800

Raman shift (cm^{-1})

Fig. 1
Raman spectra of the films deposited by (a) C$_2^-$ (200eV) and (b) CN$^-$ (200eV). The peak assignments are shown in the text.

Isotopically mass-separated positive and negative ions, ^{12}C$_2^-$ (200eV), ^{12}C^{14}N$^-$ (200eV) and C$_2^-$ (200eV) + N$^+$ (50-400eV) which energies were well-defined were irradiated on Si substrates, respectively. The energy and energy spreads of the beams were monitored by an energy analyzer. The energy spreads of beams, full widths at half maximum were about 7eV. The base pressure of a deposition chamber was 2×10^{-7} Pa. The pressure during depositions was 4×10^{-7} to 2×10^{-6} Pa. The features of the deposition apparatus were described elsewhere [16]. The substrates were ultrasonically cleaned with acetone, then stripped of their oxide using HF+H$_2$O (HF:H$_2$O = 1:5) for 1 min before deposition. Films were deposited on 30×40 mm^2, (100) oriented silicon substrates at ambient conditions (~300K). The film size was 1.5-2cm in diameter depending on beam energy.

In order to investigate properties of films such as a structure, bonding states, composition and optical properties, following measurements were carried out. Raman scattering was performed on backscattering from the sample using an argon-ion laser operating at 514.5nm at a power of 100mW. The infrared absorption spectra were obtained using a Nicolet 60SXR FTIR spectrometer. RBS measurements were performed by using a 1.8MeV helium ion beam from a Van de Graaff accelerator on Si(100) channeling condition to reduce silicon background signals. Optical constants, (n, k) were determined in the wavelength range of 250 to 1800nm by Spectroscopic ellipsometry (SE), where n is a real part of refractive index and k is the imaginary part known as the extinction coefficient.

RESULTS AND DISCUSSION

The deposited films were in-situ investigated by high energy electron diffraction (RHEED) method. Diffuse electron reflections or ambiguous ring patterns were observed in all films. It indicates that all the films are amorphous. RBS spectra showed clear signals of

C atoms in the films deposited by C_2^- (200eV) and those of both C and N atoms in the films deposited by CN^- (200eV), C_2^- (200eV) + N^+ (50-400eV). There were no apparent signals observed except Si, C and N atoms in all the films. From the spectra, the composition ratios (N/C) of the films deposited by CN^-, C_2^- + N^+ ions were estimated. In the case of CN^- deposition, the N/C were 0.4. It means that only 40 at % of arrival N ions were included in the films, although the arrival ratio, N/C was 1. In the case of C_2^- + N^+ deposition, the N/C ranged in 0.3 to 0.5 because of spatial distributions of intensities of both beams on a substrate.

On FTIR spectra, there were no distinct structures except for air absorption background around 1600cm^{-1} in the C film [13]. On the other hand, a peak around 2150cm^{-1} and a broad absorption band from 800cm^{-1} to 1800cm^{-1} were observed on the CN films [13]. The former peak is assigned to the C-N triple bond [17], and the latter is assigned to the Raman active G and D bands [18]. No other features have been observed in the both films. The absence of C-H modes (e.g. CH_2 stretching mode, 2920cm^{-1}) [13] indicates that our C and CN films are hydrogen-free on the detectable level of FTIR measurement.

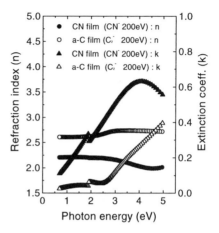

Fig. 2
Variation of refraction index (n) and extinction coefficient (k) as a function of photon energy for the films deposited by CN^- (200eV) and C_2^- (200eV).

Raman spectra of C and CN films deposited by C_2^- (200eV) and CN^- (200eV), respectively, are shown in Fig. 1 by thick lines. The spectra were deconvoluted by a Gaussian fitting method, represented by thin lines. The signals from Si substrate around 1000cm^{-1} were neglected from the fitting procedure. The spectrum in Fig. 1 (a) was fitted by two Gaussian peaks around 1550cm^{-1} and 1350cm^{-1}. The peaks around 1550cm^{-1} and 1350-1390cm^{-1} of both films were ascribed to the G and D bands which were also observed in amorphous carbon (a-C) [19], although these were small shift of 20-30cm^{-1}. The G band is a characteristic of graphitic carbon and the D band is a characteristic of disordered sp^2-bonded carbon. However, the spectrum in Fig. 1 (b), for the case of the CN film, was not able to be fitted by two Gaussian peaks. The additional Gaussian peak around 1215cm^{-1} was necessary to fit the spectrum. The peak will be ascribed to stretching modes of C-N single bond [20], since C-H modes were not observed on FTIR measurement.

The main bands of the Raman spectra for both films are the D and G bands, which indicates that a basic structure of both films is very similar to that of a-C films. However, the relative intensity ratio of the D band to the G band is smaller for the case of (a) than the case of (b). The extent of disorder in graphite lattice is known to be reflected on D band intensity [18,21]. In the point view of ion beam collision, the radiation effects should be similar for the case of (a) and (b), because kinetic energies of both ion beams were the same and the atomic masses were almost the same each other on a film formation process. Thus it is considered that an inclusion of N atoms during film formation by CN⁻ ions caused an increase of disorder level which resulted in the increase of the D band intensity.

The variation of optical constants, (n, k), with photon energy for the C and CN films deposited by CN⁻ (200eV) is shown in Fig. 2. The spectra have been observed separately in the two energy ranges of 0.6-1.9eV and 1.9-5eV with two different light sources. Thus discontinuities of data around 1.8eV are observed. In the case of the a-C film (open circles), n is almost constant, 2.6, although there is a very small increase at around 3eV. In the case of the CN film (closed circles), n is around 2.2 up to 3eV, and then it decreases gradually to 2.0 at 4.5eV. As photon energy decreases, k of the a-C film (open triangles) also decreases rapidly to nearly zero at 2.5eV, which indicates the films become transparent in the visible light region. On the other hand, k of the CN film (closed triangles) has the maximum value, 0.62, at 4.0eV and decreases with decreasing photon energy. The k, however, still remain 0.1 at 0.7eV, indicating opaque in the visible light region unlike the case of a-C.

In order to estimate the optical band gap of the deposited films, the $(\alpha\hbar\omega)^{1/2}$, where α is absorption coefficient determined by the n and k, are plotted as a function of photon energy ($\hbar\omega$) as shown in Fig. 3. The optical band gaps are obtained by extrapolating the linear plots to energy axis. The values were about 2.3eV for the a-C film and 0.8eV for the CN film deposited by CN⁻ (200eV). The value, 2.3 eV for the a-C film is quite high among other reports for a-C [22], although it is less than that of diamond, 5.3eV. According to the linear relation between sp^3 content and the optical band gap [22], sp^3 content of our a-C film is estimated to be 85%.

The optical band gap of an a-C film decreases with increasing sp^2 content in the films because electrons around absorption edge consist of sp^2 component. In graphite which consists of 100% sp^2, for example, it possesses zero band gap. Because the highest valence band arising from the bonding π orbits overlaps by 0.036 eV with lowest conduction band arising from the anti-bonding $\pi*$ orbits. The narrowing of optical gap in the CN film deposited by CN⁻ (200eV) against the a-C shows that incorporation of N atoms in carbon network induces an increase of sp^2 content in the film. However, the optical band gap in the CN film is not zero but 0.8eV, which suggests that some sp^3 components also exist.

We also investigated optical constants at λ=675nm of CN films formed by simultaneous deposition of C_2^- and N^+ ions. The energy of C_2^- ions was kept at 200eV and the energy of N^+ ions was varied from 50 to 400eV. The measured areas were about 1mm² in diameter and the composition ratios were determined by RBS. In the region, composition ratio of N/C was about 0.4 and film thickness of

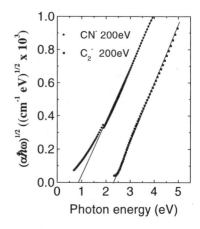

Fig. 3
$(\alpha\hbar\omega)^{1/2}$ versus photon energy plotted for the films deposited by CN⁻ (200eV) and C_2^- (200eV).

310

each sample was 40-90nm depending on the beam energy. Fig. 4 shows a plot of (n, k) versus N^+ ions kinetic energy. Both n and k increase 1.5 to 2.4 and 0.3 to 1.3, respectively, with increasing the kinetic energy. The increase in k with N^+ kinetic energy indicates that sp^2 content in the films increase with the energy.

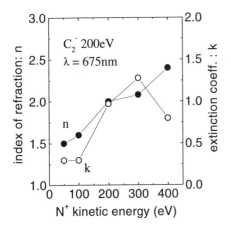

Fig. 4
Variation of optical constants, (n, k) at λ = 675nm of the film deposited by C_2^- (200eV) + N^+ (50-400eV), as a function of N^+ ion's kinetic energy.

CONCLUSIONS

Carbon and CN films were prepared by using isotopically, mass-separated, hyper-thermal (50-400eV) ion species such as $^{12}C_2^-$, $^{12}C^{14}N^-$ and $^{12}C_2^- + ^{14}N^+$ under high vacuum condition, $10^{-7} - 10^{-6}$ Pa. Structures of deposited films were amorphous carbon like. The deposited C and CN films were hydrogen-free on the detectable level of FTIR measurement. The composition ratio of N/C was 0.4 in the case of CN films deposited by CN^- ions, although the arrival ratio of N/C is 1. In the case of $C_2^- + N^+$ ions, The composition ratio of N/C was 0.3-0.5.

Optical band gap was about 2.3eV for an a-C film and 0.8 eV for a CN film. The value of 2.3eV for a-C film is a quite high value comparing to that of a-C films prepared by other methods. The gap of 0.8eV for a CN film suggests that sp^2 component increased with introducing N, but sp^3 component also exists in the film.

Optical constants at λ=675nm of films formed by simultaneous deposition of C_2^- (200eV) and N^+ (50 to 400eV) ions. The values of n and k at λ=675nm increased 1.5 to 2.4 and 0.3 to 1.3, respectively with increasing N^+ kinetic energy. The increase in k with N^+ kinetic energy indicates that sp^2 content in the films increases with the energy.

ACKNOWLEDGEMENTS

The authors thank Professor M. Wada and Mr. K. Yamashita for the operation of our negative ion source.

REFERENCES

[1] B. F. Coll and M. Chhowalla, Surf. Coat. Technol. **79**, 76 (1996).
[2] A. A. Voevodin, S. J. P. Laube, S. W. Walck, J. S. Solomon, M. S. Donley and J. S. Zabinski, J. Appl. Phys. **78**, 4123 (1995).
[3] P. Yang, Z. J. Zhang, J. Hu, C. M. Lieber, Mat. Res. Soc. Symp. Proc. Vol **438**, 593

(1997).

[4] J. Ishikawa, Y. Takeiri, K. Ogawa and T. Takagi, J. Appl. Phys. **61**, 2509 (1987).

[5] J. Kulik, G. D. Lempert, E. Grossman, D. Marton, J. W. Rabalais, Y. Lifshitz, Phys. Rev. B **52**, 15812 (1995) .

[6] J. L. Robertson, S. C. Moss, Y. Lifshitz, S. R. Kasi, J. W. Rabalais, G. D. Lempert and E. Rapoport, Science, **24**, 1047 (1989).

[7] A. Y. Liu, M. L. Cohen, Phys. Rev. **B41**, 10727 (1990).

[8] O. Matsumoto, T. Kotaki, H. Shikano, K. Takemura, S. Tanaka, J. Electrochem. Soc. **141**, L16 (1994).

[9] K. J. Clay, S. P. Speakman, G. A. J. Amaratunga, S. R. P. Silva, J. Appl. Phys. **79**, 7227 (1996).

[10] M. Y. Chen, D. Li, X. Lin, V. P. Dravid, Y.-W. Chung, M.-S. Wong, W. D. Sproul, J. Vac. Sci. Technol. **A11**, 521 (1993).

[11] D. Marton, A. H. Al-Bayati, S. S. Todorov, K. J. Boyd, J. W. Rabalais, Nucl. Instrum. Meth. Phys. Res. **B90**, 277 (1994).

[12] H. C. Hofsäss, C. Ronning, U. Griesmeier, M. Gross, Mat. Res. Soc. Symp. Proc. Vol. **354**, 93 (1995).

[13] N. Tsubouchi Y. Horino, B. Enders, A. Chayahara, A. Kinomura, K. Fujii, Mat. Res. Soc. Symp. Proc. Vol.**438**, 605 (1997).

[14] X.-A. Zhao, C. W. Ong, Y. C. Tsang, Y. W. Wong, P. W. Chan, C. L. Coy, Appl. Phys. Lett. **66**, 2652 (1995).

[15] D. M. Teter, Russell J. Hemley, Science **271**, 53(1996).

[16] Y. Horino N. Tsubouchi, K. Fujii, T. Nakata, T. Takagi, Nucl. Instrum. Meth. Phys. Res. **B106**, 657(1995).

[17] H.-X. Han, B.J. Feldman, Solid State Commun. **65**, 921(1988).

[18] K. J. Clay, S. P. Speakman, G. A. J. Amaratunga, S. R. P. Silva, J. Appl. Phys. **79**, 7227 (1996).

[19] S. Kumar, T.L. Tansley, Thin Solid Films, **256**, 44(1995).

[20] A. Bousetta, M. Lu, A. Bensaoula, A. Schultz, Appl. Phys. Lett, **65** (6), 696 (1994).

[21] J. H. Kaufman, S. Metin, D. D. Saperstein, Phys. Rev. **B39**, 13053 (1989).

[22] J. Robertson, Phil. Mag. **B76**, 335 (1997).

EFFECT OF ION ENERGY ON STRUCTURAL AND CHEMICAL PROPERTIES OF TIN OXIDE FILM IN REACTIVE ION-ASSISTED DEPOSITION (R-IAD)

Jun-Sik Cho*, Won-Kook Choi*, Ki Hyun Yoon**, J. Cho*, Young-Soo Yoon*, Hyung-Jin Jung*, and Seok-Keun Koh*
* Thin film Technology Research Center, Korea Institute of Science and Technology, Korea
**Department of Ceramic Engineering, Yonsei University, Korea

ABSTRACT

Tin oxide films were deposited on *in-situ* heated Si (100)substrates using reactive ion-assisted deposition and the effect of average impinging energy of oxygen ions on the crystalline structure and the stoichiometry of deposited films were examined. The transformation from SnO phase to SnO_2 phase of the films was dependent on the change of the average impinging energy of oxygen ion (E_a), and the relative arrival ratio of oxygen to tin. Perfect oxidation of SnO_2 was performed at $E_a = 100$, 125 eV/atom at as low as 400 Å substrate temperature. The composition (N_o/N_{Sn}) of films increased from 1.21 to 1.89, and was closely related to the average impinging energy of oxygen ion. The surface morphology of the films was also investigated by scanning electron microscopy.

INTRODUCTION

SnO_2 films are widely used in a variety of application such as opto-electronic devices, solar-cells, antireflection coatings and gas sensors [1-4]. Several methods, *e.g.* reactive evaporation [5], chemical vapor deposition [6] and sputtering [7], each with its own advantages and disadvantages, have been employed to obtain SnO_2 films. Chemical vapor deposition(CVD) processing of tin oxide films is, although inexpensive, being substituted by physical vapor deposition(PVD) processing which is more controllable and needs lower processing temperature, leading to thin films of better purity and outstanding optical and electrical properties. Although many reports have been published on the preparation of SnO_2 films by traditional PVD techniques, no one has successfully deposited pure SnO_2 films without a post annealing treatment at higher than 550 ℃ in oxygen ambient or air. Generally, in the case of evaporation or sputtering using SnO_2 as evaporant or sputtering target, the SnO_2 phase decomposes in gaseous state and mainly SnO phase condenses onto the substrate. These films have to be oxidized to SnO_2 subsequently by post annealing treatment in oxygen ambient or air at temperatures higher than 550 ℃ [5,7].

Recently many thin films including dielectric oxides, fluorides, transparent conductors and nitrides have been grown successfully by ion-assist deposition [8-10], however, there have been few reports about deposition of SnO_2 films using this deposition method. This ion based-deposition technique has many benefits such as an elaborate control of film properties including composition and structure, an enhanced adhesion between film and substrate, and a reduction of substrate temperature for crystallization of films, etc. [11]. In particular, a major advantage of the ion-assisted deposition is that ion energy and ion flux are decoupled, allowing for independent variation of either parameter. Thus the stoichiometry of films is easily controlled by varing ion flux, and additional energy added to the growing film by bombardment with energetic ions during the deposition results in lower minimum substrate temperature for thin film formation.

In this paper, reactive ion-assisted deposition is used to fabricate pure tin oxide films that are transformed from SnO phases to SnO$_2$ phases having different oxygen content by changing the average oxygen ion energy impinging on a tin atom, *i.e.*, the relative arrival ratio of oxygen ions to tin atoms without post annealing treatment. We also investigate the effect of the average oxygen ion energy and *in-situ* heating process on the composition and the structure of the deposited films using Auger electron spectroscopy (AES), glancing angle x-ray diffraction (GXRD), and scanning electron microscopy (SEM).

EXPERIMENTAL

Tin oxide films were deposited on Si(100) substrates at 300 - 500 ℃ substrate temperature by reactive ion-assisted deposition. The 99.99% tin metal was evaporated by a partially ionized beam. The deposition rate was fixed at 0.5 Å/sec and the thickness of the films was about 2000 Å. The 99.9% oxygen gas was ionized using the 5-cm grid cold hollow cathode ion gun and its flow rate introduced in ion gun was 3 sccm (*ml*/min.). The ionized oxygen was accelerated at constant ion-beam potential of 500 V and the current density of ionized oxygen was changed from 3.5 to 17.5 μ Acm^{-2} with discharge voltage. In order to measure the current density of ionized oxygen, the faraday cup biased -30 volt to remove electron acceptance was located below the substrate holder. The relative arrival ratio of oxygen to tin was converted into an average impinging energy(E_a) which was changed from 25 to 125 eV/Sn atom. Table 1 and Figure 1 showed the deposition parameters of films with the average impinging energy(E_a) and schematic diagram of reactive ion-assisted deposition system. The base pressure of vacuum chamber was 1 - 2 $\times$ 10^{-5} Torr and the working pressure was 1 - 2 $\times$ 10^{-4} Torr.

TABLE I . The deposition parameter of films by reactive ion-assist deposition.

Sample	Average impinging energy (eV/atom)	Relative arrival ratio of oxygen to tin (O/Sn)	Discharge voltage(V)	Current density (μ Acm^{-2})
S25	25	0.025	400	3.5
S50	50	0.05	420	7.0
S75	75	0.75	440	10.5
S100	100	0.1	450	14.0
S125	125	0.125	460	17.5

RESULTS AND DISCUSSION

In order to examine crystallinity of grown tin oxide films, glancing angle x-ray diffraction was performed. Sn phases and SnO phases are identified in the films deposited at E_a=25 and 50 eV/atom at 300 ℃ substrate temperature and other films deposited under these conditions appear to be amorphous (not shown here). Figure 2 shows the GXRD patterns of the films

314

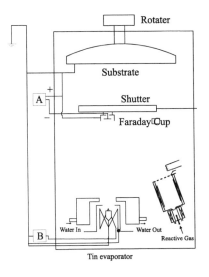

Fig. 1 Schematic diagram of reactive ion-
assisted deposition system

deposited at 400 ℃ substrate temperature with average impinging energies(E_a). The film deposited at E_a=25 eV/atom(curve (a)) shows the mixed phases of β-Sn and polycrystalline SnO which have β-Sn(200), β-Sn(220), β-Sn(211), SnO(101), SnO(200), and SnO(112). For the film at E_a =50 eV/atom(curve (b)), the intensity of the β-Sn peaks, especially the β-Sn(200) decreases abruptly and that of the SnO(110) and the SnO(002) peaks increases, which may be due to increasing the content of oxygen arriving at the surface of the film. From curve (c), it is noteworthy that the film deposited at E_a =75 eV/atom shows the different peaks with respect to the previous two cases.

The characteristic GXRD peak occurs at 34.1° corresponding the SnO$_2$ (101) phase and all SnO peaks except for the SnO (002) peak vanished.

It is also observed that the line width of the SnO (002) peak broadens and the position of the peak shifts to lower angle. The broad nature of the GXRD peak may be due to the reduction of grain size and the strain in the film. The shift of the SnO (002) peak confirms that mechanical strain is present in the film. The main cause occurring the mechanical strain may be the mixed phases of SnO and SnO$_2$ in the film[12]. From this result, it can be said that the transformation from SnO phase to SnO$_2$ phase takes place at E_a=75 eV/atom and the film has the mixed phase of SnO and SnO$_2$. The films deposited at E_a=100 and 125 eV/atom(curve (d), (e)) represent only the SnO$_2$ peaks at 26.8°, 34.1° and 51.8° corresponding to (110), (101) and (211) planes, respectively and then the films oxidized to SnO$_2$ perfectly. As the substrate temperature increases to higher than 400 ℃, no change in the position and the intensity of the GXRD peaks were observed compared with those of the films deposited at 400 ℃ substrate temperature. Figure 3 shows SEM micrographs of the films deposited with E_a's as noted. As shown in Fig. 3(a), the crystallites of the film deposited at E_a=25 eV/atom are composed of two type of the crystallites; one is that of β-Sn which has a length of about 2000 Å and a width of about 500 Å (part A) and the other is that of SnO which has the length of about 1800 Å and the width of about 1200 Å (part B). In the case of the film deposited at E_a=50 eV/atom(Fig. 3(b)), the crystallites of β-Sn disappear and those of SnO become dominant. This result has a good agreement with the GXRD pattern in Fig. 2. The interesting feature that the shape of crystallites is different from the previous films is observed in the film deposited at E_a=75 eV/atom (Fig. 3(c)). This appearance of the crystallites may be closely related to the mixed phases of SnO and SnO$_2$ phases.

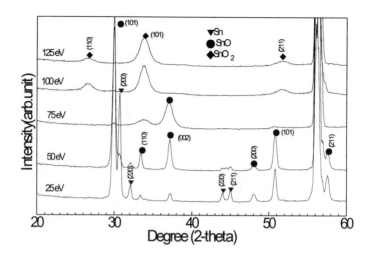

Fig. 2 GXRD patterns of tin oxide films deposited at various E_a's values at 400 ℃ substrate temperature.

From the surface micrographs of the films deposited at E_a=100 and 125 eV/atom, it can be said that these films consist of uniform granular crystallites which are typical grain shapes of SnO_2. It is also observed that the grain size of the film deposited at E_a=125 eV/atom is about 400 Å which is larger than that of the film deposited at E_a=100 eV/atom (200 Å) and this increase of grain size may be due to the enhancement of the surface mobility of tin atom which results from the increased energy of the assisting oxygen ions. All films deposited were analyzed using Auger electron spectroscopy(AES) and the atomic ratio of oxygen to tin (O/Sn) was calculated using the differential peak-to-peak height of the O $KL_{2,3}L_{2,3}$ and Sn $M_4N_{4,5}N_{4,5}$ transition lines corrected by intensity ratios computed for standard SnO_2 powder and Sn metal. The calculated O/Sn ratio of the films is represented in Fig. 4. As shown in Fig. 4, the O/Sn ratio of the films is increased from 1.21 to 1.89 with the average impinging energy, *i.e.*, the relative arrival ratio of oxygen to tin. This trend is significantly connected with the GXRD and SEM results and therefore the oxidation of deposited tin oxide films was affected by the initial oxygen content.

CONCLUSIONS

Tin oxide films were deposited on *in-situ* heated substrate by reactive ion-assisted deposition. The polycrystalline SnO_2 thin films could be deposited at lower temperature (400 ℃) than previously reported post annealing temperature (550 ℃) and the atomic ratio of the films could be easily controlled by the relative arrival ratio of oxygen to tin. The deposited SnO_2 thin films had a uniform surface and the granular crystallites with size of 200 ~ 400 Å.

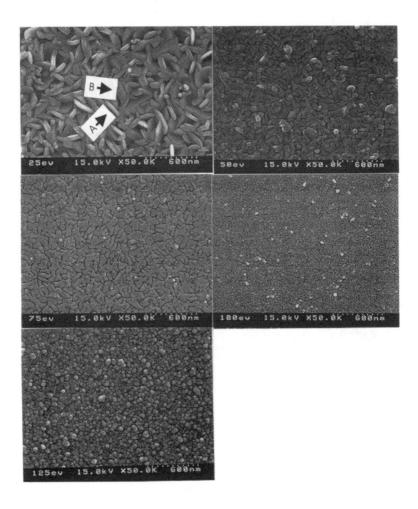

Fig.3 SEM micrographs of tin oxide films deposited at various Ea's values at 400 ℃
substrate temperature : (A) 25 eV/atom, (B) 50 eV/atom, (C) 75 eV/atom,
(D) 100 eV/atom, and (E) 125 eV/atom.

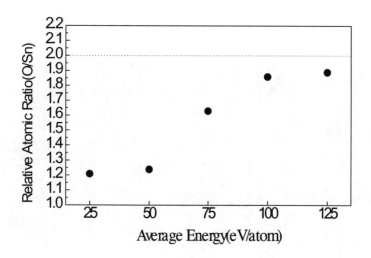

Fig. 4 Relative atomic ratios of tin oxide films deposited at various Ea's values at 400 ℃ substrate temperature

REFERENCES

1. K. L. CHOPRA, S. MAJOR and D. K. PANDYA, Thin Solid Films, **102** (1983) 1.
2. V. K. JAIN and A. P. KULSHRESHTHA, Sol. Energy Mater., **4** (1981) 151.
3. J. SANZ MAUDES and T. RODRIGUZ, Thin Solid Films, **69** (1980) 183.
4. K. D. SCHIERBAUM, U. WEIMAR and W. GÖPEL, Sensors and Actuators **B7** (1992) 709.
5. M. H. MASHUSUDHANA REDDY, S. R. JAWALEKAR and A. N. CHANDORKAR, Thin Solid Films, **169** (1989) 117.
6. M. S. MURTY, G. K. BHAGAVAT and S. R. JAWALEKAR, Thin Solid Films, **92** (1982) 347.
7. M. R. SOARES, P. H. DIONISIO, I. J. R. BAUMVOL and W. H. SCHREINER, Thin Solid Films, **214** (1992) 6.
8. P. J. MARTIN AND R. P. NETTERFIELD, Thin Solid Films, **137** (1986) 207.
9. J. A. DOBROWOLSKI, F. C. HO AND A. WALDORF, Appl. Opt., **26(24)** (1987) 5204.
10. M. SATOU AND F. FUJIMOTO, Jap. J. Appl. Phys., **22** (1983) 171.
11. J.J. CUOMO AND S. M. ROSSNAGE, "Handbook of Ion Beam Processing Technology" (Noyes Publications, New Jersey, 1989) p. 387.
12. M. H. M. REDDY, S. R. JAWALEKAR AND A. N. CHANDORKAR, Thin Solid Films, 169 (1989) 117.

ORIENTATION AND ELECTRICAL RESISTIVITY OF ALUMINUM OXIDE FILMS PREPARED BY ION BEAM ASSISTED PLASMA CVD

H. NAKAI, O. HARASAKI, J. SHINOHARA
Research Institute, Ishikawajima-Harima Heavy Industries Co., Ltd., 1-15, Toyosu 3-chome, Koto-ku, Tokyo 135, Japan, hiroshi_nakai@ihi.co.jp

ABSTRACT

Aluminum oxide (Al_2O_3) films were prepared on a nickel based superalloy (Inconel 718) by ion beam assisted plasma CVD at the multiple incident angles of the ion beams to a substrate. Crystal structure of these films was analyzed by X-ray diffraction (XRD). Electrical resistance of the films was measured at temperatures ranging from room temperature to 850℃. The incident angle of the ion beams influenced the orientation of Al_2O_3 crystals. The net integrated intensity was calculated in order to compare the intensity among the different thickness samples whose thickness is thinner than X-ray penetration depth. The electrical resistivity depends on the relative X-ray intensity of (110) and (300) of α-Al_2O_3.

INTRODUCTION

It is well-known that aluminum oxide (Al_2O_3) has an excellent electrical insulating property at high temperature. However, Al_2O_3 films, prepared by a conventional deposition process, are not suitable to form electrical insulating layers for electric devices such as thin film sensors[1] at a high temperature. Because the resistivity of insulating films decreases considerably with temperature. We are carrying out research to produce dense insulating films with closely packed grains by ion beam assisted plasma CVD in order to improve the insulating property at high temperature.

Previously, the authors have reported on the Al_2O_3 films prepared by this process[2-3]. It was found that the Al_2O_3 films prepared by this process showed superior adhesion strength compared with that by a conventional plasma CVD and simultaneous irradiation of the ion beams during plasma CVD enhanced crystallization and/or growth of fine grain structure which was not obtained by plasma and thermal CVD at lower temperature[3]. α-Al_2O_3 crystals were obtained by ion beam assisted plasma CVD at the substrate temperature of 800℃ with irradiation of 10 keV krypton ions. The film was remarkably crystallized and consisted of fine grain compared with the films by the conventional plasma CVD[3]. This film showed higher electrical resistivity at high temperature than the films prepared by conventional CVD at the same substrate temperature, and it was suggested that the film prepared by ion beam assisted plasma CVD has preferential orientation[3]. The incident angle of the ion beams to the substrate influenced the orientation of Al_2O_3 crystals.

In this paper, the effect of the incident angle of the ion beams during ion beam assisted plasma CVD on the orientation of Al_2O_3 crystals is mainly described, and the relationship between crystallinity and electrical resistivity is discussed. The crystallinity of these films was analyzed by X-ray diffraction (XRD) using CuKα. The net integrated intensity was calculated in order to compare the intensity among the different thickness samples whose thickness is thinner than X-ray penetration depth.

EXPERIMENT

Al_2O_3 films were deposited on the nickel based superalloy (Inconel 718) by ion beam assisted plasma CVD at the multiple incident angles of the ion beams. The evaporant, $Al[OCH(CH_3)_2]_3$ powder, was heated to 120℃ in an oven and the vapor was carried to a reaction chamber by pure argon gas. The substrate temperature was controlled by the resistance heater under the substrate. Plasma was produced by microwave excitation combined with electronic cyclotron resonance. The ion-beam current-density at a substrate position was measured under the same pressure as the deposition and adjusted to the prescribed values before deposition. The substrate was simultaneously irradiated by ion beams during ion beam assisted plasma CVD. An outline of this deposition system, its specifications and deposition conditions of Al_2O_3 films were described elsewhere[2-3]. The incident angle of the ion beams was defined as shown in Fig. 1.

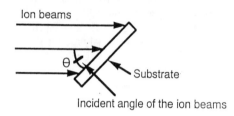

Ion beams

θ

Substrate

Incident angle of the ion beams

Fig. 1.   The incident angle of the ion beams. It was defined the angle of θ in the figure.

The crystallinity of these films was analyzed by X-ray diffraction(XRD) using CuK α. The integrated intensity of diffraction depends on the film thickness and the incident angle of the X-ray when the film thickness is thinner than X-ray penetration depth. Therefore, the integrated intensity was normalized using the film thickness and incident angle. The net integrated intensity was calculated as follows in order to compare the intensity among the different thickness samples. Kuratani et.al reported that the relation between orientation and internal stress of Ni films using the same normalization of the intensity of the X-ray diffraction[4].

The energy(dI_d) was thought to be diffracted from this beam by a layer of a powder of thickness dx, located at a depth x below the surface.

$$dI_d = (I_0K/\sin\theta)\exp(-2\mu x/\sin\theta)dx \qquad (1)$$

Where μ is the linear absorption coefficient of the powder compact. K is a constant consisting of the volume fraction of the specimen containing particles having the correct orientation for reflection and the fraction of the incident energy which is diffracted per unit volume. The total diffraction intensity from a film is obtained by integrating over a film of thickness d:

$$I_d = (I_0K/2\mu)(1-\exp(-2\mu d/\sin\theta)) \qquad (2)$$

When the sample has an infinite thickness, the total diffraction intensity is independent of θ. The integrated intensity is given by

$$I_\infty = I_0K/2\mu \qquad (3)$$

The net integrated intensity from the samples, or the intensity when the sample has an infinite thickness, is therefore

$$I_\infty = I_d/(1-\exp(-2 \mu d/\sin \theta))$$ (4)

The integrated intensity of the substrate was calculated by the same theory taking into consideration the absorption of the film.

The electrical resistance was measured as follows. Gold electrodes were sputter deposited on the sample and connected in series with a standard resistance. DC voltage(3 V) was applied to this circuit, and the electrical resistance was calculated from the measured voltage of the standard resistance. The sample was heated in an electrical furnace, and this measurement was carried out in the range of room temperature to 850°C.

RESULTS AND DISCUSSION

Fig. 2 shows the XRD patterns of the Al_2O_3 films prepared at the multiple incident angles of the ion beam. X-ray diffraction peaks of α-Al_2O_3, γ-Al_2O_3 and the substrate were observed in all films. Several peaks are shown in Fig. 2 and numbered from 1 to 7 for convenience in comparing the peak intensities. The relationship between the peak number and the diffraction angles from the experimental results is shown in Table 1. The diffraction angles and the index of α-Al_2O_3 and γ-Al_2O_3 shown in the JCPDS cards were listed in Table 1 as well as the diffraction angles of the substrate. The diffraction angles of the substrate were obtained using the XRD measurement of the substrate before the deposition.

Peak number 4 corresponded to (113) of α-Al_2O_3 and/or the diffraction from the substrate. Peak number 5 is the diffraction from the substrate. Fig. 3 shows the relationship between the net integrated intensity of peak numbers 4 and 5. The intensity of peak number 5 is directly proportional to the intensity of peak number 4. If peak number 4 corresponded to the diffraction from (113) of α-Al_2O_3, the relative intensity of peak numbers 4 and 5 should depend on the film thickness, and the linear relation would not be obtained. A similar relation was also observed for the other substrate peaks, number 8 and 9. Therefore peak number 4 corresponded most closely to the peak from the substrate.

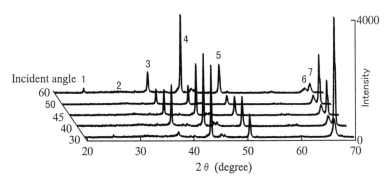

Fig. 2.   The XRD patterns of the Al_2O_3 films prepared at the multiple incident angles of the ion beams.

Table I The relationship between the peak number and the diffraction angles from the experimental results. Diffraction angles of α-Al$_2$O$_3$ and γ-Al$_2$O$_3$ are JCPDS cards data. Diffraction angles of the substrate are obtained by the measurement of the substarate without the films.

Peak number	Experimental results 2θ(deg.)	α-Al$_2$O$_3$ 2θ(deg.)	Index	γ-Al$_2$O$_3$ 2θ(deg.)	Index	Substrate 2θ(deg.)
1	25.4	25.58	012			
2	32.1			31.93	220	
3	37.6	37.78	110	37.60	311	
4	43.5	43.36	113			43.55
5	50.7					50.70
6	66.8	66.52	214	67.03	440	
7	67.8	68.21	300			
8	74.6					74.6
9	95.8					95.8

Peak number 6 corresponded to (214) of α-Al$_2$O$_3$ and/or (440) of γ-Al$_2$O$_3$. The integrated intensity of the peak was compared with peak number 2, which corresponded to (220) of γ-Al$_2$O$_3$, in order to determine whether this peak corresponded to (440) of γ-Al$_2$O$_3$. Fig.4 shows the relationship between the net integrated intensity of peak numbers 2 and 6. The intensity of peak number 6 is directly proportional to the intensity of peak number 2. The slope of a liner approximation equation is about 5. This value corresponded to the ratio of the relative intensity of (440) γ-Al$_2$O$_3$ to the intensity of (220) γ-Al$_2$O$_3$ shown in the JCPDS card. Therefore, peak number 6 is corresponded to the peak from γ-Al$_2$O$_3$.

As a results, It was considered that the peaks that corresponded to α-Al$_2$O$_3$ among the main diffraction peak from the sample are numbers 1, 3 and 7. The peak number 1, (012) of α-Al$_2$O$_3$, was observed in the patterns from incident angles of 30 and 60 degree in Fig. 2, but these peaks were significantly weak in the other patterns, while the peak number 7, (300) of α-Al$_2$O$_3$, is strong in the patterns from incident angles of 40, 45 and 50 degree.

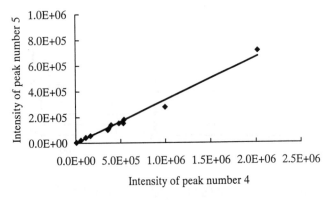

Fig. 3. The relationship between the integrated intensity of peak numbers 4 and 5.

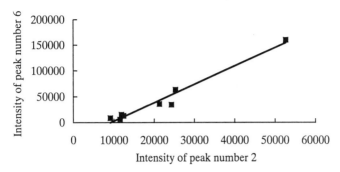

Fig. 4. The relationship between the integrated intensity of peak numbers 2 and 6.

The peak numbers 3 and 7, (110) and (214) of α-Al_2O_3, indicated the difference at the point of peak intensity among these samples.; then the peak intensity from (214) and (300) of α-Al_2O_3, was compared in order to determine the relation between the orientation of α-Al_2O_3 and the electrical resistivity.

Fig. 5 shows the relationship between the incident angle of the ion beam and the electrical resistivity of the films measured at 800℃. As indicated here, the electrical resistivity of the films is significantly affected by the incident angle of the ion beams. It is considered that the differences in the orientation of the Al_2O_3 crystals lead to the differences in the electrical resistance. In addition, it was confirmed that the Al_2O_3 film prepared by plasma CVD at the substrate temperature of 800℃ with irradiation of 10 keV krypton ions at an angle of 40 to 45 degree displayed the highest electrical resistivity through out this experiment.

Fig. 6 shows the relationship between the relative peak intensity I_7/I_3 and the electrical resistivity measured at 800℃. Where I_3 and I_7 are the X-ray intensity of peak numbers 3 and 7, respectively. The electrical resistivity varies according to the relative peak intensity I_7/I_3. The electrical resistivity depends on the ratio of the peak (300) intensity to the peak (110) intensity. Additional analysis on the relationship between the size of the grain boundary and the conductivity theory for ceramics are necessary in order to explain the reasons behind high electrical resistance in greater detail.

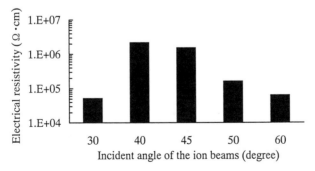

Fig. 5. The relationship between the incident angle of the ion beams and the electrical resistivity of the films measured at 800℃

323

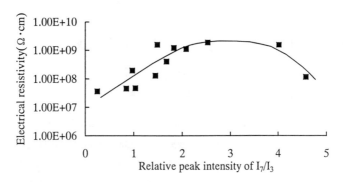

Fig. 6. The relationship between the relative peak intensity I_7/I_3 and the electrical resistivity measured at 800℃.

CONCLUSIONS

Aluminum oxide (Al_2O_3) films were prepared by ion beam assisted plasma CVD at the multiple incident angles of the ion beams to a substrate. Crystal structure of these films was analyzed by X-ray diffraction (XRD). The net integrated X-ray intensity was calculated in order to compare the intensity of the different samples with varying thickness. The electrical resistivity depends on the ratio of the peak (300) intensity to the peak (110) intensity.

ACKNOWLEDGMENTS

Part of this work was conducted in the program; "Advanced Chemical Processing Technology", consigned to the Advanced Chemical Processing Technology Research Association from the New Energy and Industrial Technology Development Organization, which is carried out under the Industrial Science and Technology Frontier Program enforced by the Agency of Industrial Science and Technology, the Ministry of International Trade and Industry. We appreciate their advice and support.

REFERENCES

1. F. Ida, J. Shinohara, T. Kunikyo and S.C.P. Cook, Proc. 18th AIAA Aerospace Ground Testing Conference,(1994)2537

2. H. Nakai,H. Kuwahara, J. Shinohara, T. Kawaratani, T. Sassa and Y. Ikegami, Nucl. Instrum. Methods Phys. Res. Sect.B.112(1996)280

3. H. Nakai, J. Shinohara, T. Sassa and Y. Ikegami, Nucl. Instrum. Methods Phys. Res. Sect.B.121(1997)125

4. N. Kuratani, Y.Murakami, O. Imai, A. Ebe, S. Nishiyama and K.Ogata, J.Vac.Sci. Technol. A15(1997)

Part IV

Optical Materials and Nanoclusters

MESOSCALE ENGINEERING OF NANOCOMPOSITE NONLINEAR OPTICAL MATERIALS

R. F. Haglund, Jr.,[1] C. N. Afonso,[2] L. C. Feldman,[1] F. Gonella,[3] G. Luepke,[1] R. H. Magruder,[1] P. Mazzoldi,[3] D. H. Osborne,[1] J. Solis[2] and R. A. Zuhr[4]

[1] Vanderbilt University, Nashville TN 37235
[2] Instituta de Optica, CSIC, Madrid, Spain
[3] CNR-INFM, Universitá di Padova, Padova, Italy
[4] Oak Ridge National Laboratory, Oak Ridge, TN 37831

Abstract

Complex nonlinear optical materials comprising elemental, compound or alloy quantum dots embedded in appropriate dielectric or semiconducting hosts may be suitable for deployment in photonic devices. Ion implantation, ion exchange followed by ion implantation, and pulsed laser deposition have all been used to synthesize these materials. However, the correlation between the parameters of energetic-beam synthesis and the nonlinear optical properties is still very rudimentary when one starts to ask what is happening at nanoscale dimensions. Systems integration of complex nonlinear optical materials requires that the mesoscale materials science be well understood within the context of device structures. We discuss the effects of beam energy and energy density on quantum-dot size and spatial distribution, thermal conductivity, quantum-dot composition, crystallinity and defects — and, in turn, on the third-order optical susceptibility of the composite material. Examples from recent work in our laboratories are used to illustrate these effects.

Introduction and Motivation

Research groups around the world have reported the fabrication of metal quantum-dot composites (MQDCs) by such varied techniques as ion implantation,[1] ion exchange followed by ion implantation,[2] sol-gel synthesis,[3] sputtering[4] and pulsed laser deposition.[5] The motivation cited in these reports has been to synthesize materials suitable for all-optical switching, to take advantage of the short duration and high pulse repetition frequency of optical signals. Invariably cited in all these publications is the large third-order nonlinearity $\chi^{(3)}$ of these materials, which produces either strong nonlinear absorption $\beta \propto \Im m[\chi^{(3)}]$ or nonlinear refraction $n_2 \propto \Re e[\chi^{(3)}]$, or both.

However, simply having a large value of $\chi^{(3)}$ is insufficient to justify deployment of these materials in all-optical switching technology. A much more complex set of problems is associated with optimizing the materials properties of MQDCs to provide an adequate nonlinear optical response within the constraints set by device functionality and operating conditions. In this context, the mesoscale properties of the MQDCs which affect high-frequency performance, such as thermal conductivity and defect properties, are of special concern.

Here we address a number of mesoscale materials properties which affect the size of the third-order nonlinearity, such as tunability, relaxation time, and high-frequency operation. Several beam-assisted processing techniques can produce either waveguide or layered structures; the challenge lies in adapting these techniques to produce self-assembling or patterned structures. Comparisons of the third-order nonlinear optical properties of MQDCs made by ion implantation, ion exchange followed by ion implantation, and pulsed laser deposition suggest that the greater regularity of layered structures, as found in pulsed laser deposition, leads to significant improvement. We also consider prospects for producing laterally structured or patterned nonlinear materials.

Physical and Optical Characteristics of Quantum-Dot Composites

The arenas of concern for mesoscale engineering of a MQDC are illustrated in Figure 1, and are circumscribed by the processes of optical excitation and relaxation which occur when the dot is irradiated by a short laser pulse. The physical and electronic structure of the quantum dot governs the third-order nonlinear response, particularly the quantum-confined component, and the dissipation of the excitation energy through electron-electron and electron-phonon interactions. The interface properties determine the transfer of thermal energy from the dot to the surrounding matrix, and are especially critical for relaxation rates under high-frequency pulsed irradiation. The mesoscale structure of the composite also determines whether or not percolation effects occur. The quantum properties of the dot, and

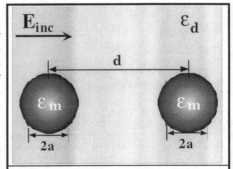

Figure 1. Schematic of a metal quantum-dot composite, showing dots of radius $2a$ spaced at a distance d from each other. In composites of the type discussed here, a is of order 2-5 nm while d ranges from 10-30 nm.

classical field effects due to the embedding of the quantum dot in an appropriate matrix, both contribute to the overall nonlinear optical response and can be optimized by materials processing.

Quantum-confinement effects come into play for metal quantum dots smaller than the mean free path of bulk conduction-band electrons, typically below dot diameters less than 10-13 nm.[6] The energy spectrum for an electron in a spherical box, with bandgap energy E_{gap}, principal and angular momentum quantum numbers (n, ℓ), momentum (k) and a band index b are[7]

$$E_{bn\ell} = \frac{E_{gap}}{2} \pm \frac{\hbar^2 k_{n\ell}^2}{2m_b} \tag{1}$$

In semiconductors, the differing band effective masses and the size dependence of the gap energy produce the well-known sensitivity of the optical response to size inhomogeneities. For metals, on the other hand, the band gap vanishes, and the effective masses of electrons and holes are roughly equal except for strongly localized electrons, such as the d-electrons. The energy-level spacing is roughly equal to the Fermi energy divided by the number of conduction-band electrons.

The relaxation time of metal quantum dot composites, measured with ultrafast lasers, is on the order of a few picoseconds,[8] even near the surface plasmon resonance,[9] with a relatively weak dependence on laser pulse energy.[10] The chief mechanisms for relaxing the excitation include electron-electron scattering (~10 fs); electron dephasing by collisions with the dot walls (~50-80 fs depending on quantum-dot radius),[11] electron-phonon coupling in the metal dot (~1-3 ps) and electron-phonon coupling with the surrounding dielectric (~100-200 ps).

The optical response of a metal quantum-dot composite is characterized by the dielectric polarization (dipole moment per unit volume). The ith component of the polarization **P** induced by an applied optical field $E = E_o \cdot exp[i\omega t]$ can be expanded up to third order in a power series as

$$P_i = \sum_j \chi_{ij}^{(1)} E_i + \sum_{j,k} \chi_{ijk}^{(2)} E_j E_k + \sum_{j,k,l} \chi_{ijkl}^{(3)} E_j E_k^* E_l \tag{2}$$

where the summation indices refer to Cartesian coordinates in the material as well as the polarization of the applied optical field. In isotropic materials, the second-order term vanishes.

Classical confinement effects, in a mean-field theory, result from the oscillations of the conduction-band electrons against the ionic background charge and the restoring force of the surface, thereby altering the field in the vicinity of the dot. For a MQDC with metal dots having a complex dielectric constant $\varepsilon_{qd}(\omega) = \varepsilon_1(\omega) + i\varepsilon_2(\omega)$ occupying a volume fraction $p \ll 1$, and a host medium with a real dielectric constant ε_h, the first-order susceptibility $\chi^{(1)}$ is given by

$$\chi^{(1)}_{eff} = p \cdot \chi^{(1)}_{qd} \cdot \left| \frac{3\varepsilon_h}{(\varepsilon_1 + 2\varepsilon_h)^2 + \varepsilon_2^2} \right|^2 \equiv p \cdot \chi^{(1)}_{qd} \, |f_c|^2 \tag{3}$$

where the quantity f_c measures the local field enhancement due to the dielectric polarization. This expression has a resonance at the frequency ω_r for which $\varepsilon_1(\omega_r) + 2\varepsilon_h(\omega_r) = 0$. This *surface plasmon resonance* can be shifted substantially by altering the shape and size of the nanocrystals, or by changing the combination of metal and dielectric, or by introducing a core-shell structure.

The classical confinement effect on the third-order susceptibility can be derived by applying Maxwell's equations to the nonlinear susceptibility to first order in the electric field.[12] If the nanoparticle radius a satisfies the inequality $(\omega n_o a/c) \ll \delta$, where δ is the skin-depth for the metal, and if $a \ll \delta$, the field is essentially constant over the entire quantum dot. Under these circumstances, the nonlinear optical susceptibility $\chi^{(3)}$ is[13]

$$\chi^{(3)}_{eff} = p \cdot \chi^{(3)}_{qd} \cdot \left| \frac{3\varepsilon_h}{(\varepsilon_1 + 2\varepsilon_h)^2 + \varepsilon_2^2} \right|^2 \cdot \left(\frac{3\varepsilon_h}{(\varepsilon_1 + 2\varepsilon_h)^2 + \varepsilon_2^2} \right)^2 \equiv p \cdot \chi^{(3)}_{qd} \, |f_c|^2 \cdot (f_c)^2 \tag{4}$$

where f_c again measures the local field enhancement due to the polarization. A self-consistent treatment using a jellium model for the metal[14] yields the same result for this special case.

The design of nonlinear optical materials and structures is constrained by the fact that the optical susceptibilities $\chi^{(1)}_{qd}$ and $\chi^{(3)}_{qd}$ of the metal quantum dot behave in opposite ways. The former increase, while the latter decrease, with quantum-dot radius. Hence, it is generally desirable to keep quantum-dot volume small. Also, whereas the amplitude of the plasmon resonance due to dielectric confinement effect is proportional to the square of the field-enhancement factor for the first-order susceptibility, the third-order susceptibility is proportional to the fourth power of f_c.

Working optical devices may be based either on dispersion or absorption. The first-order susceptibility is related to the linear refractive index n_o and absorption coefficient α_o by:

$$n_o = \Re e[1 + \chi^{(1)}] \qquad \alpha_o = \frac{\omega}{n_o c} \Im m[\chi^{(1)}] \tag{5}$$

For a material without a preferred symmetry axis, the composite third-order susceptibility has an analogous relationship to the nonlinear refractive index and nonlinear absorption coefficient:[15]

$$n_2 = \frac{12\pi}{n_o} \Re e[\chi^{(3)}] \qquad \beta = \frac{96\pi^2\omega}{n_o^2 c^2} \Im m[\chi^{(3)}] \tag{6}$$

These third-order nonlinear optical properties, and their dependence on the micro- and mesoscale materials properties of the composite, are the focus of the remainder of the paper.

Figures of Merit for χ(3) Devices

Virtually every linear electro-optic coupling, modulation or switching device can be reconfigured for all-optical operation by inserting appropriate nonlinear optical materials in waveguide or layered geometries.[16] Schematic representations of some possible nonlinear all-optical switching devices are shown in Figure 2. Quantum-dot composites appear to fit neatly into the fabrication of all of these device geometries, though the application is particularly obvious in the Mach-Zehnder interferometers, for example, where one simply has to incorporate the nonlinear material in one arm of the interferometer.

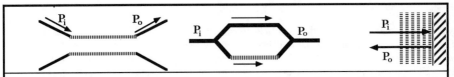

Figure 2. Schematic diagrams of nonlinear, all-optical switching devices which can be made using metal quantum-dot composite materials. The arrows indicate the direction of power flow; the vertically or horizontally hatched lines show the topology of the nonlinear material. Left to right: Full-wave directional coupler (waveguide); Mach-Zehnder interferometer (waveguide); distributed feedback reflector (multilayer structure).

The constraints imposed on mesoscale optical properties by photonic-device structure and functionality can be expressed in a convenient, albeit approximate, way by figures of merit (FOMs). *Optical* figures of merit relate the magnitude of the third-order nonlinearity to the linear response and nonlinear optical response. For purely dispersive or absorptive nonlinearities, the optical figures of merit (*not* dimensionless!) off- or on-resonance, respectively, are:[17]

$$f_{disp} = \frac{\omega \cdot n_2}{n_0} \qquad\qquad f_{abs} = \frac{\beta}{\alpha_o \cdot \tau} \qquad\qquad (7)$$

The relaxation time τ_{mat} in the expression for f_{abs} is the limiting energy relaxation time of the resonant (long) or non-resonant (short) excitation process. These two quantities measure, respectively, the change in refractive index or absorption per unit time per unit intensity, thus permitting comparison of materials with widely differing excitation and relaxation mechanisms.

A defect-related dimensionless materials figure of merit arises from limits imposed by two-photon absorption (TPA) on all-optical switching; it is defined by[18]

$$f_{TPA} = \frac{2\lambda \cdot \beta}{n_2} < 1 \qquad\qquad (8)$$

and expresses the capacity of the nonlinear material to induce a phase shift without dropping below the critical operating intensity of the device. Other optical properties of the nonlinear composites besides TPA,[19] such as the presence of optically active defects,[20] may force the systems designer to accept smaller third-order susceptibilities in order to optimize overall system performance.

Functional figures of merit relate device properties, such as phase shifts, to the mesoscale materials properties. For example, in a nonlinear directional coupler or Mach-Zehnder interferometer, switching occurs if $w \equiv \Delta\beta_{sat}L/\lambda > 2$, where L is the interaction length, λ is the wavelength, and $\Delta\beta_{sat}$ is the change in the guided-wave index at saturation, which usually equals the saturation in the material index of refraction, Δn_{sat}. However, high throughput also requires that $L\alpha < 1$, where α is the linear absorption coefficient. Combining these two conditions yields a dimensionless figure of merit for this particular device:[21]

$$W = \frac{\Delta\beta_{sat} \cdot L}{\lambda} \times \frac{1}{L\alpha} = \frac{\Delta n_{sat}}{\lambda\alpha} \tag{9}$$

Note that this particular figure of merit not only includes a referent to the device type, but also its operating point, through the inclusion of the saturated value of the change in refractive index.

The *operational* figure of merit is effectively the number of switching operations which can be performed in one thermal relaxation time. In all-optical switching schemes, one must have a phase change of π over some reasonable length (say, 0.1-1 cm for waveguide integrated-optical devices). For a device operating on nonlinear dispersion, this phase change must satisfy $\phi = 2\pi n_2 I_o L/\lambda$. On the other hand, the thermal loading produced by absorption of a single pulse of duration τ_{laser} at the intensity is $\Delta Q = \alpha L I_o a \tau_{laser}$, where α is the absorption coefficient and $V=aL$ is the volume of the waveguide in which the pulse is absorbed. The figure-of-merit defining the performance of a given material at high pulse-repetition frequencies is then:[22]

$$f_{thermal} = \frac{\text{Index change required for switching}}{\text{Thermal index change}} = \frac{n_2 C_p \rho}{\alpha\tau(dn/dT)} \tag{10}$$

where C_p is the heat capacity at constant pressure, ρ is the density, τ is the slower of the thermal and electronic relaxation times, and dn/dT is the thermo-optic coefficient. Since typical thermo-optic coefficients tend to fall in the range 10^{-4} K^{-1}, the thermal refractive index change, which relaxes only very slowly in many insulators, may mask the fast electronic relaxation processes associated with n_2 in the worst cases.

A comparison of semiconductor and metal quantum-dot composites in glass with a fused silica fiber shows that, by this figure-of-merit, optical fibers appear to be the "best" nonlinear material. However, using several meters of optical fiber may not always be compatible with many desirable device applications! Metal QDCs and semiconductor QDCs in glasses appear to have comparable figures of merit in many cases, as shown in Table 1.

Table 1: Figures of merit for various nonlinear materials

Material	$f_{disp}\left(\frac{W}{cm^2 \cdot s}\right)$	$f_{abs}\left(\frac{W}{cm^2 \cdot s}\right)$	f_{TPA}	W	$f_{thermal}$
GaAs MQW	10^7	$3\cdot10^{-4}$	6	~ 2.5	$3\cdot10^2$
CdSe:glass	10^3	66	30	0.3	$7\cdot10^4$
Cu:glass	10	50	5	0.2 - 2.0	$6\cdot10^3$
Silica fiber	10^{-5}	10^8	9	$> 10^3$	$4\cdot10^7$

Experimental Details

The drive to generate functional nonlinear optical materials is the classic synthesis *vs* materials characteristics problem. We shall compare three different processes which are currently in use in many laboratories to synthesize quantum-dot composites. We have omitted sol-gel processing, ion-beam-assisted deposition and classical sputter deposition processes for reasons of space.

Synthesis by ion implantation

Ion implantation can affect nonlinear device performance either by changing the optical properties of the matrix (index of refraction, for example) or the electronic structure of the metal quantum dot. A large variety of nanocomposites incorporating elemental and compound metal and semiconductor nanocrystals have been synthesized by ion implantation. Using energies ranging from 100 keV to a few MeV, nanocomposite layers up to a few hundred nm thick and as much as half a micron deep have been prepared. However, the process of quantum-dot nucleation and growth is poorly understood in these materials, although there is some evidence that the dots are nucleated at defect sites and grow by thermal diffusion of the implanted ions, driven by the thermal energy supplied by ion deposition. The only successful control of size distribution demonstrated so far has been achieved at the expense of deposition speed, by reducing the ion-beam current density. At low implantation current densities,

Synthesis by ion exchange and ion implantation

Planar waveguides of order 5 μm thick can also be generated in one surface of a sodium silicate (soda-lime) glass by immersing the glass in a bath of molten $AgNO_3$ or $CuNO_3$ salts held at the eutectic point (600-700 C, depending on the salt), initiating the exchange of Na^+ for Ag^+ or Cu^+, and leading to the formation of a layer of Ag or Cu ions. After ion exchange, He ions at 1.8 MeV energy (range 6.5 μm) at doses up to $1 \cdot 10^{16}$ cm^{-2} were implanted; the energy deposited locally by the light ions in transit is sufficient to initiate nucleation of the metal nanocrystals. Optical absorption spectra taken before and after implantation showed that the surface plasmon resonance for silver formed after the implantation, indicating that the deposition of energy by the He ions was sufficient to initiate aggregation of the dispersed Ag or Cu ions. Optical absorption spectra after implantation show a strong peak at wavelengths corresponding to the surface plasmon resonance.

Synthesis by pulsed laser deposition

The most recent developments in MQDC synthesis have been using pulsed laser deposition with multiple targets. In one recently published paper,[5] Cu nanocrystallites 4±1 nm in diameter were generated in a ten-period structure with Al_2O_3 as a matrix, by sequential laser ablation of a Cu metal target and a pressed-powder Al_2O_3 target. The laser used was an ArF laser, operating at a wavelength of 193 nm; typical fluences were of order 2 J·cm^{-2}. Particularly surprising was the fact that the spacing of the nanocrystals in each layer was highly uniform, measuring from the transmission electron microscopy 8±1 nm. An analysis of the ablation plume from the Cu target showed the presence of Cu_6 and Cu_7 clusters, as measured by matrix-assisted laser desorption-ionization mass spectrometry. Assuming that the clusters indeed arrive at the surface in this configuration, it would appear that cluster diffusion on the surface, perhaps nucleating at defect sites in the deposited alumina film (which is polycrystalline), leads to growth limited only by the surface temperature. Recent unpublished results in our laboratory have shown the presence of Cu nanoclusters in complex ternary oxides as well.

The relative merits of these three synthesis techniques are compared in Table 2, where we have noted the possible effects of each process on the critical mesoscale properties of MQDCs. Not all of these capabilities have been fully demonstrated in every synthetic technique.

Table 2. Comparison of mesoscale materials characteristics for various beam syntheses.

Mesoscale properties	Ion Implantation	Ion Exchange and Implantation	Pulsed Laser Deposition
Control of size distribution	Marginal	Moderate	Good
Small quantum-dot size	Marginal	Good	Excellent
Core-shell structures	Yes	No	Possible?
Waveguides	Too shallow	Yes	Yes
Lateral structuring	With great difficulty	Probably not	Possible?
Defect density	High	Low	Low
High thermal conductivity	Sometimes	Marginal	Yes

Nonlinear optics experiments

Nonlinear optical measurements were carried out using a tunable, synchronously pumped, mode-locked and cavity-dumped dye laser with an average output power of 100-250 mW at a pulse repetition frequency of 0.76-19 Mhz. The dye laser was pumped by a continuous-wave, mode-locked and frequency-doubled Nd:YAG laser producing 100 ps pulses at a pulse repetition frequency of 38 Mhz. The pulse duration of the dye laser output beam was measured by standard autocorrelation techniques to be 6 ps (using Rhodamine 6G) or 0.7 ps (using Pyhrromethene 570). The focusing lens in both experiments was 15 cm, leading to a peak focal-spot intensity on the order of $2 \cdot 10^8$ W·cm^{-2}.

The third-order susceptibility of the MQDCs was measured directly by degenerate four-wave mixing in the forward, phase-matched geometry. The beam from the dye laser was split into three beams: two strong pump beams, each with approximately 45% of the total intensity, and a single weak probe beam with the remaining laser output. The pump beams were aligned onto the sample and precisely overlapped in space and time; the probe beam was reflected from a corner cube on a movable translation stage, and was varied in time to sweep through the probe beams as they struck the sample. The phase-conjugate signal, strong enough to be visible to the naked eye, was identified from the direction consistent with conservation of momentum, and was routed to an uncooled photomultiplier detector the output of which was fed to a dual gated photon counter.

The nonlinear index of refraction n_2 and nonlinear absorption β of the samples were measured using a single-beam technique, the so-called Z-scan, in which phase distortion is converted to amplitude distortion as a well-characterized Gaussian beam is directed through a fixed, long-focal-length lens to the sample, located near the focal plane. As the sample is scanned along the optical axis of the system through the focal plane, the intensity on the sample is varied. A Z-scan measurement of both near- and far-field intensities yields the nonlinear index and nonlinear absorption from a single measurement.

For a thin nanocluster layer embedded in a transparent dielectric and for a high-repetition-rate laser, the relative far-field transmitted power is determined by both the optical response of the metal particle and the surrounding dielectric, and by the two-dimensional thermal relaxation processes through which the absorbed laser energy flows out of the illuminated region. In the limit of small nonlinearities, the power transmitted to the detector is given by

$$P_{det} = \left[\frac{1}{1 + \beta I_o L \cdot \frac{1 - exp(-\alpha L)}{\alpha}} \right] \cdot \left[1 - \frac{2C_1\zeta}{(1+\zeta^2)^2} + \frac{2C_2\zeta}{(1+\zeta^2)} \right]$$

(11)

where C_1 and C_2 are propertional to the electronic and thermal components of the nonlinear index, respectively, and ζ is a normalized z-coordinate along the optical axis. The differing z-dependences of the electronic and thermal terms in Eq. (11) arise because the electronic component of the nonlinear index is proportional to intensity, while the thermal nonlinear index is proportional to the temperature rise and hence to the total power.

Mesoscale Engineering Issues

We have seen that the nonlinear performance of MQDCs depends in a complicated way on the optical properties of the quantum dots and the matrix — as in the mean-field enhancement factor f_c. But the figures of merit also include the material relaxation factors and the thermal properties of the composite. Engineering the materials for even the simplest devices requires an understanding of which parameters control which property or group of properties. In this section, we examine several different synthesis variables and their effects on nonlinear optical performance.

In trying to produce third-order nonlinear optical materials for practical applications, there are several interrelated characteristics of the MQDCs, each imposing unique constraints on the quantum dots and the matrix in which they are embedded. These issues may be summarized as follows:

- Spectral and temporal response of the metal quantum dot, controlled chiefly by composition and size;

- Spectrum and amplitude of the surface plasmon resonance, controlled by the physical structure of the quantum dot and the dielectric functions of metal and matrix;

- Thermal relaxation time of the nanocomposite, governed primarily by the thermal transport properties of the matrix and the size of the quantum dot.

As will be demonstrated in the succeeding paragraphs, the constraints imposed by mesoscale materials properties can be altered in a number of interesting ways using energetic beams to control the microscopic (*i.e.*, atomic scale) properties of the quantum dot, interface and matrix.

Quantum-dot optical response

Semiconductor quantum dots achieve wavelength tunability by varying the quantum-dot size: As the size of the dots is progressively reduced below the Bohr exciton radius, the bandgap energy shifts progressively toward the blue; moreover, the electron and hole effective masses are also size dependent. While the oscillator strength is large for any specific optical transition, the bands are nearly parabolic, meaning that there is relatively little anharmonicity, and the increase in transition probability comes almost exclusively from enhanced oscillator strength. In metal quantum dots, on the other hand, tunability is low, because energy levels are spaced almost equidistant. The anharmonicity is high, on the other hand, because the confining potential is very deep and nearly a square well. The only resonant enhancement is provided by either quantum confinement effects on the intraband transition, or by the surface plasmon resonance.

The spectral and temporal response of metal quantum dots is significantly affected by whether the optical transition is an intraband or interband transition. The holes created in interband transitions are, for noble metals, in the d band and are effectively screened by the s-p conduction and electrons. Also, because the d orbitals are already relatively strongly localized, the electronic structure of the quantum dots controls their frequency response as well as the relative size of the third-order susceptibility.

We measured the intrinsic response time of quantum dots excited via the intraband transition in Au:SiO$_2$, and compared it to the response time on the interband transition for Cu:SiO$_2$, both at a laser wavelength of 575 nm. The experiment was a four-wave mixing measurement, in which the phase-conjugate signal for the two pump and probe beams produces a coherent signal proportional to the relaxation time of the grating created by the pump beams. The measured autocorrelation width of the laser pulse was 730 fs. The phase-conjugate signals show distinct differences in the two cases: On the interband transition for Cu quantum dots, the dephasing signal is symmetrical with respect to zero time

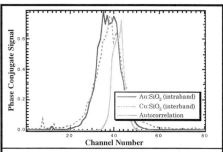

Figure 3. Degenerate four-wave-mixing signal for Au:SiO$_2$ and Cu:SiO$_2$ MQDCs, showing the finite relaxation time for the nanocrystals compared to the laser pulse (730 fs).

delay (channel 40), indicating a constant relaxation rate probably due to electron-phonon coupling; the measured relaxation time is approximately 2.2 ps. In the case of the Au:SiO$_2$ sample, the phase-conjugate signal shows a pronounced asymmetry, and a flat top with a width of 1.2 ps. This is probably due to Pauli blocking: electrons excited out of conduction-band levels near the Fermi surface have fewer states available into which they can scatter — a consequence of the small size of the nanocrystal. This causes the electron-phonon relaxation rate to slow down as the energies of the scatter electrons drop down closer to the Fermi level. Both of these times are much shorter than the thermal relaxation time, so that it is the latter which is the rate-limiting process in the high-frequency response of the quantum dots.

Surface plasmon resonance

The increased nonlinear response made possible by the surface plasmon resonance (SPR) is a primary advantage of metal quantum dots for nonlinear optical applications. As shown by Eq. (3), the enhancement f_c in the optical response from the surface plasmon is governed by the relationship between the dielectric functions of the metal and the host dielectric. Changing that relationship alters the strength and frequency characteristics of the resonance, as shown in Figure 4. Here Cu and Ag ions were coimplanted in a fused-silica matrix, with beam energies adjusted to insure that the implantation depth was equal for both species. The driving force for the alloying is believed to be the heat deposited in the matrix by the second implantation. Alloy nanoclusters were formed, and the resulting dielectric functions differed in each case. As Fig. 4 shows, the strength of the plasmon resonance near 410 nm (the position of the SPR for pure Ag nanocrystallites) was clearly

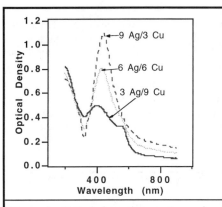

Figure 4. Optical absorption spectrum of an Ag/Cu:SiO$_2$ nanocomposite, showing the effect of the increasing fraction of Cu relative to Ag on the plasmon resonance.

altered by the co-implantation. Other strategies which can potentially increase the strength of the plasmon resonance include embedding the nanocrystals in a *nonlinear* dielectric, because the intrinsic dipole moment then adds to the local field enhancement produced by the plasmon resonance. Interestingly, the nonlinear response was also altered, with modest decreases in the nonlinear index but with a significant change in the nonlinear absorption. Such a change, of course, will alter some of the figures of merit governing the performance of the material.

The central frequency and strength, as well as the width, of the plasmon resonance can also be altered by creating core-shell nanostructures. In this case, a proper match of the boundary conditions across the core, shell and matrix interface can produce an even larger field enhancement than is possible for a single-element nanocrystal in a dielectric. This effect is shown in Figure 5, in which Ag was co-implanted with several different metals to produce core-shell structures in the two cases (Sb and Cd) which are not miscible with Ag.

Thermal Transport Properties

The performance figure-of-merit introduced earlier probably poses the greatest single materials challenge to the construction of practical third-order metal-quantum-dot composites. At high frequencies, the residual thermal loading can be substantial, even at relatively low powers; the problem is compounded by the fact that the dielectric medium in which the quantum dots are embedded is typically a poor conductor of heat. Previous studies of the nonlinear index of refraction showed that the thermal component of Eq. (18) comes to dominate the electronic component at relatively low pulse repetition frequencies. For example, Figure 6 shows how the near-field and far-field Z-scans vary with pulse repetition frequency for Au nanocrystallites in fused silica, synthesized by ion implantation at a total ion dose of approximately 10^{17} Au ions·cm^{-2}. At a pulse repetition frequency above 10 MHz, the dominance of the thermal component is shown by the sudden rise in the peak-to-valley ratio of the far-field Z-scan [Figure 6(a)] and by the shift from nonlinear absorption to nonlinear saturation [Figure 6(b)] in the near-field Z-scan.

This problem can be overcome by embedding the quantum dots in a medium with higher thermal conductivity than is typical for glasses.

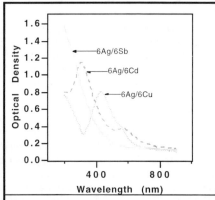

Figure 5. Optical density *vs* wavelength for a sample implanted with two differing ion species showing the shifting of the surface plasmon resonance with composition.

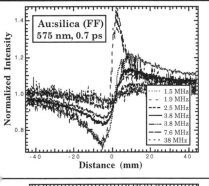

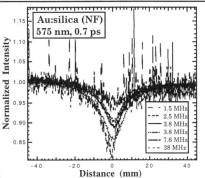

Figure 6. Z-scans showing far- and near-field transmission (upper and lower panels, respectively) of an Au:SiO$_2$ MQDC, as a function of pulse repetition frequency.

In another experiment, Cu quantum dots were embedded in Al$_2$O$_c$ by pulsed laser deposition, as described above, and the near-field and far-field scans were used to determine the relative size of thermal and electronic effects on the nonlinear index and absorption. These data were then compared with earlier published data on the relative sizes of the thermal and electronic components of the nonlinear index, and with the nonlinear absorption. The Cu:Al$_2$O$_3$ composite was found to have a somewhat lower thermal to electronic component than the Cu:SiO$_2$ composite. However, the improvement in the nonlinear absorption was much more dramatic: the saturable absorption decreased by two orders of magnitude in the Cu:Al$_2$O$_3$ material! This suggests that embedding the nanocrystals in even higher-conductivity matrix materials should lead to significant improvements in the high-pulse-repetition-frequency performance of these nanocomposites. We have recently begun experiments aimed at doing just that, by using pulsed-laser-deposition synthesis to nucleate and grow Cu quantum dots in indium-tin oxide, a transparent, electrically conductive oxide with a conductivity over two orders of magnitude larger than that of alumina.

In recent computational studies of the thermal conductivity of nanocluster composites, it was found that the transport of heat across the boundary cannot be adequately described by standard Fourier heat-conduction theory once the size of the nanocrystals drops below the mean-free path of the heat carriers. As it turns out, the thermal conductivity is reduced by as much as a factor of two over bulk values, so that as the nanocrystallites are reduced in size, the heat transport tends to be described not by continuum Fourier heat conduction theory, but by a microscopic Boltzmann transport formalism. It remains to be seen precisely what the tradeoff is between the increased nonlinear-optical performance derived from smaller cluster sizes and the penalty for high-frequency operation incurred by the reduced thermal conductivity of the smaller nanoclusters.

Conclusions

In summary, we have described a number of the microscopic and mesoscale materials properties of metal quantum-dot composites which can be influenced by energetic laser and ion beams. Nonlinear optical experiments show clearly that these properties are directly influenced by these atomic-scale properties. Hence, future studies of materials will need to take these properties into account if they are to produce meaningful progress toward nanocomposites suitable for photonics applications. In particular, the question of thermal transport is critical and needs to be addressed.

Acknowledgement

Research at Vanderbilt is partially supported by the Army Research Office under contract DAAH04-93-G-0123; research at Padua is supported by the Institute of Materials Physics of the Italian National Research Council (INFM-CNR). G. Luepke is supported by a Feodor-Lynen Fellowship of the Alexander von Humboldt Foundation. The Oak Ridge National Laboratory is partially supported by the U. S. Department of Energy, Division of Basic Energy Sciences, under contract DE-AC05-96OR22464 with Lockheed-Martin Energy Systems, Inc.

References

1. C. Buchal, S. P. Withrow, C. W. White and D. B. Poker, *Ann. Rev. Mat. Sci.* **24** (1994) 125.
2. P. Mazzoldi, G. W. Arnold, G. Battaglin, F. Gonella and R. F. Haglund, Jr., J. Nonlinear Opt. Phys. and Matls. **5** (1996), 285.
3. G. De, L. Tapfer, M. Catalano, G. Battaglin, F. Gonella, P. Mazzoldi and R. F. Haglund, Jr., Appl. Phys. Lett. **68** (1996) 3820.
4. H. B. Liao, R. F. Xiao, J. S. Fu, P. Yu, G. K. L. Wong and P. Sheng, Appl. Phys. Lett. **70** (1997), 1-3. I. Tanahashi, Y. Manabe, T. Tohda, S. Sasaki and A. Nakamura, J. Appl. Phys. **79** (1996) 1244.
5. J. M. Ballesteros, R. Serna, J. Solís, C. N. Afonso, A. K. Petford-Long, D. H. Osborne and R. F. Haglund, Jr., Appl. Phys. Lett. **71** (1997) 2409.
6   F. Hache, D. Ricard and C. Flytzanis, J. Opt. Soc. Am. B **3** (1986) 1647.
7. S. Schmitt-Rink, D. A. B. Miller, and D. S. Chemla, Phys. Rev. B **35** (1987) 8113.
8. T. Tokizaki, A. Nakamura, S. Kaneko, K. Uchida, S. Omi, H. Tanji and Y. Asahara, Appl. Phys. Lett. **65** (1994) 941.
9. M. Perner, P. Bost, U. Lemmer, G. von Plessen, J. Feldmann, U. Becker, M. Mennig, M. Schmitt and H. Schmidt, Phys. Rev. Lett. **78** (1997) 2192.
10. J.-Y. Bigot, J.-C. Merle, O. Cregut and A. Daunois, Phys. Rev. Lett. **75** (1995) 4702.
11. R. F. Haglund, Jr., Li Yang, D. H. Osborne, H. Hosono, C. W. White and R. A. Zuhr, Appl. Phys. A **62** (1996) 403.
12. D. Stroud and V. E. Wood, J. Opt. Soc. Am. B **6** (1989) 778.
13. D. Stroud and P. M. Hui, Phys. Rev. B **37** (1988) 8719.
14. F. Hache, D. Ricard and C. Girard, Phys. Rev. B **38** (1988) 7990.
15. M. J. Weber, D. Milam, D. and W. L. Smith, W. L., Opt. Eng. **17** (1978) 463.
16. G. I. Stegeman, E. M. Wright, N. Finlayson, R. Zanoni, R. and C. T. Seaton, *IEEE J Lightwave Tech.* **6** (1988) 953.
17. C. Flytzanis, F. Hache, M. C. Klein, D. Ricard and Ph. Roussignol, *Prog. Optics* **2?** (1991) 323.
18. V. Mizrahi, K. W. DeLong, G. I. Stegeman, M. A. Saifi and M. J. Andrejco, *Opt. Lett.* **1** (1989) 1140.
19. J. W. Aitchison, M. K. Oliver, E. Kapon, E. Colas, and P. W. E. Smith, *Appl. Phys. Lett* **56** (1990), 1305.
20. K. W. DeLong, V. Mizrahi, G. I. Stegeman, M. A. Saifi, and M. J. Andrejco, *Appl. Phys. Lett.* **56** (1990) 1394.
21. G. I. Stegeman and R. H. Stolen, *J. Opt. Soc. Am. B* **6** (1989) 652.
22. S. R. Friberg and P. W. Smith, *IEEE J. Quantum. Electron.* **23** (1987) 2089.

SILVER NANOCLUSTER FORMATION IN IMPLANTED SILICA

G. MARIOTTO* and F.L. FREIRE Jr.**
* Istituto Nazionale per la Fisica della Materia and Dipartimento di Fisica, Università di Trento, 38050 Povo, TN, Italy
** Departamento de Fisica, PUC-Rio, Rio de Janeiro, 22452-970, RJ, Brazil

ABSTRACT

Samples of fused silica were implanted at room temperature with 300 keV-Ag^+ for doses ranging from 0.8×10^{16} to 14×10^{16} ions/cm^2. A multi-technique approach including Rutherford backscattering spectrometry (RBS), x-ray diffraction (XRD), optical absorption and Raman scattering spectroscopies has been used to characterize silver precipitate. The Ag-depth profiles of samples implanted with doses higher than 6×10^{16} Ag^+/cm^2 show a bi-modal distribution, with the appearance of a secondary maximum near the surface. XRD spectra indicated the formation of silver nanocrystals of ~10 nm in size within the heavily implanted samples. Optical absorption has been used to monitor the effects of ion doses on the optical properties of the metal clusters in the UV-Vis region. A single broad absorption band, due to surface plasmon resonance, is peaked at about 400 nm for low implantation doses. For doses higher than 4.3×10^{16} Ag^+/cm^2, a second broad band originates at higher wavelengths, peaking at 625 nm for the highest dose. The evolution of optical spectra is tentatively discussed in terms of the formation of silver particle aggregates with no longer spherical shape. An estimate of the mean size of silver nanoclusters of about 5.5 nm is obtained from low-frequency Raman scattering due to acoustic vibrations localized at the cluster surface. The discrepancies in the metal particle size obtained from XRD and Raman scattering measurements are discussed with respect to optical absorption data.

INTRODUCTION

Recent increase in research activities on semiconductor and metal nanocrystals embedded in solid transparent matrices is strongly motivated by the potential applications in the field of non-linear processing devices and by the interest in fundamental physics on the excited states of these composite systems. Metal nanoclusters dispersed in transparent matrices present highly enhanced optical non-linearity due to quantum confinement of electrons, which strongly affects the optical properties of these composite systems. Therefore, as far as the opto-electronic applications are concerned, a very crucial step is the accurate control of cluster size during the synthesis process. Ion implantation provides a suitable route to precipitate metal colloids or semiconductor particles in dielectric matrix [1]. The major advantages of ion implantation in metal composite fabrication lie in its capability of incorporating much higher local concentration of metal particles in the host matrix compared to the melt quenching method. Furthermore, it allows for the depth-distribution control of the precipitated nanostructures. Basic properties of metal and semiconductor nanocrystals formed by high-dose ion implantation in fused silica and sapphire have been recently reviewed by White et al. [2].

Since the pioneering work of Arnold and Borders [3] most of the experimental investigations carried out on metal colloid composites were addressed to the characterization of structural and optical properties of metal nanoparticles in glass matrices [4,5]. In particular, a number of publications reported on transmission electron microscopy (TEM) measurements of metal cluster sizes and distribution in fused silica under various implantation and annealing conditions [6-7].

Mat. Res. Soc. Symp. Proc. Vol. 504 © 1998 Materials Research Society

Recently, low-frequency Raman scattering from acoustic vibrations, localized at the nanocluster surface, has been proven to be an efficient non-destructive tool alternative to TEM, to study the size of very small silver particles dispersed in dielectric matrices [8-11]. The analysis of low-frequency Raman scattering data in terms of Lamb´s theory, which gives the vibrational frequencies of a homogeneous elastic body with a spherical form [12], provides an estimate of the silver cluster mean size, in reasonable agreement with TEM determinations.

In this work, a multi-technique approach, including RBS, XRD, optical absorption and low-frequency Raman scattering, has been adopted to characterize the silver colloid particles precipitated by ion implantation in fused silica. Our data suggest the formation of simple spherical silver clusters at low implantation doses and of complex aggregates of precipitated silver clusters at implantation doses equal or higher than 7.9×10^{16} Ag^+/cm^2.

EXPERIMENTAL PROCEDURES

Silver ions were implanted at energy of 300 keV-Ag^+ into commercial fused silica (Herasil) samples, held at room temperature. The implantation doses ranged from 0.8×10^{16} to 14×10^{16} Ag^+/cm^2 and the current density was ~1 $\mu A/cm^2$. No post-implantation thermal annealing was performed and the samples were stored in air. Under white light, samples implanted at low doses show a pale yellowish color, which tends to become more grayish when the ion dose increases, indicating the formation of metal colloids. RBS spectra were obtained using 2.0 MeV He^+ beams at a scattering angle of 165°.

X-ray diffraction patterns were recorded in the glancing incidence geometry (incidence angle 1.5°) using the Cu-K_α radiation of a ω-diffractometer (Siemens model D5000). The detector was equipped with a Soller slit and a LiF monochromator. The diffractograms were scanned between 34° and 46° in 2θ in order to detect both the (111) and the (200) diffraction lines of silver.

Optical extinction spectra were carried out in the wavelength range 200 nm to 800 nm using a Cary 14 dual beam spectrophotometer. The spectra were recorded at room temperature in the standard differential mode, with an unimplanted fused silica sample as reference.

Low-frequency Raman spectra were measured at room temperature using a commercial equipment consisting of an Ar^+-ion laser, of a double monochromator (Jobin Yvon, model Ramanor HG2-S), equipped with holographic gratings (2000 lines/mm), and of a photon counting system. Two polarization settings were used in order to probe the symmetry character of the vibrational modes causing the light scattering, with the electric field of the incident beam parallel (V) or perpendicular (H) to the scattering plane. The scattered light was filtered by a polarizing-plate with vertical axis (V). These settings allowed to record of VV and HV spectra, alternatively.

RESULTS AND DISCUSSION

RBS measurements were used to determine the depth-profile of Ag^+ ions implanted at 300 keV under different doses in the range 0.8 to 14×10^{16} ions/cm^2 in silica. Part of the RBS spectra of implanted samples is shown in Fig. 1. The silver depth profiles, as determined from RBS measurements, from samples implanted with a dose lower than 7.9×10^{16} Ag^+/cm^2, indicate a maximum concentration at R_p = 83 nm, with a full width at half maximum ΔR_p = 26 nm. The full width at half maximum of silver depth profiles in the heavily implanted samples is somehow narrower than those of samples implanted with low ion doses, presumably due to the formation of metal precipitate grown at high ion doses. Furthermore, the silver depth profiles of samples implanted at ion doses higher than 7.9×10^{16} Ag^+/cm^2 clearly show a bi-modal distribution, with a

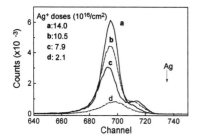

Figure 1- RBS spectra of implanted fused silica implanted with 300 keV-Ag$^+$ at different ion doses. Only the energy region around the silver peak is shown. An arrow indicates the position of Ag atoms at the sample surface.

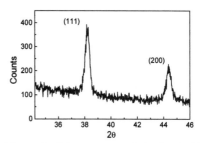

Figure 2- XRD patterns of the fused silica sample, implanted with 7.9×10^{16} Ag$^+$/cm^2, showing both the (111) and the (200) diffraction lines of crystalline silver with cubic structure.

secondary maximum close to the depth where the maximum deposition of energy by the incident ions occurs.

The formation of silver crystalline aggregates at high doses is revealed by x-ray diffraction patterns. Figure 2 shows a XRD spectrum obtained at glazing angle from the sample implanted with 7.9×10^{16} Ag$^+$/cm^2. The presence of silver nanocrystals is revealed by the XRD patterns through the occurrence of sharp reflections from the (111) and the (200) planes of silver with cubic structure, peaking at 38.1° and 44.3°, respectively. Assuming an isotropic distribution of metal particle shape, the Debye-Scherrer analysis of the line broadening of XRD peaks of Fig. 2 indicates an average diameter of ~10 nm for the metal particles in this sample. Similar mean sizes were obtained from XRD patterns of samples implanted with higher ion doses, but no silver diffraction line was observed from our fused silica samples at lower ion doses.

The extinction spectra of implanted samples are shown in Fig. 3. The formation of silver colloids even at the lowest dose of 0.8×10^{16} Ag$^+$/cm^2 is revealed by the weak extinction occurring at the surface plasmon energy (~3.1 eV). The extinction spectra show a very strong intensity increase with the ion dose accompanied by a drastic change of their shape. In fact, the absorption spectra of samples implanted at doses not higher than 4.3×10^{16} Ag$^+$/cm^2 consist of a single symmetric band peaked at about 400 nm. They broaden and increase in intensity on the side of higher wavelengths when the ion dose rises, and turn out into a clearly revolved doublet, with the second band peaked at about 625 nm at the highest dose. Similar double-band optical absorption spectra were observed in SiO$_2$ glass implanted at room temperature with 7.6×10^{16} Ag$^+$/cm^2 of 150 keV [7]. They were discussed in terms of surface plasmon excitation of two types of silver colloids, the sizes of which were estimated to be lower than 10 nm and about 50 nm, according to calculations performed by Arnold and Borders [3]. However, TEM data did not show the presence of any 50 nm-sized silver cluster in this implanted SiO$_2$ glass [7].

We interpret our optical data with a structural model, which assumes the occurrence of silver clusters for low implantation doses, and the formation of complex aggregates from precipitated silver clusters, with chain-like or even cluster-like clumping characters, for high implantation doses. The optical absorption spectra of our samples, in fact, closely recall the experimental observations of Granquist et al. [13], which were reasonably well simulated using the Maxwell-Garnett formalism [14]. Therefore, the evolution of extinction spectra versus the implantation

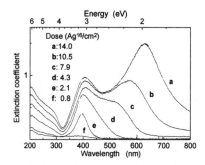

Figure 3- Extinction spectra of fused silica samples Ag-implanted at six different doses.

dose suggests that silver precipitation in fused silica may result into the formation of bigger metal particles, having no longer spherical symmetry.

Figure 4 shows the VV polarized Raman spectra recorded under excitation of the 488.0 nm laser line from two Ag-implanted silica samples (top panels) and from an unimplanted sample (bottom panel). The spectrum obtained from the not-implanted sample fits very well that of fused silica [15]. Raman spectra of both implanted samples show a paramount low-frequency scattering band, extending up to 200 cm^{-1} far from the excitation energy in both anti-Stokes and Stokes sides, which is superimposed on an apparently flat background extending up to about 600 cm^{-1}. A 20 times magnified plot of the region 200 cm^{-1} to 1200 cm^{-1} for sample implanted with 4.3×10^{16} Ag^+/cm^2 (mid panel) undoubtedly reveals that the underlying background originates from silica matrix. The most significant change is due to the additional band peaked at about 980 cm^{-1}; this band is related to stretching vibrational mode of terminating Si-OH units in silica [16]. Its

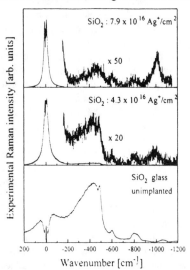

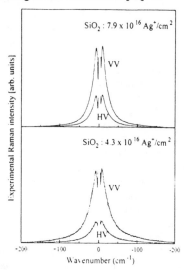

Figure 4- VV-polarized Raman spectra of silver implanted fused silica samples. Both anti-Stokes component below 200 cm^{-1} and the overall Stokes spectrum till 1200 cm^{-1} are shown.

Figure 5- Polarized (VV) and depolarized (UV) low-frequency Raman scattering from two silver implanted silica samples. Both anti-Stokes and Stokes components are reported.

presence indicates that some hydration occurred in the implanted sample and that hydrogen decorates the chemically reactive defects produced by the implantation in silica [17]. Similar considerations can be inferred from the 50 times magnified plot for sample implanted under higher dose (top panel), where the relative intensity of the band at 980 cm^{-1} suggests even a more important hydrogen decoration of the internal surface defects, due to ion radiation damage. Obviously, the reduced spectral intensity from the glass matrix observed in these implanted samples results from the strong absorption of exciting radiation by the colloidal surface plasmons. On the other hand, the excitation of surface plasmons localized in the silver clusters enhances the low-frequency Raman scattering observed from implanted samples.

Figure 5 shows both the low-frequency (below 200 cm^{-1}) anti-Stokes and Stokes components of the VV and the HV polarized Raman spectra recorded under identical excitation conditions from two implanted samples. Despite the remarkably different intensity (about a factor 4 between the two samples), low frequency Raman scattering from both samples peaks at the same frequency ($\sim$8.5 cm^{-1} far from the laser energy, independent of the polarization), showing also a similar spectral shape. Moreover, low-frequency Raman scattering from silver implanted silica is depolarized, and shows a constant depolarization ratio throughout the overall spectral region below 200 cm^{-1}. This conclusion can be easily drawn by considering the shapes of HV and VV spectra of both samples. In other words, the depolarization ratio $\rho = I_{HV}/I_{VV}$, where I_{HV} and I_{VV} are the spectral intensity observed in HV and in VV polarization, is independent of the frequency, and is $\approx$0.3 [18]. All our samples implanted at higher doses show similar spectral characteristics, with respect to the peak position, bandshape and depolarization ratio.

Recent, independent observations [9] of low frequency depolarized Raman scattering from silica films, containing co-sputtered silver particles, show spectral characteristics (peak frequency in the region between 10 and 20 cm^{-1}, same spectral shape in VV and HV polarization, and depolarization ratio $\rho \approx$0.27) similar to those described above. These results were discussed in terms of acoustic vibrations confined at the surface of the silver clusters. The mean size of metal particles - supposed to be spherical in shape - has been directly derived from the observed peak frequency. In fact, according to Lamb's theory [12], the frequencies of all vibrational modes of a homogeneous elastic sphere with a free surface scale as the inverse of the linear dimensions of the sphere. In addition, group-theory arguments suggest that the only Raman active modes of a sphere are the symmetric (l=0) and the quadrupolar (l=2) spheroidal modes [19]. Among them only the surface quadrupolar vibrational modes give a Raman depolarized spectrum. Using the results of calculations carried out by Fuji et al. [9], we estimate an average diameter of 5.5 nm for silver clusters precipitated in samples implanted with 4.3x10^{16} Ag$^+$/cm^2 or with higher doses. Finally, the depolarization ratio we obtained for the same samples is even closer to 1/3, i.e. the value expected for silver clusters in fused silica on the basis of theoretical calculation [20]. Therefore, the above findings definitively suggest that low frequency Raman scattering from our implanted fused silica samples originates from spheroidal quadrupolar vibrations localized at the surface of the silver cluster of 5.5 nm in size.

CONCLUSIONS

Silver precipitates in fused silica implanted with 300 keV-Ag$^+$ at some doses, ranging from 0.8x10^{16} to 14x10^{16} Ag$^+$/cm^2, were characterized by a multi-technique approach including RBS, XRD, optical absorption and low-frequency Raman scattering. Both, the depth profiles and distribution widths of the implanted ions were provided by RBS. XRD patterns clearly indicate the formation of silver crystals of the order of 10 nm in heavily (doses higher than 7.9x10^{16} Ag$^+$/cm^2) implanted samples. Optical absorption spectra of samples implanted under low doses (lower than 4.3×10^{16} Ag^+/cm^2) consist of a broad symmetric band, peaked at the expected surface plasmon resonance energy for silver nanoclusters, which evolve toward a double-band structured spectra for higher implantation doses. On the other hand, Raman scattering from acoustic vibrations of silver particles shows spectral features that are independent of the ion dose. The average size of silver clusters, estimated from the peak position of low-frequency Raman scattering from samples underwent implantation doses between 4.3×10^{16} and 7.9×10^{16} Ag^+/cm^2 is of the order of 5.5 nm.

The discrepancy between the cluster sizes detected by Raman scattering and x-ray diffraction should be not fortuitous, but is probably due to the fact that these two analyzing tools are probing structures characterized by different sizes: the former one probing the precipitated silver clusters, while the latter one probes the metal aggregates formed only at the highest implantation doses. The existence of these types of particles is, in fact, compatible with the observed optical extinction spectra, which suggest the existence of both spherical clusters and aggregates with chain-like or even cluster-like clumping characters in samples implanted with the highest doses.

ACKNOWLEDGMENTS

This work has been partly supported by the Italian CNR and Brazilian CNPq within a bilateral cooperation project. The authors are indebted to Prof. I.J.R. Baumvol (UFRGS-Brazil) for ion implantation. They are also grateful to Dr. M. Cremona for his help in XRD experiments.

REFERENCES

1. Ch. Buchal, S.P. Withrow, C.W. White and D.B. Poker, Annu. Rev. Sci. **24**, 125 (1994).
2. C.W. White, J.D. Boudai, J.G. Zhu, S.P. Withrow, D.M. Hembree, D.O. Henderson, A. Ueda, Y.S. Tung and R. Mu, Mat. Res. Soc. Symp. Proc. **396**, 377 (1996).
3. G.W. Arnold and J. A. Borders, J. Appl. Phys. **48**, 1488 (1977).
4. H. Hosono, Y. Abe and N. Matsunami, Appl. Phys. Lett. **60**, 2613 (1992).
5. R.H. Magruder III, Li Yang, R.F. Haglund, Jr., C.W. White, Lina Yang, L. Dorsinville and R.R. Alfano, Appl. Phys. Lett. **62**, 1730 (1993).
6. H. Hosono, H. Fukushima, Y. Abe, R. A. Weeks and R.A. Zuhr, J. Non-Cryst. Solids **143**, 157 (1992).
7. N. Matsunami and H. Hosono, Appl. Phys. Lett. **63**, 2050 (1993).
8. G. Mariotto, M. Montagna, G. Viliani, E. Duval, S. Lefrant, E. Rzepka and C. Mai, Europhys. Lett. **6**, 239 (1988).
9. M. Fuji, T. Nagareda, S. Hayashi and K. Yamamoto, Phys. Rev. B **44**, 6243 (1991).
10. M. Ferrari, F. Gonella, M. Montagna and C. Tosello, J. Appl. Phys. **79**, 2055 (1996).
11. F.L. Freire Jr., N. Broll and G. Mariotto, Mat. Res. Soc. Sym. Proc. **396**, 385 (1996).
12. H. Lamb, Proc. London Math. Soc. **13**, 187 (1882).
13. C.G. Granquist, N. Calander and O. Hunderi, Solid State Commun. **31**, 249 (1979).
14. C.G. Granquist and O. Hunderi, Phys. Rev. B **16**, 3513 (1977).
15. C.A.M. Mulder and A.A.J.M. Damen, J. Non-Cryst. Solids **93**, 387 (1987).
16. G.Mariotto,M.Montagna,G.Viliani,R.Campostrini and G.Carturan, J.Phys.C **21**, L797 (1988).
17. C. Burman and W.A. Lanford, J. Appl. Phys. **54**, 2312 (1983).
18. G. Mariotto and F.L. Freire Jr., to be published.
19. E. Duval, Phys. Rev. B **46**, 5795 (1992).
20. M. Montagna and R. Dusi, Phys. Rev. B **52**, 10080 (1995).

NANOCRYSTAL GROWTH AT HIGH DOSE RATES
IN NEGATIVE COPPER-ION IMPLANTATION INTO INSULATORS

N. Kishimoto, V.T. Gritsyna*, Y. Takeda, C.G. Lee, N. Umeda**, T. Saito
National Research Institute for Metals, 1-2-1 Sengen, Tsukuba, Ibaraki 305, Japan,
kishin@nrim.go.jp
*Kharkov State University, Kharkov 310077, Ukraine
**Tsukuba University, 1-1 Ten-nodai, Tsukuba, Ibaraki 305, Japan

ABSTRACT

Nanoparticles of Cu were fabricated by negative-ion implantation, leading to spontaneous formation at high beam fluxes. Negative ions, alleviating surface charging, exhibit significant merits in carrying out low-energy implantation at high dose rates. The kinetic processes were studied by measuring dose-rate dependence of colloid formation and resultant optical properties. Negative-Cu ions of 60 keV were implanted into silica glasses at high-current densities, up to 260 $\mu A/cm^2$, fixing the total dose at 3.0×10^{16} ions/cm^2. Spherical nanocrystals of Cu atoms formed within a narrow region, near the projectile range of Cu ions. Simultaneously, much smaller particles spread out beyond a depleted zone, deeper than the projectile range. The nanocrystal growth and optical properties were greatly dependent on the dose rate and the specimen boundary condition. The growth process is explained by a droplet-model based on surface tension and radiation-induced diffusion. Beam-surface interactions also play an important role in the mass transport from the beam flux to the interior solid.

INTRODUCTION

Nanoparticles of a metal, embedded in dielectrics, have recently attracted much attention to develop novel electronic and optical properties. The metal-colloid systems exhibit nonlinear [1] and fast optical response [2], which may become important for opto-electronics. Various fabrication methods, from conventional thermal precipitation to state-of-the-art atom manipulation, have so far been applied for colloid formation. The fabrication methods now expand their variety, not only to improve structural control for various materials, but also to form colloids of compounds, etc. Our efforts have been directed to a colloid system distributed within a thin thickness, i.e., 2-dimension-like distribution. The 2D-configuration may become important for both electronic and optical devices in future. To attain such a special distribution, we have employed ion implantation of a relatively low energy, 60 keV. When implanting at such a low energy into insulators, positive ions are subjected to surface charging via secondary-electron emission, and the specimen will be charged up to the acceleration voltage or the breakdown voltage [3], resulting in inaccurate implantation or material damage. Consequently, we have employed negative Cu ions of the low energy.

One of the important aspects for the colloid performance is compatibility between atomic-scale control and reduction of internal and interfacial defects. Although negative ions may cause intra-solid processes identical to positive ions, high-current phenomena in themselves are not well understood. Dose-rate dependence of plasmon absorption was observed even at the higher energy, 160 keV of Cu [4], where the absorption greatly increased with current density, up to 7.5 $\mu A/cm^2$. Previously, we observed in 60 keV-Cu⁻ implantation that optical absorption was unusually dependent on the current density, up to 260 $\mu A/cm^2$ [5,6]. The colloid formation processes have both features of high-temperature equilibrium [7] and flux-dependent dynamical processes. In this paper, we discuss the material kinetics over the higher dose-rate range, exploring a possibility to control the colloid formation process.

EXPERIMENTAL PROCEDURES

Negative Cu ions of 60 keV were produced by a plasma-sputter-type ion source [8] and were implanted into silica-glass substrates. The total Cu⁻ current of 3 - 4 mA, at the maximum, was homogenized by a fast scanner and the current density was 260μA/cm². In measuring dose-rate dependence, the total dose was fixed to 3.0×10^{16} ions/cm². Depth profiles and sputtering yields for 60 keV Cu were roughly estimated with the TRIM code[9]. The projectile range and the straggling width obtained are 45 nm and ±15 nm, respectively. The fixed total dose corresponds to about 10 % at the projectile peak.

Substrates used were optical-grade silica glasses (KU-1: 820 ppm OH⁻, other impurities ~6.3 ppm) of 15 mm in diameter and 0.5 mm in thickness. The silica glasses were sufficiently transparent up to the ultra-violet region.

To carry out the high-current implantation, a mask method was applied. A metal (Cu) mask, having either 2 mmφ-holes (2.2 mmφ × 37) or a 12 mmφ hole, was mounted on the specimen surface to efficiently dissipate the beam load. The Cu beam reached the specimen through the holes. Specimen temperature was monitored by a thermocouple and an infrared pyrometer.

Optical absorption was measured in a photon energy range from 1.4 to 6.5 eV using a dual beam spectrometer. Cross-sectional electron microscopy (TEM) was conducted to evaluate microstructures in the Cu-implanted region. After the ion implantation, a thin Cr film (5 nm) as a marker layer was coated onto the implanted surface. Thickness of the cross-section was measured with a tilting method for the Cr layer. Surface morphology of the implanted substrates was also observed by atomic- force microscopy (AFM) in the tapping mode.

RESULTS AND DISCUSSION

Ion implantation has advantages of spatial controllability and non-equilibrium nature for a variety of material phases. However, the ion techniques inevitably suffer from the drawback of radiation-induced damage. To seek the good compatibility, we have focused on optimization of implantation kinetics, studying dose-rate dependence on the microstructures and optical properties. After developing the mask method, we have managed to carry out high-current implantation up to 260μA/cm². High-current implantation has demonstrated merits not only for efficient implantation but also for nanocrystal formation without thermal annealing [5,6]. It has, however, been realized that optical spectra of absorption and reflection are greatly dependent on Cu dose rate [4,5]. The dose-rate dependence of absorption spectra consists of two major features. i.e., variation in the spectral shape of surface plasmon and change in the optical density of the whole spectrum [5]. The former and the latter variations are respectively associated with colloid morphology and beam-surface interactions. Previously, we showed that, at least, beam-surface interactions played an important role in the mass transport of Cu atoms into the solid [5].

Negative ions probably alleviate Coulomb-repulsion effects around the substrate, at least, better than positive ions. Even if it is the case, negative ions could not accomplish simple implantation independent of the dose rate. As will be shown later, a major factor to affect the implantation is a boundary condition of the irradiation, especially the surface area exposed to the incident ions. Up to the present, an optimum condition for the largest absorption spectrum has been found to be about 10^2 μA/cm² with a 2 mm-φ

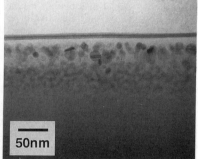

50nm

Fig. 1 Cross-sectional TEM image of a specimen, Cu⁻-implanted at 95 μA/cm² to 3×10^{16} ions/cm². A dark layer on the surface is Cr coating as a marker.

mask. The highest current density 260 μA/cm² gave less efficient implantation.

Focusing on a typical case of 10² μA/cm², we discuss characteristics of the nanoparticle growth in high-current implantation. Figure 1 shows a TEM image of the specimen, Cu⁻-implanted at a dose rate of 95μA/cm² to 3 × 10¹⁶ ions/cm² (2 mm-φ mask). Spheres of Cu nanoparticles of 15-20 nmφ form within a narrow depth, ~50 nm, near the surface. It was verified by dark-field imaging that most of the Cu spheres were of fcc single crystals, randomly oriented. Between the larger spheres, there are no discernible strain contrasts of defect clusters nor the smaller Cu particles. The pronounced sphere growth without other fine clusters exhibits a high-current characteristics that radiation-enhanced diffusion and energy-spike effects are dominant around the projectile range. The pronounced roundness of the nanocrystals definitely indicates a mechanism dominated by the surface tension of the spheres. Besides those single nanocrystals, some of the larger spheres included twin-like parallel boundaries, most of which had mirror symmetry with respect to the mid-axis of the sphere. This tendency may be due to the hydrostatic stress of surface tension during the crystal growth.

Thus, our objective to create a thin-layered colloid distribution was, to some extent, accomplished by the low-energy implantation. The distribution is, however, not very narrow at this moment and is further improved as will be described later.

Simultaneously with the nanoparticles of 15-20 nmφ, fine black dots spread in the deeper region. The finer dots also seemed to be of Cu single crystals, though dark-field images could not be acquired for the smallest dots. The total implanted zone at 95μA/cm² ranged 1.5 times deeper than predicted by the TRIM code. This wide distribution supports the radiation-enhanced diffusion of Cu atoms. Consequently, the whole particle-distribution becomes bimodal, not only for the depth profile but also for the size distribution.

Figure 2 shows another TEM image of the specimen at a dose rate 10 μA/cm² to 3 × 10¹⁶ ions/cm² (2 mm-φ mask). Even at this lower dose-rate, spontaneous colloid formation and a bimodal depth profile were qualitatively similar to those at 95μA/cm². Roundness of the nanospheres was evident for the larger spheres around the projectile range. However, the bimodal zone-separation along the depth is less pronounced, as compared to the case of 95μA/cm². The bimodal distribution

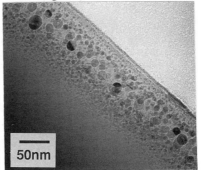

Fig. 2 Cross-sectional TEM image of a specimen, Cu⁻-implanted at 10 μA/cm² to 3 × 10¹⁶ ions/cm².

and enhanced diffusion become less dominant, with decreasing the dose rate. As was previously reported [6], no bimodal separation apparently occurred at 3 μA/cm² (2 mmφ-mask) and fine dots of 1-5 nmφ spread out.

To our knowledge, there have been no experiments reported which use the combination of high currents and low energy as in our case. Up to now, the dose-rate dependence has attracted attention in some cases [4,5], but it has been often believed [7] that colloid formation processes are analogous to a high-temperature equilibrium. Namely, thermal-spike effects of heavy ions may attain a kind of thermal equilibrium[7], i.e., chemical reactions including the precipitation quickly saturate and the colloid size and density tend to be determined by a total amount of implants, irrespective of the dose rate. The results of our experiment are in favor of the kinetic aspect, showing strong dose-rate dependence.

For the quantitative discussion, size distribution of the nanocrystals and atomic number density of the Cu atoms are evaluated from the cross-sectional images for the two cases of 95μA/cm² and 10 μA/cm².

Figures 3 (a) and 3(b) show the size distributions of the nanospheres for the two cases, where the abscissa stands for number density of colloids, as a function of the particle diameter. Since there are copious small particles at greater depths, population in the smaller

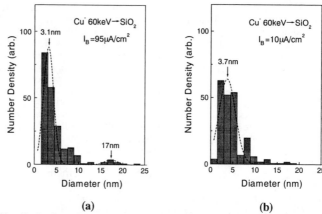

(a)   **(b)**

Fig. 3   Size distributions of nanocrystals for two cases of Cu⁻-implanted at 95μA/cm² ; left (a)
and 10 μA/cm² ; right (b). The dotted lines are Gaussian curves fitted with the histogram.

is more impressive as for the number density. However, the bimodal size-distribution is
clearly seen for the higher current density (a).   The average diameters fitted with each
Gaussian are 3.1 nmφ and 17 nmφ for the two peaks.   For the lower-current case (b), the
bimodal distribution is less pronounced and the average diameter yields as 3.7 nmφ.

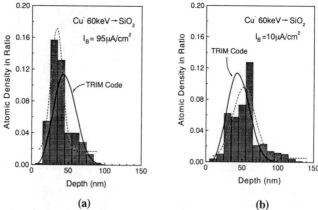

(a)   **(b)**

Fig. 4   Depth profiles of Cu atomic density in ratio, retained in the nanocrystals for
two cases of Cu⁻-implanted at 95μA/cm² ; left (a) and 10 μA/cm² ; right (b).
The dotted lines are Gaussian curves fitted with the histogram.

Figures 4(a) and 4(b) show the Cu atomic density in ratio, which is derived from the
integrated volume of Cu spheres by slicing the depth into strips of 10 nm. An assumption of
the mass density equal to pure Cu is justified also by TEM. This data treatment is practically
not easy, because statistically sufficient particles should be counted in the superimposed TEM
images. However, if appropriate thickness of the specimen is selected, this method could give
reliable information of the absolute amount of Cu atoms. The depth profiles of Cu atoms are

compared to those translated from the TRIM code[9]. For the higher current density (Fig. 4(a)), the peak is located at about 30 -35 nm in depth, smaller than the projectile range predicted, i.e., 45 nm. The peak shift is attributable to formation of a depleted zone between the lateral spheres and the fine-dot zone. The mechanism is explained as follows: The TRIM calculation conveys that the ionization energy is highest at the projectile range, and the radiation-induced diffusion promotes escaping motion of Cu atoms from the projectile range to the surface or the deeper area. Since the energy deposition of ionization keeps a high level in the direction to the surface, the nanocrystal growth dominantly occurs in a region shallower than the projectile peak. However, a surface-sputtering effect is also possibly responsible for the peak-shift mechanism. Although the sputtering phenomena at the high current densities are not

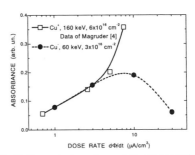

Fig .5 Dose rate dependence of absorbance at the plasmon energy with the 12 mmφ mask, for the specimens, Cu$^-$-implanted to 3 × 10^{16} ions/cm^2. Data by Magruder III et al, for positive 160 keV-Cu ions, are included for comparison.

known, a rough estimation with the TRIM-code predicts that the sputtering up to 10 nm may occur during the Cu irradiation up to 3 × 10^{16} ions/cm^2. For the lower current density (Fig. 4(b)), the whole distribution is inhomogeneous but the Gaussian curve fitted with the data roughly agrees with the TRIM-prediction. Namely, the colloid formation of the larger spheres is relatively suppressed and the small precipitates spread towards the deeper region.

It has been thus concluded that dose-rate dependence for the nanocrystal growth definitely exists even for negative-ion implantation. The important factors are energy-spike effects and radiation-induced diffusion, leading to the surface-tension-controlled growth.

Furthermore, the dose-rate dependent processes are drastically affected by the boundary condition of the surface area. Figure 5 shows the dose-rate dependence of absorbance at the plasmon peak for the 12 mm-φ mask. Although the dose-rate dependence for the 2 mmφ mask gave unstable variation [5], the mask prolonged the capable region up to the highest current density, 260 μA/cm^2. On the contrary, the 12 mmφ-mask showed an earlier drop in the absorbance, below 100 μA/cm^2. The change in the boundary condition resulted in reduction in the optimum dose-rate. Although the wider implantation is practically preferred, the applicable dose rate for the 12-mmφ is restricted lower. Dose-rate dependence of positive Cu ions of 160 keV, given by Magruder III [4] et al is also included in Fig. 5, where the values are normalized at 1 μA/cm^2 to each other. Their divergent variation appears to be moderated by using the negative ions. However, these results cannot simply be compared with each other, because the boundary condition of the specimen has a great influence on the eventual morphology.

Cu-implanted at 10μA/cm^2
to 3 x 10^{16} ions/cm^2

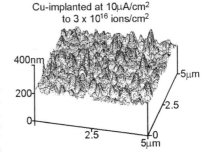

Fig. 6   AFM image of the specimen surface, Cu$^-$ implanted at 10 μA/cm^2 with the 12 mmφ-mask, to 3 x 10^{16} ions/cm^2.

It is suggested by this experiment that beam-surface interactions give important effects on the inward mass transport. One of the possible mechanisms is the surface sputtering. Surface morphology of a specimen, Cu$^-$-implanted at 10 μA/cm^2 with the 12 mmφ- mask, is given in Fig. 6. After the implantation, the surface morphology changes to a rugged texture, whose roughness amounts to several tens nanometers, while the same current density did not cause such surface damage in the case of the 2 mmϕ-mask. It is informative that the rugged layer very easily drops off leaving the flat surface, even by a soft touch of a cotton pin. This process leads to recession of the specimen surface during the irradiation.

Those tendencies in Figs. 5 and 6 indicate that usage of the wider mask would result in a shift in dose-rate dependence toward the low current density. The mechanism is attributable to an increase in low dissipation of the beam load and enhanced sputtering. Here, it is noted that temperature measured by an infrared pyrometer was below 500 K.

Figure 7 shows a cross-sectional TEM image of a specimen, Cu⁻-implanted at 45 μA/cm^2 with the 12 mmϕ-mask. The photograph shows a marvelous linear arrangement, i.e., a 2D-like distribution of Cu nanoparticles. A similar arrangement within a narrow thickness was also obtained at a higher current density ~100 μA/cm^2 using the smaller 2 mmϕ-mask. Moreover, an interesting feature of Fig. 7 is absence of the fine-

Fig. 7 Cross-sectional TEM image of a specimen, Cu⁻-implanted at 45 μA/cm^2 with the 12 mmϕ-mask.

dot zone in the deep region. It is considered that a balance between the surface recession and the enhanced growth resulted in the 2-D like arrangement.

It is thus demonstrated that changes in Cu dose rate and the boundary condition alter the morphology of Cu nanoparticles and also that 2D-like distribution is attainable by taking advantage of the relevant kinetic processes.

CONCLUSIONS

Negative-Cu ions of 60 keV have been successfully implanted into silica glasses up to 260 μA/cm^2. Spherical nanocrystals of a Cu single phase formed near the projectile range and, simultaneously, smaller particles spread beyond a depleted zone. Nanocrystal growth and the optical properties were significantly dependent on the Cu dose rate, and is dominated by surface tension and radiation-induced diffusion. The low-energy implantation at 60 keV fabricated spherical colloids within a narrow layer near the surface. Change to the wider mask resulted in an earlier drop of absorbance, in the dose-rate dependence. Controlling the dose-rate dependence enables us to handle the morphology of the Cu nanoparticles. Changing the beam-energy dissipation also provides us with a controllable parameter for the nanoparticle fabrication.

REFERENCES

[1] F. Hache, D.Richard, C. Flytzanis and U. Kreibig, Appl. Phys. A **47**, 347 (1988).
[2] T. Tokizaki, A. Nakamura, S. Kaneko, K. Uchida, S. Omi, H. Tanji and Y. Asahara, Appl. Phys. Lett. **65**, 941(1994).
[3] J. Ishikawa, H. Tsuji, Y. Toyota, Y. Gotoh, K. Matsuda, T. Tanjo and S. Sasaki, Nucl. Instrum and Meth. **96** (1995) 7.
[4] R.H. Magruder III, R.F. Haglund, Jr, L.Yang, J.E. Wittig and R.A. Zuhr, J. Appl. Phys., **76**, 708 (1994).
[5] N. Kishimoto, V.T. Gritsyna, K. Kono, H. Amekura and T. Saito, Nucl. Instr. and Meth., **B127/128**, 579(1997).
[6] N. Kishimoto, V.T. Gritsyna, K. Kono, H. Amekura and T. Saito, Mater. Res. Soc. Symp. Proc. Vol. 438 (1997) 435.
[7] H. Hosono, H. Fukushima, Y. Abe, R.A. Weeks and R.A. Zuhr, J. Non-Cryst. Solids, **143**, 57(1992).
[8] Y. Mori, A. Takagi, K. Ikegami, A. Ueno and S. Fukumoto, Nucl. Instr. and Meth., **B37/38**, 63 (1989).
[9] J.F. Ziegler, J.P. Biersack and U. Littmark, The Stopping and Range of Ions in Solids, (Pergamon Press, New York, 1985), Chap 8.

THE OPTICAL PROPERTIES OF ION IMPLANTED SILICA

C.C. SMITH*, D. ILA** AND E.K. WILLIAMS**, D.B. POKER*** AND D.K. HENSLEY***
*NASA, MSFC, AL, 35812
**Center for Irradiation of Materials, Alabama A&M University, Normal, AL 35762-1447,
***Solid State Division, Oak Ridge National Laboratory, Oak Ridge, TN 37831

ABSTRACT

We present the results of our investigation of the change in the optical properties of silica, "suprasil", after keV through MeV implantation of copper, tin, silver and gold and after annealing. Suprasil, name brand of silica glass produced by Hereaus Amersil, which is chemically highly pure with well known optical properties. Both linear and nonlinear optical properties of the implanted silica were investigated before and after thermal annealing. All implants showed strong optical absorption bands in agreement with Mie theory. For implants with a measurable optical absorption band we used Doyle's theory and the full width half maximum of the absorption band to calculate the predicted size of the formed nanoclusters at various heat treatment temperatures. These results are compared with those obtained from direct observation using transmission electron microscopic techniques.

INTRODUCTION

The production of colored glass devices was motivated mostly by desired changes in the linear properties of glasses. In recent years, attention has been directed to both linear and nonlinear properties of the silica glasses caused by the surface plasmon resonance optical absorption which depends on the index of refraction of the host substrate and the electronic properties of the metal colloids. These glasses were prepared classically by mixing selected metals powders with molten glass then cooling to form a homogeneous glass. Recently, ion implantation has been used to introduce nonlinear optical properties [1-8] in layers near the surface of optical materials. To form nanoclusters after ion implantation below the saturation fluences the material must be treated either by thermal annealing, laser annealing or by post-implantation ion bombardment [8]. An attractive feature of ion implantation is that the linear and nonlinear properties occur in a well defined space in an optical device.

The colors due to formation of small metallic particles or colloids embedded in dielectrics are associated with optical sbsorption at the surface plasmon resonance frequency. For clusters with diameters much smaller than the wavelength of light (λ), Mie theory [9] can be used to calculate the absorption coefficient (cm^{-1}) of the composite:

$$\alpha = \frac{18\pi Q n_o^3}{\lambda} \frac{\varepsilon_2}{(\varepsilon_1 + 2n_o^2)^2 + \varepsilon_2^2}, \qquad (1)$$

where Q is the volume fraction occupied by the metallic particles, n_o is the refractive index of the host medium, and ε_1 and ε_2 are the real and imaginary parts of the frequency dependent dielectric constant of the bulk metal. Equation (1) is a Lorentzian function with a maximum value at the surface plasmon resonance frequency (ω_p), where

$$\varepsilon_1(\omega_p) + 2n_o^2 = 0, \qquad (2)$$

where $\varepsilon_1 = n^2 - k^2$ with n and k being the optical constants of the bulk metal. Using the tabulated optical constants the theoretical values for the wavelength of the absorption bands for Au, Ag, Cu, and Sn are tabulated in table 1, where $n_o = 1.46$ for suprasil.

Using the "characteristic" optical absorption band, the average radius of the metal spheres, is determined from the resonance optical absorption spectrum and using the Doyle theory [10] according to the equation $r = V_f/\Delta\omega_{1/2} =$, where V_f is the Fermi velocity of the metal and $\Delta\omega_{1/2}$ is the full width at half maximum (FWHM) of the absorption band due to the plasmon resonance in small metal particles.

EXPERIMENTAL PROCEDURES

The silica glass used in this work is Suprasil provided by Hereaus Amersil, Inc. It contains 150 ppm OH, 0.05 ppm of Ti, Na, Ca and Al and less than 0.01 ppm other metals. Samples 10 x 10 x 0.5 mm were implanted with 3.0 MeV gold, 0.16 MeV Silver, 1.5 MeV silver, 2.0 MeV copper, or 0.35 MeV tin ions at a current density less than 2 $\mu A/cm^2$. Ion fluences varied from 1 x 10^{16} ions/cm^2 to 8 x 10^{17} ions/cm^2. We studied and measured the threshold fluence, above which spontaneous formation of nanoclusters was observed by optical means.

The heat treatment for the samples was done in air at temperatures between 100°C and 1200°C for 1 hour at each heat treatment temperature. After each heat treatment, an optical absorption spectrum was measured. From these spectra we calculate the size of the metallic clusters which are responsible for the optical absorption band.

The average radius of metal spheres, small compared with the wavelength of light, is then determined using Doyle theory, the measured optical photospectrometry, and the RBS results from each implanted sample.

RESULTS and DISCUSSIONS

The higher the atomic number and fluence of the bombarding ion, the more the damage and modification of the optical properties [8,9] of the silica. This was observed by absorption spectrometry during all of our bombardments. Except for the effects attributable to the metal clusters, these other effects did not produce any appreciable characteristic absorption band for suprasil and were reduced with increase in the heat treatment temperature.

With heat treatment the near neighbor clusters coalesce and the host accommodates to the volume reduction. RBS and TEM measurements confirm that the depth profile of the metal clusters formed after heat treatment is almost identical to, or slightly narrower than, that of the atoms initially implanted by ion bombardment.

Figure 1 shows the optical absorption spectra for 1.5 MeV Ag implanted in silica at 2.0 x10^{16} Ag/cm^2 at various heat treatment temperatures. The prominent absorption resonance at 410 nm increases until it disappears for heat treatment temperatures above 1000°C.

Figure 2 shows the optical absorption spectra for 160 keV Ag implanted to a fluence of 2 x 10^{17}/cm^2. At this energy and fluence the atomic concentration of Ag, as computed with SRIM96 [11], is ten times that of the 1.5 MeV implant. A second absorption peak appears at ~620 nm. This peak decreases with heat treatment and disappears completely at 500°C. The second peak is due to nonspherical Ag clusters [12] . A strong possibility is that there are many pairs of clusters, "dumbbells", that coalesce into spheres upon heat treatment. The absorption spectra from the keV implant began to decrease in height at a slightly lower temperature than those from the MeV implant. The copper and gold implanted samples behave similarly. Figure 3 shows the optical absorption spectra for 2.0 MeV Cu implanted suprasil at as implanted at 24°C and after various annealing temperatures. For a heat treatment temperature beyond a

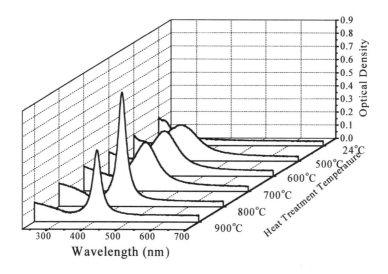

Figure 1. Resonance optical absorption of silica implanted with 0.2×10^{17} Ag/cm^2 at 1.5 MeV and heat treated for 1 hour to the temperatures shown.

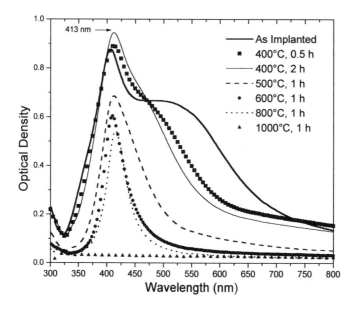

Figure 2. Optical absorption spectra from silica implanted with 2×10^{17} Ag/cm^2 at 160 keV.

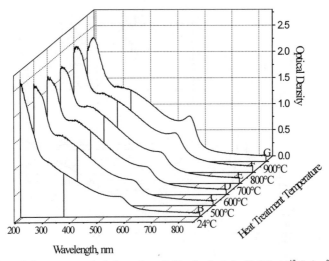

Figure 3. Resonance optical absorption of silica implanted with 2.0 x10^{17} Cu/cm^2 at 2.0 MeV and heat treated for 1 hour to the temperatures shown.

Table 1. Specie, energy and fluence of the ions implanted in suprasil as well as the theoretical and

Ion Specie	Ion Energy (MeV)	Ion fluence (cm^{-2})	Absorption Wavelength Theory/Experiment (nm)	Average Radius (nm)
gold	3.0	1.2 x 10^{17}	520/526	2.3
silver	1.5	6.0 x 10^{16}	410/407	4.7
silver	0.16	2.0 x 10^{17}	410/409	2.6
copper	2.0	2.0 x 10^{17}	575/570	4.4
tin	0.350	8.0 x 10^{17}	225/227	0.2

critical temperature the resonance optical absorption disappears as metal atoms evaporate from the clusters as shown in Figs. 1 and 2 for silver implanted silica. The critical temperature correlates only roughly with the bulk metal melting temperature.

Figure 4 is a typical TEM result for Ag implanted suprasil, showing the distribution of implanted species as well as the distribution of nanoclusters throughout the implanted layer. The TEM measurements were used to compare the avarage cluster size predicted by Doyle theory with a direct measurement, as well as to study the type and crystallinity of the formed

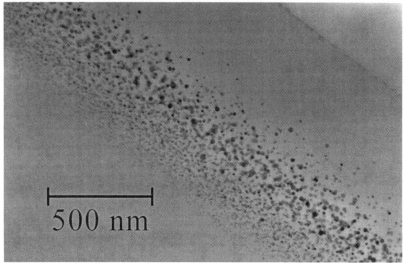

Figure 4. TEM micrograph of Ag implanted into Suprasil 1.

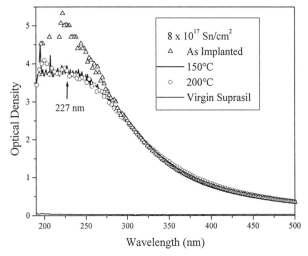

Figure 5. Optical absorption spectra of 350 keV Sn implanted into suprasil-1 to a fluence of 8 x 1017 Sn/cm2 in the As Implanted state and after 1 h anneals at 150° and 200°C.

clusters. Typical nanocluster sizes, as measured by TEM, were about twice that calculated using Doyle theory.

Figure 5 shows the optical absorption spectrometry result from 350 keV tin implanted suprasil before and after annealing at 150°C, and 200°C for one hour. An absorption peak due to clusters was observed for Sn implanted at 8 x 10^{17} Sn/cm^2. Annealing caused an increase in the FWHM of the absorption spectra, mostly due to diffusion of Sn into the silica substrate.

Using the Doyle theory [10] and the resonance optical absorption bandwidths, average cluster radii in the three implanted samples are calculated for each heat treatment temperature.

Table 1 is a list of the implantation parameters, energy and fluence, as well as the theoretical and experimental values for the peak wavelength of the plasmon band and the calculated average radius.

CONCLUSION

Samples of silica implanted with MeV ions of Au, Ag, Sn, or Cu and subjected to careful heat treatment contain metal nanoclusters whose radii can be obtained from the resulting characteristic resonance optical absorption bands at the same regin predicted by Mie theory. The calculated average nanoclusters radius, using Doyle theory, is sligthly off from those measured by TEM which could be due to the sensitivity of the optical techniques to the density of the nanoclusters. In our case, there are less of the large clusters (2 to 50 nm) than clusters below 2 nm. The Optical measurement can be used as a nondestructive tool for qualitative measurement of the nanocluster formation.

ACKNOWLEDGMENTS

This project was supported by the Center for Irradiation of Materials at Alabama A&M University and Alabama EPSCoR-NSF/ALCOT Grant No. OSR-9553348. The work at ORNL was sponsored by the Division of Materials Science, U.S. Department of Energy, under Contract DE-AC05-96OR22464 with Lockheed Martin Energy Research Corp.

REFERENCES

1. G. W. Arnold, J. Appl. Phys. **46**, 4466 (1975).
2. G. W. Arnold and J. A. Bordes, J. Appl. Phy. **48**, 1488 (1977).
3. R. H. Magruder III, R. A. Zuhr, D. H. Osborne, Jr., Nucl. Inst. & Meth. in Phys. Res. **B99**, 590 (1995).
4. Y. Takeda, T. Hioki, T. Motohiro, S. Noda and T. Kurauchi, Nucl. Instr. Meth. **B91**, 515-519 (1994).
5. C. W. White, D. S. Zhou, J. D. Budai, R. A. Zuhr, R. H. Magruder and D. H. Osborne, Mat. Res. Soc. Symp. Proc. **Vol. 316**, 499 (1994).
6. K. Fukumi, A. Chayahara, M. Adachi, K. Kadono, T. Sakaguchi, M. Miya, Y. Horino, N. Kitamura, J. Hayakawa, H. Yamashita, K. Fujii and M. Satou, Mat. Res. Soc. Symp. Proc. **Vol. 235**, 389-399 (1992).
7. D. Ila, Z. Wu, R. L. Zimmerman, S. Sarkisov, C. C. Smith, D. B. Poker, and D. K. Hensley, Mat. Res. Soc. Symp. Proc. **Vol. 457**, 143 (1997).
8. D. Ila, Z. Wu, C.C. Smith, D. B. Poker, D. K. Hensley, C. Klatt, and S. Kalbitzer, Nucl. Instr. Meth. in Phys. Res. **B127**, 570 (1997).
9. G. Mie, Ann. Physik **25**, 377 (1908).
10. W. T. Doyle, Phys. Rev. **111**, 1067 (1958).
11. F. Ziegler, J. P. Biersack and U. Littmark, The Stopping and Range of Ions in Solids (Pergamon Press Inc., New York, 1985).
12. F. L. Freire and G. Mariott, Mat.Res. Soc. Symp. Proc. Vol. 504, 1997, (in press).

MECHANISMS OF FORMATION OF NONLINEAR OPTICAL LIGHT GUIDE STRUCTURES IN METAL CLUSTER COMPOSITES PRODUCED BY ION BEAM IMPLANTATION

S.S. Sarkisov *, E.K. Williams **, M. J. Curley *, C.C. Smith **, D. Ila **, P. Venkateswarlu *, D.B. Poker ***, D.K. Hensley ***
* Department of Natural and Physical Sciences, Alabama A&M University, 4900 Meridian Drive, Normal, AL 35762, sergei@caos.aamu.edu
** Center for Irradiation of Materials, Alabama A&M University, 4900 Meridian Drive, Normal, AL 35762
*** Solid State Division, Oak Ridge National Laboratory, Oak Ridge, TN 37831

ABSTRACT

We report the results of characterization of linear and nonlinear optical properties of a light guide structure produced by MeV Ag ion implantation of LiNbO$_3$ crystal (z-cut) in relation to the mechanisms of formation.

INTRODUCTION

Ion implantation has been shown to produce a high density of metal colloids in glasses and crystalline materials. The high-precipitate volume fraction and small size of metal nanoclusters formed leads to values for the third-order susceptibility much greater than those for metal doped solids [1]. This has stimulated interest in use of ion implantation to make nonlinear optical materials. On the other side, LiNbO$_3$ has proved to be a good material for optical waveguides produced by MeV ion implantation [2]. Light confinement in these waveguides is produced by refractive index step difference between the implanted region and the bulk material. Implantation of LiNbO$_3$ with MeV metal ions can therefore result into nonlinear optical waveguide structures with great potential in a variety of device applications. We describe linear and nonlinear optical properties of a waveguide structure in LiNbO$_3$-based composite material produced by silver ion implantation in connection with mechanisms of its formation.

EXPERIMENT

The sample was made of 1-mm thick LiNbO$_3$ (z-cut) implanted with 1.5-MeV Ag ions to a dose of 2.0x10^{16} cm^{-2}. Implantation was done at room temperature. Fig.1 presents the distribution of the implanted silver along the depth of the sample calculated by the Monte-Carlo simulation program TRIM96 [3]. The position of the peak of the distribution defines the depth of the implanted layer at 0.41 μm and FWHM of the distribution gives an estimate of 0.24 μm for the thickness of the layer.

Light transmission along the implanted layer (treated at 500^{0}C) was studied using the prism coupling technique [4]. The technique is based on light tunneling from a prism with high refractive index (rutile, index is 2.8643) to the implanted light guiding layer through a small air gap (Fig. 2). The phase matching condition for light coupling is

$$n_p \sin\theta_m = N_m ,$$ (1)

where n_p is the prism refractive index; θ_m is the angle of incidence; N_m is the propagation

357

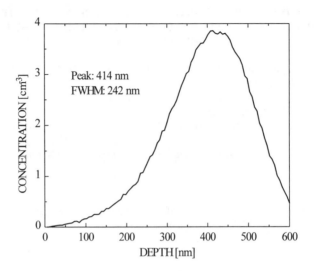

Fig. 1. The concentration of the implanted Ag ions versus the depth of the LiNbO₃ sample calculated by the Monte-Carlo method. The energy of the Ag ions is 1.5 MeV, the fluence is 2.0×10^{16} cm⁻².

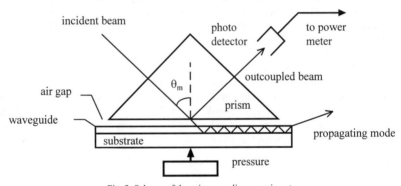

Fig. 2. Scheme of the prism coupling experiment.

number for the m-th order mode in the waveguide. Energy channeling to a guided mode leads to the drop of the intensity of the light outcoupled from the prism which is monitored with a photo detector. Measuring the intensity dependence on the incidence angle, it is possible to obtain θ_m and then to calculate N_m. The implanted layer refractive index n_f and thickness h_f can be found from the solution of the system of dispersion equations at least for two modes. For s - polarization (TE - modes) these equations can be written as:

$$\frac{2\pi h_f}{\lambda}(n_f^2 - N_m^2)^{1/2} - \tan^{-1}\left(\frac{N_m^2 - n_c}{n_f^2 - N_m^2}\right)^{1/2} - \tan^{-1}\left(\frac{N_m^2 - n_s^2}{n_f^2 - N_m^2}\right)^{1/2} - m\pi = 0 \tag{2}$$

where λ is the light wavelength, n_s and n_c are the substrate and the cladding (air) refractive indices respectively . Similar equations can be formulated also for TM-modes. The prism coupler Metricon 2010 was used in our measurements. Fig. 3 shows experimental results which

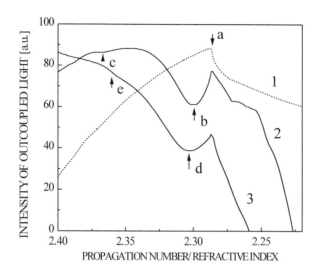

Fig. 3. Intensity of the light outcoupled from the coupling prism versus the propagation number of modes injected into the light guiding implanted layer. Curve 1 corresponds to virgin LiNbO$_3$ crystal, where point (a) marks the threshold of total internal reflection which determines the refractive index at 2.2867. Curves 2 and 3 correspond to two different locations in the implanted crystal. Points (b) and (c) mark the dips related to the zeroth order and the first order propagating modes in the first location with propagation numbers $N_0 = 2.3652$ and $N_1 = 2.3003$ respectively. Points (d) and (e) mark the modes $N_0 = 2.3938$ and $N_1 = 2.3029$ for the second location. The average refractive index and thickness of the implanted layer in the first location is calculated to be 2.3887 and 0.748 μm respectively while the same parameters for the second location are 2.4254 and 0.6228 μm. All the measurements are taken at 633 nm.

indicate that the light guiding layer with the index greater than that of the substrate (2.2867) is spread along the depth up to 0.75 μm and therefore goes beyond the limits of the 0.24-μm-wide nuclear stopping region where the implanted Ag ions were originally located (Fig. 1). This is indicative for silver ion diffusion out of the original location during heat treatment accompanied by the increase of the refractive index of the crystal.

The linear optical absorption spectra of the implanted sample are shown in Fig. 4. The spectrum right after implantation exhibits the prominent single surface plasmon peak near 430 nm which is the signature of nanometer size clusters of metallic silver formed in the host. The radius of the clusters can be estimated by using the relation $R = V_f/\Delta\omega_{1/2}$, where V_f and $\Delta\omega_{1/2}$ are Fermi's velocity of silver (1.39×10^8 cm/s [5]) and the half width of the absorption peak, respectively [6]. In our case $R \approx 1.1$ nm. Heat treatment after the implantation at 500^{0}C for 1 hour lead to the red shift of the absorption peak to 550 nm (Fig. 4) without significant change of $\Delta\omega_{1/2}$ (and, correspondingly, of the radius R). Shang, et. al. proposed to explain this effect as a result of the volume fraction increase due to Ag precipitation near the surface [7]. Another

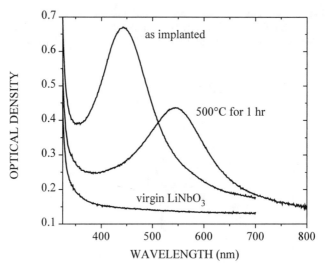

Fig. 4. Optical absorption spectra of the Ag implanted LiNbO₃ sample.

likely contributing factor is that the heat treatment is working to remove implantation damage [8]. LiNbO₃ implanted with 190 KeV Ag at 1×10^{17}-cm^{-2} fluence has been shown to undergo full epitaxial regrowth after heat treatment at 400^0C but an anneal of 800^0C for 1 h is necessary for full removal of the implantation damage [9]. At the end of range ions implanted into LiNbO₃ have been shown [2, 8] to decrease the index of refraction by over 5 percent. Reducing the host index by 5 to 10 percent would result in a shift of the expected absorption peak for Ag from 520 nm to approximately 480 nm.

The nonlinear refractive index of the sample after heat treatment was characterized using the Z-scan technique [10]. The laser source was a tunable dye laser (with laser dye Rhodamine 6G) pumped by a frequency-doubled mode-locked Nd:YAG laser (76-MHz pulse repetition rate). The tuning range was 555 to 600 nm. The average power of the laser radiation applied to the sample varied from 100 to 350 mW. The laser pulse duration (FWHM) was 4.5±0.8 ps. The beam diameter was 3.0 mm. Estimated laser peak power density in the sample placed near the focus of the lens (125-mm focal distance) in the Z-scan experiment was 0.025 to 0.088 GW/cm². The closed aperture Z-scan demonstrated typical behavior of a nonlinear refractive medium with positive nonlinear refractive index [10]. The open aperture Z-scan showed saturation of nonlinear absorption of the sample (optical transmission peak) at the distance z = 0 from the focus. The measured nonlinear refractive index of the sample is plotted against the wavelength in Fig.5 (circles) . It repeats the linear absorption spectrum fraction of which is depicted by a solid line 1. This is a typical picture when the surface plasmon resonance contributes to the intrinsic nonlinear response of nanoclusters originated from interband, intraband, and hot electron photo excited transitions [11].

The nonlinear refractive index of the silver implanted LiNbO₃ sample was compared against Cu-implanted silica sample as a reference. The sample was prepared at the conditions

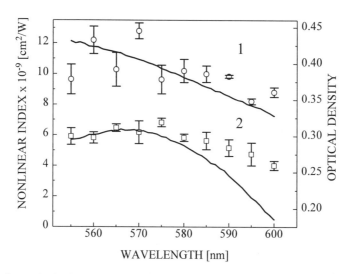

Fig. 5. Nonlinear refractive index of the Ag:LiNbO$_3$ sample (circles) and Cu:silica reference sample (squares) versus wavelength. Solid lines 1 and 2 represent optical absorption spectra of Ag:LiNbO$_3$ and Cu:silica respectively.

similar to those in Ref. 1. The energy of the Cu ions and the fluence were 2.0 MeV and 1.0×10^{17} cm^{-2} respectively. Implantation was performed at room temperature. Heat treatment was done at 1000^{0}C in the air for 1 hour. The nonlinear refractive index of the reference sample was characterized with the Z-scan technique at the same conditions as the Ag-implanted LiNbO$_3$ sample. It is plotted together with the absorption spectrum in Fig. 5 (squares and solid line 2 respectively). The nonlinear refractive index of the reference also repeats the shape of the Cu surface plasmon resonance peak in full agreement with Ref. 1. The nonlinear index of the Ag-implanted sample is approximately two times greater than the index of the reference. The optical density is greater as well. At the same time, the fluence used to implant the silver sample is 5 times less than the fluence used for the reference. We believe that this enhancement of the nonlinear refractive index and the optical density can be attributed to the high volumetric density of Ag nanoclusters in the ion-implanted layer remaining even after intensive heat treatment and possible recrystallization of the nuclear damage region in the LiNbO$_3$ matrix. The absolute values of the nonlinear index of both samples ($(8.21 \pm 0.20) \times 10^{-9}$ cm^2/W $\leq n_2 \leq$ $(12.82 \pm 0.52) \times 10^{-9}$ cm^2/W and $(4.02 \pm 0.30) \times 10^{-9}$ cm^2/W $\leq n_2 \leq (6.82 \pm 0.28) \times 10^{-9}$ cm^2/W for silver and copper respectively) are at least one order of magnitude greater than those for similar materials reported in the literature [1] possibly due to the cumulative thermal self-focusing effect at relatively high pulse repetition rate (76 MHz versus 3.8 MHz in Ref. 1).

CONCLUSIONS

We have fabricated a planar nonlinear optical waveguide using implantation Ag ions in LiNbO$_3$. The implanted composite light guiding layer exhibits a Kerr-type nonlinear susceptibility. The nonlinear refractive index for the Ag:LiNbO$_3$ composite compares well to other metal colloid composites prepared by ion implantation. It is particularly twice as high as the index for Cu:silica composite.

ACKNOWLEDGMENTS

This research was supported by the Department of Natural and Physical Sciences, the Center for Irradiation of Materials, Alabama A&M University, Alliance for Nonlinear Optics (NASA Grant NAG5-5121), and the Division of Material Sciences, U.S. Department of Energy (Contract DE-AC05-96OR22464 with Lockheed Martin Energy Research Corp.).

REFERENCES

1. R.F. Haglund, L. Yang, R.H. Magruder, J.E. Wittig, K. Becker and R.A. Zuhr, Opt. Lett. **18**, 373 (1993).

2. S.S. Sarkisov, E.K. Williams, D. Ila, P. Venkateswarlu, D.B. Poker, Appl. Phys. Lett. **68**, 2329 (1996).

3. J.F. Zeigler, J.P. Biersack and U. Littmark, The Stopping and Range of Ions in Solids (Pergamon Press, New York, 1985).

4. R.Ulrich and R.Torge, Appl. Opt. **12**, 2901 (1973).

5. A.E. Hugs and S.C. Jian, Adv. Phys. **28**, 717 (1979).

6. G.W. Arnold, J. Appl. Phys. **46**, 4466 (1975).

7. D.Y. Shang, Y. Saito, R. Kittaka, S. Toniguchi and A. Kitahara, J. Appl. Phys. **80** , 6651 (1996).

8. E.K. Williams, PhD thesis, Alabama A&M University, 1996.

9. D.B. Poker and D.K. Thomas, Nucl. Inst. and Meth. **B 39**, 716 (1989).

10. M. Sheik-Bahae, A.A. Said, T.H. Wei, D.J. Hagan, E.W. Van Stryland, IEEE J. Quantum Electronics, **26**, 760 (1990).

11. R.F. Haglund, Jr. Quantum-Dot Composites for Nonlinear Optical Applications, in Handbook of Optical Properties. Volume II. Optics of Small Particles, Interfaces, and Surfaces, edited by R.F. Hummel and P. Wismann (CRC Press, Boca Raton, 1997) p. 198.

STUDY OF THE EFFECTS OF MEV AG, CU, AU AND SN IMPLANTATION ON THE OPTICAL PROPERTIES OF LINBO₃

E.K. WILLIAMS[*], D. ILA[*], S. SARKISOV[*], M. CURLEY[*], D.B. POKER[**], D. K. HENSLEY[**], C. BOREL[***]
*Dept. of Natural and Physical Sciences, PO Box 1447, Normal, AL 35762-1447,
**Solid State Division, Oak Ridge National Laboratory, Oak Ridge, TN 37831,
***Université Claude Bernard, Lyon, France.

ABSTRACT

We present the results of characterization of linear absorption and nonlinear refractive index of Au, Ag, Cu and Sn ion implantation into $LiNbO_3$. Silver was implanted at 1.5 MeV to fluences of 2 to 17 x 10^{16}/cm^2 at room temperature. Gold and copper were implanted to fluences of 5 to 20 x 10^{16}/cm^2 at an energy of 2.0 MeV. Tin was implanted to a fluence of 1.6 x 10^{17}/cm^2 at 160 kV. After heat treatment at 1000°C a strong optical absorption peak for the Au implanted samples appeared at ~620 nm. The absorption peaks of the Ag implanted samples shifted from ~450 nm before heat treatment to 550 nm after 500°C for 1h. Heat treatment at 800°C returned the Ag implanted crystals to a clear state. Cu nanocluster absorption peaks disappears at 500°C. No Sn clusters were observed by optical absorption or XRD. The size of the Ag and Au clusters as a function of heat treatment were determined from the absorption peaks. The Ag clusters did not appreciably change in size with heat treatment. The Au clusters increased from 3 to 9 nm in diameter upon heat treatment at 1000°C. TEM analysis performed on a Au implanted crystal indicated the formation of Au nanocrystals with facets normal to the c-axis. Measurements of the nonlinear refractive indices made using the Z-scan method showed that Ag implantation changed the sign of the nonlinear index of $LiNbO_3$ from negative to positive.

INTRODUCTION

Introducing metal colloids into a glass or other transparent material matrix has been used to change their color for decoration and more recently for fabricating optical filters and other fast optical devices. In recent years, ion implantation has been used to introduce nonlinear optical properties [1-10] in layers near the surface of optical materials. To form nanoclusters after ion implantation the material may need to be thermally annealed or laser annealed [11]. An attractive feature of ion implantation is that the metal colloids can be placed in a well defined space near the surface and that by using focused ion beams, point quantum confinement may be accomplished.

It has long been known that small metallic particles or colloids embedded in dielectrics produce colors associated with optical absorption at the surface plasmon resonance frequency [12,13]. For a collection of spherical clusters with diameters much smaller than the wavelength of the incident light, the absorption coefficient (cm^{-1}) of the composite can be calculated from [12]:

$$\alpha = \frac{18\pi Q n_o^3}{\lambda} \frac{\varepsilon_2}{(\varepsilon_1 + 2n_o^2)^2 + \varepsilon_2^2} , \tag{1}$$

where Q is the volume fraction occupied by the metallic particles, n_0 is the refractive index of the host medium, and ε_1 and ε_2 are, respectively, the real and imaginary parts of the frequency-dependent dielectric constant of the bulk metal. Equation 1 is maximized at the surface plasmon resonance frequency (ω_p) when $\varepsilon_1 = -2n^2$.

Using tabulated values [14,15] of ε_1 for Cu, Sn, Ag and Au as a function of the photon wavelength, and using $n_0 = 2.2$, Eq. 1 predicts that the wavelengths for the surface plasmon resonance frequencies should be about 520 nm for Ag and near 620 nm for Cu and Au and at ~350 nm for Sn.

If the metal spheres are small compared with the wavelength of the incident light, their radius estimated from the optical absorption spectrum is $r = A_m v_f / \Delta\omega_{1/2}$ [1,16], where v_f is the Fermi velocity of the metal and $\Delta\omega_{1/2}$ is the full width at half maximum of the absorption band and A_m is a constant that varies for different metals. For Ag we have taken $A_m = 1$ and for Au we have used $A_m = 1.5$ [16].

EXPERIMENTAL PROCEDURES

Single crystal, 1 mm thick $LiNbO_3$ wafers were obtained from Crystal Technology Inc., (Palo Alto, CA). We used 1.5 MeV Ag and 2.0 MeV Au and Cu with beam currents of 2 to 3 $\mu A/cm^2$ and the samples were maintained at room temperature. Ag was implanted to fluences ranging from $2 \times 10^{16}/cm^2$ to $1.7 \times 10^{17}/cm^2$ while Au and Cu were implanted from $5 \times 10^{16}/cm^2$ to $2.0 \times 10^{17}/cm^2$. Sn was implanted at 160 keV to fluences up to $1.6 \times 10^{17}/cm^2$.

Optical absorption spectrometry was done soon after implantation and immediately after heat treatment using a Cary model 3e spectrophotometer. Heat treatments were done in air at temperatures of 300°C to 1000°C for 0.5 to 14 hours. Average radii of the metal clusters were measured from these spectra.

To investigate the effects of the implantation on the third order nonlinear optical coefficient, n_2, we used the z-scan technique [17, 18] employing a tunable pulsed dye laser pumped by a frequency doubled mode locked Nd:YAG laser run to give a wavelength of 575 nm and a pulse width of 4.5 ps at a repetition rate of 76 MHz. The average power was 70 to 350 mW, giving a power density applied to the sample in the range of 0.02 to 0.1 GW/cm^2.

RESULTS AND DISCUSSION.

The damage and modification of the optical properties are directly proportional to the atomic number and fluence of the bombarding ion [19, 20]. The surfaces of all of the Ag implanted samples were noticeably textured by the implantation. The higher fluence samples even functioned as poor quality diffraction gratings. The Au implanted $LiNbO_3$ also showed some surface texturing but not to the same extent as the Ag implanted samples. The texturing was still present after 10 hours at 1000°C.

The absorption spectra for Cu and Sn implanted $LiNbO_3$ are shown in Figure 1. The peak due to Cu clusters disappears after 500°C heat treatment for one hour. After 400°C for 1 h in air the Sn spectrum is the same as that of a virgin crystal. RBS analysis of the Sn implanted sample shown no significant diffusion at these low temperatures. XRD analysis did not indicate the presence of Sn crystals.

The absorption spectra from a Ag implanted sample at a fluence of $5 \times 10^{16}/cm^2$ is shown in Figure 2. The absorption peaks decreased in height and shifted to longer wavelengths, with the greatest shift occurring during the initial 500°C heat treatment. After 800°C heat treatment

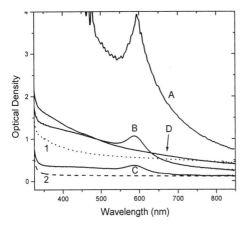

Figure 1. Optical absorption spectra from 2.0 Mev
Cu, A) 2×10^{17}/cm^2, B) 5×10^{17}/cm^2, C) 2×10^{16}/cm^2,
D) 2×10^{17}/cm^2 heated to 500°C and from 160 keV
Sn, 1.6×10^{17}/cm^2: 1) As Implanted, 2) 400°C, 1 hr.

Table I. Nonlinear index of refraction as
measure by z-scan method for varoius
fluences of 1.5 MeV Ag in LiNbO$_3$.

Fluence, ions/cm^2	Temp °C	Optical Density at 575 nm	Nonlinear index, n_2, 10^{-8} cm^2/W
2×10^{16}	500	0.39	0.96
2×10^{16}	800	0.15	0.35
5×10^{16}	500	1.0	2.90
5×10^{16}	800	0.2	0.38
17×10^{16}	500	1.87	6.81

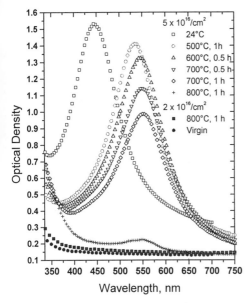

Figure 2. Optical absorption spectra of 1.5 MeV
Ag implanted LiNbO$_3$ after various heat treatments.
The calculated radius increased from 1.1 nm at 24°C
to 1.6 nm at 700°C.

the 2×10^{16} Ag/cm^2 implanted sample's
absorption spectrum was very close to
that of a virgin crystal. The change in
height of the spectra may be due to a
reduction in the number of Ag clusters
as Ag diffuses atomistically into the
substrate. The wavelengths of the Ag
absorption peaks and the radii of the
clusters are shown in Table I for each
fluence and heat treatment. The radii are
initially the same for all three fluences,
then increase from 30 to 60 percent after
heat treatment. Another increase in
radius is seen just before the absorption
spectra disappear.

In the as-implanted state the
absorption peaks were at a wavelength
shorter than that expected from theory
(~520 nm). This blue shift may be
caused primarily by the end of range
damage, which has been shown [23] to
decrease the index of refraction by over
5 percent. From Eq. 1, changing the host
index by 5 to 10 percent would result in a
shift of the expected absorption peak of
20 to 40 nm. Heat treatment works to
remove much of this implantation

damage and restores the refractive index. LiNbO$_3$ implanted with Ag 1×10^{17}/cm^2 at 190 keV has been shown to undergo full epitaxial regrowth after heat treatment at 400°C [24], but an anneal of 800°C for 1 h is necessary for full removal of the implantation damage. Prism coupling results indicate that for a Ag fluence of 2×10^{16}/cm^2 and heat treatment at 500°C the refractive index is increased by over 5 percent, sufficient to move the plasmon resonance peak position from the expected 520 nm to the experimentally observed 542 nm

The absorption spectra for a 5×10^{16} Au/cm^2 implanted sample is shown in Fig. 3. No absorption peak is visible prior to heat treatment. There was little change in the spectrum from the 5×10^{16} Au/cm^2 sample after 0.5 and 14 h at 500°C but for higher fluence samples (9×10^{16}/cm^2 and 2.0×10^{17}/cm^2) the 500°C for 0.5 h heat treatment resulted in an absorption spectrum typical of Au. At 700°C and above, the peak positions shift to longer wavelengths and dramatically increase in height. Shang, et. al. [21] observed a decrease in peak height at 600°C

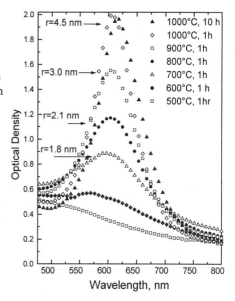

Figure 3. Optical absorption spectra from 5 x 10^{16} Au/cm^2 at 2.0 MeV into z-cut LiNbO$_3$. The radius calculated from each spectrum is indicated.

for Au implanted into LiNbO$_3$ at 23 keV. In their case heat treatment at 600°C for up to 8 hr resulted in a blue shift and a large decrease in the peak height with the expectation that the absorption spectrum would disappear with further heat treatment, as is the case for Ag in LiNbO$_3$. For the two higher fluence samples mentioned above, we observed a slight decrease in peak height at 650°C. Analysis of the absorption spectra indicates that the clusters begin to grow significantly larger at 800 to 900°C and appear to reach a maximum size at 1000°C.

TEM analysis of the 5×10^{16}/cm^2 Au implanted sample after heat treatment at 1000°C for 10 h shows (Figure 4a) that the Au clusters are in a ~380 nm thick band, 100 nm below the surface. Figure 4b shows faceting of the Au crystals normal to the c-axis. Very few of the crystals are smaller than 9 nm in diameter, many are approximately 9 to 11 nm in diameter and they range in size up to 40 nm in diameter. This agrees reasonably well with the cluster sizes determined from the absorption spectra, since in that method the width of the resonance is inversely proportional to the cluster size. The absorption spectrum from mixture of large and small clusters should differ little from the spectrum of small clusters only.

The results of the z-scan for the Ag implanted samples are shown in Table 1. The thickness of the colloidal layer used in the calculation of n$_2$ (240 nm for 1.5 MeV Ag) was estimated using SRIM96 [25]. For both Ag and Au the magnitude of the nonlinear index increased with increasing fluence and optical density. For the low fluence (2×10^{16}/cm^2) Ag implanted sample heat treated to 800°C, the absorption spectrum is nearly identical to that of the virgin substrate yet a positive nonlinear index is seen. This is not the intrinsic nonlinear index of

LiNbO$_3$, which is negative, as was confirmed by a z-scan of an unimplanted area. The nonlinear indices reported here are at least one order of magnitude greater than those reported for other silver clusters in the literature [26]. This indicates the presence of the cumulative thermal self-focussing effect which can still occur for the short pulses used in this work due to the relatively high pulse repetition rate (76 MHz versus 3.8 MHz in Ref. [26]). For the Au implanted LiNbO$_3$ a positive n_2 was seen prior to heat treatment as well as after cluster formation was observed.

CONCLUSIONS

The Cu, Sn, Ag and Au implanted LiNbO$_3$ reacted differently to heat treatment as observed in the optical absorption spectra. Silver clusters disappeared at a heat treatment temperature of 800°C, Cu clusters disappeared at 500°C and Sn clusters were not observed. Heat treatment of the Au implanted samples was necessary to see the expected absorption spectrum from Au clusters. Au clusters grew at heat treatment temperatures up to 1000°C and formed crystals with facets normal to the c-axis. For Ag implanted LiNbO$_3$ the Ag absorption peak appears without heat treatment (Figure 2) and disappears upon heat treatment at 800°C but enhancement of the nonlinear index remains after reduction of the absorption.

ACKNOWLEDGMENTS

This research was supported by the Dept. of Natural and Physical Sciences, Center for Irradiation of Materials, Alabama A&M University, and the Division of Materials Sciences, U.S. Dept. of Energy, under contract DE-AC05-96OR22464 with Lockheed Martin Energy Research Corp. and Alliance for Nonlinear Optics (NASA Grant NAGW-4078).

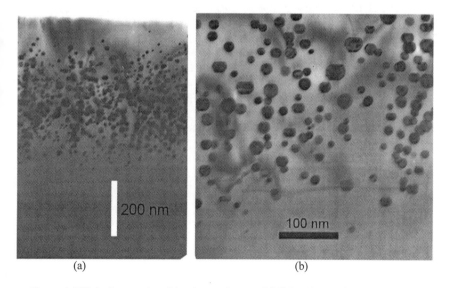

(a)   (b)

Figure 4. TEM micrographs of Au clusters in z-cut LiNbO$_3$. The Au fluence was 5 x 10^{16}/cm^2, heat treatment was 1000°C for 10 hr. The facets in the Au nanocrystals are normal to the c-axis.

REFERENCES.

1. G. W. Arnold, J. Appl. Phys. **46** (1975) 4466.

2. R. H. Magruder III, R. A. Zuhr, D. H. Osborne, Jr., Nucl. Inst. and Meth. **B 99** (1995) 590.

3. Y. Takeda, T. Hioki, T. Motohiro, S. Noda and T. Kurauchi, Nucl. Instr. and Meth. **B 91,** (1994) 515.

4. C. W. White, D. S. Zhou, J. D. Budai, R. A. Zuhr, R. H. Magruder and D. H. Osborne, Mat. Res. Soc. Symp. Proc. **Vol 316,** (1994)499.

5. K. Fukumi, A. Chayahara, M. Adachi, K. Kadono, T. Sakaguchi, M. Miya, Y. Horino, N. Kitamura, J. Hayakawa, H. Yamashita, K. Fujii and M. Satou, Mat. Res. Soc. Symp. Proc. **Vol 235,** (1992) 389.

6. D. Ila, Z. Wu, R. L. Zimmerman, S. Sarkisov, C.C. Smith, D. B. Poker, and D. K. Hensley, Mat. Res. Soc. Symp. Proc. **Vol 457** (1997) 143.

7. D. Ila, Z. Wu, R. L. Zimmerman, S. Sarkisov, Y. Qian, D. B. Poker, and D. K. Hensley, Mat. Res. Soc. Symp. Proc. **Vol 438** (1997) 417.

8. Q. Qian, D. Ila, K. X. He, M. Curley, D. B. Poker, Mat. Res. Soc. Symp. Proc. **Vol 396** (1996) 423.

9. D. Ila, Z. Wu, C.C. Smith, D. B. Poker, D. K. Hensley, C. Klatt, and S. Kalbitzer, Nucl. Instr. and Meth. B 127/128 (1997) 570.

10. Y. Qian, D. Ila, R. L. Zimmerman, D. B. Poker, L. A. Boatner, and D. K. Hensley, Nucl. Instr. and Meth. B 127/128 (1997) 524.

11. F. Gonella, G. Mattei, P. Mazzoldi, G. W. Arnold, G. Battaglin, P. Calvelli, R. Poloni, R. Bertoncello, and R. F. Haglund, Jr., Appl. Phys. Lett. **69** (20) 3101.

12. G. Mie, Ann. Physik **25** (1908) 377.

13. J. A. Creighton, D. G. Eadon, J. Chem. Soc. Faraday Trans. **87** (1991) 3881.

14. D. R. Lide, Ed., *CRC Handbook of Chemistry and Physics, 76th Ed.* (CRC Press, Boca Raton, 1995).

15. E. D. Palik, Ed., *Handbook of Optical Constants of Solids* (Academic Press, San Diego, 1985).

16. U. Kreibig and M. Vollmer, *Optical Properties of Metal Clusters (Springer Series in Materials Science, Vol25)* (Springer Verlag, Berlin, 1995)

17. M. Sheik-bahae, A. A. Said, T. H. Wei, Y. Y. Wu, D. J. Hagan, M. J. Soileau and E. W. van Stryland, SPIE Vol. 1148 Nonlinear Optical Properties of Materials, 41.

18. M. Sheik-bahae, A. A. Said, T. H. Wei, D. J. Hagan and E. W. van Stryland, IEEE J. Quantum Electronics **26** (1990) 760.

19. E. R. Schineller, R. P. Flam and D. W. Wilmot, J. Opt. Soc. Am. **58** (1968) 1171.

20. P. D. Townsend, Nucl. Instr. and Meth. **B 46** (1990)18.

21. D. Y. Shang, H. Matsuno, Y. Saito, and S. Suganomata, J. Appl. Phys. **80** (1996) 406.

22. D. Y. Shang, Y. Saito, R. Kittaka, S. Taniguchi and A. Kitahara, J. Appl. Phys. **80** (1996) 6651.

23. P. D. Townsend, Rep. Prog. Phys. 50 (1987) 501.

24. D. B. Poker and D. K. Thomas, Nucl. Inst. and Meth. B 39 (1989) 716.

25. J. F. Zeigler, J. P. Biersack and U. Littmark, *The Stopping and Range of Ions in Solids* (Pergamon Press, NY, 1985)

26. R. F. Haglund, L. Yang, R. H. Magruder, J. E. Wittig, K. Becker and R. A. Zuhr, Opt. Lett. 18 (1993) 373.

ELECTRON PARAMAGNETIC RESONANCE STUDY OF ION-IMPLANTED PHOTOREFRACTIVE CRYSTALS

A. Darwish[a], D. Ila[b], E. K. Willams[b], D. B. Poker[c] and D. K. Hensley[c],

a Center for nonlinear optics and materials and Laser Matter Research Lab, Al A&M University, PO BOX 245, Normal Al, 35762
b Center for Irradiation of Materials, Normal, Al 35762;
 c Solid state Division, Oak Ridge National Laboratory, Oak Ridge, TN 37831.

ABSTRACT

The effect of the ion implantation (Fe) on $LiNbO_3$, MgO, and Al_2O_3 crystals is studied using electron paramagnetic resonance (EPR). EPR measurements on these crystals were performed as a function of fluence at room temperature. The fluence was $1x10^{14}$ and $1x10^{16}$ ions/cm^2. The unpaired carrier concentration increases with increasing fluence. The photosensitivity of these crystals was determined by observing in situ the effect of the laser illumination on the EPR signal and measuring the decay and the growth of the EPR signal. The EPR signal of Fe^{3+} was found to decrease in both MgO, and Al_2O_3; and was found to increase in $LiNbO_3$. This indicated that in case of MgO, and Al_2O_3 Fe^{3+} will transfer into Fe^{2+}/Fe^{4+}, but in case of $LiNbO_3$ Fe^{2+}/Fe^{4+} will transfer into Fe^{3+}; increasing the EPR signal. This was found primary due to some Fe^{2+} and Fe^{4+} ions, which is not intentionally doped on the $LiNbO_3$ crystal but exist as a defect on the crystal .

INTRODUCTION

$LiNbO_3$ is a ferroelectric crystal with a variety of applications. Dopants play an important role in developing favorable properties for optical purposes. Transition metals (e.g. Fe, Mn) increase the photorefractive sensitivity. The properties and applications of lithium niobate have been widely studied. The high level of interest in lithium niobate is due to its combination of useful dielectric, elastic and opto-electronic properties. Its optical switching speed is as fast as 50 ps. In consequence, high speed switches, modulators, waveguide arrays, filters and polarization converts have been built based on lithium niobate.
Electron paramagnetic resonance measurements can provide critical information about structural and transport properties of the implanted samples as they give information on the spin mobility and degree of localization. EPR also can determine the effect of the laser illumination on the spin concentration and the charge carrier on the crystals. Also, it can identify the ions that participate on the processes of the charge transfer.
We have reported an EPR investigation of $LiNbO_3$: Fe^{3+}:Mn^{4+} and the effect of the irradiation, which produced a new center [1]. Although there have been many studies of impurity states in $LiNbO_3$ by EPR[2-4], the analysis of Fe impurity is especially of importance because this impurity appears to be responsible for optical refractive index damage and phase holographic storage in $LiNbO_3$.

Mat. Res. Soc. Symp. Proc. Vol. 504 © 1998 Materials Research Society

EXPERIMENTAL

X-cut(the C-axis perpendicular to the ab plane) single crystals of $LiNbO_3$, MgO, and Al_2O_3, were polished by EPI technique and were implanted with 160 keV Fe^+ at room temperature at fluences of 1×10^{14} /cm^2 and 1×10^{16}/cm^2. Also, $LiNbO_3$ crystals were implanted with 2.0 MeV Mn at room temperature at fluences of 2.5×10^{14} /cm^2 and 2.5×10^{16}/cm^2. Electron Paramagnetic Resonance was taken using a Bruker X-band 300ESP spectrometer. The modulation frequency was 100 kHz and the microwave frequency was 9.3 GHz. The sample was placed in an optical while acquiring the spectrum. Diphenyl picryl-haydrazyl (DPPH) with a g value of 2.0036 was taken as a g-marker and a reference for the calculation of the spin concentration. A HeNe laser with 35mW power was used as light source.

RESULTS AND DISCUSSION

The EPR spectra for the three crystals were taken at room temperature and the laser illumination was done in situ to observe the decay and the growth of the EPR signal i.e. the charge transfer due to the laser illumination. The EPR spectra are shown in figures 1 to 5. For $LiNbO_3$: Mn^{2+} as shown in figure 1; there is one group of sextet due to Mn^{2+} (s=5/2) and two strong lines due to Fe^{3+}. This assignment is based on the fact that the sextet-lines are due to hyperfine interaction of the d electrons of Mn with the s=5/2, ^{55}Mn nucleus. One strong set of sextets are observed, at θ =0° and 90°, the other weak two are overlapping with the Fe broad line which decreases the resolution of the hyperfine structure splitting. It is clear that the crystal contains some Fe^{3+} ions which is unintentionally doped but the trace of Fe^{3+} does exist on the starting materials as impurity. Fe^{3+} is playing the main role in the photorefractive processes in these crystals.

There are three possible sites for paramagnetic impurities in $LiNbO_3$, a Li site, a Nb site and a structural vacancy. Arguments have been advanced which favor both Li and Nb sites as possible candidates for Fe^{3+} impurity. EPR studies of Cr^{3+} in $LiNbO_3$ have shown that both Li and Nb sites are possible candidates for the Cr^{3+} impurity.

To interpret the EPR for the $LiNbO_3$: Fe crystal, Fe^{3+} (S =5/2) was found in an axially symmetric site and is described by the spin Hamiltonian [7]

$$H = g\beta H \bullet S + B^0_2 O^0_2 + B^0_4 O^0_4 + B^3_4 O^3_4$$

In this expression

$$O^0_2 = 3S^2_z - S(S+1),$$
$$O^0_4 = 35 S^4_z - 30S(S+1) S^2_z + 25 S^2_z - 6S(S+1),$$
$$O^3_4 = \tfrac{1}{4} [S^Z (S^3_+ + S^3_-) + (S^3_+ + S^3_-) S_Z]$$

At θ =0, the energy eigenvalues are given by [6]

$$E(\pm 5/2) = \pm g_\| \ \beta H + B^0_2 + 90 B^0_4 + \{[\pm(3/2) g_\| \ \beta H + 9 B^0_2 - 30 B^0_4]^2 + 90 \mid B^3_4\mid^2\}^{1/2}$$
$$E(\pm 3/2) = \pm 3/2 \ g_\| \ \beta H - 2B^0_2 - 180 B^0_4,$$
$$E(\pm 1/2) = \pm g_\| \ \beta H + B^0_2 + 90 B^0_4 - \{[\pm(3/2) \ g_\| \ \beta H + 9 B^0_2 - 30 B^0_4]^2 + 90 \mid B^3_4\mid^2\}^{1/2}$$

The crystal field parameters at room temperature were found to be $B^0_2 = 540$ G, $B^0_4 = -0.922$ G and $B^3_4 = 10$ G

In the case of MgO_3 the EPR, spectra are shown in figures 2, 3, 4 and 5 for low and high fluences. The hyperfine splitting for the Fe^{3+} is resolved even with low fluence as shown in figures 4 and 5. In these crystals, we expect to observe Fe^+. Iron in natural abundance mostly consists of Fe^+ (except when the line is narrow). The nuclear moment of ^{57}Fe is one of the smallest of any nuclide for which $I=1/2$ and this will give a large deviation of g value from the free-spin g factor 2.0023. Also, the magnitude of the EPR signal may be considered large due to isoelectronic ion. In our observation there is no intense line was observed with large g value, so it is safe to consider only the lines of the hyperfine structure to be due to Fe^{3+}.

In the case of $ALO_3:Fe^{3+}$ the EPR spectrum is shown in figure 6. Fe^{3+} goes in the cubic site (signal I) as well as interstitial (signal II).

EFFECT OF THE LASER ILLUMINATION:

By illuminating with the HeNe laser during the EPR measurements , we have observed that the signal of Fe^{3+} decreased in both MgO, and Al_2O_3; and was found to increase in $LiNbO_3$. This indicating that in case of MgO, and Al_2O_3 Fe^{3+} will transfer into Fe^{2+}/Fe^{4+}, but in case of $LiNbO_3$ Fe^{2+}/Fe^{4+} will transfer into Fe^{3+}; increasing the EPR signal. This was found primary due to some Fe^{2+} and Fe^{4+} ions, which unintentionally doped on the $LiNbO_3$ crystal. By turning, the laser off one can see that the EPR signal gains back its strength almost on the same time. On the other hand, the process of the charge transfer from Mn^{2+} was not strong enough to be measured. This is due to the overlapping of the Fe line and the broad cubic Fe line, which decreased the resolution of the Mn lines.

The spin concentration measurements were done by comparing the integrated intensity of the EPR signal with that of a known standard DPPH sample. The DPPH sample and the sample under study were placed in a double cavity to acquire their EPR signal at the same time and under the same experimental conditions. The relation between the two EPR signals and their spin concentration can be written as:

$$S_x = P/R \qquad (1)$$

as

$$P = S_{std} A_x R_x \text{ Scan}_x^2 G_{std} M_{std} (g_{std})^2 S(S+1)_{std} \text{ and}$$
$$R = A_{std} R_{std} \text{ Scan}_{std}^2 G_x M_x g_x^2 s(s+1)_x$$

where S_{std} is the concentration of the standard, A is the area under the EPR curve or the integration of the absorption signal, G is the gain, M is the modulation amplitude, s is the spin value, Scan is the amplitude of the magnetic field sweep, R is the multiplicity, and g is the g-value. From the integrated intensity, the analysis given using equation 1 yields spin concentration of $2 \times 10^{18}/cm^3$ for the of $2.5 \ 10^{14} /cm^2$ fluence and $10^{22}/cm^3$ for the $1 \times 10^{17} /cm^2$ fluence.

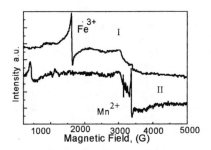

Figure 1. EPR spectrum for $LiNbO_3:Mn^{2+}$ high fluence
Z-axis perpendicular (I) / parallel to the magnetic field (II).

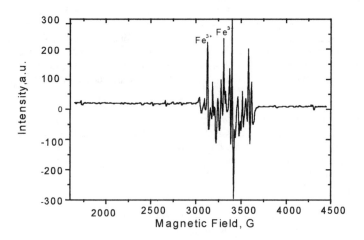

Figure 2. EPR spectrum for $MgO:Fe^{3+}$ high fluence
Z-axis parallel to the magnetic field.

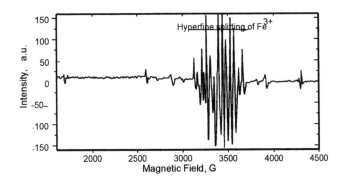

Figure 3.   EPR spectrum for MgO:Fe^{3+} high fluence
Z-axis perpendicular to the magnetic field.

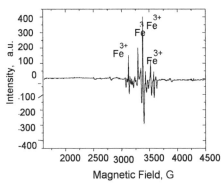

Figure 4.   EPR spectrum
for MgO:Fe^{3+} low fluence
Z-axis parallel to the
Magnetic field.

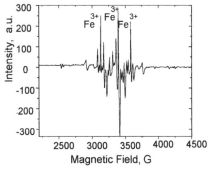

Figure 5.   EPR spectrum for
MgO:Fe^{3+} low fluence Z-axis
perpendicular to the magnetic
field.

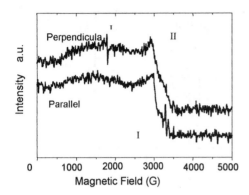

Figure 6. EPR spectrum for $AlO_3:Fe^{3+}$ high fluence Z-axis perpendicular / parallel to the magnetic field.

CONCLUSIONS

The higher number of the spin concentration for the high fluence crystal accounted for the Fe^{3+} impurities on the crystal as well as the implanted Mn ions. The difference on the decay time for Fe in both crystals with high and low fluence is found to be due to the long spin lattice relaxation time for Fe^{3+} at room temperature and the structure of the $LiNbO_3$ crystal. The spin concentration in these crystals was measured and the effect of the Fe^{3+} ions was studied.

ACKNOWLEDGMENTS

Research sponsored by the Center for Nonlinear Optics and Materials and the Center for Irradiation of Materials of Alabama A&M University, and by the Division of Materials Science, U.S. Department of Energy, under contract DE-AAC05-96OR22464 with Lockheed Martin Energy Research Corp.

REFERENCES
1. A. Darwish, D. McMillen, T. Hudson, and P. Banerjee, "Investigations of the charge transfer and the photosensitivity in single and double doped $LiNbO_3$ single crystals; an optical-electron paramagnetic resonance study (Part I)", J . Opt. Eng., Tech Digest, Vol. 2362, 29-31, (1997).
2. A.A. Mirzakhanyan, "The splitting in zero field of ground state levels of the Ni^{2+} ion in $LiNbO_3$," Sov. Phys. Solid State 23, 8, (1981).
3. F. Jermann and J. Otten ,"Light induced charge transport in $LiNbO_3$:Fe at high light intensities," J. Opt. Soc. Am. B, Vol. 10,11,2085-2092, (1993).
4. H. Towner and H. Story, "EPR studies of crystal field parameters in $Fe^{3+}:LiNbO_3$", The J. of Chem. Phys., Vol. 56, 7, (1972).
5. L.E. Halliburton and C. Chen, "ESR and optical point defects in Lithium Niobate," Nuc. Inst. and Meth. in Phys. Res., B1, 344-347, (1984).
6. P. Gunter and J.P. Hunignard, Photorefractive Materials and Their Applications I and II ,(Springer-Verlag, Heidelberg, 1988, 1989).
7. F. Huixinan, W. Jinke, W. Huafu, and X. Yunxia, "EPR studies of Fe in Mg-Doped $LiNbO_3$ crystals", J. Phys. Cem . Solids Vol. 51, No. 5, 397-400, (1990).

FORMATION OF NANOCLUSTER COLLOIDS OF TIN, GOLD AND COPPER IN MAGNESIUM OXIDE BY MeV ION IMPLANTATION

R. L. ZIMMERMAN *, D. ILA *, E. K. WILLIAMS * and S. SARKISOV *, D. B. POKER ** and D. K. HENSLEY **,
*Center for Irradiation of Materials, Department of Natural and Physical Sciences, Alabama A&M University, P. O. Box 1447, Normal, AL 35762-1447, ila@cim.aamu.edu
**Solid State Division, Oak Ridge National Laboratory, Oak Ridge, TN 37831

ABSTRACT

We have implanted ions of Sn, Au and Cu at energies between 160 keV and 2.0 MeV into single crystals of MgO (100) at room temperature. The formation of nanoclusters was confirmed using photospectrometry, in combination with Mie's theory, which was indirect but nondestructive. Using Doyle's theory, as well as Rutherford Backscattering Spectrometry (RBS), we correlated the full width half maximum of the absorption bands to the estimated size of the metallic nanoclusters between 1-10 nm. These clusters were formed both by over implantation and by a combination of threshold fluence of the implanted species and post thermal annealing. The changes in the estimated size of the nanoclusters, after annealing at temperatures ranging from 500°C to 1000°C, were observed using photospectrometry.

INTRODUCTION

The classical treatment of small spheres imbedded in an optical material of index n_o shows that an imposed electric field produces in each sphere a dipole moment proportional to the field and to the factor

$$\frac{\varepsilon - n_o^2}{\varepsilon + 2n_o^2},$$ (1)

where ε is the dielectric constant of the material of the spheres. For spheres of conducting material $\varepsilon = \varepsilon_1 + j\varepsilon_2$, where the real component ε_1 is negative and the imaginary component ε_2 is proportional to the conductivity which causes energy loss from a time varying electric field, such as that in visible light. Mie [1] derived the optical absorption coefficient α of a material with a volume fraction Q occupied by metal spheres whose radii are small compared with incident light of wavelength λ

$$\alpha = \frac{18\pi Q 2 n_o^3}{\lambda} \frac{\varepsilon_2}{(\varepsilon_1 + 2n_o^2)^2 + \varepsilon_2^2} \quad cm^{-1}.$$ (2)

A minimum occurs in the denominator of Equation 2 when

$$\varepsilon_1(\lambda_\rho) + 2n_o^2 = 0$$ (3)

and causes a maximum absorption of light at a characteristic wavelength λ_p, the so called surface plasmon resonance [2-5]. Metallic colloids embedded in a dielectric transmit only part of the visible spectrum, a phenomenon used since ancient times to decorate glassware. Figure 1 shows the values of ε_1 and ε_2 for gold, silver, copper and tin derived from the electronic constants of the bulk metals [6,7]. The downward trend of ε_1, linear in wavelength, together with the slow variation of ε_2, leads us to expect quasi Lorentzian optical absorption peaks centered at a wavelength λ_p characteristic of the index of refraction of the host and of the permittivity of the

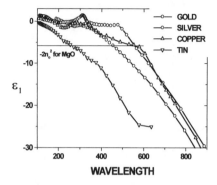

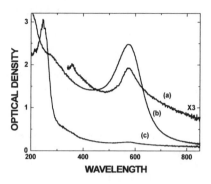

Figure 1 The permittivity for gold, silver, copper and tin from Refs {11,12]

Figure 2 Optical absorption spectra for (a) copper, (b) gold and (c) tin.

metal. Using n_0=1.73 for MgO and the values of ε_1 from Figure 1, Equation (3) predicts wavelengths of 564, 551 and 250 nm for the surface plasmon resonance for colloidal Cu, Au and Sn in MgO, respectively. Figure 2 shows experimental optical absorption spectra for gold, tin and copper, with the observed wavelengths for maximum optical density in reasonable agreement with Equation 2. Ion implantation induces additional optical absorption in the MgO substrate that must be removed by careful heat treatment of the implanted samples.

Doyle [8] has improved the Mie theory by showing that for spheres whose size is less than the mean free path of the conduction electrons the plasmon resonance is broadened not by the conductivity, represented by ε_2, of the bulk material but by the radius r of the spheres. For the full width at half maximum $\Delta\lambda$ of the peak determined from an optical absorption measurement, in accordance with Doyle

$$r = \frac{v_f \lambda_p^2}{2\pi c \Delta\lambda}, \qquad (4)$$

where v_f is the electron velocity corresponding to the Fermi energy of the metal. λ_p depends on the substrate and the element implanted in it and $\Delta\lambda$ is related to the size of the nanoclusters.

These changes in the optical properties of insulating materials can be made to occur following the implantation of metal using MeV ion accelerators. Small metal clusters form in the implanted layer, either spontaneously or by heat treatment, and absorb light at the surface plasmon resonance wavelength [9-11]. The nonlinear optical parameters of the implanted layer have been reported to be larger than the host dielectric [12].

The thermal behavior of ion implanted material differs significantly from optical material prepared with a uniform distribution of colloidal material. The host dielectric may be a single crystal, as well as a glass or ceramic. Secondly, the initial distribution of metal atoms is easily controlled by the energy and fluence of the incident ion beam. And there is no *a priori* restriction on the chemical affinity or stoichiometry of the implanted ions. Subsequent thermal treatment controls the formation of nanocrystals of the metal. The metal atoms aggregate into single crystals of desired size aligned with the host crystal [12].

Because most of the crystal initially has no metal atoms, heating above a critical temperature causes the clustered atoms to disperse irreversibly. In the present paper we compare the thermal behavior of copper, tin and gold colloids implanted near a (100) surface of MgO single crystals.

EXPERIMENTAL PROCEDURES

Implantations were made into epi-polished 1x1 cm MgO (100) single crystals, a commercial product of Commercial Crystal Laboratories, Inc., using low fluxes to maintain the MgO crystals at room temperature. Room temperature optical absorption measurements were made for each of the implanted samples before any heat treatment and repeated after each heat treatment. The heat treatment was performed in air. Table I summarizes the samples and their treatment. Prior experience has shown that tin atoms are dispersed at relatively low temperatures [11] and in the experiments reported here the samples implanted with tin were not heat treated. Heat treatment of the copper implanted samples at 500°C also resulted in dispersal of the guest atoms and the disappearance of the optical absorption resonance. Persistent heat treatment of the gold implanted samples up to 1000°C did not destroy the optical absorption resonance characteristic of gold.

Because implantation introduces crystal defects in the MgO, each of which has a characteristic optical absorption, virgin MgO crystals were bombarded with $1.5 \times 10^{17} \, cm^{-2}$ helium ions of energy 1.4 MeV. The damage resulted in crystal defects whose thermal properties were studied.

The depth and distribution of the implanted ions shown in Table I were calculated from SRIM [13] and verified by Rutherford Backscattering Spectrometry (RBS). The alpha particle channeling technique was employed to detect evidence that the metal nanocrystals were aligned with the MgO substrate.

Table I Treatment of MgO samples

Implantation Material	Implantation Energy	Fluence ($10^{16} \, cm^{-2}$)	Implantation Depth (μm)	Heat treatment (all fluences)
Gold	2,000	40 20 8.0 4.0	0.32 ± 0.05	None and 500 to 1000°C 100°C intervals for 1 hr
Copper	2,000	70 19 8.0	0.80 ± 0.12	None and 500°C, 1 hour
Tin	160 320	10 3.0	0.047 ± 0.013 0.089 ± 0.021	None

The fluences used in these implantations were chosen to reach a concentration of guest atoms between 10 to 100 times less than that of the host atoms; i. e., the ratio of Au to Mg of 0.1 to 0.01 for gold implanted MgO. For the 0.1 ratio, the mean separation of the guest atoms is about three times that of the host atoms. We used low current densities such that no significant temperature increase occurred during implantation. When clusters of high density metal form, the volume in the substrate occupied by the metal atoms decreases.

Heat treating the sample further reduces the strains and the charge imbalance in the substrate. This change in the host material effects the diffusion of the implanted atoms. In our tests, the implanted atoms move to the lower energy metallic state of the clusters whose volume fraction approaches an equilibrium which depends on the initial implantation parameters, i.e., total

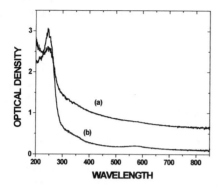 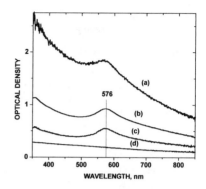

Figure 3 Optical absorption of MgO implanted with 3x10^{16} cm^{-2} Sn at 320 keV (a) and 1x10^{17} cm^{-2} Sn at 160 keV (b).

4. Spectra of MgO implanted with 2.0 MeV Cu at fluences of (a) 7.0x10^{17}, (b) 1.9x10^{17} and 8.0x10^{16} ions cm^{-2} (c) without and (d) with thermal treatment at 500°C for one hour.

fluence and substrate temperature. With an initial mean separation of the implanted atoms of only a few nanometers virtually all will diffuse to a cluster and not reach the surface of the layer. Moreover, other experiments have shown that the damaged region in the host may itself offer lower energy states to individual guest atoms and inhibit diffusion loss from, and enhance diffusion into, that region. With heat treatment the near neighbor clusters coalesce which is an Ostwald ripening process and the host accommodates to the volume. RBS measurements confirm that the depth profile of the metal clusters formed after heat treatment is almost identical to or slightly narrower than that of the atoms initially implanted by ion bombardment.

RESULTS AND DISCUSSION

Figure 3 shows the optical absorption spectra from Sn implanted MgO crystals. From Equation 4, and the position and width of the optical absorption peak, we infer that nanoclusters have formed with average radii of 1.5±0.3 nm for the 160 keV Sn fluence of 1x10^{17} cm^{-2} and 1.3±0.3 nm for the 320 keV Sn fluence 3x10^{16} cm^{-2}. The uncertainty is owing to a significant contribution to the optical absorption by crystal defects induced by the ion implantation. Studies of radiation induced optical absorption in MgO have been reported elsewhere [12,14,15]

Figure 4 shows the optical spectra of three MgO(100) crystals implanted with copper at room temperature with fluences of (a) 8.0 x 10^{17}, (b) 1.9 x 10^{17} and (c) 8.0 x 10^{16} ions cm^{-2}. The optical absorption bands at 576 nm show that metal nanoclusters form during implantation at room temperature. Heat treatment of each sample at 500°C for one hour caused this absorption band to disappear as shown in Figure 4 (d) for the sample implanted with 8.0 x 10^{16} ions cm^{-2}. Evidently, such heat treatment causes irreversible dispersion of the Cu atoms from the implantation layer in the MgO crystal, a phenomenon shown also by silver implanted MgO, but only after heat treatment at temperatures above 1100°C [12,16]. Evidently the thermal stability of nanoclusters formed during implantation or by post bombardment thermal treatment is not related directly to the melting temperature of the bulk metals. The position and width of the optical absorption peaks in Figure 4 are almost independent of the implanted fluence and yield the value 1.3 ± 0.2 nm for the Cu nanoclusters.

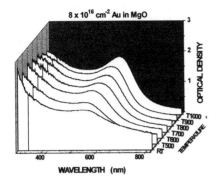

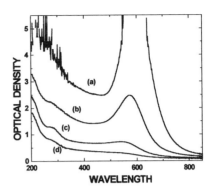

Figure 5 Optical spectra of a single MgO(100) crystal implanted with 2 MeV Au ions to a fluence of 8×10^{16} cm^{-2} and heat treated for one hour at each of the temperatures shown.

Figure 6 Optical spectra of four samples of MgO implanted with 2 MeV Au ions to fluences of (a) 4×10^{17}, (b) 2×10^{17}, (c) 8×10^{16} and (d) 4×10^{16} cm^{-2}. Each has been annealed at 1000°C for 10 hours.

Figure 5 shows the optical density as a function of wavelength for a single MgO sample implanted with 2.0 MeV gold at a fluence of 8×10^{16} ions cm^{-2}. Spectra as implanted at room temperature and after heat treatment for one hour at several temperatures are shown. Evidence of a broad absorption band at about 580 nm appears before any heat treatment. At increasing temperatures the band becomes more pronounced, narrows and shifts slightly to shorter wavelengths. Samples implanted with other fluences respond similarly to heat treatment. Figure 6 shows the spectra for four samples implanted with differing Au fluences all annealed at 1000°C for 10 hours. Further annealing at 1000°C does not alter the optical properties of Au implanted MgO. The gold nanoclusters are stable and, except for strains in the implantation layer, the MgO crystal defects have disappeared. Dispersal of the Au atoms require heat treatment at temperatures above 1200°C. Spectra such as those shown in Figure 6 yield average gold radii of 1.4 ± 0.2 and 2.1 ± 0.2 nm for the Au fluences 8×10^{16} and 2×10^{17} Au ions cm^{-2}, respectively.

Surprisingly, the average radius of stable gold nanoclusters increases significantly more than that expected [8] from the cube root of the fluence increase . Coalescence of small clusters by a nearby larger one could explain this phenomenon. The greater initial separation of nanoclusters in samples implanted with low fluence may permit less coalescence during thermal treatment.

RBS measurements accomplished on implanted MgO crystals that are aligned along a principal crystal plane show that metal clusters are single crystals also aligned in accordance with their host. Figure 7 shows two RBS measurements with equal fluences of 3 MeV α particles, one along the 100 plane of MgO and the other in a random direction. The former shows a significant channeling both in the gold nanocluster and in the MgO crystal. Evidence obtained by RBS that metallic nanocrystals are single crystals aligned with their host and confirmed by TEM have been reported elsewhere [16].

CONCLUSION

The implantation of metal ions into a single crystal of MgO(100), which is then submitted to careful heat treatment, creates a well defined layer of material with new optical properties. The optical absorption at the plasmon frequency of implanted gold may be used to calculate the average radius of the metal nanoclusters as a function of thermal treatment temperatures and

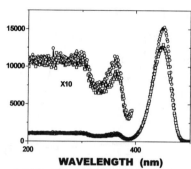

WAVELENGTH (nm)

Figure 7 RBS measurements of an MgO crystal implanted with 4×10^{17} Au ions/cm^2 along a [100] plane and in a random direction (round symbols).

times. Subsequent heat treatment of the gold implanted MgO causes metallic nanoclusters to grow to an equilibrium average radius that depends on the implantation fluence. An increase of the fluence of implanted gold causes an increase in the equilibrium radius of the gold nanoclusters. No such increase would be expected if the nucleation site density is proportional to the defect density caused by damage during implantation. In fact we observe an increase greater than the cube root fluence dependence expected for a constant density of initial nucleation sites. Possibly coalescence of smaller nanoclusters with a single larger nanocluster occurs more with samples implanted with high fluences.

Gold, copper and tin atoms diffuse in MgO during room temperature implantation to form nanoclusters. Subsequent heat treatment causes Sn and Cu to disperse irreversibly into the host, unlike gold and silver [12,16]. This striking difference is not attributable to differences in bulk melting temperatures. The close match of the lattice constants of Au and Ag with those of MgO may give nanoclusters of those metals relatively more thermal stability.

ACKNOWLEDGMENTS

Research supported by the Center for Irradiation of Materials, Alabama A&M University, EPSCoR-NSF Grant No. EHR-9108761/STI-9108761 and the U. S. Department of Energy, under contract DE-AC05-96OR22464 with Lockheed Martin Energy Research Corp.

REFERENCES

1. G. Mie, Ann. Physik **25**, 377 (1908).
2. G. W. Arnold, J. Appl. Phys. **46**, 4466 (1975).
3. G. W. Arnold and J. A. Borders, J. Appl. Phy. **48**, 1488 (1977).
4. G. Abouchacra, G. Chassagne, and J. Serughetti, Radiation Effects **64**, 189 (1982).
5. G. Fuchs, G. Abouchacra, M. Treilleux, P. Thévenard, and J. Serughetti, Nucl. Instr. and Meth. **B32**, 100 (1988).
6. David R. Lide, Ed., Handbook of Physics and Chemistry **76**, CRC Press, New York 1995, pp **12**-129-131
7. G. Jéséquel, J. C. Lemonnier and J. Thomas, J. Phys. F: Metal Phys. 7, 2613 (1977).
8. W. T. Doyle, Phys. Rev. **111**, 1067 (1958).
9. Y. Qian, D. Ila, K. X. He, M. Curley, D. B. Poker and L. A. Boatner, Proc. Mat. Res. Soc. **396**, 423 (1996).
10. A. E. Hughes and S. C. Jian, Adv. Phys. **28**, 717 (1979).
11. Y. Qian, D. Ila, R. L. Zimmerman, D. B. Poker, L. A. Boatner and D. K. Hensley, Nucl. Inst. and Meth. **B127/128**, 524 (1997).
12. D. Ila, Z. Wu, R. L. Zimmerman, S. Sarkisov, Y. Qian, D. B. Poker, and D. K. Hensley, Proc. Mat. Res. Soc. **438**, 417 (1997).
13. J. F. Ziegler, J. P. Biersack and U. Littmark, The Stopping and Range of Ions in Solids (Pergamon NY 1985).
14. J. Narayan, Y. Chen and R. M. Moon, in Defect Properties and Processing of High-Technology Nonmetallic Materials, J. H. Crawford, Jr., Y. Chen and W. A. Sibley, Eds., MRS Symp. Proc. **24** (Elsevier 1984), 101
15. S. Clement and E. R. Hodgson, Defect Properties and Processing of High-Technology Nonmetallic Materials, Y. Chen, W. D. Kingery and R. J. Stokes, Eds., (MRS 1986) 191
16. D. Ila, R. L. Zimmerman, E. K. Williams, C. C. Smith, S. Sarkisov, D. B. Poker and D. K. Hensley, KK6.25, Fall Meeting MRS, 1997

CHANGE IN THE OPTICAL PROPERTIES OF SAPPHIRE INDUCED BY ION IMPLANTATION

D. ILA[a*], E. K. WILLIAMS[a], S. SARKISOV[a], D. B. POKER[b] and D. K. HENSLEY[b]
a) Center for Irradiation of Materials, Alabama A&M University, Normal AL 35762
b) Solid State Division, Oak Ridge National Laboratory, Oak Ridge TN

ABSTRACT

We have studied the formation of nano-crystals, after implantation of 2.0 MeV gold, 1.5 MeV silver, 160 keV copper and 160 keV tin into single crystal of Al_2O_3. We also studied the change in the linear optical properties of the implanted Al_2O_3 before and after subsequent annealing by measuring the increase in resonance optical absorption. Applying Doyle's theory and the results obtained from Rutherford backscattering spectrometry (RBS) as well as the full width half maximum of the absorption band from Optical Absorption Photospectrometry (OAP), we measured the average size of the metallic clusters for each sample after heat treatment. The formation and crystallinity of the nanoclusters were also confirmed using transmission electron microscopy (TEM) technique.

INTRODUCTION

In general the interaction of energetic ions with ceramics (insulator) is different from their interaction with metals [1,2]. Since the process of ion implantation is inherently a nonequilibrium process, it can be used to introduce very low solubility impurities into sapphire thus creating charge-compensating defects [3,4].

Sapphire is from R3C space group, rhombohedral, where each Al^{+3} ion has six O^{2-} ions, three at 0.1858 nm and three at 0.1971 nm, instead of all at equal distances on the vertices of an octahedron [5]. In this paper, we have used keV and MeV ion implantation to introduce metals, such as gold, silver, tin and copper into chemically ultra-pure sapphire crystals. Then we studied the optical properties of the ion implanted crystals before and after thermal annealing at each heat treatment temperature.

EXPERIMENTAL PROCEDURES

We have used 6 mm x 6 mm epi-polished, highly pure c-cut single crystals of Al_2O_3, c-axis perpendicular to the surface of the crystals. Then individual Al_2O_3 crystals were implanted at room temperature with 2.0 MeV gold ions, 1.5 MeV silver ions, 160 keV tin ions, or 160 keV copper ions separately at fluences between 2.0 x 10^{16} ions/cm^2 to 20 x 10^{16} ions/cm^2. The implantations were performed at a current density less than 2 $\mu A/cm^2$ to avoid the premature formation of metal clusters due to the ion beam heating, and with target holder maintained at room temperature.

The heat treatments were done in air at temperatures between 673°K to 1323°K for one hour. After each heat treatment, an optical absorption spectrum at room temperature was measured. Using these spectra, the RBS results, as well as TRIM simulations [6], we calculate the size of the metallic clusters which are responsible for the optical absorption band. The RBS was done using 3.0 MeV alpha particles. The ion channeling was done along the c-axis of the crystals. Optical absorption measurements were performed at room temperature using a Cary 13E spectrophotometer capable of measuring absorption in UV and visible portions of the spectrum (i.e., from 190 to 900 nm). For all these measurements, the unimplanted part of the sample was used as a reference. For cross-section TEM analysis the samples were prepared by lapping and polishing with diamond, and a final polishing with colloidal silica. Final thinning was done by Ar ion milling (4.5 keV, 18°, sector milling) for five minutes per side. The TEM specimen preparation sequence preserved the original implanted surface for depth measurements.

The average radius of metal nanoclusters is small compared with the wavelength of light and is determined [7] from the resonance optical absorption spectrum according to the equation

$$r = (v_f/\Delta\omega_{1/2}),$$

where v_f is the Fermi velocity of metal and $\Delta\omega_{1/2}$ is the full width at half maximum of the absorption band due to the plasmon resonance in small metal particles. The value of the FWHM is determined from the absorption band ($\Delta\omega_{1/2} = 2\pi c \, \Delta\lambda/\lambda_p^2$) where $\Delta\lambda$ is the full width at half maximum wavelength of the plasmon band and λ_p represents the observed peak wavelength of the plasmon band.

RESULTS and DISCUSSIONS

The structure of the implanted sapphire is complex and the amount and the type of damage at various depths strongly depends on the orientation of the ion beam relative to the crystallographic axes [8,9]. The radiation induced damage was observed by RBS channeling, TEM and by absorption photospectrometry. At higher annealing temperatures the radiation induced damage was reduced except in the layer where the density of the clustered nanocrystals was peaked. In that layer the (006) lattice spacing is highly distorted, measuring approximately 5% larger than for the single crystal sapphire substrate.

Figure 1 compares the optical absorption spectra of four sapphire crystals implanted either with gold, silver, copper, or tin after subsequent heat treatment at 1323°K for Au implanted, 773°K for Ag implanted, 673°K for Cu implanted, and at 773°K for Sn implanted crystals. The

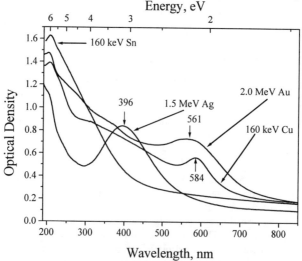

Figure 1, the absorption optical density for sapphire crystals implanted with Ag at 8 x 10¹⁶/cm², Au at 12 x 10¹⁶/cm², Cu at 20 x 10¹⁶/cm² and Sn at 8 x 10¹⁶/cm² after heat treatment.

absorption spectra for sapphire implanted with each ion species follows the Mie's theory predicted shape [10].

To observe the effects of the MeV ion beam induced changes in accordance with the shape predicted by the Mie theory in the optical absorption spectra of sapphire, we bombarded a c-cut pristine sapphire with 3.0 MeV alpha particles at $1.0 \times 10^{16}/cm^2$ and $20 \times 10^{16}/cm^2$. These crystals were optically studied before and after annealing at temperatures between $673^\circ K$ to $1173^\circ K$. Typically, the innert ion bombarded sapphire crystals showed absorption bands at 257 nm, 230 nm, and 203 nm. These bands were reduced to one absorption band at 203 nm with a 50% reduction in its absorption optical density after annealing at $673^\circ K$, and further reduced an additional 50% when heat treated at $1173^\circ K$.

The information obtained from alpha particle bombarded sapphire crystals helped us to distinguish between optical absorption band due to ion beam induced damage in the sapphire crystals and the optical absorption bands due to formation of nanoclusters. The typical absorption peak for Au implanted sapphire is at 561 nm, for Cu is at 583 nm, for Ag is at 396 nm. The absorption band for Sn implanted sapphire is in the ultraviolet range, approximately 6 eV, which is in the same region as the absorption band from the ion beam induced defects in sapphire.

The room temperature implanted sapphire generally displayed a high absorption in the ultra-violet region decreasing in the infra-red region. As the implanted crystals were annealed, the optical absorption due to the ion beam induced damage to the sapphire substrate decreased and the optical absorption due to formation of nanocrystals increased. Figure 2, the optical absorption photospectra for 2.0 MeV Au implanted sapphire at $12 \times 10^{16}/cm^2$ as implanted at room temperature and after heat treatment at $973^\circ K$ and at $1323^\circ K$, shows the thermal history of the formation of the optical absorption band at 561 nm due to formation of gold nanoclusters after annealing.

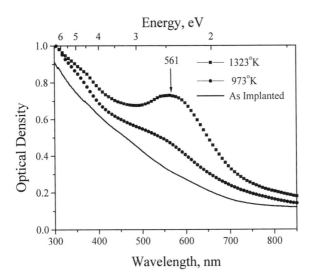

Figure 2, the optical density for 2.0 MeV gold implanted epi-polished c-cut sapphire crystal at $12 \times 10^{16}/cm^2$ at as implanted and annealed at $973^\circ K$ and at $1323^\circ K$.

The ion implanted sapphire crystals suffer damage throughout the range of the implanted ions as well as slightly beyond the range. The distribution of the nanoclusters observed by RBS extends toward the surface and continues beyond the TRIM predicted range with maximum density of implanted ions, which is 280 nm for Au, 370 nm for Ag, 43 nm for Sn, and 63 nm for Cu. Figure 3 is a TEM distribution of nanoclusters near the depth where the density of implanted species is maximum. This figure shows a high density of very fine (~2 nm) Au nanocrystals. The large crystals are formed in the middle of the distribution of small nanocrystals. This image and the electron diffraction pattern, as well as bright field imaging, show that the Au nanoclusters are crystalline at this depth and have a texture with <111> parallel with the <0001> of the sapphire substrate. The size of the gold nanocrystals was found to be between ~200-600 nm, with a few scattered small nanocrystals outside this range. The TEM results show that the Au nanocrystals were mostly in the same orientation as the host substrate and they were strongly faceted with principal planes aligned with the crystal planes of the sapphire host.

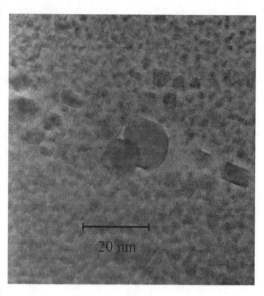

Using Doyle's theory [7] and the resonance optical absorption bandwidths, average cluster radii in the three metal implanted samples are calculated for each heat treatment temperature. Since we could not separate the measured absorption band for Sn implanted sapphire from the optical absorption due to ion beam induced damage to the substrate, we could not use Doyle's theory for Sn implanted sapphire. Table 1 shows ion

Figure 3, transmission electron micrograph of the 2.0 MeV gold implanted sapphire crystal, annealed at 1323°K at a depth of 300nm.

Ion Species	Fluence (ions/cm²)	Energy (MeV)	Temperature (°K)	Peak (nm)	Ave. Radius (nm)
Cu	20 x 10¹⁶	0.16	673	564	2.3
Ag	8 x 10¹⁶	1.5	773	396	0.63
Sn	8 x 10¹⁶	0.16	773	none	--------
Au	1.2 x 10¹⁷	2.0	1323	561	1.5

Table 1, showing ion species and their implantation energies, the calculated average cluster size and heat treatment temperature.

species and their implantation energies, the calculated average cluster size, and heat treatment temperatures. The average cluster size was calculated by two methods: 1) using the FWHM of the absorption band and the Doyle's theory, and 2) using the Mie's theory for the absorption coefficient, which is also a function of the dielectric constant of the implanted species, the index of refraction of the host material, and the volume fraction occupied by the implanted species. RBS results from the implanted sapphire were used in measuring the volume fraction of the implanted species.

CONCLUSION

We have induced a strong optical absorption band in sapphire after implantation by Cu, Ag, Au, or Sn and subsequent annealing. The positions of the absorption bands are in agreement with that predicted by Mie's theory. We have also excluded the possibility that these absorption bands could be due to the ion beam induced damage to the substrate by bombarding a pristine sapphire crystal and studying its thermal behavior. The optical absorption bands are proven to be due to formation of nanocrystals with orientations governed by the host sapphire crystal and some have also shown a strong indication of faceting.

ACKNOWLEDGMENTS

This project was supported by the Center for Irradiation of Materials at Alabama A&M University and   is based on work supported by, the U.S. Army Research Office under contract/grant No. DAAH04-96-1-0127. The work at ORNL was sponsored by the Division of Materials Science, U.S. Department of Energy, under Contract DE-AC05-96OR22464 with Lockheed Martin Energy Research Corp.

*Corresponding author: Tel +1 205 851 5866, FAX +1 205 851 5868, e-mail ila@cim.aamu.edu

REFERENCES

[1]   P. Mazzoldi and G. Arnold (eds.), "Ion Beam Modification of Insulators" (Elsevier, Amsterdam, 1987).

[2]   D. M. Parkin, in C. J. McHargue, R. Kossowsky and W. O. Hofer (eds.), "Structure-Property Relationships in Surface Modified Ceramics", (Vol. 170, NATO ASI Series, E. Kluwer, Dordrecht, 1989), 47.

[3]   C. J. McHargue, P. S. Sklad, C. W. White, G. C. Farlow, A. Perez and G. Marest, J. Mater. Res. **6** (1991) 2145.

[4]   L. J. Romana, P. Thevenard, S. Ramos, B. Canut, L. Gea, M. Brunel, L. L. Horton and C. J. McHargue, J. Surface Coatings **51** (1992) 410.

[5]   M. L. Kronberg, Acta Met. **5** (1951) 507

[6]   F. Ziegler, J. P. Biersack and U. Littmark, The Stopping and Range of Ions in Solids (Pergamon Press Inc., New York, 1985).

[7]   W. T. Doyle, Phys. Rev. **111**, (1958) 1067.

[8]   C. J. McHargue, P.S. Sklad and C. W. White, Nucl. Instr. and Meth. in Phys. Res. **B46**, (1990) 79.

[9]   G. C. Farlow, P. S. Sklad, C. W. White, C. J. McHargue and B. R. Appleton, in: Defect Properties and Processing of High Technology Nonmetallic Materials, eds. Y. Chen, W. D. Kingerly and R. J. Stokes (Materials Research Society, Pittsburgh, 1986), 387-394.

[10] G. Mie, Ann. Physik **25**, (1908) 377.

CONTROL OF OPTICAL PERFORMANCE FROM Er-DOPED ALUMINA SYNTHESIZED USING AN ECR PLASMA

J.C. Barbour and B.G. Potter
Sandia National Laboratories, Albuquerque, NM 87185-1056

ABSTRACT

Hydrogen in deposited optical ceramics can modify the optical properties, and therefore the role of the hydrogen needs to be understood to control its effects. Erbium-doped amorphous alumina films were deposited using simultaneous electron beam evaporation of aluminum and erbium while bombarding the sample with 30 eV O_2^+ ions from an electron cyclotron resonance (ECR) plasma. The hydrogen content was measured, using elastic recoil detection, as a function of isochronal annealing treatments. The data was fit to a simple trap-release model in order to determine an effective activation energy for the thermal release of H from alumina and Er-doped alumina. The intensity of the ion-beam stimulated luminescence from these samples was monitored in the visible and near infrared regions as a function of the thermal treatments. In order to gain a better understanding of the influence of hydrogen, the ionoluminescence (IL) data from samples containing hydrogen were fit with a simple linear equation.

INTRODUCTION

The optical performance of Er-doped alumina can be modified through atomic-level engineering of the local structural environment and through modification of the composition, in particular the hydrogen content. We have shown in an earlier paper [1] that the shape and lifetime, τ, of the Er photoluminescence (PL) peak for 1.54 μm emission differs between the alpha, gamma, and amorphous phases of alumina. The PL spectrum for Er in the amorphous phase (a-Al_2O_3) was nearly identical in shape and peak positions to that for Er in γ-Al_2O_3 and differed from that for Er in α-Al_2O_3. However, τ was different for each of these phases. The difference in local environment may account for the difference in τ between the α and γ phases, but the presence of hydrogen in a-Al_2O_3 was thought to decrease the lifetime for the amorphous phase. In this work, we will examine the thermal release of hydrogen from electron cyclotron resonance (ECR) deposited alumina and the effect of the hydrogen release on the intensity of the luminescence spectra. For simplicity in this experiment, visible and near infrared (IR) light emission from Er-doped a-Al_2O_3 will be stimulated by an ion beam (ionoluminescence) rather then through laser excitation. Low temperature deposition of this important class of materials is needed for integrated optical amplifier components on III-V semiconductors, and therefore the formation of these different phases at low temperature was studied using an electron cyclotron resonance (ECR) plasma.

Previously [2], we showed that ECR O_2 plasmas combined with electron beam evaporation could form Al_2O_3 in either the amorphous (a-Al_2O_3) or the FCC γ-phase by varying the ion energy and temperature during deposition. An ECR-Al_2O_3 layer deposited at $\leq 400°C$ without a bias is amorphous, while an ECR-Al_2O_3 layer deposited at 400°C with a bias ≥ 140 V is polycrystalline γ-Al_2O_3. This paper examines the effect of H content on luminescence from the amorphous phase and therefore the sample was left floating and a deposition temperature of only 150-160°C was used in this experiment.

EXPERIMENT

Both Er-doped and undoped amorphous alumina films (300 nm thick) were deposited in order to measure the effect of Er on the hydrogen thermal-release energy. The films were deposited at 150-160°C onto Si substrates using an electron-beam evaporation source downstream from a horizontally positioned ECR plasma source. The ECR plasma allows for operation in the low-pressure regimes needed for the metal deposition. Microwaves (80-130 Watts) at a frequency of 2.45 GHz were injected into the ECR chamber to excite the O_2 gas, forming neutral atoms and molecular ions. The energetic O_2^+ (~30 eV) stream down from this plasma generation region onto the layer and substrate. The ECR system base pressure was ≈0.3x10^{-8} Torr, and the operating pressure was 3-5x10^{-5} Torr with an O_2 gas flow rate of 2.5 sccm. Evaporation rates for Al and Er were measured simultaneously using two quartz crystal monitors, and films were grown at 0.2 nm/s for Al while simultaneously evaporating Er at 0.01-0.03 nm/s.

Hydrogen from the chamber walls is introduced into the films during deposition as a result of the interaction between the energetic plasma and the chamber. The hydrogen content of the films was measured, using elastic recoil detection, as a function of isochronal annealing treatments. This data was fit to a simple trap-release model in order to determine an effective activation energy for the thermal release of H from alumina. The composition of the films was determined using 2.8 MeV He$^+$ Rutherford Backscattering spectrometry (RBS) at a scattering angle of 164°, and by using 16 MeV Si^{+3} Elastic Recoil Detection (ERD) at a scattering angle of 30°. Two isochronal annealing times were examined in this experiment in order to test the effects of kinetics on the trap-release model: 30 min. and 4 hrs. The samples were annealed in a vacuum ≤ 5x10^{-8} Torr over the temperature range of 400 to 700°C.

The intensity of ion-beam stimulated luminescence (IL) from the samples annealed for 4 hrs. was then monitored at room temperature in the visible and near infrared regions as a function of the thermal treatments. The IL spectra examined in this paper were obtained from samples containing 4 at.% Er. A 2.8 MeV He$^+$ ion beam was used as an IL excitation source with the beam incident at an angle of 45° from the sample normal. The range of these ions in Al_2O_3 is 6.4 μm, and therefore the He$^+$ ions traverse the 300 nm thick films coming to rest deep in the Si substrate and lose energy in the films primarily through electronic stopping. The emitted IL light was focussed into an optical fiber bundle using a lens 1.25 cm from the sample and with the lens optic axis oriented along the sample normal. The collected light was guided into an Instruments SA HR320 monochromator containing a 150 lines/mm grating and the IL spectrum was then detected with a thermoelectrically-cooled CCD area-array photodetector. The resolution of this detection system was ≅ 2 nm. This equipment allowed six samples to be mounted and analyzed without breaking vacuum or changing the excitation/light-collection geometry, thereby allowing relative changes in emission intensity to be examined. The incident ion current was held constant at 200 nA/cm^2 during the ionoluminescence experiments. The IL spectra were collected for times of 10 s and 120 s at repeated intervals over the course of a day in order to examine possible beam-damage effects. The intensity of the IL spectra was unaffected by ion beam fluences up to 3x10^{15} He/cm^2.

RESULTS AND DISCUSSION

As stated above, the samples examined in this experiment were amorphous alumina containing hydrogen incorporated during the deposition. Figure 1 shows the H composition-depth profile from the 4 at.% Er-doped alumina in the as-deposited condition (solid line), and after annealing for 30 min. at: 400°C, (O), 500°C (■), 600°C (dashed line), and 700°C (+). The large hydrogen peak shown at a depth of approximately 300 nm corresponds to the H remaining on the Si substrate after the predeposition cleaning treatment. Separate experiments showed that hydrogen was incorporated into the film as a result of O_2 plasma interactions with the chamber walls even when the Si substrate had a negligible surface hydrogen peak. These profiles show that the H profile changes very little after annealing at 400°C but does change significantly after annealing at 500°C. At 500°C, the Si-interface hydrogen peak and the alumina-surface hydrogen peak are both decreased. In addition, an apparent concentration gradient is established within the alumina layer. The sample annealed at 600°C shows a similar concentration gradient but for a lower H content. This gradient could result from re-trapping of hydrogen in the film, slowing the thermal release of H, and thereby modifying the apparent activation energy. Alumina is a fairly ionic material and therefore as the hydrogen is released from the sample and the local electronic density is modified, the structure of the amorphous alumina is expected to rearrange in order to maintain charge neutrality of the matrix. The

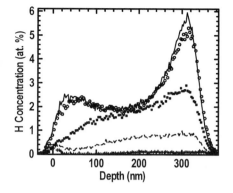

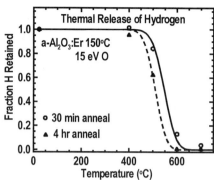

Fig. 1. Hydrogen profiles determined for the as-deposited (solid line) amorphous Al_2O_3 containing 4 at.% Er, and after annealing for 30 min. at: 400°C (O), 500°C (■), 600°C (dashed line), and 700°C (+).

Fig. 2. Fraction of H retained after isochronal annealing treatments. The solid and dashed lines were determined using the model given in the text with E_{eff} = 2.7 eV.

effects of local relaxation and atomic rearrangements may be one method for slowing hydrogen through re-trapping. Thus the activation energy for thermal release of H determined in these experiments will be an effective activation energy (E_{eff}) which also reflects the effects of re-trapping. After a 700°C anneal, the sample contains very little hydrogen within the film but does retain a small surface H peak. Similar H concentration-depth profiles were obtained for undoped alumina films (not shown) but with a smaller H concentration than shown for the Er-doped film of Fig. 1. The annealing behavior of the undoped film was the same as that of the Er-doped film.

Previously, the thermal release of H from ECR-deposited SiO_2 films was characterized [3] and a distributed trap energy model was developed to determine the average energy for release of hydrogen. The model accurately described the energetics for the thermal stability of H and predicted the release of H from SiO_2 and fluorinated-SiO_2. This model will be used in this paper to determine the effective thermal release energy of H from ECR-deposited alumina. Re-trapping of H will be neglected to first order, but experiments measuring the release of H as a function of time at a given temperature will be examined below to check the magnitude of the effect from that assumption. Figure 2 shows the experimentally determined total hydrogen concentrations as a function of annealing temperature for two isochronal annealing times: O 30 min. and ▲ 4 hrs. The data are given for the integrated H concentration over the total depth of the sample and are shown as individual data points. The lines (dotted and solid) are determined for each of these samples using the initial, as-deposited H concentration, and the effective activation energy in the model described below. This figure shows that the onset temperature for release of hydrogen for different isochronal annealing times is approximately the same but the temperature to reach 50% retained H is lowered by ≈50°C for the longer annealing time (8 times longer).

For a single trap energy, the probability (Π) of H thermally escaping a trap at an annealing temperature T is given by an attempt frequency times a Boltzmann factor: $\nu \exp(-E_B/kT)$, where ν is the order of 10^{13} s^{-1} for thermal vibrations. After having filled the traps to saturation (assumed to occur during deposition and supported by the fact that annealing below the deposition temperature does not alter the H concentration), the release of H from a trap with energy E_B is governed by a first order rate equation in which the concentration of H in traps at time t is: $d[H]/dt = -\Pi [H]$. The solution to this equation is:

$$[H(T)] = [H_0(T)] \exp(-\nu t \exp(-E_B/kT)), \qquad (1)$$

where [$H_o(T)$] is the initial H concentration before annealing at temperature T. If a single energy is considered for E_B, then the curves for the thermal release of H as a function of annealing temperature would appear as a step function in which nearly all of the H is released at one temperature. A distribution of trap energies would cause these curves to be rounded and produce a sum over the exponential term in (1). In order to determine the breadth of the possible distribution of trap energies, a Gaussian distribution of nine trap energies, E_B, centered around E_{eff} was used to represent the possible distribution of bonds and state of relaxation of the amorphous Er-doped Al_2O_3. The standard deviation of the distribution, $\sigma=0.1$ eV, was minimized to obtain a good fit to the data. The effective activation energy for thermal release of H was determined by fitting a sum over equations of the form shown in (1) to the data using the measured initial H concentration and t=1800 s an t=14400 s. The E_{eff} determined for both the 30 min. and 4 hr. annealing treatments shown in fig. 2 was 2.7 eV. This activation energy is the same as that determined for undoped a-Al_2O_3. Thus, the large quantities of Er in the samples used in this experiment has a negligible effect on the effective release energy for hydrogen.

The validity of neglecting H re-trapping is also born-out by the fact that both annealing times were fit by the same activation energy. The annealing times were chosen for this experiment to increase the possible effective diffusion distance by more than a factor of 4, if a diffusion limited process was predominant. Figure 2 shows that further loss of H with extended annealing can occur, although the amount of loss is small. Moreover, the data for the extended anneals can still be fit well with an activation energy of 2.7 eV. Equation (1) shows that the dependence on time in determining the effective activation energy for thermal release of H from ECR-deposited oxides is smaller than the dependence on temperature, and therefore the value of E_{eff} is changed very little by using longer annealing times. The reduced temperature to reach 50% retained H is therefore modeled well by the time dependence invoked in this model.

Once the H thermal-release behavior was characterized, the effects on the IL spectra of varying the H content from thermal annealing was examined. The intensity of the ion-beam stimulated luminescence from these samples was monitored in the visible and near infrared regions as a function of the thermal treatments. Figure 3 shows the visible region of the IL spectra at 297 K from samples annealed for 4 hr. at 400°C (solid line), 500°C (Δ), 600°C (dashed line), and 700°C (+). Respectively, these samples contain:

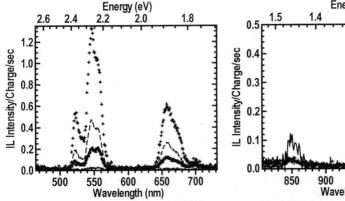
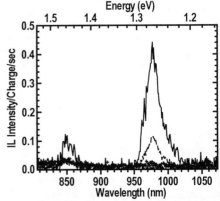

Fig. 3. Visible region of the IL spectra at 297 K from samples annealed for 4 hr. at 400°C (solid line), 500°C (Δ), 600°C (dashed line), and 700°C (+). Respectively, these samples contain: 1.81 at.% H, 1.35 at.% H, 0.02 at.% H, and 0 at.% H.

Fig. 4. Near Infrared region of the IL spectra for the samples annealed at 500°C (+), 600°C (dashed line), and 700°C (solid line). The atomic transitions giving rise to all peaks (Fig.3 & 4) are identified in the text.

1.81 at.% H, 1.35 at.% H, 0.02 at.% H, and 0 at.% H. Figure 4 shows the near infrared region of the IL spectra for the samples annealed at 500°C (+), 600°C (dashed line), and 700°C (solid line). Overall, figs. 3 and 4 show that the luminescence intensity increased with decreasing H content (increasing temperature for thermal annealing). The atomic transitions for Er^{+3} in Al_2O_3 which give rise to the peaks in these figures were identified by comparison to possible fluorescing levels cited in the literature for the free ion and for Er^{+3} in Y_2O_3, $ErCl_3$ and Er_2O_3 [4-6]. The peaks in fig. 3 correspond to the following atomic-level transitions: $^2P_{3/2}$ to $^4I_{9/2}$, at 524 nm; $^4S_{3/2}$ to $^4I_{15/2}$, at 546 nm; $^2H_{9/2}$ to $^4I_{13/2}$, at 556 nm; $^4F_{9/2}$ to $^4I_{15/2}$, at 659 nm; $^2G_{9/2}$ to $^4I_{9/2}$, at 668 nm; and $^2H_{9/2}$ to $^4I_{11/2}$, at 697 nm. The peaks in fig. 4 correspond to the following transitions: $^4S_{3/2}$ to $^4I_{13/2}$, at 846 nm; $^4F_{7/2}$ to

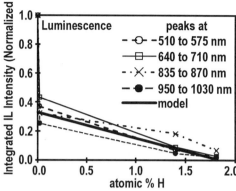

Fig. 5. The integrated IL signal from four regions of the visible and near IR regions are plotted as a function of hydrogen content in the samples (symbols connected by dashed lines). A linear fit of data is shown as a heavy solid line.

$^4I_{11/2}$, at 964 nm; and $^4I_{11/2}$ to $^4I_{15/2}$, at 988 nm. Luminescence from the well known $^4I_{13/2}$ to $^4I_{15/2}$ radiative transition (1.54 μm infrared light) was not examined in this experiment. The peak shapes and relative peak heights changed little as a function of annealing whereas the intensity of each peak increased significantly with annealing temperature. The constant peak shapes indicate that the modification to the sample by thermal annealing and loss of hydrogen alter the local environment of the Er in the alumina matrix very little, however the increased intensity indicates that the optical activity of Er^{+3} and radiative lifetimes are increased. Similar results for intensity and peak shape are expected for the $^4I_{13/2}$ to $^4I_{15/2}$ transition.

In order to gain a better understanding of the influence of hydrogen, the integrated ionoluminescence intensity from samples containing different levels of hydrogen was plotted as a function of H content. There were nine fluorescent transitions observed in the wavelengths ranging from visible to near infrared. Not all of the IL peaks arising from these transitions were easily separable and therefore the integrated intensity was determined for groups of closely spaced transitions. In this way, four integrated IL intensities were measured, and then each was normalized by the integrated intensity from the sample containing 0 at.% H and plotted as shown in fig. 5. First, this figure shows that the integrated intensity for each group of peaks drops quickly with the introduction of hydrogen into the sample. Second, the introduction of H into the sample causes the normalized IL signal to decrease linearly with the increase in H content, for each of the integrated peaks. The data were fit with a simple linear equation: I/Io = (1-n*H/Er)-K, (2) where H/Er is the ratio of hydrogen atoms to Er atoms in the sample, n is the fraction of H which reduces the Er optical activity, and K is a constant related to a reduction in intensity resulting from an increased non-radiative lifetime when even a small amount of hydrogen is present in the sample. A fit of the data to equation (2) gives a value of n=0.72 for all of these peaks, demonstrating the strong dependence of the luminescence efficiency on the hydrogen content. The fact that n=0.72 for a wide range of luminescence transitions suggests that the reduction in optical activity of an Er ion in alumina resulting from the presence of a H atom in alumina is nearly one to one. One mechanism for loss of energy from the radiative transitions is through phonon assisted non-radiative transitions. A decreased integrated IL intensity with increased H concentration is consistent with a transfer of energy to O-H infrared vibrationl modes or even

Er-H vibrational modes. Further experiments will determine if the relationship given in (2) holds for Er in different phases of alumina and for samples containing lower Er concentrations.

CONCLUSIONS

Optical performance of Er in alumina can be modified by controlling the phase and hydrogen content of the host matrix. Ion irradiation and deposition temperature during growth modify both of these properties by producing: 1) high H-content amorphous alumina at low deposition temperatures without simultaneous ion irradiation; 2) low H-content amorphous alumina at high deposition temperatures without simultaneous ion irradiation; and 3) low H-content γ-alumina at high deposition temperatures with simultaneous ion irradiation. One method to release all of the hydrogen from the low-temperature deposited alumina is to anneal the material at 700°C for 4 hrs. The effective activation energies for thermal release of H from undoped amorphous alumina or Er-doped amorphous alumina are the same and equal to 2.7 eV. By thermal annealing, the H content of Er-doped amorphous alumina decreases which causes an increase in the ionoluminescence intensity. This increased luminescence can be modeled by a simple linear relationship: $I/Io = (1-0.72*H/Er)-K$, where H/Er is the atomic fraction of hydrogen to erbium and K is an offset resulting from an increased non-radiative lifetime for the Er in the presence of hydrogen. This equations shows that the fraction of H which reduces the Er optical activity is nearly one-to-one. Therefore, each additional hydrogen atom in an optical device is very efficient at curtailing the optical efficiency of a device and thus synthesis and processing methods are required which can eliminate the incorporation of hydrogen for processing temperatures below 700°C.

REFERENCES and ACKNOWLEDGMENTS

We thank K. Minor and D. Buller for their technical assistance. Sandia is a multiprogram laboratory operated by Sandia Corporation, a Lockheed Martin Company. This work was supported by the U.S. DOE under contract DE-AC04-94AL85000: Office of Basic Energy Sciences (Division of Materials Sciences/Metals and Ceramics) and a Laboratory Directed Research and Development Program.

[1] J.C. Barbour, B.G. Potter, D.M. Follstaedt, J.A. Knapp, and M.B. Sinclair, in Materials Modification and Synthesis by Ion Beam Processing, edited by D. Alexander, B. Park, N. Cheung, and W. Skorupa (Mater. Res. Soc. Proc. 438, Pittsburgh, PA, 1997) p. 465.
[2] J.C. Barbour, D.M. Follstaedt, and S.M. Myers, Nucl. Instrum. Methods B 106, 84 (1995).
[3] J.C. Barbour, C.A. Apblett, D.R. Denison, and J.P. Sullivan, in Silicon Nitride and Silicon Dioxide Thin Insulating Films, edited by M.J. Deen, W.D. Brown, K.B. Sundaram, and S.I. Raider (The Electrochem. Soc. Proc. 97-10, Pennington, NJ, 1997) p. 520.
[4] H.M. Crosswhite and H.W. Moos, Optical Properties of Ions in Crystals, (Interscience Publishers, John Wiley & sons, New York, NY, 1967) 552 pp.
[5] G. H. Dieke, Spectra and Energy Levels of Rare Earth Ions in Crystals, (Interscience Publishers, John Wiley & sons, New York, NY, 1968) 401 pp.
[6] C. A. Morrison and R. P. Leavitt, in Handbook of the Physics and Chemistry of Rare Earths, Vol. 5, edited by K. A. Gschneider, Jr. and L. Eyring (North-Holland Publishing Co., New York, NY, 1982) p. 461.

Vanadium Oxide Precipitates in Sapphire Formed by Ion Implantation

Hiroaki Abe [1,2,*], Hiroshi Naramoto [2] and Shunya Yamamoto [2]
1. Department of Physics and Astronomy, Arizona State University, Tempe, AZ 85287-1504
2. Department of Materials Development, Japan Atomic Energy Research Institute-Takasaki, Watanuki 1233, Takasaki, Gunma 370-12, Japan

ABSTRACT

Some vanadium oxides undergo phase transformations which give rise to attributable for large variations in optical and electronic properties. Since one can expect dynamic rearrangement of implanted species and a distorted lattice due to implantation, we investigated the precipitation process under ion implantation at high temperature. Sapphire samples were implanted with 300-keV V^+ ions at temperatures from 470 to 1070 K in a transmission electron microscope interfaced with ion accelerators. Evolution of vanadium oxide precipitates was observed simultaneously. Damage evolution such as dislocation loops and voids were observed at fluences of the order of 10^{18}-10^{19} and 10^{20} ions/m^2, respectively. At implantation fluence of the order of 10^{21} ions/m^2, dot and plate contrast was observed in addition to radiation damage. Electron diffraction analysis reveals that hexagonal and monoclinic V_2O_3, tetragonal and monoclinic VO_2, and V_7O_{13} precipitates were formed in the substrate depending on the surface normal of the substrate. Some of precipitates were thermally unstable phases. Crystallographic relationship between matrix and the precipitates was investigated as well as the swelling effect both in the substrate and in the precipitates. Temperature dependence reveals precipitation starts at temperature higher than 670 K.

INTRODUCTION

Vanadium oxide has more varieties in its chemical forms than any other transition metal oxides due to the relatively large number of possible valence states. It has been shown that some of vanadium oxides show abrupt changes in some physical properties, such as electrical conductivity, magnetic susceptibility and optical properties at transition temperatures (T_t). In VO_2 the transition is a first-order semiconductor-to-metal (monoclinic-to-tetragonal) transition with $T_t = 341K$ [1,2]. The substance is expected to be useful for the application of optical switching, modulators and polarizers at submillimeter-wavelength region [3]. On the other hand, V_2O_3 undergoes a first-order insulator-to-metal transition (monoclinic-to-corundum) with T_t =150-162K [4,5].

Investigations have been made to form buried vanadium oxide precipitates by ion implantation into sapphire, and have revealed various kinds of vanadium oxides in transparent sapphire substrate [6,7]. Experiments were successful to produce V_2O_3 [6] and VO_2 [7] precipitates in sapphire by co-implantation with vanadium and oxygen ions at ambient temperature followed by annealing at temperatures between 973 and 1273 K. Due to implantation-induced defects, amorphization in substrate material (sapphire) was observed at 36K [8], while no amorphization was detected at room temperature up to the fluence of 3.5×10^{20} ions/m^2. The results are consistent with the study of irradiation-induced amorphization in sapphire [9], which shows that the critical dose for amorphization depends on the irradiation temperature, and is about 4 and 8 dpa (displacement per atom) at temperature lower and higher than 130 K, respectively. Also the critical temperature for amorphization varies from 200 to 250 K depending on ion flux. The annealing behavior of the implanted sapphire [8] was described as follows; (1) damage recovery in Al sublattice is grater in sapphire implanted to higher doses, (2) implanted species (V ions) are partially stabilized at substitutional position at Al sublattice and some are pushed out to surface due to the regrowth process from the undamaged region. At annealing temperatures above 973K, annihilation of F-centers (oxygen vacancies) and V-centers (aluminum vacancies) has sufficiently occurred [10]. Irradiation at this temperature also showed void-superlattice formation [11,12] indicating vacancy migration and it induces swelling up to about 4% [12] depending on irradiation condition, but much larger than thermal expansion (~0.7% from RT to 1073K). Therefore, it is quite reasonable that one expects dynamic recovery of defects, void formation, swelling and other effects during implantation at high temperature. We will report a preliminary result on the precipitation of vanadium oxides in sapphire implanted at high temperature. Crystallographic orientation and swelling due to implantation will be revealed.

EXPERIMENTS

Single crystals of sapphire wafers (c-, a- and r-planes) were supplied by Kyocera Co Ltd. They were cut into 3 mm diameter disks with an ultrasonic cutter, and mechanically thinned to get thickness between 80 and 100 μm. The disks were dimpled to get minimum thickness of 20 μm. The thinning and dimpling processes were made until the optical grade polish was achieved. The disks were then subjected to ion thinning with 7 keV Ar$^+$ ions whose incident angle was 20° from the surface of disks until perforation occurred (about 12 hrs). The final polishing was performed for 1 h with 4 keV Ar$^+$ ions whose incident angle was 12°.

Alumina thin films were also used in this work. The samples were prepared by electron-beam deposition technique from 99.99% Al$_2$O$_3$ bulk materials. Thickness of films was about 160 nm estimated from measurement with quartz oscillators, which were calibrated with Rutherford backscattering spectroscopy (RBS) measurements, about 4 cm away from the sample. Thickness was also confirmed by electron energy loss spectrum [13]. The effect of radiation damage will be expected much more than implantation effects since its thickness is comparable to the range of ions which will be described later.

The electron-transparent samples were subjected to implantation with 300 keV V$^+$ ions in the TEM-Accelerators Facility at JAERI-Takasaki [14], so as to simultaneously observe the microstructural evolution due to the implantation and without changing the temperature. Microstructural evolution was observed through bright-field and center-dark-field image as well as electron diffraction from the area of 0.2 μm in diameter on a specimen. Current and flux of ions, measured by the Faraday cup (250 μmφ) at the sample position, were 1-3 nA and 1-3x10^{17} ions/m^{2}s, respectively. The implantation temperatures were 470, 670, 870 and 1070 K, monitored with a thermocouple about 2 mm away from samples. The estimated beam heating assuming simple geometry was about 10 to 50K during implantation. However, implantation also arises crack formation and bend of the sample. The phenomena were much severe in a-plane and r-plane sapphire presumably due to anisotropy in thermal expansion and mechanical properties. Therefore, changes in sample thickness, in crystallographic orientation and in beam heating effect must be expected during implantation.

The range of ions was calculated using the TRIM-95 [15] code assuming a displacement energy of 40 eV for all species. The range and straggling of 300 keV vanadium ions in sapphire are 154 nm and 42 nm, respectively.

RESULTS AND DISCUSSION

Tiny defects, typically 3 nm in diameter or less, were observed in the as-ion-milled sample as shown in figure 1 (a). The growth of the defects was not observed at 1070 K. However, due to irradiation with electrons, some of defects grew so as to show black-and-white contrast in thick regions, as shown in the upper right corner. In contrast, voids were formed in thin regions, which is previously reported [16,17]. Due to ion implantation up to 0.43 dpa (displacement per atom), some of the tiny defects grew into black contrast features but majority of the defects were annihilated as shown in figure 1 (b). The area and volume density of the defects were roughly estimated as 1.1x10^{15} m^{-2} and 7.3x10^{21} m^{-3}, respectively, hereafter the range of ions was taken as thickness for the estimation of the volume density. Right and middle side part of the figure is the area irradiated with electrons prior to the ion implantation. High density but small defect clusters are obvious, presumably because the high density of nucleation was achieved due to the electron irradiation. Void contrast, as mentioned above, was also observed in the lower part of the figure. With increasing ion fluence, as shown in figure 1 (c), defect clusters grew into dislocation loops whose density was evaluated as 5.4x10^{14} m^{-2} and 3.8x10^{21} m^{-3}. The loops show edge-on or double-arc contrast whose fractions are roughly estimated as 80 % and 20 %, respectively. The former one is of interstitial-type loops on the habit plane of (0001) with Burger's vector $\mathbf{b}$=1/3[0001] [11]. The latter one, on the contrary, is presumed to be of interstitial type loops with the habit plane of {10$\bar{1}$0} and Burger's vector $\mathbf{b}$=1/3<10$\bar{1}$0> [11]. The figure 1 (c) was taken slightly overfocused condition so that voids show black dot contrast, whose size and density are about 2.8 nm, 4.7x10^{16} m^{-2} and 3.0x10^{23} m^{-3}.

respectively. They distributed randomly in the figure, though aligned along the c-axis, quite similar to the void superlattice [12] and bubble arrays [11]. Precipitation of vanadium oxide was detected in electron diffraction patterns (described later) with the ion fluence of the order of 10^{21} ions/m^2. Figure 1 (d) shows the center-dark-field image of the sample. Here, the objective-lens aperture was set on one of the precipitate-related reflections which were so close to the matrix's ones that both could not be divided. Two-and-half dimension technique [18] revealed that the white dot contrast was attributable to the implantation-induced precipitates. The size and the density of the precipitates were between 5.5 and 12.5 nm in diameter, 2.2×10^{15} m^{-2} and 1.4×10^{22} m^{-3}, respectively. In the figure both tangled dislocations which have weak contrast and void superlattice were also observed.

Figure 2 shows change in diffraction patterns in 3 kinds of sapphire implanted with vanadium ions at 1070 K. Spots due to precipitation can be seen. The result shows that the precipitates have the identical orientation relationship depending on the sample. To index the spots related to the precipitates, we took into account the plane distance and angles for most of all the possible vanadium oxides, aluminum oxides, aluminum-vanadium oxides and both metals, except for V_4O_7 whose crystallographic information was not available. The estimated experimental error is no more than 0.1 % in plane distance measurement and about 0.3 degrees in plane angle measurement. The results are summarized in table 1 together with the lattice expansion or compression due to implantation which will be described later.

The vanadium oxides formed were V_2O_3, VO_2 and V_7O_{13}. Some of them, hexagonal V_2O_3 (h-V_2O_3), tetragonal VO_2 (t-VO_2) and V_7O_{13} are the only phases stable at 1070K. V_7O_{13} belongs to the triclinic V_nO_{2n-1} Magnéli series with n=4, 5, 6, 7 and 8, though no sign of other oxides was detected. Monoclinic V_2O_3 (m-V_2O_3) and m-VO_2 are the low temperature phase which does not exist in the V-O system at temperatures examined in this work. Transition temperatures of the phases are 341 K and 150-162K for monoclinic-to-tetragonal transition in VO_2 and monoclinic-to-hexagonal transition in V_2O_3, respectively. In a-plane sapphire, t-VO_2 phase was the only phase detected. The precipitates show moiré fringes which is formed by lattice mismatch (~14%) and orientation (2.5°). Compression of the substrate was detected only in this case. In r-plane sapphire, two or three kinds of precipitates were observed; i.e. t-VO_2, V_7O_{13} and m-V_2O_3, latter two of which were not distinguished in this experiment because of the quite similar diffraction patterns. It is more likely to determine V_7O_{13} rather than m-V_2O_3, because V_7O_{13} is much more stable than m-V_2O_3 in the phase diagram and from swelling data as shown in table 1. Tetragonal VO_2 phases both in a-plane and in r-plane sapphire were recognized as different type. Effects of surface, such as sublimation and lattice relaxation, are presumed to form the different precipitates in a- and r-plane sapphire. The fluence (6.5×10^{20} ions/m^2) at which precipitation was observed in a-plane Al_2O_3 was almost half of that in the others (1.2-2.0×10^{21} ions/m^2), so the fluence effect is the another factor to be concerned in the precipitation process.

In c-plane sapphire, two kinds of precipitation were observed; (1) m-VO_2 phase and (2) m-VO_2 and h-V_2O_3 phases. They result in a large difference in swelling rate. The diffraction pattern indicates m-VO_2 precipitates have strong g(VO_2)=002 which is parallel to g(Al_2O_3)=11$\bar{2}$0, but weak diffraction of g(VO_2)=400 which is 2.5° away from g(Al_2O_3)=$\bar{1}$2$\bar{1}$0. As shown in table 1, the c-plane sapphire that has both VO_2 and V_2O_3 precipitates is thought to be more stable because of the lower swelling rate, which is comparable to the thermal expansion (~0.7%). The possible reason for the difference is the ion flux and fluence. The implantation condition for the substance which has VO_2 and V_2O_3 was higher flux and higher fluence; 4.1×10^{17} ions/m^{2}s and 2.0×10^{21} ions/m^{2}s. On the contrary, the flux and fluence of ions for the other one are 1.1×10^{17} ions/m^{2}s and 1.5×10^{21} ions/m^2, respectively. High flux implantation is attributable for enhancement of dynamic rearrangement of the species due to beam heating. Also we can suspect precipitation of V_2O_3 starts at rather high fluence and/or high temperature. Further investigation is required to clarify the precipitation process.

Temperature dependence of the precipitation process is examined in c-plane sapphire. Damage accumulation was observed at temperatures examined, though size and density of the defects are temperature dependent. At 473 K, void formation was observed but precipitation of V-oxide was

Table 1. Results of 300 keV V^+ ion implantation in sapphire at 1070K. Crystallographic orientations between sapphire and observed vanadium oxide and amount of swelling both of sapphire and of V-oxide are shown. Hexagonal (corundum), monoclinic and tetragonal phases are denoted as h, m and t, respectively. See text for details.

sapphire	observed oxide	crystallographic relationship		swelling (%) sapphire / V-oxide
a-plane	t-VO$_2$	$(11\bar{2}0) \mathbin{/\mkern-5mu/} (\bar{1}30)$,	$[0006]$ 2.5° $[313]$,	-0.6 / +0.5
			$[1\bar{1}02]$ 2.5° $[310]$ [2]	
r-plane	t-VO$_2$ [1]	$(20\bar{2}1) \mathbin{/\mkern-5mu/} (001)$,	$[\bar{1}2\bar{1}0] \mathbin{/\mkern-5mu/} [400]$	+2.0 / +3.5
	m-V$_2$O$_3$ [1,3]	$(20\bar{2}1) \mathbin{/\mkern-5mu/} (\bar{3}24)$,	$[\bar{1}10\bar{2}] \mathbin{/\mkern-5mu/} [211]$	+2.0 / -2.5
	V$_7$O$_{13}$ [1,3]	$(20\bar{2}1) \mathbin{/\mkern-5mu/} (401)$,	$[\bar{1}10\bar{2}] \mathbin{/\mkern-5mu/} [104]$	+2.0 / -0.5
c-plane	m-VO$_2$	$(0001) \mathbin{/\mkern-5mu/} (010)$,	$[11\bar{2}0] \mathbin{/\mkern-5mu/} [002]$	+2.5 / +0.3
c-plane	m-VO$_2$ [1]	$(0001) \mathbin{/\mkern-5mu/} (305)$,	$[11\bar{2}0] \mathbin{/\mkern-5mu/} [040]$	+0.8 / +2.3
	h-V$_2$O$_3$ [1]	$(0001) \mathbin{/\mkern-5mu/} (1\bar{1}02)$,	$[30\bar{3}0] \mathbin{/\mkern-5mu/} [11\bar{2}0]$	+0.8 / +0.9

1) Both phases were observed.
2) Angles between two directions are shown.
3) Either or both oxides are formed since both oxides could not be distinguished.

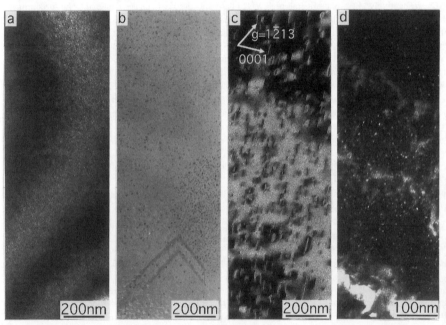

Figure 1. Bright field image (a)-(c) and center-dark-field image (d) of r-plane sapphire implanted with 300 keV V^+ ions at 1070K. Ion fluences are (a) 0, (b) 4.9x10^{18}, (c) 2.5x10^{20} and (d) 1.2x10^{21} ions/m^2, respectively.

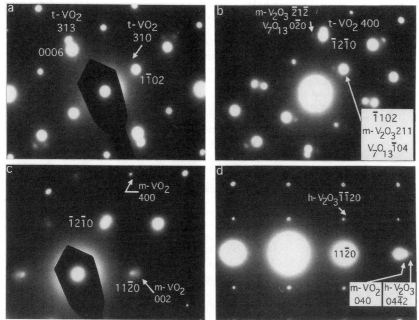

Figure 2. Selected-area electron diffraction patterns of sapphire; (a) a-plane, (b) r-plane and (c,d) c-plane. Implantation fluences are (a) 6.5×10^{20}, (b) 1.2×10^{21}, (c) 1.5×10^{21} and (d) 2.0×10^{21} ions/m^2.

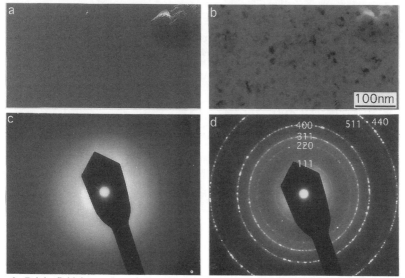

Figure 3. Bright-field image (a and c) and selected-area electron diffraction patterns (b and d) of electron-beam-deposited alumina. Implantation fluences are 0 for (a) and (b), 1.3×10^{21} ions/m^2 for (c) and (d), respectively.

not detected up to the fluence of 1.8×10^{21} ions/m^2. At temperature above 673 K, both voids and precipitates were formed. The observed precipitates were m-VO_2.

Thin films of electron-beam-deposited alumina were irradiated with 300 keV V$^+$ ions. Thickness of the films was comparable to the ion range. But presumably due to sublimation or evaporation from the surface, no sign of precipitation was detected. The film readily transformed into polycrystal gamma-phase alumina when we started irradiation ($<5 \times 10^{18}$ ions/m^2). Further irradiation was attributable for the grain growth. Figure 3 shows the change in microstructure (a, c) and electron diffraction (b, d) in the alumina thin film. Precipitation of γ-alumina is characteristic of the deposited thin films but was not observed in the crystalline sapphire samples at all temperature examined in this work. The swelling of the implanted area was 1.3 %, while it was from 0.3 % to 1.1 % in the non-implanted area kept at the same temperature for the same time, where precipitation of γ-alumina was also observed. Both of damage effects and implantation effects were suspected because of the comparable thickness to the ion range.

CONCLUSIONS

We have examined vanadium implantation in sapphire at high temperature in the microscope interfaced with the accelerator. Damage evolution, void formation and V-oxide precipitation were observed as increasing implantation fluence. Crystallographic orientation and swelling were extensively investigated. Precipitation depends strongly on the surface normal of the substrate, and some of the precipitates were low temperature phases thermally unstable. The swelling due to implantation varies from -0.6 to 2.5 %, where only a-plane sapphire shows compression.

One of the authors (HA) acknowledges Dr. J. M. Zuo at Arizona State University for the fruitful discussion.

REFERENCES
* Corresponding author, e-mail: abe@taka.jaeri.go.jp
[1] F.J. Morin, Phys. Rev. Lett. 3 (1959) 34.
[2] H.W. Verleur, A.S.Barker and C.N. Berglund, Phys. Rev. 172 (1968) 788.
[3] J.C.C. Fan, H.R.Fetterman, F.J. Bachner, P.M. Zavracky and C.D. Parker, Appl. Phys. Lett. 31 (1977)11.
[4] D.B. McWhan, A. Menth, J.P. Remeika, W.F. Brinkman and T.M. Rice, Phys. Rev. B7 (1973) 1920.
[5] D.B. McWhan and J.P. Remeika, Phys. Rev. B2 (1970) 3734.
[6] L.A. Gea, L.A. Boatner, J. Rankin and J.D. Budai, Mater. Res. Soc. Symp. Proc. 354 (1995) 268.
[7] L.A. Gea and L.A. Boatner, Appl. Phys. Lett. 68 (1996) 3081.
[8] H. Naramoto, Y. Aoki, S. Yamamoto and H. Abe, Nucl. Instrum. Methods B127/128 (1997) 599.
[9] H. Abe, S. Yamamoto and H. Naramoto, Nucl. Instrum. Methods B127/128 (1997) 170.
[10] G.P. Pells, J. Am. Ceram. Soc. 77 (1994) 368.
[11] M.D. Rechtin, Rad. Eff. 42 (1979) 129.
[12] F.W. Clinard, G.F. Hurley and L.W. Hobbs, J. Nucl. Mater. 108&109 (1982) 655.
[13] R.F.Egerton, Electron Energy Loss Spectroscopy in the Electron Microscope p.293 (Plenum, New York, 1986).
[14] H. Abe, H. Naramoto, K. Hojou and S. Furuno, JAERI-Research 96-047 (1996).
[15] J.F. Ziegler, J.P. Biersack and U. Littmark, The Stopping and Range of Ions in Solids (Pergamon, Oxford 1985).
[16] Y. Tomokiyo, T. Manabe and C. Kinoshita, Microsc. Microanal. Microstruct. 4 (1993) 331.
[17] Y. Tomokiyo, T. Kuroiwa and C. Kinoshita, Ultramicroscopy 39 (1991) 213.
[18] W.L. Bell, J. Appl. Phys. 47 (1976) 1676.

ION BEAM SYNTHESIS of CdS, ZnS, and PbS COMPOUND SEMICONDUCTOR NANOCRYSTALS

C. W. WHITE,* J. D. BUDAI,* A. L. MELDRUM,* S. P. WITHROW,* R. A. ZUHR,*
E. SONDER,* A. PUREZKY,* D. B. GEOHEGAN,* J. G. ZHU,** and
D. O. HENDERSON***
*Oak Ridge National Laboratory, Oak Ridge, TN
**New Mexico State University, Las Cruces, NM
***Fisk University, Nashville, TN

ABSTRACT

Sequential ion implantation followed by thermal annealing has been used to form encapsulated CdS, ZnS, and PbS nanocrystals in SiO_2 and Al_2O_3 matrices. In SiO_2, nanoparticles are nearly spherical and randomly oriented, and ZnS and PbS nanocrystals exhibit bimodal size distributions. In Al_2O_3, nanoparticles are facetted and oriented with respect to the matrix. Initial photoluminescence (PL) results are presented.

INTRODUCTION

The unusual properties of compound semiconductor nanocrystals compared to bulk materials has stimulated considerable interest in exploring new methods to fabricate these materials [1-3]. Ion implantation is a very versatile technique which has been used recently to form a wide range of elemental and compound semiconductor nanocrystals encapsulated in a variety of host materials [4-5]. In this method, individual constituents of the desired compound are implanted sequentially at energies chosen to give an overlap of the profiles. If the individual constituents are insoluble in the matrix, then thermal annealing (or implantation at elevated temperatures) leads to precipitation, and the compound forms if the constituents have a strong chemical affinity for each other. In this paper, we demonstrate that this method can be used to form CdS, ZnS, and PbS nanocrystals encapsulated in matrices of SiO_2 and Al_2O_3. Nanocrystals of CdS have been fabricated previously by chemical methods in colloidal solution [6], by arrested precipitation [7], or encapsulated in glass through heat treatment of doped glasses [8,9] and their size-dependent optical properties have been investigated. ZnS nanoparticles have been prepared previously in colloidal solution [6] or powder form and when doped with impurities (Mn, Cu) have attracted interest recently as high-quantum efficiency luminescent materials [10-11]. PbS nanocrystals have been proposed as photocatalysts for energy applications and have been synthesized in inverse micells [12].

EXPERIMENTAL

Substrates used in this work were fused silica (SiO_2), thermally oxidized silicon wafers (SiO_2/Si) with an oxide thickness of ~850 nm, or c-axis oriented Al_2O_3 single crystals. To form compound semiconductor nanocrystals in these substrates, the individual constituents were implanted sequentially at energies chosen to give an overlap of the profiles. Implants were carried out at room temperature in SiO_2 and at 550°C in Al_2O_3 in order to keep the matrix crystalline during implantation. Stoichiometry is determined by the implanted dose. Both Gaussian profiles resulting from single energy implants as well as flat profiles resulting from multiple energy implants have been used. Following implantation, samples were annealed at 1000°C for 1 h in flowing Ar + 4% H_2 to induce precipitation and nanocrystal formation. Samples were characterized using a variety of techniques including Rutherford backscattering, x-ray diffraction (Cu-Kα), and transmission electron microscopy (TEM). PL spectra were obtained from selected samples using 457 nm or 248 nm excitation.

RESULTS

X-ray diffraction results demonstrating the formation of CdS nanocrystals in SiO_2 and Al_2O_3 are shown in Figs. 1 and 2. In SiO_2 (Fig. 1), diffuse scattering from SiO_2 is observed in addition to strong lines arising from CdS. In Al_2O_3, there are a multitude of diffraction lines arising from CdS, in addition to the expected diffraction from the matrix. The position and relative intensity of x-ray lines characteristic of the hexagonal wurtzite structure of CdS and the two possible cubic structures, zincblende (ZB) and rocksalt (NaCl), are indicated in Figs. 1 and 2 (from powder files). Detailed analysis of the x-ray diffraction results show that most of the CdS nanocrystals in both SiO_2 and Al_2O_3 have the hexagonal structure. In addition, the CdS nanocrystals in Al_2O_3 are oriented with the matrix having their (002) planes parallel to the c-planes of Al_2O_3. These nanoparticles exhibit strong in-plane alignment also. Therefore, oriented CdS nanoparticles with hexagonal structure are produced in Al_2O_3 if implantation is carried out at elevated temperatures. By contrast, we find that the cubic structure is produced if implantation is done at low temperature where we form the amorphous phase during implantation [13].

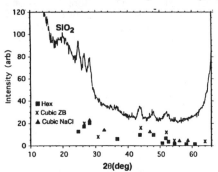 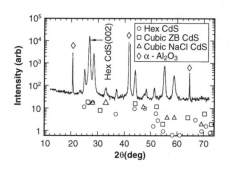

Fig. 1. X-ray diffraction showing the formation of CdS nanocrystals in SiO_2/Si. Equal concentrations (5.3×10^{21}/cm³) of Cd and S were implanted to a depth of ~180 nm.

Fig. 2. X-ray diffraction from CdS nanocrystals in Al_2O_3. Equal doses (4.3×10^{16}/cm²) of Cd (450 keV) and S (164 keV) were implanted at 550°C.

Figures 3 and 4 are x-ray diffraction results demonstrating the formation of ZnS (Fig. 3) and PbS (Fig. 4) in SiO_2 and Al_2O_3 as a result of implantation and annealing. ZnS has hexagonal and cubic structures, and the ZnS nanocrystals produced in SiO_2 and Al_2O_3 by ion implantation are a mixture of the two structures. In SiO_2, the nanocrystals are randomly oriented, but in Al_2O_3, they are oriented with respect to the matrix having the (002) planes of the hexagonal ZnS and the (111) planes of cubic ZnS parallel to the c-planes of Al_2O_3. In Al_2O_3, the ZnS nanoparticles also exhibit strong in-plane orientation.

Figure 4 shows that cubic PbS nanocrystals are formed in both SiO_2 and Al_2O_3 as a result of ion implantation and annealing. In SiO_2, the nanoparticles are randomly oriented, but in Al_2O_3, they are oriented with their (002) planes parallel to the c planes of Al_2O_3.

Figures 5 and 6 are cross-section TEM micrographs showing the microstructures in the near-surface region of SiO_2 and Al_2O_3 samples containing CdS, ZnS, and PbS nanocrystals. In Fig. 5, CdS nanocrystals are nearly spherical and have an average diameter that is slightly larger than 100 Å. Nanocrystals of smaller size are formed when the Cd + S concentration is reduced. By contrast, ZnS and PbS nanocrystals exhibit pronounced bimodal size distributions with a few large nanocrystals (several hundred angstroms diameter) in addition to a large number of smaller nanocrystals. The large nanocrystals form in well-defined spatial regions, being predominantly located in the vicinity of the end-of-range in the case of ZnS, and near the peak of the implanted profile for the case of PbS. These larger particles presumably develop as a result of diffusion

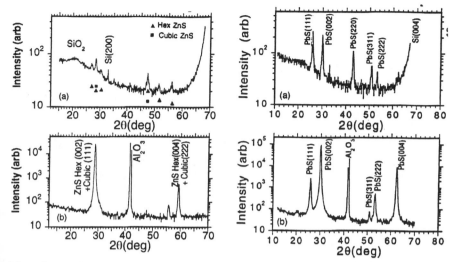

Fig. 3. X-ray diffraction from ZnS nanocrystals formed in SiO$_2$/Si (a) and Al$_2$O$_3$ (b). In (a) equal concentrations (7.5×10^{21}/cm^3) of Zn and S were implanted to a depth of 250 nm. In (b) equal doses (6×10^{16}/cm^2) of Zn (280 keV) and S (150 keV) were implanted at 400°C.

Fig. 4. X-ray diffraction from PbS nanocrystals formed in SiO$_2$/Si (a) and Al$_2$O$_3$ (b). In (a) equal doses (5×10^{16}/cm^2) of Pb (320 keV) and S (82 keV) were implanted. In (b) equal doses (5×10^{16}/cm^2) of Pb (850 keV) and S (180 keV) were implanted at 550°C.

controlled Ostwald ripening, but why they form in such localized regions is not understood and is currently under investigation.

High-resolution transmission electron micrographs showing CdS, ZnS, and PbS nanoparticles in Al$_2$O$_3$ are shown in Fig. 6. In contrast to the case of SiO$_2$ where nanocrystals are nearly spherical, in Al$_2$O$_3$ the nanocrystals exhibit pronounced faceting. The average size of these nanoparticles is also greater than in the case of SiO$_2$.

These nanoparticles give rise to strong optical absorption and in some cases strong photoluminescence. In this paper, we show some of the PL results. Figure 7 shows PL arising from CdS nanocrystals in SiO$_2$. After annealing (1000°/1 h), the peak in the PL spectra at 510 nm

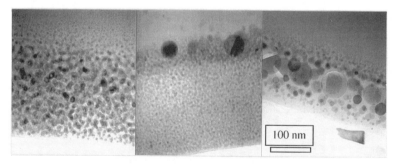

Fig. 5. Cross-section micrographs showing CdS (left), ZnS (center), and PbS (right) nanocrystals in SiO$_2$. Implantation and annealing condiitions are given in the captions of Fig. 1 (CdS), Fig. 3a (ZnS), and Fig. 4a (PbS).

Fig. 6. High-resolution cross-section micrographs showing ZnS (left), PbS (center), and CdS (right) nanocrystals in Al₂O₃. Implantation and annealing conditions are given in the captions of Fig. 2 (CdS), Fig. 3b (ZnS), and Fig. 4b (PbS).

is very close to the bandgap of CdS (2.43 eV) and we attribute this peak to recombination of electron-hole pairs in the nanocrystals. There is also a broad PL band at lower energies which probably arises from surface-related defects, as others [9] have discussed.

Figure 8 shows the PL spectra arising from CdS nanocrystals in Al₂O₃. In the as-implanted state there is little or no PL, but after annealing, there is intense bandedge PL arising from recombination of electron-hole pairs in the CdS nanoparticles. PL from the nanoparticles is blue shifted relative to that of a standard, and this may be the result of quantum confinement, but more likely results from lattice strain. In Al₂O₃, there is little or no PL at lower energies and this suggests that surface defects are relatively less important than when these particles are encapsulated in SiO₂ (see Fig. 7).

ZnS nanoparticles also give rise to strong PL as demonstrated in Fig. 9. In this case, the PL occurs at energies less than the bandgap, and this suggests that defects or surface states are responsible for the emission. The spectra is similar to that reported in ref. 12 for ZnS nanoparticles synthesized by chemical means where subbandgap PL was attributed to a deficiency of S in the nanoparticles.

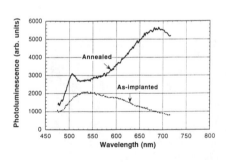

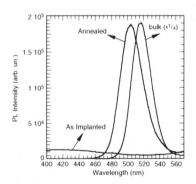

Fig. 7. PL spectra (excited at 457 nm) arising from encapsulated CdS nanocrystals in SiO₂ formed by the implantation of equal doses (1 × 10¹⁷/cm²) of Cd (450 keV) and S (164 keV).

Fig. 8. PL spectra (excited at 248 nm) from CdS nanocrystals in Al₂O₃. Implant conditions are the same as those in Fig. 2. PL spectra from bulk CdS is shown for comparison.

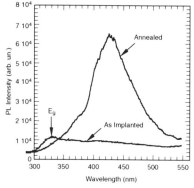

Fig. 9. PL spectra (excited at 248 nm) from ZnS nanocrystals in SiO_2. Equal concentrations ($1.5 \times 10^{21}/cm^3$) of each constituent were implanted to a depth of ~250 nm prior to annealing. E_g denotes the energy of the bulk bandgap of ZnS.

CONCLUSIONS

Sequential ion implantation and thermal annealing has been used to form CdS, ZnS, and PbS nanoparticles in SiO_2 and Al_2O_3. In SiO_2, the nanoparticles are nearly spherical and randomly oriented. Bimodal size distributions are observed for ZnS and PbS in SiO_2. In Al_2O_3, nanoparticles are oriented with respect to the lattice and when implantation is carried out at elevated temperature, CdS has the hexagonal structure. ZnS is a mixture of hexagonal and cubic structures and PbS has the cubic structure in both Al_2O_3 and SiO_2. Strong PL is observed from CdS nanoparticles in both SiO_2 and Al_2O_3. ZnS nanoparticles in SiO_2 show strong PL but at energies smaller than the bandgap.

ACKNOWLEDGMENT

Oak Ridge National Laboratory is managed by Lockheed Martin Energy Research Corp. for the U.S. Department of Energy under contract number DE-AC05-96OR22464.

REFERENCES

1. "Microcrystalline and Nanocrystalline Semiconductors," ed. by L. Brus, R. W. Collins, M. Hiroshi, F. Koch, and C. C. Tsai, Mat. Res. Soc. Proc. **358** (MRS, Pittsburgh, 1995).

2. "Advances in Microcrystalline and Nanocrystalline Semiconductors–1996," ed. by A. P. Alivisatos, P. Fauchet, I. Shimizu, R. Collins, T. Shimuder, and J. C. Vial, Mat. Res. Soc. Proc. **452** (MRS, Pittsburgh, 1997).

3. A. P. Alivisatos, Science **271**, 933 (1996).

4. C. W. White, J. D. Budai, S. P. Withrow, J. G. Zhu, S. J. Pennycook, R. A. Zuhr, D. M. Hembree, Jr., D. O. Henderson, R. H. Magruder, M. J. Yacaman, G. Mondragon, and S. Prawer, Nucl. Instrum. Methods in Physics Res. B **127/128**, 545 (1997).

5. J. D. Budai, C. W. White, S. P. Withrow, R. A. Zuhr, and J. G. Zhu, Mat. Res. Soc. Proc. **452**, 89 (1997).

6. R. Rossetti, R. Hull, J. M. Gibson, and L. E. Brus, J. Chem. Phys. **82**, 552 (1985).

7. C. B. Murray, D. J. Norris, and M. G. Bawendi, J. Am. Chem. Soc. **115**, 8706 (1993).

8. V. Sukumar and R. H. Doremus, Phys. Stat. Sol. (b) **179**, 307 (1993).

9. B. G. Potter and J. H. Simmons, Phys. Rev. B **37**, 10838 (1988).

10. R. N. Bhargava, D. Gallagher, X. Hong, and A. Nurmikko, Phys. Rev. Lett. **72**, 416 (1994).

11. A. Khosravi, M. Kundu, L. Jatwa, S. K. Deshpande, U. A. Bhagwat, M. Sastry, and S. K. Kulkarni, Appl. Phys. Lett. **67**, 2702 (1995).

12. D. E. Bliss, J. P. Wilcoxon, P. P. Newcomer, and G. A. Samara, Mat. Res. Soc. Proc. **358**, 265 (1995).

13. J. D. Budai, C. W. White, S. P. Withrow, M. F. Chisholm, J. G. Zhu, and R. A. Zuhr, *Nature* **390** 384 (1997).

ION BEAM EFFECTS ON THE FORMATION OF SEMICONDUCTOR NANOCLUSTERS

S. SCHIESTEL, C.A. CAROSELLA, R. STROUD, S. GUHA, C.M. COTELL, K.S. GRABOWSKI, US Naval Research Laboratory, 4555 Overlook Ave., Washington, D.C. 20375

ABSTRACT

Silicon rich silica films were deposited by coevaporation of silica and silicon with and without simultaneous ion bombardment. The Si-silica ratios are correlated to changes of index of refraction and shifts in an asymmetric stretching mode IR absorption. Photoluminescence between 550 and 600 nm is observed for all films which is attributed to SiO_2 defects. After annealing this photoluminescence peak shows a shift to 750 nm and an increase in intensity, indicating the formation of silicon nanoclusters. This effect is more pronounced for samples prepared without ion beam treatment. ERD measurements show a correlation of photoluminescence with the presence of hydrogen in the films. The microstructure of these films were investigated by TEM. Photoluminescence from the Si nanoclusters in the films is optimized when the arrival rate of Si/silica is slightly less than 0.4.

INTRODUCTION

The discovery of photoluminescence of porous silicon has stimulated enormous interest in the synthesis of nanocrystalline silicon for optoelectronic applications [1]. Many researchers are exploring techniques that would produce nanocrystals of Si with processes that are compatible with semiconductor technology. They have already demonstrated that silicon and other semiconductor nanoclusters are formed by ion implantation and subsequent annealing [2-4]. Ion implantation has obvious disadvantages imposed by the limitation of thicknesses of potential photoluminescent layers, the non uniform distribution of nanoclusters and the limited density of nanoclusters. We have shown that ion beam assisted deposition (IBAD) can overcome these difficulties for the production of noble metal clusters in various insulating matrices such as SiO_2 [5]. The energetics of the IBAD process resulted in a homogeneous, high density distribution of metal clusters. The same behavior is expected for the synthesis of semiconductor clusters. This paper will explore preliminary work in this area.

EXPERIMENT

Silicon nanocluster containing films were deposited by coevaporation of silicon and silica from two e-beam hearths in an ultrahigh vacuum chamber ($1 \cdot 10^{-8}$ mbar). The system is both cryo-pumped and turbo-pumped during depositions. One set of samples was simultaneously bombarded with an argon ion beam from an 8 cm. ion gun. The energy of the argon ions was 100 eV; the current density was 35 $\mu A/cm^2$. All films were deposited at room temperature. The deposition rate of silica was 0.2 nm/s; silicon deposition rate was varied from 0.04 to 0.16 nm/s. After deposition the samples were post-annealed in vacuum or nitrogen between 500 °C and 900 °C. The composition of the films was determined by Rutherford back scattering analysis (RBS) and elastic recoil detection of hydrogen (ERD), using He ions accelerated in a 3 MeV tandem Van de Graaff. The index of refraction was measured by preparing thick samples (>500 nm) on silicon substrates and analyzing the interference in the transmission and reflection curves vs. wavelength. The data was fit to the Selmeir dispersion formula. The change of optical properties, which indicate indirectly the presence of silicon clusters, were investigated by UV- and IR-spectroscopy. The photoluminescence spectra were measured with a single monochromator (HR 460) and a cooled charged coupled device (CCD) array. The size of the silicon clusters was determined by HR-TEM using a 300 keV transmission electron microscope (Hitachi H9000).

RESULTS

Index of refraction

The evaporation of pure silica results in a transparent film with a index of refraction of 1.47, in excellent agreement with the literature [6]. Simultaneous ion bombardment during the silica evaporation leads to a slightly higher index of refraction of 1.5, resulting from a densification of the silica due to ion beam mixing. The coevaporation of silicon and silica may result in a dissolution of silicon in silica or in a formation of silicon monoxide SiO, especially under argon ion bombardment where additional reaction energy is available. Heats of formation suggest that the first one is expected. In both cases the expected index of refraction can be calculated for different SiO_2-Si and SiO_2-SiO mixtures from Maxwell-Garnet theory:

$$n_{eff} = (\varepsilon_{eff})^{1/2} \tag{1}$$

$$\varepsilon_{eff} = \varepsilon_m \cdot [1+2 f_v (\varepsilon_e - \varepsilon_m)/(\varepsilon_e + 2 \varepsilon_m)]/ [1-f_v (\varepsilon_e - \varepsilon_m)/(\varepsilon_e + 2 \varepsilon_m)] \tag{2}$$

where n_{eff} and ε_{eff} are the index of refraction and the dielectric constant of the mixture, ε_m the dielectric constant of the matrix, ε_e the dielectric constant of the cluster element and f_v the volume fraction of the clusters. The difference in index of refraction for Si and SiO is large enough to distinguish between the SiO_2-Si and the SiO_2-SiO mixtures. In Fig. 1 the indices of refraction for two different silicon/silica arrival rates (0.2 and 0.4) are compared to the ones of pure SiO_2 and pure SiO. The index of refraction of SiO is obtained from the literature [6]. Assuming that all evaporated Si would form SiO, the resulting calculated indices of refraction at 800 nm are 1.52 for the 0.2 Si/SiO_2 arrival rate and 1.58 for 0.4 arrival rate. The measured refractive indices for the Si-SiO_2-mixtures are higher (1.6 and 1.85 respectively) suggesting therefore the formation of Si-SiO_2. The index of refraction of samples which were deposited under simultaneous argon ion bombardment do not much differ from those that were only e-beam evaporated. The slightly lower indices might be due to lower silicon concentration as consequence of preferential sputtering during ion bombardment.

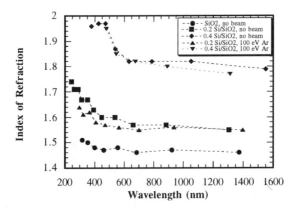

Figure 1. Index of refraction vs. wavelength for Si/SiO_2 thin films prepared with no ion beam and with a 100 eV Ar ion beam. The SiO_2 film has an index near 1.47. Two arrival rates of Si/SiO_2 are shown: 0.2 and 0.4. Modeling of the data with Maxwell-Garnet theory suggests that the Si is not oxidized in the process, even in the presence of the Ar ion beam.

IR spectroscopy

An asymmetric stretching motion of O atoms in Si-O-Si bonds results in a broad aborption band at 1075 cm^{-1}. Pai et al. [7] demonstrated that the frequency of this IR band monotonically scales with the film composition SiO$_x$, shifting to lower frequency with decreasing x. IR absorption spectra of pure SiO$_2$ and Si-SiO$_2$-mixtures are shown in Fig. 2. Fig. 2a shows a slight shift of the band peak (1080 cm^{-1}) in the direction of x < 2 for the SiO$_2$ deposited with an ion beam. Calculation of x from the IR data for this case gives x=1.8. The peak shifts back to 1110 cm^{-1} after annealing in nitrogen for 1 hour at 900 °C, indicating that x=2. RBS analyses of stoichiometry confirm these results.

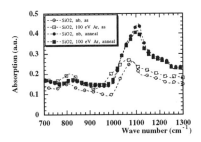 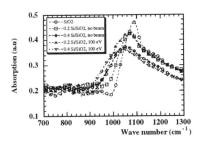

Fig. 2a, left: IR-spectra of evaporated silica, with an without an Ar ion beam. The shifts in the IR peaks due to the ion beam indicates that SiO$_x$ (x=1.8) forms and then reacts to give nanoclusters of Si after annealing. Fig. 2b, right: IR-spectra of different SiO$_2$-Si mixtures after deposition, with and without an ion beam. The regular shifts to lower frequency of the peaks with increasing Si/SiO$_2$ arrival ratio is summarized in Table 1. Ion beams also change the stoichiometry slightly.

Previous studies [9] of SiO$_x$ films show that annealing of substochiometric SiO$_x$ leads to the formation of mesoscopic Si particles in a SiO$_2$ matrix as follows:

$$2 \text{ SiO}_x \text{ ---> } x \text{ SiO}_2 + (2-x) \text{ Si} \tag{3}$$

Therefore, we expect IBAD of silica to form silicon clusters by partial decomposition of the silica. Fig. 2b shows that coevaporation of silica and silicon leads to larger IR shifts depending on the Si/SiO$_2$ arrival rates. Table 1. summarizes the data of Fig. 2b, comparing the IR peak positions with RBS data on the same samples.

Table 1. The stoichiometry of Si/SiO$_2$ films as measured by IR and RBS. The hydrogen content of the films, as measured by ERD is also tabulated. An argon ion beam (100 eV, 35 mA/cm^2) causes a small change in stoichiometry, but significantly reduces the hydrogen content of the films.

Si Evap. Rate (nm/s)	Freq (cm^{-1})	x in SiO$_x$ (IR)	x in SiO$_x$ (RBS)	y in SiO$_x$H$_y$ (ERD)
.04	1063	1.75	2.05	0.135
.08	1048	1.52	1.57	0.15
.04 with ion beam	1056	1.65	1.65	0.09
.08 with ion beam	1041	1.45	1.13	0.026

The binding state of hydrogen that is measured by ERD is still unclear . There are no Si-H bands and no water contamination observed in the IR spectra. Hydrogen is ubiquitous in vacuum systems, however care was taken to bake out the system before deposition.

<u>Photoluminescence and TEM</u>

The above analyses suggest that the deposited films contain silicon nanoclusters and thus they should exhibit photoluminescence (PL). Silica samples, deposited with and without an Ar ion beam, have a weak PL peak centered at 550 nm which is attributed to defects in the Si-O structure [3]. This behavior was also found for the Si-SiO$_2$ mixtures for a Si/SiO2 arrival rate 0.2. The sample evaporated with a Si/SiO2 arrival rate of 0.4 (without ion beam) shows a

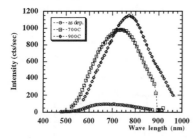 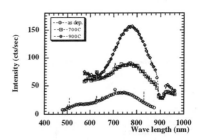

Fig. 3a, left: PL-spectra of films made at a Si/SiO2 arrival rate of 0.4 without ion beam, shown as deposited and after 700 °C and 900 °C vacuum anneals for one hour. Fig. 3b, right: the same conditions, except for the presence of an Ar ion beam. The PL intensity is normalized for film thickness.

broad, low intense PL band at 700 nm after deposition, without any annealing. Annealing of this sample at 700 °C leads to an increase in intensity and a small shift of the peak position to 750 nm, indicating the formation of silicon nanoclusters. If the post-deposition annealing is performed at 900 °C the intensity increases even further and the peak shifts to 780 nm.

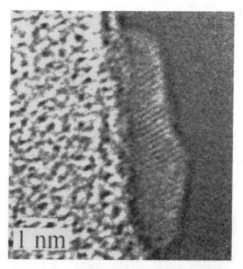

Figure 4. TEM of a Si cluster in the 900 °C annealed film whose PL appears in Fig. 3a. The cluster is in the right-center of the micro graph. To the left is the amorphous Si O$_2$.

The TEM image of this sample (Fig. 4) shows crystalline silicon clusters 1.4 to 4.0 nm in size, within larger 0.1 μm silica grains . This range of cluster size and the position of the PL peak are in good agreement with other papers.

The IBAD sample is shown in Fig. 3b. Two low intense peaks at 550 nm and 740 nm are observed after deposition. Creation of defects during the IBAD process is expected and these defects are almost totally removed by annealing. Annealing of this sample also increases the PL intensity and shifts the position to 780 nm. The PL intensity of the IBAD sample is about a factor of 7 smaller than the one evaporated without IBAD, even though the film characterization suggests more Si in the IBAD samples. One explanation might be the different hydrogen concentration in both samples. The evaporated sample contains about 6 times more hydrogen than the IBAD samples. Although it is still not known how hydrogen contributes to the mechanism of photoluminescence, the dependence of photoluminescence on the presence of hydrogen was also found by other authors [4].

Films were also prepared at the higher Si/SiO_2 arrival rate of 0.8. The PL from these samples for the 700 nm peak was quite weak. Annealing the films in oxygen at 900 °C for various time periods demonstrate that the clusters gradually oxidize. TEM of these samples show that the silicon clusters are originally too big, about 10 nm, to promote PL, but during the oxidation the clusters shrink in size and the PL increases. Eventually the clusters become too small and the PL once again falls. Similar oxidation studies of Si/SiO_2 at lower arrival rates suggest that a Si/SiO_2 arrival rate of slightly less 0.4 optimizes the PL. We have also observed that the films made with this technique are quite unstable, prepared with or without an ion beam. PL measured a few months after sample preparation show a pronounced decreased signal.

CONCLUSIONS

The simultaneous deposition of Si and silica by vacuum evaporation can produce photoluminescence from nanoclusters of Si crystals in an amorphous silica matrix. The largest PL is obtained for an arrival rate for Si/silica of slightly less than 0.4, e.g. 0.08 nm/sec of Si for 0.2 nm/sec of silica. Higher Si/silica arrival rates produce nanoclusters that are too large to give PL. The presence of hydrogen in the film enhances the PL signal. Ion beams effectively limit the take-up of hydrogen during the growth process of the IBAD films. This is probably due to the removal of OH ions from the surface of the growing films via the argon ion beam. Annealing in hydrogen could potentially enhance the PL signal. Indirect experimental evidence presented here suggests that IBAD-processed films do indeed have Si nanoclusters. Direct verification awaits more TEM work . The observation that the PL fades away with time (months) may be related to the grain structure of the amorphous silica matrix produced by the vacuum evaporation process without IBAD. Grain boundaries could provide paths for internal oxidation of the Si nanoclusters.

REFERENCES

1. L.T. Canham, Appl. Phys. Lett 57, p. 1046 (1990)

2. C.W. White, J.D. Budai, S.P. Withrow, J.G. Zhu, S.J. Pennycook, R.A. Zuhr, D.M. Hembree Jr. , D.O. Henderson, R.H. Magruder, M.J. Yacaman, G. Mondragon, S. Prawer, Nucl. Instr. Meth. Phys. Res. B 127/128, p. 545 (1997)

3. S. Guha, M.D. Pace, D.N. Dunn and I. Singer, Appl Phys. Lett. 70(10) 1997

4. K.S: Min, K.V. Scheglov, C.M. Yang, H. Atwater, M.L. Brongersma and A. Polman, Appl. Phys. Lett. 69, p. 2033 (1996)

5. S. Schiestel, C.M. Cotell, C.A. Carosella, K.S. Grabowski and G.K. Hubler, Nucl. Instr. Meth. Phys. Res. B 127/128, p. 566 (1997)

6. Handbook of Optical Constants, ed by E. D. Palik (Academic Press, Orlando, FL 1985), pp 753 and 767.

7. P.G. Pai, S.S. Chao, Y. Tagaki and G. Lucovsky, J. Vac. Sci Technol. A 4 (3), p. 689 (1986)

8. S.Hayashi and K. Yamamoto, J.of luminescence, 70, p.352 (1996)

9. L. Nesbit, Appl. Phys. Lett. 46, p.38 (1985)

FORMATION OF Pb INCLUSIONS IN Si BY ION IMPLANTATION

V.S. TOUBOLTSEV[1], E. JOHNSON[1], U. DAHMEN[2], A. JOHANSEN[1], L. SARHOLT[1] AND S.Q. XIAO[2]

1. Ørsted Laboratory, Niels Bohr Institute, Universitetsparken 5, DK 2100 Copenhagen Ø, Denmark.
2. National Center for Electron Microscopy, E.O. Lawrence Berkeley National Laboratory, UC Berkeley, CA, USA

Abstract

Si<110> single crystals were implanted at a temperature of 835 K with 150 keV Pb^+ ions to a fluence of $1 \cdot 10^{20}$ m^{-2} corresponding to an average concentration of 2-3 at%. The implanted samples have been studied by Rutherford Backscattering (RBS)/channeling and transmission electron microscopy (TEM) techniques. In as-implanted samples the main fraction of implanted Pb was located on substitutional sites in the Si matrix thus providing a highly supersaturated solution of Pb in Si. Spontaneous precipitation of Pb, giving rise to formation of nanosized Pb inclusions, was found to take place only in the peak region of the implantation. TEM analysis showed that the Pb precipitates had sizes from about 2 to 20 nm and that they grew in parallel cube orientation relationship with the host matrix. The shape of the inclusions was found to be approximately cuboctahedral with poorly developed {111} and {100} facets.

In-situ RBS/channeling heating/cooling experiments on both as-implanted samples and samples previously furnace-annealed at 1175 K showed a distinct melting/solidification hysteresis of the Pb inclusions around the bulk melting point for Pb at 600 K. These results were verified by in-situ TEM heating/cooling experiments on as-implanted samples.

Introduction

For the past two decades structural, thermodynamic and diffusion properties of crystalline nanoparticles completely embedded in a metallic matrix have been investigated in detail in a variety of heterogeneous alloys of immiscible elements produced by both ion implantation and rapid solidification (see Refs. 1-6 and references therein). In contrast, nanosized elemental inclusions in semiconductors have not received the same degree of attention. This is due to the fact that such inclusions, to our knowledge, can be produced in silicon or germanium crystals only by ion implantation of elements with a negligible solubility, leading to spontaneous segregation and precipitation. In microelectronic devices the precipitation process is of importance for the redistribution of dopants during implantation and annealing [7]. Lead is a good choice for such embedded inclusions in a silicon matrix since there is no appreciable mutual solubility of the elements in the solid state [8].

Here we present results of an investigation of Pb inclusions produced by ion implantation into single crystal Si. The resulting inclusion and defect microstructures, and the melting and solidification of the inclusions were studied in parallel by RBS/channeling and TEM. In RBS/channeling measurements melting is directly associated with increased dechannelling in liquid inclusions, whereas in TEM observations, melting results in the disappearance of moiré fringes. Thus RBS provides a global measurement of the behavior of inclusions as an ensemble, while TEM gives a local measurement of individual inclusions.

Experimental

Pb^+ ions with an energy of 150 keV were implanted into Si<110> single crystals in a random direction to a fluence of $1 \cdot 10^{20}$ m^{-2} using an ion beam current density of about 20 $nAmm^{-2}$. To avoid irradiation induced Si amorphization the implantations were performed at a temperature of

835 K. Subsequent RBS/channeling measurements using a 500 keV He^{2+} beam supplemented by TEM analysis were carried out to analyze the microstructure, the channeling characteristics and the thermal behavior of the implanted Pb atoms embedded in the Si matrix. RBS/channeling spectra were recorded as a function of temperature in series of heating/cooling cycles, either in-situ on as-implanted samples or ex-situ on samples after vacuum furnace annealing at 1175 K for 1 hour. Both implantations and RBS analyses were performed at a pressure of 10^{-5} Pa. Single crystals for TEM studies implanted under similar conditions were annealed in-situ up to 1230 K in a JEOL 200CX microscope using a Gatan double tilt heating holder. In both the RBS and TEM measurements the sample temperature was estimated to be within an accuracy of 2-3 degrees.

Results and discussion

Figure 1 shows the Pb depth profiles in the implanted samples before and after annealing. In the as-implanted state the Pb atoms are seen to have a bimodal depth distribution with a poorly developed peak at a depth around 20 nm and the main peak at a depth of 60 nm. The experimental depth profile is seen to be much wider than the range distribution of Pb in Si

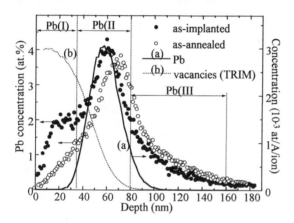

obtained from a TRIM calculation [9] (given in Fig.1 by solid line), although the position of the main peak around 60 nm coincides with the TRIM peak. Such a discrepancy is quite common as the TRIM code does not take into account irradiation-induced diffusion processes taking place during the implantations. In many cases TRIM predicts correctly only the average position of the implanted depth profile while it underestimates the range straggling [10]. In this view, the broad experimental profile is ascribed to radiation-enhanced diffusion which is faster than thermal diffusion alone. The damage distribution calculated by

Figure 1. Depth distributions of 150 keV Pb implanted into Si<110> at 835 K shown for the sample before and after annealing at 1175 K for 1 hour. The arrows indicate the axes for the different distributions.

TRIM as the vacancy distribution for 150 keV Pb ions in Si is shown by the dashed line in figure 1. Since oversized Pb atoms diffuse in Si via vacancies [11], redistribution of the implanted Pb is believed to take place at the vacancy concentration gradients driving Pb atoms towards the surface. This outdiffusion is believed to cause the partly resolved peak on the shallow flank of the Pb profile measured on as-implanted samples.

Post-implantation annealing at 1175 K is seen in figure 1 to change the Pb depth profile significantly. After annealing, the near-surface region of the implanted layer has been depleted of Pb and the surface peak on the flank of the as-implanted profile has disappeared. The profile has accordingly become narrower as compared with the as-implanted state and at the same time the main peak of the profile has been shifted about 10 nm towards larger depths. The depletion of Pb is tentatively ascribed to enhanced outdiffusion of Pb mediated by the high density of irradiation damage existing in the surface region of the implanted layer. This provides an easy diffusion path and will during the early stage of annealing lead to partial Pb segregation to the surface followed by direct vaporization.

Figure 2 shows Pb and Si <110> axial channeling dips, i.e. the backscattered yields for each element normalized to the random RBS spectra as functions of tilt angle around the <110> axis, for both the as-implanted sample (a) and after furnace annealing to 1175 K (b). For the as-implanted sample, figure 2a shows angular scans with channeling dips for Pb in the three depth regions indicated in figure 1 subdivided as follows: 0-35 nm (PbI), 35-80 nm (PbII), and 80-160 nm (PbIII). The angular scan for Si was measured as an average over the entire implantation zone. The measured values for the critical angles $\psi_{1/2}$ and the minimum yields χ_{min} obtained from figure 2 for Pb and Si are given in Table 1. Distinct channeling dips were observed for both Si and Pb within each of the selected depth regions. It is obvious that implantation of Pb at a temperature of 835 K prevents amorphization and leads to spontaneous recovery of the Si matrix even after extensive irradiation by such heavy ions as Pb$^+$. The critical angles $\psi_{1/2}$ for PbI and PbIII are seen in Table 1 to be very close to the value for Si, while $\psi_{1/2}$ for PbII is somewhat larger corresponding to a wider channeling dip in figure 2a. In contrast, after annealing at 1175 K which caused a significant redistribution of the implanted Pb, the channeling dip for Pb shown in figure 2b is substantially wider than for Si (see Table 1), although it is averaged over the implanted layer as a whole (a depth region of 0-160 nm). The similarity between the channeling dips for PbI, PbIII and Si in as-implanted samples, and the nearly identical values for the corresponding critical angles show that the main fraction of Pb atoms in these two depth regions are located on substitutional sites in the Si lattice. Since He^{2+} ions backscattered by Pb atoms in inclusions and clusters smaller than about 1 nm do not contribute to the channeling yield [12], regions I and III of the implanted layer are believed to be highly supersaturated with substitutional Pb in a Si matrix that is essentially free of Pb clusters or precipitates larger than $\approx$ 1 nm in size. In contrast, the larger width of the angular Pb scan in region II shows the presence of precipitated Pb.

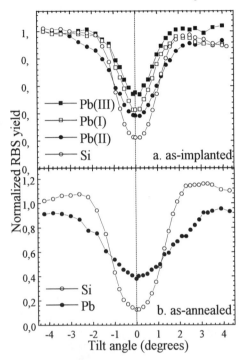

Figure 2 Axial channeling dips in Si implanted with 150 keV Pb shown for the sample before (a) and after annealing at 1173 K for 1 hour (b). The three Pb dips in (a) refer to the three depth intervals in figure 1 while the Si dip is an average measure from all three depth intervals.

A plan-view TEM analysis of as-implanted samples provides direct evidence for the existence of precipitates in the implanted layer. The TEM micrograph in figure 3 taken from an orientation close to Si <110> shows a fairly homogeneous distribution of Pb inclusions with sizes ranging from 1 nm up to 20 nm. A diffraction analysis showed that the precipitates grow in parallel cube alignment with the Si matrix which gives rise to characteristic moiré images reflecting the difference in lattice parameter between Pb (f.c.c., a_{Pb} = 0.495 nm) and Si (d.c., a_{Si} = 0.548 nm). Although the detailed shape of the inclusions is rather irregular, their overall shape was found to be cuboctahedral with poorly developed {111} facets truncated on the tips by {100} facets. The larger Pb inclusions were frequently found to have distorted moiré fringes (figure 3) indicating the existence of residual strains. It is the highly perfect topotaxial alignment of the Pb precipitates with the Si matrix that gives rise to the wider channeling dip for Pb in region II (see figure 2a). During annealing at 1175 K the main fraction of implanted Pb in the entire implantation zone

Table I

Critical angles and minimum RBS yields

	as-implanted				as-annealed	
	PbI	PbII	PbIII	Si	Pb	Si
$\psi_{1/2}$	0.83	1.20	0.85	0.95	1.70	1.20
χ_{min}	0.43	0.37	0.54	0.22	0.39	0.12

can be redistributed by diffusion and is believed to form crystalline inclusions. In this view, the wide angular scan dip for Pb in figure 2b is unambiguously due to the channeling in these precipitates.

In figure 4 the minimum RBS yields for Pb, i.e., the normalized yields in the channeling direction, have been plotted as a function of temperature. For the as-implanted sample, data have been plotted separately for regions I, II and III, whereas for a sample that was subsequently annealed for 1h at 1175 K, an average over the implantation profile is shown. Within the accuracy of our RBS measurements, the minimum RBS/channeling yields corresponding to the PbI and PbIII regions from the as-implanted sample show no distinct change with temperature except for a slight monotonic increase. This is presumably due to increased thermal vibrations of both the Si matrix atoms and the substitutional Pb atoms that will inevitably deteriorate the channeling of the He^{2+} ions. Since He^{2+} ions backscattered by Pb atoms in inclusions and clusters smaller than about 1 nm do not contribute to the channeling yield [12], regions I and III of the implanted layer are believed to be highly supersaturated with substitutional Pb in a Si matrix that is essentially free of Pb clusters or precipitates larger than $\approx$ 1 nm in size. In contrast, the larger width of the angular Pb scan in region II shows the presence of precipitated Pb.

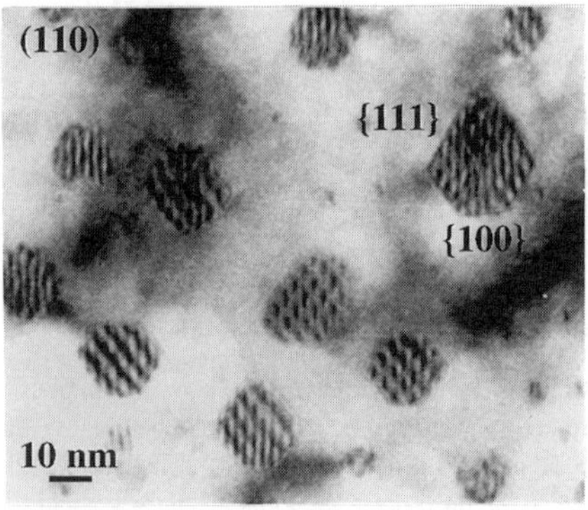

This confirms our interpretation of the angular scans (figure 2a) that the PbI and PbIII depth regions contain no precipitates detectable by RBS. A step-wise jump of the minimum yield in the PbII region within a temperature interval of 600-625 K is associated with melting of Pb inclusions in the depth range around the peak of the implant distribution (35-80 nm). Upon cooling, solidification of the inclusions sets in around 580 K and is completed around 550 K. At a temperature as high as 855 K when all inclusions in the PbII region are anticipated to be liquid, the minimum RBS yield is still less than 0.7. This value clearly indicates that even in the PbII region a substantial fraction of the Pb atoms is located substitutionally in the Si host lattice, thus providing for the large degree of channeling.

Figure 3 TEM micrograph showing nanosized Pb inclusions in a Si<110> single crystal after implantation with 150 keV Pb^+ ions at 835 K. In the <110> orientation the Pb display larger {111} and smaller {100} facets, both seen edge-on.

Heating/cooling experiments carried out in-situ in the electron microscope on as-implanted samples also showed a distinct melting/solidification hysteresis. In general, the temperature intervals where melting and solidification take place were found to be in reasonable accordance with those measured by RBS/channeling. However, both melting and solidification temperatures of individual inclusions were found to depend on their size. Large inclusions were observed to melt first, while the smallest inclusions solidified first.

Melting of the inclusions occurs most likely by heterogeneous nucleation at the interfaces with the surrounding matrix. In completely embedded inclusions melting requires a break-down of the solid/solid interfaces and hence an energy contribution that is comparable to or larger than for a free crystal. The smallest inclusions have the largest fraction of atoms located at the interface and are thus most prone to show size dependent deviations in their properties. Significant superheating is only seen for Pb and In inclusions in Al which have atomically smooth low-energy facets [2, 13]. The Pb inclusions in Si have less well-defined facets which can explain the smaller observed superheating. Furthermore, melting of the Pb inclusions in Si is most likely also influenced by pressure-broadening. This is due to the fact that the volume expansion and the shape change associated with melting of Pb in this case can not be accommodated quickly by Si diffusion, since the diffusion in Si at 550-650 K is much slower than in Al.

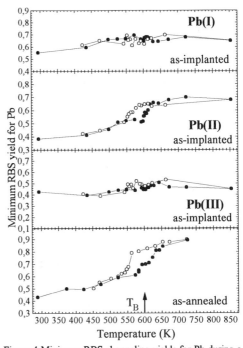

Figure 4 Minimum RBS channeling yields for Pb during a heating/cooling cycle on Si<110> implanted with 150 keV Pb shown before and after annealing. (•) indicates heating and (o) indicates cooling.

Upon cooling, solidification of the Pb inclusions is associated with substantial undercooling which also exhibits a clear dependence on size. The TEM results showed that smaller inclusions solidified first while larger inclusions required larger undercooling. At first sight this behavior contradicts heterogeneous nucleation theory because the larger inclusions should provide more nucleation sites and thus solidify with less undercooling. However, if the inclusion size is of the order of the critical nucleus then smaller inclusions would be more likely to solidify by spontaneous fluctuation involving the entire inclusion in the right orientation relationship with the matrix[1]. TEM heating/cooling experiments showed that there is a tendency for the liquid Pb inclusions completely embedded in the Si matrix to retain some of the properties normally attributed to those in the solid state. Upon melting, the inclusions were found to remain faceted, retaining their approximately cuboctahedral shape unchanged up to a temperature as high as 1230 K. In contrast, in an Al matrix, where the lattice vacancies are in copious supply at temperatures around 600 K, nanosized Pb inclusions were found to become spherical when melting [1,2].

The heating/cooling cycle measured after furnace annealing at 1175 K can be seen in figure 4 to show substantial changes in the melting/solidification hysteresis in comparison with the behavior of the PbII region in the as-implanted samples. In the annealed sample, melting of the Pb inclusions begins around 585 K and is completed around 650 K. On cooling, solidification sets in around 570 K followed by a steep decrease in the minimum RBS/channeling yield down to 530 K. The RBS/channeling yield measured at the highest temperature of 725 K gave a value of 0.9, justifying our interpretation of the angular scans that after annealing the main fraction of the implanted Pb is confined to the precipitates. During annealing, precipitation of new small inclusions as well as thermal coarsening of already existing inclusions can provide for changes in the size distribution of the precipitates.

Conclusions

It has been shown that ion implantation of 150 keV Pb^+ ions into Si<110> crystals can give rise to formation of dense distributions of nanosized crystalline Pb inclusions in a highly supersaturated substitutional solution of lead in silicon. The Pb inclusions were found to grow in topotaxy with the host matrix following a parallel cube orientation relationship. The shape of the inclusions was approximately cuboctahedral with poorly developed {111} and {100} facets. Heating/cooling experiments using both RBS/channeling and TEM showed a distinct hysteresis in melting and solidification of the inclusions. Large inclusions were observed to melt first, while smaller inclusions solidified first. Annealing at 1175 K was found to induce a significant redistribution of the implanted Pb. It is accompanied by diffusion-assisted lead precipitation from the supersaturated solution and thermal induced coarsening of the as-implanted inclusion.

Acknowledgments

This work is supported by the Danish Natural Sciences Research Council and the Director Office of Energy Research, Office of Basic Energy Sciences, Materials Sciences Division of the U.S. Department of Energy under Contract No. DE-ACO3-76SFOOO98.

References

[1] H..H. Andersen and E. Johnson, Nucl. Instr. and Meth. B 106 (1995) 480.

[2] S.Q. Xiao, E. Johnson, S. Hinderberger, A. Johansen, K.K. Bourdelle and U. Dahmen, J. Microscopy 180 (1995) 61.

[3] K.I. Moore, D.L. Zhang and B. Cantor, Acta Metal. Mater. 38 (1990) 1327.

[4] D.L. Zhang and B. Cantor, Phil.Mag. A 62 (1990) 557

[5] D.L. Zhang, K. Chattopadhyay and B. Cantor, J.Mater.Sci. 26 (1991)1536

[6] E. Johnson, A. Johansen, N.B. Thoft, H.H. Andersen and L. Sarholt-Kristensen, Philos.Mag. Lett. 68 (1993) 131.

[7] S. Solmi, F. Baruffaldi and M. Derdour, J. Appl. Phys. 71 (1992) 697.

[8] R.W. Olesinski and G.J. Abbaschian, Binary Alloy Phase Diagrams, edited by T.B. Massalski, J.L. Murray, L.H. Bennett and H. Baker (Metals Park, Ohio: ASM) (1986) 1845.

[9] Ziegler, J.F., Biersack, J.P., and Littmark, U., 1985, The Stopping and Ranges of Ions in Solids (Pergamon, New York).

[10] H.H. Andersen, J. Bottiger and H. Wolder-Jorgensen, Appl. Phys. Lett. 26 (19??) 678.

[11] W. Frank, Crucial Issues in Semiconductor Materials and Processing Technologies, edited by S. Coffa, F. Priolo, E. Rimini, and J.M. Poate, NATO ASI Series E: Applied Sciences, Vol. 222, Kluwer Academic Publishers, Dordrecht-Boston-London 1992, p. 383.

[12] K.K. Bourdelle, V.A. Khodyrev, A. Johansen, E. Johnson, and L. Sarholt-Kristensen, Phys. Rev. B, 50 (1194) p.82

[13] H. Saka, Y. Nishikawa and T. Imura, Philos. Mag. A57 (1988) 895.

IN-SITU OBSERVATION OF ATOMIC PROCESSES IN Xe NANOCRYSTALS EMBEDDED IN Al

K. Mitsuishi, M. Song, K. Furuya *, R. C. Birtcher, C. W. Allen **, and S. E. Donnelly †
* National Research Institute for Metals, Sakura, Tsukuba 305, Japan
** Materials Science Division, Argonne National Laboratory, Argonne, IL60439, USA
† Department of Physics, University of Salford, Manchester, UK

ABSTRACT

Self-organization processes in Xe nanocrystals embedded in Al are observed with in-situ high-resolution electron microscopy. Under electron irradiation, stacking fault type defects are produced in Xe nanocrystals. The defects recover in a layer by layer manner. Detailed analysis of the video reveals that the displacement of Xe atoms in the stacking fault was rather small for the Xe atoms at boundary between Xe and Al, suggesting the possibility of the stacking fault in Xe precipitate originating inside of precipitate, not at the Al/Xe interface.

INTRODUCTION

Recently, nanocrystals have attracted fundamental and applied interest in materials research because various properties such as, electrical, optical, and magnetic properties, largely depend on size and/or shape of the crystals. The behavior of nanometer size precipitates created by implantation of inert gases in materials has been studied extensively for more than 35 years because of problems associated with the development of fusion and fission reactors[1].

When it is implanted at room temperature, the rare gas Xe precipitates as nanocrystals less than 4nm diameter and mesotactically aligned with the surrounding matrix. The lattice parameter of the precipitates varies with the precipitate size and the combination of gas and metal matrix [2, 3, 4]. The shape and topotactical alignment of the precipitates have been established by high-resolution electron microscopy (HRTEM) [5, 6].

However, little is known about the movement of individual atoms, or structural changes at the atomic level that are necessary to know when one tries to understand the self-organization processes. Therefore, the motivation for following the movement of individual atoms in-situ is very strong. Although HRTEM equipped with a video system can be one of the most promising tools for this purpose, there is an obstacle to surmount that the Xe lattice image is overlapped with Al fringes, because the Xe nanocrystals are embedded in the Al matrix. Therefore, in order to extract the Xe image, it is required to perform some sort of image processing such as FFT, which interfere with in-situ observation.

In this paper, we use the specific off-Bragg condition to extract the Xe lattice image in-situ, and observe self-organization process of Xe, such as inducement and recovery of the defect under electron irradiation.

EXPERIMENT

Specimen preparation

The Al specimens were cut from 5-9's starting material and electronically polished using and electrolyte of HNO_3: $CH_3OH= 1:2$ at 253 K and implanted with 30 keV Xe at room temperature to a dose of 3×10^{20}ions/m^2. Specimens were then annealed at 523 K for 0.5 h in a vacuum to remove radiation damage in the Al matrix and to consolidate the Xe within the precipitates.

Mat. Res. Soc. Symp. Proc. Vol. 504 © 1998 Materials Research Society

Templier et al.[7] have reported that such annealing results in sharper precipitate image and more intense electron diffraction from the solid Xe in Al.

HRTEM observation

HRTEM observations were performed with a JEM ARM-1000 high-voltage TEM, operated at a voltage of 1 MV with a spatial resolution of 0.13 nm, with the incident electron beam tilted slightly from a $<110>_{Al}$ direction. The defocus values were also chosen to realize specific off-Bragg condition to extract the Xe lattice image.

The electron beam current density was $1.3 \times 10^6 A/m^2$ to stimulate the system.

Models for simulation

Xe precipitates in an Al lattice were modeled as a tetradecahedron with faces parallel to $\{111\}_{Xe}$ and $\{100\}_{Xe}$ in Al cavities faceted on these crystal planes[8, 9]. The lattice parameter of the FCC Xe crystal was chosen to be 50% larger than that of Al, consistent with electron diffraction data. Relaxation of interface atoms is not included.

Multi-slice image simulations were carried out to simulate images from the JEM ARM-1000 at 1000keV,where the spherical aberration coefficient C_s is 2.6 mm, the spread of defocus δ is 10 nm and the beam semi-divergence α is 0.3 mrad. The defocus value was varied from +28 to -160 nm. Scherzer defocus for structure imaging is about -56 nm.

RESULTS

Off-Bragg condition

Fig.1 shows a HRTEM image of Xe precipitates in Al with the selected area diffraction pattern. The incident electron beam is parallel to $< 011 >$. Atoms appear as black dots in this image taken near Scherzer defocus. Two-dimensional Moiré patterns along one $< 100 >$ and two $< 111 >$ directions are seen. Both the Moiré patterns of HRTEM and the spacing of the diffraction spots in the SAD pattern indicate that the Xe precipitates are FCC and are mesotacitic with the Al

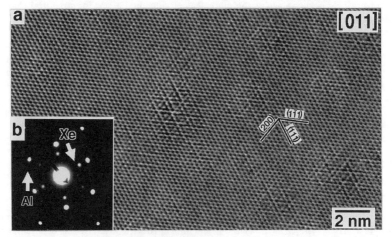

Figure 1: High-Resolution TEM (HRTEM) image of Xe precipitates in Al, observed near Scherzer defocus (a) with selected area diffraction pattern (b). The incident electron beam is parallel to $< 011 >$.

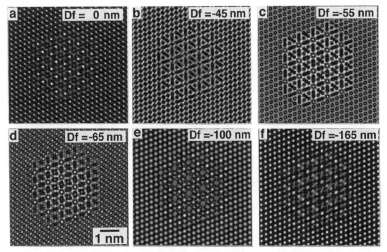

Figure 2: The results of multi-slice image simulations of a Xe precipitate in Al matrix at different defoci (Df) with [011] direction. a) Df = 0 nm, b) Df= -45 nm, c) Df=-55 nm (Scherzer defocus), d) Df= -65 nm, e) Df= -100 nm and f) Df= -165 nm.

matrix with a lattice spacing about 50% larger than that of Al. A similar observation has been reported by Donnelly et al[10].

In fig. 2, the calculated images of Xe precipitate in [011] direction are shown. The simulated image at Scherzer defocus is in qualitative agreement with the HRTEM image in fig.1. As in fig. 1, the shape of the precipitate is not clear although the image itself is very clear, because the Xe

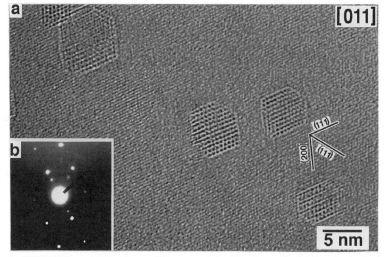

Figure 3: HRTEM image obtained in an off-Bragg condition The electron beam direction is tilted about 3 degree toward a [1$\bar{1}$1] from [011] zone axis, and the defocus value is taken as -76 nm. Al lattice fringes are weak and Xe precipitate can be observed clearly.

lattice image is overlapped with Al matrix lattice fringes. In this case, it is impossible to perform in-situ observation of recovery processes in the Xe.

To extract the Xe precipitate image from that of the Al matrix, a specific off-Bragg condition is used. The diffraction spots of Xe precipitates are spread in reciprocal space while Al matrix spots are more localized, so that the appropriate selection of illumination angle may weaken the Al lattice fringe and strengthen the contrast of Xe precipitate. The optimum defocus value is also determined by the contrast transfer function in effect detuning the Al matrix image in favor of that of the Xe. The resultant image, therefore, is mainly constructed from Xe spots.

Fig.3 shows the HRTEM image obtained by the specific off-Bragg condition that the electron beam direction is illuminated about 3 degree to a [1$\bar{1}$1] direction from [011] direction, and the defocus value is taken as -76 nm. Al lattice fringes are weak and the Xe precipitate can be observed clearly. It should be noted that the interpretation of such images needs some care because the defocus is not at Scherzer. It is confirmed that the black spots in fig. 3 correspond to the Xe atoms by multi-slice calculations of specific off-Bragg condition [11].

Defect inducement

Fig. 4 is video-captured images of a defect introduced in the Xe precipitate. Fig. 4 (a) shows the frame of immediately before its introduction, and fig. 4 (b) is just afterwards. The white arrows indicate the plane {111} the defect lies on. The defect is parallel to the < 110 > projected direction. The first few atomic lines of Xe from the bottom are not displaced, and the mismatch of the Xe line occurred at forth atom row from right side, coming to the gap between third and forth line of left. The displacement of Xe atoms which forms the stacking fault depends on the atom location, and surprisingly, the displacement was rather small for the Xe atoms at boundary between Xe and Al. This suggests the possibility of the defect in the Xe precipitate originating within the precipitate, not from the Al/Xe interface. By the argument of surface energy of Xe precipitates in Al matrix, in general, it is believed that the shape of the Xe precipitates is strongly confined by Al at the interface [12, 13]. However, it is not clear in this instance that the defect is caused by the boundary.

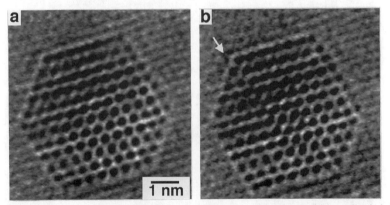

Figure 4: Video captured images of the defect introduction in a Xe precipitate; (a) immediately before the event, (b) immediately afterward. The white arrows indicate the plane of the defect.

Because the Xe precipitates are FCC, this defect is stacking fault type defect. To confirm this, a simple calculation of stacking fault in a Xe precipitate neglecting again the interface relaxation, was performed. Fig. 5 shows the result of calculation for the defocus value $\Delta = -76$ nm. The calculated image of the defect shape coincides well with the experimental regardless of the shapes of interface. Therefore, the defect can be considered as stacking fault type defect.

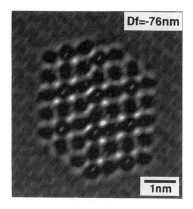

Recovery process

The successive recovery of the defect is shown in fig.6 where white arrows indicate the layers of atoms moved and the black arrows indicate the direction of movement. The numbers are the time, measured from the first observation of the defect in seconds. The recovery occurred within a layer unit. Since the time intervals of events in each layer are long, a few

Figure 5: Simulated image of the stacking fault type defect induced in Xe nanocrystal. The defect is in {111}.

seconds, each step of the process is spread over many frames of the video and is visible at normal video recording speed.

In fig. 6 (d), opposite movement of the layer previously recovered is observed, and in fig. 6 (e), several layers recover at the same time. This illustrates the delicate balance of forces involved in the recovery processes.

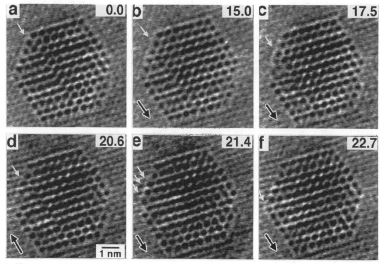

Figure 6: The successive images of the defect recovery. The white arrow indicates the layer in which the recovery occurred, and the black arrow indicates the direction of movement. The number indicates the time in seconds, measured from the first observation of the defect.

CONCLUSION

High-resolution transmission electron microscopy (HRTEM) was carried out on Al TEM thinned specimens implanted with 30keV Xe$^+$ at room temperature to a dose of 3×10^{20} ions/m^2. Self-organization processes in Xe nanocrystals embedded in Al were observed with in-situ electron microscopy technique by tilting the electron beam about 3 degree from a $< 011 >$ direction and defocusing the objective lens -76 nm. It is revealed that the displacement of Xe atoms which forms the stacking fault was rather small for the Xe atoms at the Xe/Al interfaces.

This may suggests that the stacking fault in the Xe precipitate is originates inside of precipitate for the particular Xe precipitate presented here, but it is difficult to decide only by two dimensional image of HRTEM.

Recovery occurs in a layer by layer manner rather slowly so that the process can be recorded by a standard video system.

REFERENCES

1. V. N. Chernikov, W. Kesternich and H. Ullmaire, J. Nucl. Mater. 227, 157, (1996).

2. R. C. Birtcher and C. Liu, J. Nucl. Mater. 165, 101 (1989).

3. R. C. Birtcher and W. Jäger, Ultramicroscopy 22,267 (1987).

4. R. C. Birtcher, S. E. Donnelly and C. Templier, Phys. Rev. B, 50, 764 (1994).

5. S. E. Donnelly and C. J. Rossouw, Phys. Rev. B13,485 (1986).

6. D. I. Potter and C. J. Rossouw, J. Nucl. Mater. 161, 124 (1989).

7. C. Templier, H. Garem, J. P. Riviere and J. Delafond, Nucl. Instr. and Meth. B18, 24 (1986).

8. L. Gråbæk, J. Bohr, H. H. Andersen, A. Johansen, E. Johnson , L. Sarholt-Kristensen and I. K. Robinson, Phys. Rev. B45, 2628 (1992).

9. S. Q. Xiao, E. Johnson, S. Hindenberger, A. Johannsen, K. K. Bourdelle and U. Dahmen, J. Microscopy 180, 61 (1995).

10. S. E. Donnelly, C. J. Rossouw, and I. J. Wilson, Radiat. Eff. 97, 265 (1986).

11. K. Mitsuishi, M. Song, K. Furuya, R. C. Birtcher, C. W. Allen, and S. E. Donnelly, presented at the 1997 Jpn.Phy.Soc. Meeting, Kobe,Jpn,1997(unpublished).

12. N. Ishikawa, M. Awaji, K. Furuya, R. C. Birtcher, C. W. Allen, Nucl. Instrum. Methods B127/128, 123 (1997).

13. M. S. Anderson and C. A. Swenson, J. Phys. Chem. Solids 36,145 (1975).

Part V

Polymers

MODIFICATION OF WETTING PROPERTY OF POLYCARBONATE BY MEANS OF ION BEAM IMPLANTATION AND STORAGE IN DIFFERENT GAS ENVIRONMENTS

A. POIRIER *, G.G. ROSS *, P. BERTRAND **, V. Wiertz **
*INRS-Énergie et Matériaux, Université du Québec, 1650 boul. Lionel-Boulet, Varennes (Québec), Canada, J3X 1S2, ross@inrs-ener.uquebec.ca
**PCPM, Université Catholique de Louvain, Place Croix du Sud 1, 1348 Louvain-la-Neuve, Belgium.

ABSTRACT

The wetting property of polymers is very important in different applications such as biomaterials, textiles, aerospace (fluid management and materials processing in microgravity), and thin film adhesion. Therefore, there is a strong interest in the development of a new technology for the modification at will of this property. The use of low energy ion beams allows the modification of the first surface atomic layers. Nitrogen ions of 500 eV/at. were used to bombard the surface of polycarbonate (PC) samples to a fluence of $5x10^{16}$ at/cm^2. Five different environments (oxygen, nitrogen, argon, dry air and vacuum) were used to store the samples for some hours (1 to 24 hours) after the implantation. Aging studies of the contact angle (advancing and receding) have shown that the environment gas influences the long term value of the contact angle and helps to maintain the stability of the treated surfaces with the passage of time. XPS and ToF-SIMS have been used to study the chemical effects of both N_2^+ ion irradiation and storage gas surrounding the samples. The results show faster aging in the case of the samples stored in vacuum, a harmful effect of nitrogen gas on the treatment and the formation of new chemical species for all treatments.

INTRODUCTION

The surface modification by ion irradiation is a domain of interest for several research and development sectors such as microelectronic, optic, acoustic, thin film technology, thermonuclear fusion, biomaterials and aerospace. This strong interest is due to the necessity of having specific surface properties distinguishable from that of the bulk of the materials. Therefore, it is important to understand and control the phenomena involved in the modification of the surface properties to develop new technic for the modification at will of these properties.

The liquid-solid interaction is also a domain of interest. In fact, the liquid affinity with a solid surface depends on the wetting properties of one on the other. This property can be evaluated by the measurement of the contact angle between the liquid (in our case water) and the sample surface. The contact angle is defined as the angle between the tangent at the surface of a liquid drop at its contact point with the sample surface, and the solid surface under the drop. It is convenient to define two different contact angles named advancing (ACA) and receding (RCA). The ACA is measured during the increase of the drop diameter while the RCA is obtained during the decrease of the drop diameter.

For the modification of the wetting properties, the use of low energy (~500 eV) ion beam is attractive [1, 2, 3]. Combining their nuclear and electronic energy loss, low mean range (a few nm), erosion rate and fluence, it allows the modification of the first surface atomic layers. After the modification by implantation, it is important to control the recovery of the contact angles. In fact, the wetting modification is evolving with the passage of time. In this paper, we will report the use of different types of storage gases which have an effect on the aging of polymer after ion implantation.

EXPERIMENT

Materials

Only commercial polycarbonate (LEXAN) was used in the surface modification experiments. The repeat units of this polymer is shown in Figure 1. The PC samples have been cut of a sheet in blocks of 10 x 10 x 3 mm^3 and ultrasonically cleaned in ethanol and dried with soft paper. The surfaces were also examined by means of optical microscopy.

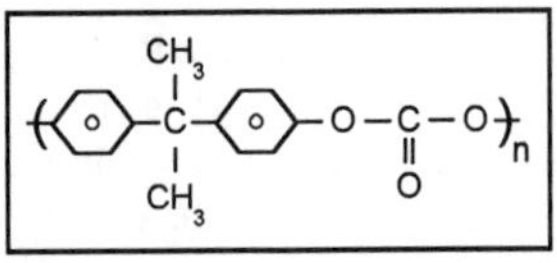

Figure 1. The repeat units of polycarbonate (PC).

Implantation

The samples were introduced into the implantation chamber where the base pressure was ~2x10^{-7} torr. All samples were implanted at room temperature with nitrogen (N$_2^+$) ions to an energy of 500 eV/atom and a fluence of 5x10^{16} atom/cm^2. The mean range and the standard deviation of the N ion distribution in the PC were calculated by means of the Monte Carlo code TRIM95 [4] and are 34 Å and 16 Å, respectively.

After the implantation, the samples were stored during 1 to 25 hours either in different gas environments (nitrogen, oxygen, dry air and argon) directly in the implantation chamber at atmospheric pressure or under vacuum. Some samples were taken out of the vacuum chamber immediately after the implantation (nitrogen was used to fill up the chamber) and put in room air.

Contamination

The samples put under vacuum were slightly contaminated by polydimethylsiloxane (PDMS). PDMS is an usual contaminant in vacuum chamber and was clearly identified only by means of ToF-SIMS. Using contact angle measurements and the relationship relating apparent contact angle to the chemical composition of a surface (cos θ' = f$_1$cosθ$_1$ + f$_2$cosθ$_2$), an average fraction of 15% PDMS was deduced. This contamination has a small effect on the contact angles (~5°) and XPS measurements and its contribution was subtracted from the contact angle measurements (adjusted contact angles) and the atomic concentration ratios obtained by XPS. However, the presence of PDMS doesn't allow the sample characterization by Tof- SIMS. So, the samples were cleaned by means of hexane to remove the PDMS. XPS measurements have insured that this cleaning procedure has no noticeable effect on the PC surface composition.

Surface analysis

I. Contact angle. Wettability, before and after treatment, was characterized by the measurement of the contact angle hysteresis (ACA and RCA) of a drop of distilled water (surface tension = 72mJ/m^2) is deposed on the sample surface [5]. The maximum volume of the drop was 10 μl and the injection and suction speed was 6 μl/min.

II. XPS. The surface composition was determined by XPS (model ESCALAB 220i-XL) equipped with a non-monochromatic Al Kα (1486,6 eV) source at a power of 4.05 W/mm^2 (15 kV, 27 mA) and a large area XL. The surface charge compensation was achieved by using a low-energy (6 eV) electron flood gun and by covering the sample with a metallic ring. The total pressure in the main vacuum chamber during analysis was typically 9x10^{-9} torr. For each sample, a detail scan of C 1s, O 1s and N 1s was performed by order. The total exposition time for each sample never exceeded

30 minutes to keep small the changes of the elemental composition due to the sample degradation under x-rays and low energy electron beam. In spite of this precaution, an increase of the contact angles (8° in advancing and 3° in receding) of the implanted and non-implanted PC was measured. According to Beamson and Briggs [6], the effect on the O/C ratio is lower than 0.3 %.

The calibration of the binding energy (BE) scale was made by setting the C 1s BE of the C-C and C-H peak at 285.0 eV. A non-linear Shirley-type background substraction was used, and the peaks were decomposed by using a least-square routine assuming a Gaussian/Lorentzian line shape. The intensity ratios were converted to atomic concentration ratios by using the sensitivity factor proposed by VG Instruments. In our case, the sensitivity factors are 1 for carbon, 1.8 for nitrogen and 2.93 for oxygen. The resolution of the apparatus is approximately 1 eV.

III. ToF-SIMS. Positive and negative ToF-SIMS spectra were obtained with a Charles-Evans time-of-flight SIMS spectrometer with the following standard operating conditions: a pulsed 15 keV gallium ion beam (530 pA DC, pulses of 5 ns, 5 kHz repetition rate) was rastered over 196 x 196 μm² area for a maximum acquisition time of 300 s, corresponding to a total ion dose of 6.5 x 10¹⁰ ions/cm² per spectrum, insuring static conditions. The secondary ions were extracted by a ±3 kV voltage and then deflected by three electrostatic analyzers in order to compensate for the initial energy distribution of ions with the same masses. The mass spectra of these ions (positive and negative, separately) were recorded by measuring their time-of-flight distribution from the sample surface to the detector. The scanned mass range was from 0 to 5000 amu. The charge compensation of the sample was achieved with a pulsed electron gun (20 eV) and a grounded stainless steel grid covering the sample.

RESULTS AND DISCUSSION

The ACA and the RCA of the non-implanted PC are 79°±2° and 59°±3°, respectively. Immediately after a N_2^+ implantation (5×10^{16} at./cm² at 500 eV/at.), the ACA and the RCA become 54°±2° and 27°±2°, respectively. Also, we note that the implantation causes the increase of the hysteresis of the contact angles from 20° before implantation to 27° after.

Figure 2 shows the evolution of the contact angles of the PC stored in vacuum or nitrogen immediately after a N_2^+ implantation. We note a different aging effect of the contact angles on each sample. However, in all cases, the aging follows a logarithmic function (between the interval from 0.1 to 25 hours). The samples stored in vacuum recover partially the values before the implantation faster than the samples stored in nitrogen.

Figure 3 shows that the storage gas environments used after an N_2^+ implantation in the PC samples have a large influence on the values of the wetting properties of the PC. In fact, oxygen and argon improve the wettability of the PC after the implantation whereas nitrogen storage gas can't keep the contact angles at the same values as immediately after the implantation. The effects of dry air and room air storage gases on wetting are

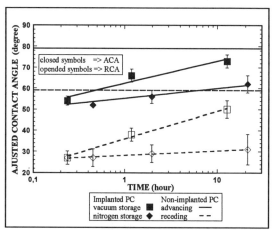

Figure 2. Aging of implanted PC under nitrogen and vacuum conditions.

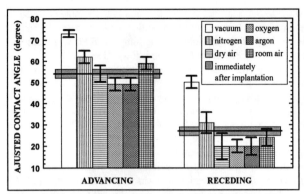

Figure 3. Contact angles of nitrogen implanted PC after 20 hour storage in different gases.

set between those, on the one hand, of oxygen and argon and, on the other hand, of nitrogen gas. Since the argon has the same effect as the oxygen, we can't assert that the presence of oxygen at the surface improves the wettability of the PC. Thus, we can only make the hypothesis that the nitrogen is harmful for the wettability after the implantation.

Figure 4 shows our experimental results on the implanted PC, aged in air, after 20 hour storage in different gas environments. Lines are used instead of distinct points (~ 25 points per line) to make easier the visualization. This figure indicates that the contact angles vary as a function of time after the implantation and tend to recover the values measured before the implantation. Also, there is clear evidence that the gas used to store the samples immediately after the implantation (during the first ~20 hours) has an influence on the long-term behaviour of the wetting properties of the PC kept in room air. Figure 4 points out that the long-term (after ~10,000 hours) contact angles of the samples stored in oxygen remain lower than the contact angles of samples stored in other gases. It seems that the oxygen helps to stabilize the contact angles at lower values than for non-implanted PC. If we look at the first 20 hours, we observe a difference between the evolution of the ACA of the implanted PC stored in oxygen and stored in argon. This observation suggests that oxygen has a positive effect on the active surface after the implantation. Therefore, the presence of oxygen and nitrogen in air can explain that the contact angles of samples stored in dry air and room air stay between those stored in nitrogen and oxygen. Finally, the decrease (in the first hours) of the contact

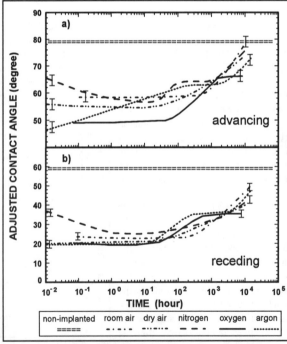

Figure 4. Aging in room air of nitrogen implanted PC after 20 hour storage in different gases.

angles of the PC implanted and stored in nitrogen (observed before the stabilisation at the same values than the implanted PC stored in room air), suggests once again that the nitrogen has a harmful effect on the improvement of wettability following the implantation by low energy N_2^+.

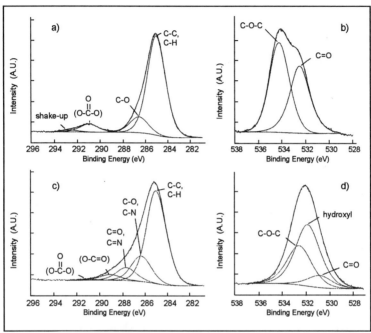

Figure 5. XPS spectra of the non-implanted PC a) C 1S line, b) O 1s line and of the implanted PC stored in oxygen (20 hours) c) C 1S line and d) O 1s line.

Figure 5a)c) shows a comparison of the C 1s XPS spectra between non-implanted and implanted PC. The peak positions have been chosen in accordance with other works [4, 7, 8, 9]. On the one hand, there is formation of two new peaks; one peak represents a combination of (C=O) and (C=N) at 287.7 eV, while the other stands for (O-C=O) at 289.1eV and a decrease of the shake-up contribution. On the other hand, we note a displacement from 286.4 eV to 286.6 eV and an increase of the C-O peak. We cannot explain these observations only by the presence of nitrogen in the first surface atomic layers (C-N), because the N 1s spectra are not enough large to explain this strong increase. However, creation of hydroxyl might be the explanation [7]. We can discuss of this phenomenon by looking at the O1s spectra (figure 5 b) and d)). A good fit of the O 1s curve has been obtained in adding a new peak (the hydroxyl contribution (C-OH)) between the carbonyl (O=C) and the ether (C-O-C) groups at 1.0 eV from the carbonyl peak [7]. This contribution can be justified by the ToF-SIMS results. Indeed, the ToF-SIMS positive and negative spectra of the implanted PC, show an increase of characteristic peaks with C-OH ending chains. These peaks are observed at 227 amu in the negative spectra and at 107 and 125 amu in the positive spectra [10,11].

The effect of a storage under vacuum immediately after the nitrogen implantation is seen on the C 1s XPS spectrum. A reorganization of the surface in agreement with the initial structure of the PC and a slight loss (diffusion either in the bulk or out of the sample) of nitrogen near the surface

(N/C decrease from 0.10 to 0.07) are observed. This loss of nitrogen is more pronounced for the amines/imines than for the amides on the N 1s spectrum. The same phenomena are noted in the ToF-SIMS negative spectrum, where it has an intensity decrease larger in the CN and NH peaks than in the CNO peak. Finally, a decrease of the characteristic peaks containing hydroxyl radicals is observed in agreement with the reorganization of the O 1s XPS spectra.

If we consider the atomic ratios, we can notice that the O/C ratios are almost the same before and after the implantation, and this is valid for all storage gases. This suggests that the low energy implantation doesn't cause an oxygen desorption. However, since the x-radiation can induce only a slight loss of oxygen [6], it seems that the low energy implantation (nitrogen) doesn't increase the oxygen proportion in the first surface atomic layers.

CONCLUSION

This work has highlighted that low energy (500 eV) ion beam (N_2^+) can be used to improve the wetting properties of polycarbonate. Furthermore, the evolution of wettability after the treatment depends on the gas used for the storage of the samples during the first hours. Indeed, oxygen storage gas used immediately after the implantation improves the effect of the treatment. No notable different aging between samples stored in dry and room air was observed. However, nitrogen has a harmful effect and increases the contact angles. Thus it seems that nitrogen (and not humidity) is the principal responsible of the aging. The improvement of the hydrophilic character of polycarbonate seems to be linked to the increase of hydroxyl radicals near the surface.

We acknowledge G. Veilleux for his help and suggestions concerning the XPS and F.Quirion for valuable discussion on wetting.

REFERENCES

1. J.F. Pageau, G.G. Ross, É. Couture, A. Poirier and F. Quirion in Ion-Solid Interactions for Materials Modification and Processing, edited by D.B. Poker, D. Ila, Y.-T. Cheng, L.R. Harriott and T.W. Sigmon (Mater. Res. Soc. Proc. **396**, Pittsburgh, PA, 1996) p. 299.
2. S.C. Park, H.J. Jung and S.K. Koh in Ion-Solid Interactions for Materials Modification and Processing, edited by D.B. Poker, D. Ila, Y.-T. Cheng, L.R. Harriott and T.W. Sigmon (Mater. Res. Soc. Proc. **396**, Pittsburgh, PA, 1996) p. 347.
3. W.K. Choi, S.K. Koh and H.J. Jung, JVST A, **14** (4), p. 2366 (1996).
4. J.F. Zeigler, J.P. Biersack and U. Littmark, The Stopping and Range of Ions in Solids, edited by J.F. Zeigler (Pergamon, New York, 1985), p.321.
5. Éric Couture. Master's thesis, Université du Québec (INRS-Énergie et Matériaux), 1995.
6. G. Beamson and D. Briggs, High Resolution XPS of Organic Polymers. The Scienta ESCA300 Database, 1st ed. (Wiley, Chichester, 1992)
7. L.J. Gerenser, J. Adhesion Sci. Technol., 7 (10), p. 1019. (1993).
8. C.Briggs, and M.P. Seah Practical Surface Analysis (2 éd.) vol.1: Auger and X-ray Photoelectron Spectroscopy, edited by M.P. Seah (Wiley, New York, 1990).
9. J.-B.Lhoest, P. Bertrand, L.T. Weng and J.-L. Dewez, Macromolecules, **28** (13), p. 4631 (1995)
10. J.G.Newman, B.A. Carlson, R.S. Michael, J.F. Moulder and T.A. Hohlt (editor), Static SIMS Handbook of Polymer Analysis, (Perkin-Elmer Corporation, Minnesota).
11. D. Briggs, A. Brown and J.C. Vickerman, Handbook of Static Secondary Ion Mass Spectrometry, (John Wiley & Sons, New York).

ION ASSISTED REACTION IN POLYMER AND CERAMICS

S. K. Koh, S. C. Choi, S. Han. H-J Jung
Thin Film Technology Research Center, Korea Institute of Science and Technology, P. O. Box
131, CHEONGRYANG, SEOUL 130-650, KOREA

ABSTRACT

Ion assisted reaction (IAR), which was firstly presented in 1995 MRS Fall meeting, has been reviewed for the surface modifications of polymer and ceramics. The reaction is assisted by energetic ions from 0.5 to 1.5 keV, doses 10^{14} to 10^{17} ions/cm^2, and blowing rate of oxygen 0 ~ 8 ml/min. Hydrophilic surfaces of polymers (wetting angle < 20^0 and surface energy 60 ~ 70 erg/cm^2) have been accomplished by the reaction, and an improvement of wettability and an increment of the surface energy are mainly due to the polar force and hydrophilic functional groups such as C=O, (C=O)-O, C-O, etc., without surface damage. The IAR was also applied on aluminum nitride in an O_2 environment and AlON on AlN is formed by the Ar$^+$ irradiation. The improvement of bond strength of Cu films on the AlN surface resulted from the interface bonds between Cu and the surface layers. Comparisons between the conventional surface treatments and the IAR are described in terms of physical bombardment, surface damage, functional group, and chain mobility in polymer.

INTRODUCTION

Surface modification by ion beam has been researched for scientific or industrial purpose and it has been applied to various field in material engineering. The usage of ion beam could change surface properties of materials and create new compounds which could not be formed in equilibrium reaction. In the surface modification of solid, one of the widely used method is direct irradiation of ion onto the solid surface. When non-reactive ion such as He$^+$, Ar$^+$, and Ne$^+$, etc., is irradiated onto the solid surface, it sputters the target atoms and destroys the surface structure, but it does not form new compound. Though irradiation of reactive ion such as a oxygen ion, nitrogen ion, etc., could form new chemical compounds, it induces the damage on the surface, which, especially in the case of polymer modification, results in reducing the density of functional groups on the surface after washing with water [1].

Ion assisted reaction (IAR) is one of the useful methods to create new chemical compounds on the solid surface, and it uses non-reactive ion instead of reactive ion. In IAR, new compounds on the surface could be formed by reaction between blowing gases near the target surface and target components. In this reaction, irradiating ion induces the activated sites or the radicals on the target surface which easily react with other foreign components. The IAR showed excellent characteristics in polymer modification [2-5].

In this article, polystyrene (PS) and polyvinylidenefluoride (PVDF) were modified by the IAR. PS is one of the most widely used polymers as a biomedical material especially as a petri-dish due to durability, low production cost, and easy shapability. Though petri-dish has been widely used as a disposable culture dish, it does not offer suitable environment to some cells like tissue cell which needs hydrophilic surface in order to affix to the substrate [6]. To offer suitable environment to such a cell, PS petri-dishes were modified by the IAR and culturing capability of modified PS petri-dishes was surveyed. PVDF has a piezoelectric characteristic. In order to apply PVDF in electrical devices, contact with a metal electrode is inevitable. Poor adhesion between the metal electrode and the substrate often induces device failure. Adhesion strength is closely related to hydrophilic surface in metal/polymer system [7]. So, PVDF was also modified by the IAR to improve adhesion strength and the relation between the adhesion strength and the hydrophilic surface was investigated. The IAR process also is applied to modify AlN ceramics and the effect of the IAR treatment was also surveyed by measuring the adhesion strength.

EXPERIMENT

The ion assisted reaction system consists of a conventional ion beam system, a reactive gas introducing system, an ion gun, and a sample holder. A cold hollow type ion gun has been used in the system. The ion energies were adjusted from 0.5 to 1.2 keV, and current of ion beam controlled by discharge voltage and current. The amount of ion irradiating the specimen was measured by a Faraday cup biased -24 voltage to remove the secondary electron, and was controlled by changing the exposure time at fixed ion beam current. Reactive gases were blown from 0 to 8 ml/min during the ion beam irradiation. More detail experimental system and methods such as sample preparation, wetting angle measurement, x-ray photoelectron spectroscopy (XPS), etc., were described in detail elsewhere [2-4].

Relation between wettable PS petri-dish surface and cell growth was investigated by growing the rat pheochromocytoma cell (PC12 cell). The PC12 cell has property to adhere very poorly to surface of plastic dish, thus shows difficulties to grow on the PS petri-dish without any treatment. The PC12 cells were seeded in RPMI1640 media which contains 5% fetal bovine serum and 10% horse serum, and cultured in the incubator with 5% CO^2. The growth of the PC12 cell was compared by counting the number of cells from each group in every 24 hour for 6 days. Counting the number of cells were performed as follows; The media were removed by aspiration and cells were rinsed with phosphate buffered saline (pH 7.3). The cells were detached from the PS petri-dishes by incubating with trypsin solution (GIBCO) for 5 min. at 37°C. Cell suspensions were pipetted several times to obtain single cell to count. Then, the cell suspensions were diluted with media and counted the number of cell by optical microscope using hematocytometer.

Adhesion between PVDF and Pt was tested by boiling methods. Pt/PVDF samples were dipped into water and boiled at 100°C for 4 hours and subsequently cooled by dipping into the tap water. After cooling, the surfaces of Pt/PVDF samples were examined by optical microscope.

AlN ceramics was also modified by the IAR. The characteristics of modified ceramic surface was studied by XPS. Cu film was deposited on modified ceramics and adhesion between Cu film and ceramics was investigated by scratch test.

RESULTS

Figures 1(a) and (b) show changes of the contact angles of the PS and PVDF as a function of irradiating ion dose with and without blowing oxygen gas at the acceleration voltage of 1.0 kV, respectively. As shown in Fig. 1(a), the contact angles of specimens irradiated by Ar^+ ions without blowing oxygen gas were reduced as much as $30° \sim 40°$ from those of untreated samples. The effect of blown oxygen gas was also surveyed. The contact angles of PS irradiated with Ar^+ ions were reduced as the oxygen gas flow rate increased. Changes of wettabilities mainly depend on the ion dose and the oxygen gas flow rate. As shown in Fig. 1(b), the minimum contact angle is $19°$ at the ion dose of 5×10^{16} ions/cm^2 and the oxygen gas flow rate of 8 ml/min which is lower than that of PS irradiated without oxygen gas under the same ion dose. When the ion dose exceeded 1×10^{17} ions/cm^2, a slight increment of the contact angle was observed. In the case of

PVDF, the contact angle was remarkably reduced at the dose of 1×10^{15} ions/cm^2 and slowly increased by further ion irradiation. In both cases, the introduction of oxygen gas makes the contact angle considerably decrease and some chemical reaction would occur between the oxygen gas and the surface of PS and PVDF.

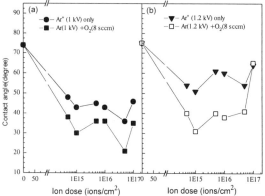

Figure 1. Changes of the contact angles of (a) the PS and (b) PVDF as a function of ion dose with and without blowing oxygen gas at the acceleration voltage of 1 kV.

Figure 2 shows the C1s core level spectra of (a) PS and (b) PVDF. In Fig. 2(a), the C-C peak and/or the C-H peak was observed at 284.6 eV in the untreated PS. In Fig. 2(b), C-F peak and C-H peak was observed at 289.2 eV and at 285.0 eV, respectively, in the untreated PVDF. After the IAR treatment of PS (Fig. 2(a)), other peaks are observed. The ridge line that appeared at higher binding energy than 285 eV should be attributed to C-O, C=O and -(C=O)-O- functional groups formed by the IAR, respectively. In the case of PVDF, the peak intensity of 289.2 eV rapidly decreased but that of 285 eV does not and only a slight decrement and peak broadening was observed. The intensity level of the ridge between 285 and 292 eV might be due to C-O, C=O, -(C=O)-O-, and C-F functional groups. The above fact means that F atoms were preferentially sputtered by ion bombardment [8] and active sites were produced by ion bombardment. The active sites reacted with other carbon atoms to form crosslinks or with the introduced oxygen atoms to form hydrophilic groups. However, as shown in Fig. 2, carbonization (~284 eV) process more rapidly occurred in PVDF than PS and the increment of the contact angle at high ion dose might be due to carbonization.

Figure 3(a) shows the growth curves of PC12 cells cultured in PS petri-dishes. Initially, 2.5×10^5 cells were seeded in each petri-dish (5-cm in diameter). For the group of non-treated petri-dishes, after one day of seeding, most of the cells were not attached to the surface of the dishes and only a few cells were attached so that they survived. While, for both groups of IAR-treated petri-dishes, most of the cells were able to attach to the dishes and survived. The reduction of cell concentration after one day of seeding means either the number of cells in day 0 is over saturated or not every cell is healthy to attach to the surface of the dishes, and those phenomena are common for culturing animal cells. The number of PC12 cells in petri-dishes treated by IAR with the doses of 5×10^{16}/cm^2 and 1×10^{15}/cm^2 increase up to 2.0×10^5 and 1.4×10^5 after 2 days and up to 1.6×10^5 and 1.9×10^5 after 3 days, respectively. After the number of viable cells reached maximum value, the population of the cells was decreased in both of the treated groups, which might be due to a depletion of media and a contact inhibition.

Figure 3(b) and (c) show the microscopic images of PC12 cells grown on petri-dishes after 2 days. Panel (b) represents the image of the cells on the non-treated petri-dishes, and panel (c) represents that on the IAR treated with dose of 5×10^{16}/cm^2. As shown in Fig. 3(b) and (c), distinct difference in cell growth was observed which might be depending on the surface hydrophilicity. PS petri-dishes modified by the IAR showed exclusively preferential culturing properties of PC12 cells.

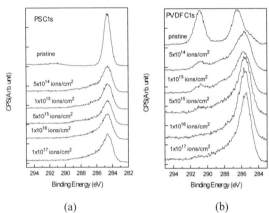

(a)   (b)

Figure 2. XPS C1s core level spectra of (a) PS and (b) PVDF after IAR treatment.

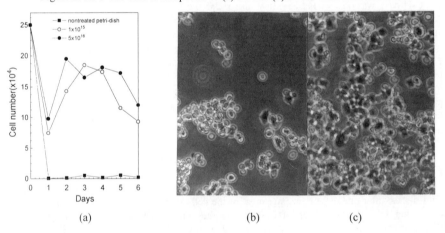

(a)   (b)   (c)

Figure 3. (a) PC12 cell growth curve on PS petri-dishes modified by the IAR methods and optical microscope images of grown cells on petri-dish. (b) before treatment, and (c) after IAR treatment with dose of 5×10^{16} in oxygen environment.

In order to examine the adhesion strength of Pt/PVDF samples, the boiling test were performed. Surface images of Pt/PVDF film after the boiling test was observed by optical microscopy. Images of the test for Pt layer on pristine PVDF is shown in Fig. 4(a). Pt layer deposited on non-treated PVDF shows the buckling phenomena on the whole area, but Pt layer deposited on the IAR treated PVDF shows the crack propagation rather than common buckling phenomena. Adhesion between Pt and non-treated PVDF is not good so thermal stress due to different thermal expansion coefficient are released by separating Pt layer from PVDF resulting in forming buckling. While, in the case of Pt film deposited on IAR treated PVDF, adhesion between Pt and PVDF is good enough to endure the thermal stress so stress at the interface is released by forming cracks.

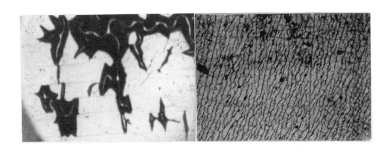

(a)   (b)

Figure 4. Optical microscope images of Pt layer after boiling test. Pt film was deposited on (a) pristine PVDF and (b) IAR treated PVDF with dose of 1×10^{15} ions/cm^2 in oxygen environment.

AlN surface was also modified by the IAR at various oxygen environments and 1000 Å Cu was deposited by ion sputtering. The result of the scratch test is shown in Fig. 5. Cu films deposited on non-treated AlN showed poor bond strength and were delaminated from the beginning stage. On the other hand, Cu film deposited on AlN modified by Ar^+ ion with dose of 1×10^{15} ions/cm^2 and oxygen flow rate of 4 ml/min showed an increased acoustic emission at the normal load of 30 Newton. These results indicate that the bond strength between Cu and AlN was significantly increased by the IAR treatment.

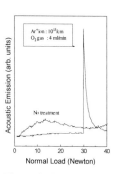

Figure 5. The variation of acoustic emission as an increase of normal load during scratch test of Cu film on the AlN substrates.

Figure 6 represents the XPS core-level spectra acquired from the sample of Cu film deposited on AlN modified by the IAR. After 20 sec. of sputtering, pure Cu (3p) peaks were observed at 75.5 and 77.5 eV. After 10 min. of sputtering, Cu3p peaks were shifted to higher binding energies and were broaden. The results mean that copper oxide was formed in this layer by the IAR. After 20 min. of sputtering, Al2p core level spectrum was obtained at the binding energy of 72.2 eV which was shifted to higher position comparing to the spectrum of AlN (71.8 eV) after

45 min.of sputtering. The Al2p spectrum obtained after 25 min. of sputtering is not due to pure AlN but due to other complex. O1s core level spectra shows the formation of oxide layer at the interface region. The O1s spectra obtained after 10 min. and 25 min. of sputtering show CuO (531.5 eV) peak and AlON (529.5 eV) peak as shown in Fig 6(b). These results confirmed that a intermediate layer such as CuO and AlON was formed between the deposited Cu and new surface layer due to chemical reaction among the irradiated AlN surface. The presence of a intermediate layer should contribute directly to improve the adhesion strength between Cu film AlN substrate.

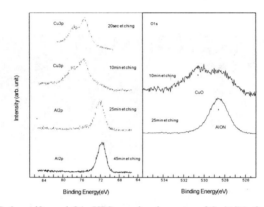

Figure 6. Cu3p, Al2p and O1s XPS core level spectra of Cu/AlN after modification.

CONCLUSIONS

We survey the IAR process. The IAR process uses low energy non-reactive ion (~ keV). So, modification of large area is easy and the cost of constructing the system is cheap. Moreover, in polymer (PS and PVDF) and ceramics(AlN) surface modification, IAR technique shows excellent results in formation of hydrophilic surface and improvement of adhesion. Further research should be done to apply this technique to various material and for industrial purpose.

REFERENCES

1. B. M. Callen, M. L. Ridge, S. Lahooti, A. W. Neumann, and R. N. S. Sohdi, J. Vac. Sci. Technol. **A13**, 2023 (1995).
2. J. S. Cho, W. K. Choi, K. H. Yoon, H. J. Jung, and S. K. Koh, J. Mater. Res. **12**, 277 (1997).
3. W. K. Choi, S. K. Koh, and H. J. Jung, J. Vac. Sci. Technol. **A14**, 2366-2371 (1996).
4. S. K. Koh , W. K. Choi, J.S. Cho, S. K. Song, Y. M. Kim, and H-J. Jung, J. Mater. Res. **11**, 2933 (1996).
5. S. K. Koh , H-J. Jung, S. K. Song, W. K. Choi, Y. S. Yoon, and J.S. Cho, USA patent, "Process for modifying surface of polymer, and polymer having surface modified by such process" 1996. 3. 19.
6. J-B. Lhoest, J. L. Dewez, and P. Bertrand, Nucl. Instrum. Methods **B105**, 322-327 (1995).
7. R. Schalek, M. Hlavacek, and D. S. Grunmmon, Mat. Res. Soc. Symp. Proc. **236**, 335 (1992).
8. E. H. Adem, S. J. Bean, C. M. Demanet, A. Le Moel, and J. P. Duraud. Nucl. Instrum. Methods **B32**, 182 (1988).

ENHANCEMENT OF ADHESION BETWEEN CU THIN FILM AND POLYIMIDE MODIFIED BY ION ASSISTED REACTION

S. C. Choi*, K. H. Kim*, H-J. Jung*, C. N. Whang**, and S. K. Koh*
*Thin Film Technology Research Center, Korea Institute of Science and Technology, P.O.Box 131 Cheongryang, Seoul, 130-650, Korea
**Department of Physics, Yonsei University, Shinchondong, Seodemoongu, Seoul, 120-749, Korea

ABSTRACT

Polyimide films are modified by ion assisted reaction method using various ion beams in various gases environments. Amount of ion and blown gases rate were changed from 5×10^{14} to 1×10^{17} and from 0 to 8 sccm, respectively. Wetting angles between water and polyimide films modified by Ar^+ ion without oxygen blowing decrease from 67° to 40° and surface free energies increase from 46 to 64 dyne/cm^2. Wetting angle of polyimide films modified by Ar^+ ion in an oxygen environment decreases to 12° and surface free energy increases to 72 dyne/cm^2. The lowest wetting angle was obtained by oxygen ion irradiation in the oxygen gas environment and its value was 7°. In the case of polyimide film modified by Ar^+ ions in an oxygen environment, the wetting angle increases up to 65° when it kept in air and that increases up to 46° when it kept in water after 5 day. In the case of polyimide film modified by O_2^+ ion in oxygen environment, however, the wetting angle of polyimide film dose not increase. X-ray photoelectron analysis shows that the chemical bonds between polyimide components are severed by ion irradiation and hydrophilic groups such as CO and C=O are formed by the reaction between newly formed radicals and blown oxygen. It was found that adhesion between Cu and polyimide modified by ion assisted reaction was improved. The main reason of the enhanced adhesion is due to the reaction between Cu and C-O or C=O groups formed by ion assisted reaction on the polyimide surface.

INTRODUCTION

Polyimide (PI) has been received considerable attention due to its thermal stability, low dielectric constant and good mechanical properties [1]. It has been used in application ranging from aerospace to microelectronics. However, one of the drawback is its low adhesion to metals. To improve adhesion between polymers and metals, various methods such as ion beam mixing, insertion of an interlayer, annealing, etc., were studied [2-4]. Though annealing process improves adhesion between PI and metals, it takes place diffusion of a metal layer [5]. Insertion method usually use Cr as a interlayer, but Cr induces pollution problem. While, ion beam modification process, usually performed at room temperature, could avoid these problems. Adhesion between polymers with metals is closely related to wettability [6,7]. To give hydrophilic functionality to polymer, we employ the ion assisted reaction (IAR) techniques, with which we already show remarkable improvement in wettability and adhesion applying to various polymers [8-11].

In this article, we examine the effects of ion assisted reaction in oxygen environment and the relation between hydrophilic groups and adhesion between Cu and PI films. The initial growing stage of Cu film on the IAR treated PI films was investigated by X-ray photoelectron spectroscopy (XPS). Though many studies on interaction between Cu and PI have been done and the results have been published, interaction between Cu and PI films which has hydrophilic surface was not studied. Interaction between Cu and PI films which has a hydrophilic surface also was studied by in-situ XPS.

EXPERIMENTS

Hydrophilic surface of PI was generated by the IAR technique. In the IAR process, the gas flow rate near the substrate was controlled by a mass flow controller from 0 to 8 ml/min(sccm). In this article, the gas flow rate was fixed at 8 sccm . Various ions were generated by a hollowed cathode type ion gun and ion beam potential was fixed at 1-kV. Ion beam current arrived at target was monitored by a Faraday cup. Ion dose was changed from 5×10^{14} to 1×10^{17} ions/cm^2. Detail experimental conditions about the IAR process were described elsewhere [11]. Chemical changes on a surface of PI after the IAR treatment were studied by XPS. Monochromatized Al K-α line (1486.8 eV) was used as an X-ray source. The take-off angle was fixed at 45° and a working pressure was 2×10^{-10} Torr. For adhesion test, Cu (1000-Å) films were deposited by ion beam sputtering. The adhesion test between Cu and polyimide was done by boiling test. Each sample was dipped into boiling water for 8 hours. After boiling, samples were dried in air and peel test using scotch tape (3M Magic Tape^{TM}) was done. In order to elucidate the relation between adhesion and interface of Cu/PI, *in-situ* XPS measurements were performed after depositing every 1-Å Cu film by ion sputter.

RESULTS

Figure 1(a) represents changes of the contact angles of water drops on the PI films as a function of ion dose. In the following contents, modification conditions of PI films are denoted as "PI$_{X,X}$", where the first and the second subscripts represent the species of the ions and the environment gases, respectively. In the case of PI film modified by ions without blowing gases, only the first subscripts are denoted. The contact angle of water drops on the pure PI was measured to be 67°. In the case of PI$_{Ar}$, the contact angle decreased to 40°. The contact angles of PI$_{Ar,N}$ or PI$_{Ar,CO2}$ were nearly the same as PI$_{Ar}$. However, the contact angles of PI$_{O,O}$ and PI$_{Ar,O}$ decreased to 7° and 12°, respectively. The contact angle of PI$_O$ does not remarkably decreased at low ion dose, but PI$_{O,O}$ decrease to 10° at the dose of 5×10^{14}. These results represent that the oxygen gas plays an important role to improve wettability of PI in the IAR process. One of the interesting thing is that, in the case of PI$_{Ar,O}$, the wetting angle increases up to 65° when it kept in air and that increases up to 46° when it kept in water after 5 day. In the case of PI$_{O,O}$, however, the wetting angle of polyimide film dose not increases. This fact means that the hydrophilic groups formed by IAR was not washed by water.

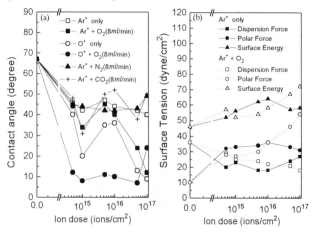

Fig. 1. Changes of (a) the contact angles and (b) the surface energy as a function of ion dose.

Figure 1(b) represents changes of the surface energy (the polar force + the dispersion force) of IAR treated PI as a function of ion dose. The reason to choose $PI_{Ar,O}$ is that, as shown in Fig 1(b), the contact angles are drastically changed as a function of ion dose. Evaluation of the exact surface energy of $PI_{O,O}$ is very difficult because the contact angle is too low to exactly measure. However, from the rough measurements, the surface energy of IAR treated $PI_{O,O}$ was evaluated to be higher than 70 erg/cm^2 though small amount of ion irradiated on PI films by the IAR. The surface free energy of PI_{Ar} gradually increases with increment of ion doses up to 1×10^{16} ions/cm^2 and slight decrement of surface free energy was observed when ion dose exceeded 5×10^{16} ions/cm^2, while, in the cases of $PI_{Ar,O}$, such a decrement was not observed. As shown above, ion beam irradiation with blowing oxygen gas is more effective to modify the PI surface than without blowing oxygen.

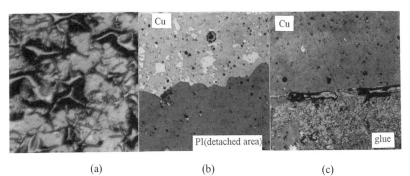

(a)   (b)   (c)

Fig. 2. Optical microscopic images of Cu/PI. (a) Cu on pure PI after boiling test (b) result of scotch tape test of Cu on the IAR treated $PI_{O,O}$ with ion dose of 1×10^{15} ions/cm^2 after boiling test and (c) result of scotch tape of Cu on the IAR treated $PI_{O,O}$ with ion dose of 5×10^{16} ions/cm^2 after boiling test

Wettability and surface energy are related to the adhesion strength to metals, adhesive and dissimilar materials with polymers. $90°$ peel test or tensile test has been commonly used for measuring the adhesion strength. In our adhesion test, $90°$ peel test could not be carried out due to the higher adhesion strength at interface of Cu/PI then the adhesion strength between Cu and epoxy resin (> 50 g/mm) to fix the samples to holder. In order to show the difference, we adopted boiling test method which has been used for adhesion test of metal/polymer system [12]. When metal/polymer samples are dipped into boiling water, large amount of stress is applied to the interface due to the different thermal expansion coefficient and boiling water continuously attacks the weak point to form hillock or crack. Figure 2 shows the results of the peel test of Cu/PI samples after dipping into boiling water for 8 hours. As shown in Fig. 2, hillock was formed on Cu film deposited on pure PI after boiling test (Fig. 2(a)) and Cu film is detached from PI by scotch tape test. The IAR treated PI shows enhanced adhesion. After boiling test, hillock was not observed at the IAR treated PI by optical microscope (Figs. 2(b),(c)) but Cu films were detached from the IAR treated PI films with low ion dose ($5 \times 10^{14} \sim 5 \times 10^{15}$ ions/cm^2) by scotch tape test (Fig. 2(b)). While Cu films deposited on IAR treated PI with high ion dose (> 1×10^{16} ions/cm^2) were not detached from the IAR treated PI films by scotch tape test and a glue layer of scotch tape remained on Cu (Fig. 2(c)). This result clearly shows enhancement of adhesion by IAR treatment.

XPS study was performed to investigate the chemical reaction on a PI surface. For this purpose, two kinds of samples were prepared: pure PI film and IAR treated $PI_{O,O}$ with the ion dose of 5×10^{16} ions/cm^2 and the flow rate of 8 sccm. Figure 3 represents C1s, O1s and N1s core level spectra of pure and IAR treated PI films. In Fig. 3, each core level spectrum was deconvoluted following the results of

R. N. Ramb *et al.*[13] and each peak position was represented in Fig. 3.

XPS spectra of the IAR treated $PI_{O,O}$ shows many changes in each core level spectrum. First, the peak appeared at 285.8 (C-N or C-C bonds in PMDA) unit decreased. Second, new peak is formed at 287.8 eV. Third, the other peak heights likes 286.6 eV (C-O-C), 285.8 eV (C-C in PMDA unit) and 288.6 eV (C=O) are changed. Reduction of peak height at 285.8 eV is mainly due to the bond scission of C-N and the destruction of phenyl ring structure in PMDA unit. After ion irradiation, the intensity of N1s decreases more than half of its original value and the shake up satellite peak (292 eV) due to the $\pi \rightarrow \pi^*$ transition disappeared. The energetic incident ions randomly break the bonds in the PI unit. If the bonds between the imide nitrogen and the carbonyl carbon in the PMDA unit are broken, -C-N· and -(C=O)· could be induced. Generation of this amide like moiety increases the intensity level at 287.8 eV (C1s), 530.4 eV (O1s) and 399.8 eV (N1s) [14]. XPS studies of PI after Ar^+ ion irradiation have been investigated by many research groups and its results were published [14]. After modification by Ar^+ ion only, the peak intensities related to O1s and N1s decrease. However, in the case of $PI_{O,O}$, the intensity of O1s increases and the intensity level at higher binding energy than 286 in C1s increases, which means that amount of C-O and C=O groups are increased, i.e., The hydrophilic groups are generated by the chemical reaction. The energetic ions could easily remove the hydrogen atoms or carbon atoms in phenyl ring [15] and the created radicals in phenyl ring react with blowing oxygen. Consequently, the shake up satellite peak (292 eV) disappeared. There are many different processes induced by ion irradiation, besides above described, such as isoimidization, carbonization, cross-linking, etc. It is impossible to discern all processes by XPS, but, from the XPS study, we could observe that the formation of hydrophilic groups such as >C-O-H, -(C=O)-O, and -(C=O)-(NH)- on the PI surface.

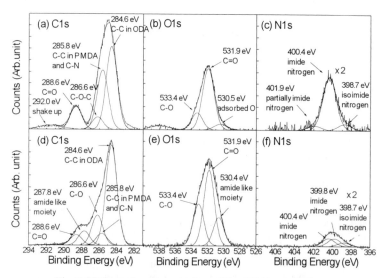

Fig. 3. XPS core-level spectra of pure PI and IAR treated $Pi_{O,O}$.

Figures 4(a), (b) and (c) show changes of C1s, O1s, N1s peaks of pure PI after depositing 1 and 3 Å of Cu films. After depositing Cu, one of the most distinct change in C1s peak shape is reduction of carbonyl peak and formation of new peaks near 287.5 eV and 283.2 eV. When metals are deposited on PI, C=O (288.5 eV) preferentially reacts with metals and its peak intensity is decreased. This change was known to be due to the charge transfer [16]. The peak appeared at 287.5 eV is due to intermediate state of Cu_2O and 283.2 eV is due to Cu_2O [5]. In O1s, as the thickness of Cu increased, the width of the peak increased and the intensity of the peak decreased due to the charge transfer and the Cu over layer. Peak broadening might be due to the formation of Cu_2O (531.9 eV) phase which increases the peak intensity of the low binding energy region. After deposition of 3-Å Cu over layer, a shoulder was developed between 532.0 and 533.1 eV which means that C=O group is more rapidly react to Cu than C-O group.

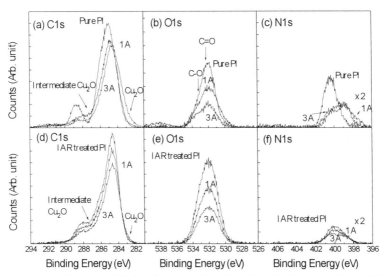

Fig. 4. XPS core level spectra of Cu/PI after deposition of Cu film. (a)~(c) on pure PI and (d)~(e) on IAR treated PI.

The XPS spectra of Cu on the IAR treated PI films show a different trend to the pure PI. As shown in Fig. 4(d), (e), and (f), attenuation of the carbonyl peak also observed in the IAR treated $PI_{O,O}$. One of the distinct difference of XPS spectra was observed in O1s. In O1s spectrum of Cu/IAR treated PI (Fig. 4(e)), the peak intensity also decreased as the thickness of Cu increased. However, the shoulder was not observed after deposition of Cu as contrast to pure PI. As shown in Fig. 4(e), peak shapes were rarely changed as the Cu film thickness increased. The decrement of peak intensity is due to Cu overlay and the interaction between the oxygen atoms and the Cu atoms. After deposition of 1-Å Cu, these adsorbed Cu atoms react with the oxygen atoms formed by the IAR treated on the PI surface, which results in rapidly decrement of the total intensity of the O1s peak. Peak shape of O1s spectra of the IAR treated PI was scarcely changed, which means ether oxygen (not C-O-C in pure PI but C-O) reacts with adsorbed Cu as much as carbonyl oxygen dose. D. E. King et al.[17] deposited Cu on an organized molecular assembly of $HS(CH_2)_{10}COOH$, 11-mercap-toundecanoic acid (MUA), which terminated with (C=O)-OH group and reported that Cu prefers to react with C-O rather than with C=O. J. L. Droulas et al. [18] examined the adhesion strength between various polymers and metals. They used Au and Al as depositing metals and polyethylene (PE), polystyrene (PS), poly (ethylene terephthalate) (PET), poly (ether ether ketone) (PEEK), and polyvinylalcohol (PVA) as the substrates. PE consists of C and H and does not contain a phenyl pendent group and PS also consists of C and H but it contains a phenyl pendent group. On the other hands, PET, PEEK, and PVA have the polar groups. PET and PEEK have carbonyl group and PVA has the ether group. They found that the adhesion strength depends on existence of oxygen groups and adhesion between metals and PVA are strongest among the above listed polymers. This fact means that the adhesion between metals with ether group (alcohol group) is stronger than with carbonyl group. So, we could infer that enhancement of adhesion is due to formation of the hydrophilic groups on the PI surface.

CONCLUSIONS

Wetting angles and surface free energies of the PI films after Ar^+ ion irradiation with dose of 1×10^{17} ions/cm^2 are changed from 67° to 40° and from 46 dyne/cm^2 to 64 dyne/cm^2, respectively. When PI films were modified by Ar^+ ions with blowing oxygen with the dose of 1×10^{17} ions/cm^2, wetting angles and surface free energies are changed from 67° to 12° and from 46 dyne/cm^2 to 72 dyne/cm^2. While, in the case of PI films modified by oxygen ion with blowing oxygen, wetting angles are lower than 10° and surface free energies were higher than 70 dyne/cm^2 though ion dose was low. Surface modification of polymer by ions with blowing oxygen is more effective to improve wettability than without blowing oxygen. XPS analysis shows that hydrophilic groups such as C-O-H, C=O are formed on PI surface by IAR. These hydrophilic groups contribute to improve wettability. From the *in-situ* deposition of Cu and XPS analysis, we found that, in pure PI, C=O group preferentially reacts with depositing Cu and, in IAR treated PI, not only C=O but also C-O group which should be generated by the IAR plays an important role improving the adhesion.

REFERENCES

1. A. M. Wilson, Polyimides-Synthesis, Characterization and Applications, K. L. Mittal, ed. (Plenum, New York, 1984) p. 715.
2. R. J. Jensen, J. P. Cummings, and H. Vora, IEEE Trans. Components Hybrids Manuf. Technol. 7, 385 (1984).
3. S. K. Koh, K. D. Pae, and R. Caracciolo, Polym. Eng. Sci. 32, 558 (1992).
4. A. Chenite, A. Selmani, and A. Yelon, J. Vac. Sci. Technol. A12, 513 (1994).
5. Lj. Atanasoska, S. G. Anderson, H. M Meyer III, Z. Lin and J. H. Weaver. J. Vac. Sci. Technol. A5, 3325 (1987).
6. S. R. Forest. M. L. Kaplan, P. H. Schmidt, T. Venkatesan and A. E. Lovinger, Appl. Phys. Lett. 41, 708 (1982).
7. L. Hultman, U. Helmersson, S. A. Barnett, J. E. Sundgren and J. E. Greene, J. Appl. Phys. 61, 552 (1987).
8. S. K. Koh, S. C. Park, S. R. Kim, W. K. Choi, H-J. Jung, and K. D. Pae, J. Appl. Polym. Sci. 64, 1913 (1997).
9. J. S. Cho, W. K. Choi, H-J. Jung and S. K. Koh, J. Mater. Res. 312, 277 (1996).
10. S. K. Koh, W. K. Choi, J. S. Cho, and H-J. Jung, J. Mater. Res. 11, 2933 (1996).
11. W. K. Choi, S. K. Koh and H-J. Jung, J. Vac. Sci. Technol. A14, 2366, (1996).
12. J. B. Ma, J. Dragon, W. V. Derveer, A. Entenberg, V. Lindberg, M. Anschel, D. Y. Shih, and P. Lauro, J. Adhesion. Sci. Technol. 9, 487 (1995).
13. R. N. Ramb, J. Baxter, M. Grunze, C. W. Kong, and W. N. Unertl, Langmuir 4, 249 (1988).
14. G. Marletta, C. Oliveri, G. Ferla, and S. Pignataro, Surf. and Inter. Anal. 12, 447 (1988).
15. R. Flisch and D-Y. Shin, J. Vac. Sci. Technol. A8, 2376 (1990).
16. R. Height, R. C. White, B. D. Silverman, and P. S. Ho, J. Vac. Sci. Technol. A6, 2188(1991).
17. D. E. King, A. W. Czanderna, and D. Spaulding, Metallized plastics 3 edited by K. L. Mittal and J. R. Susko (Plenum press, NewYork, 1989) pp.149-161.
18. J. L. Droulas, Y. Jugnet, and T. M. Due, ibid., pp.123-140.

IMPROVING THE THERMAL STABILITY OF PHOTORESIST FILMS BY ION BEAM IRRADIATION

F.C. ZAWISLAK*, IRENE T.S. GARCIA**, D. SAMIOS***, D.L. BAPTISTA*, P.F.P. FICHTNER****, E. ALVES*****, MARIA F. DA SILVA*****, J.C. SOARES******
*Instituto de Física, UFRGS, Caixa Postal 15051, 91501-970 Porto Alegre, Brasil, zawislak@if.ufrgs.br; **PGCIMAT, UFRGS, Porto Alegre, Brasil; ***Instituo de Química, UFRGS, Porto Alegre, Brasil; ****Escola de Engenharia, UFRGS, Porto Alegre, Brasil; *****Instituto Tecnológico e Nuclear, Sacavém, Portugal; ******Centro de Física Nuclear, UL, Lisboa, Portugal

ABSTRACT

The thermal stability of ion irradiated 1.7 μm thick AZ-1350J photoresist films was investigated using the RBS and ERDA techniques to measure the composition of the irradiated and annealed films. The films have been irradiated with He, N and Ar ions at energies from 380 to 760 keV and fluences between 2×10^{15} and 10^{16} ions cm^{-2}. A considerable increase in the thermal stability of the He irradiated film is observed from $\approx 200°C$ – when the non-irradiated film starts to decompose – to 400°C after the irradiation. The FTIR spectroscopy and the SEM observations were used to study the chemical structural changes and the surface morphology of the irradiated samples. The results are discussed in terms of the energy density deposited by the ions, the large loss of H during irradiation, and the resulting increase in cross-linking density.

INTRODUCTION

Ion beam implantation and irradiation of polymers has been used, during the last years, to induce modifications in electrical, optical and mechanical properties near the surface of the material. The published work shows that surface-sensitive properties such as resistance to wear, erosion, fatigue and specially hardness [1-3], are considerably improved by an adequate exposure of the polymers to ion bombardment. Because of the inherent softness of most of the polymers, the improvement of the mechanical properties is very important, specially at higher temperatures. The poor thermal stability of many polymers and polymer blends is a severe limitation in various technological applications. This problem is usually overcome by the use of stabilizers or by γ-irradiation.

It is well known that highly cross-linked polymers have an improved resistance to thermal degradation, and cross-linking produced by γ-irradiation has been used since many years. However, many polymers (as Polystyrene and Novolac based photoresists) have a relatively low sensitivity to γ-rays and to electrons. An alternative is the application of ion beams, which can be more efficient in the cross-linking process, because much higher energy density transfers are possible via ion irradiations.

In previous studies of the thermal stability of photoresists we investigated the effects of shallow implantations of heavy ions [4-6] (50 keV Bi, Au, Sn and Sb). These measurements, performed on 1.5 μm photoresist films isochronally annealed for short times of 20 minutes in the temperature range of 200 to 450°C, have shown an increase of the temperature at which the photoresist starts to decompose. In the present work we expand the investigation in two directions. First, we study the thermal degradation of the irradiated polymers as function of an isothermal annealing for periods of up to 15 hours. The thermal stability for longer periods at a fixed T is a more closely related test to the real conditions of use of the polymer. Second, as the thermal properties of a polymer are related to the bulk of the material, it seems important to investigate the effects of the energy density (eV/Å^3) deposited over the whole volume of the photoresist by the irradiation. From this point of view, a shallow implantation, modifying only the region near the surface, may not be an adequate approach.

The experimental results reported here address some of the above problems. Rutherford backscattering (RBS) and elastic recoil detection analysis (ERDA) are used to investigate changes in the stoichiometric composition of AZ-1350J photoresist films as consequence of irradiation and thermal treatments. The chemical structural changes of some of the irradiated films have been also studied using Fourier transform infrared spectroscopy (FTIR) and their surface morphology was characterized by scanning electron microscopy (SEM).

EXPERIMENTAL DETAILS AND ANALYSIS

The photoresist AZ-1350J, supplied by Shipley Europe Ltd., is composed of 70% of novolak plus 30% of diazonaftoquinone, a photoactive compound. The novolak-diazoquinone resists are presently the most important imaging materials of the semiconductor device industry. The chemical composition of the AZ1350J photoresist is $C_{6.17}H_6O_1N_{0.14}S_{0.06}$ and the density is $\rho = 1.3$ g/cm^3. The photoresist films were obtained by spin coating clean silicon wafers. The films with thickness of $\cong 1.7$ µm were baked at 100°C for 1 hour in order to eliminate the solvent and to improve the adhesion to the substrate. Samples of 1 to 2 cm^2 were subsequently irradiated with Ar^{++}, N^{++} and He$^+$ ions at energies and fluences shown in table I. The irradiations were done at room temperature using the 400 kV ion implanter of the Institute of Physics, Porto Alegre, and beam current densities <50 nA/cm^2, in order to avoid local heating and damage of the films.

The RBS technique, with 760 keV α particles from the implanter was used to probe the stoichiometry of the films in terms of C and O. The probing of H was performed via ERDA using the 3.1 MeV Van de Graaff accelerator from Sacavém (Portugal). The backscattered α particles were detected by a silicon barrier detector with energy resolution of 13 keV. The incident and exit angles of the ERDA experiment were of 79° relative to the sample normal. The ERDA detector had a 8 µm capton foil in front to prevent the detection of forward scattered α particles. The infrared spectroscopy measurements were carried out by the attenuated total reflection Fourier transform infrared (ATR-FTIR) technique in a BOMEN spectrometer model OA8, with a detector type MCT, cooled with liquid N$_2$. The IR spectra were obtained by 64 scanning accumulations with a resolution of 4 cm^{-1}. The SEM micrographs were performed in a JEOL JSM-5800 system.

Table I shows the ion beam parameters for He, N and Ar used in the irradiations of the AZ1350J films. The projected range R$_p$ and the electronic (S$_e$) and nuclear (S$_n$) stopping powers are calculated, via TRIM (1995 version) simulations [7], with $\rho = 1.3$ g/cm^3 for the AZ-1350J film. Table I also displays the total electronic (T$_e = \phi S_e$) and nuclear (T$_n = \phi S_n$) energy density transfers (in eV/Å^3) for the different ions and fluences ϕ.

TABLE I. Parameters for He, N and Ar ions implanted in the AZ-1350J photoresit (see text).

Ion	E (keV)	ϕ (at cm^{-2})	S$_e$ (eV/Å)	S$_n$ (eV/Å)	T$_e = \phi S_e$ (eV/Å^3)	T$_n = \phi S_n$ (eV/Å^3)	R$_p$ (Å)
He$^+$	380	10^{16}	28	0.08	28.0	8×10^{-2}	20500
N^{++}	760	2×10^{15}	75	1.2	15.0	2.4×10^{-1}	14700
Ar^{++}	760	2×10^{15}	75	14	15.0	2.8	9500

RESULTS

Table I shows that the irradiation with He represents a large deposition of electronic energy density of 28 eV/Å^3 and a very small nuclear energy density of 8×10^{-2} eV/Å^3. The irradiation with the heavier ions N and Ar means an enhancement of the nuclear energy deposition reaching 2.8 eV/Å^3 for 760 keV Ar ions. In addition, only the 380 keV He ions have a range larger than the 1.7 µm thickness of the film.

It is known that the novolac-diazoquinone photoresist starts to decompose at about 220°C. Previous results have shown that at 250°C there is already a large loss of oxygen and carbon and the film becomes considerably thinner as consequence of this loss of material [6]. In the experiments described in this paper we investigate the thermal stability of samples of AZ-1350J taken from the same spin coated wafer. One sample was non-irradiated, and is called virgin sample. The other samples have been irradiated as described in table I.

Fig. 1 shows the α 760 keV RBS spectra of the virgin sample annealed isothermally in various steps at 300°C for a total of 15 hours, in a vacuum better than 10^{-6} Torr. The spectra

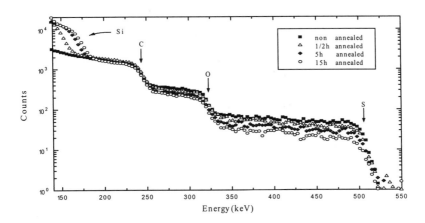

Fig. 1. 760 keV RBS spectra of the AZ-1350J virgin film and isothermally annealed at 300°C for various periods (in hours).

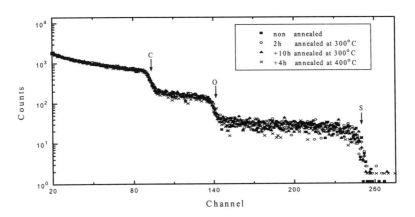

Fig. 2. 760 keV RBS spectra of an AZ-1350J film irradiated with 380 keV 10^{16} He cm^{-2}. The spectra refer to non-annealed and isothermally annealed film at 300°C and 400°C for various periods (in hours).

445

of fig. 1 correspond to different annealing times, and it is observed that after the first annealing for 30 minutes, the film starts to decompose and the RBS Si edge appears, corresponding to the film's thickness decrease by about 40%. Subsequent annealings display a continuous loss of material and a consequent reduction of the film thickness. After 15 hours of annealing the film lost about half of the material and the final thickness is ≈8000 Å. Fig. 2 represents a similar experiment but with the He irradiated film. Here we do not observe significant changes in the composition of the film up to 12 hours of annealing at 300°C, and an additional annealing of 4 hours at 400°C. This means that the irradiation with 380 keV He ions at a fluence of 10^{16} ions cm^{-2} has improved considerably the thermal stability of the film up to 400°C.

The fittings of the RBS spectra after irradiation and following the annealings (fig. 2) confirm the stoichiometic composition of the AZ-1350J within 5 to 10%. This is a typical error in RBS experiments. However, we have also to consider that the polymer loses hydrogen atoms as a consequence of the irradiation [8]. Our measurements of the relative amount of H in the virgin, and He, N and Ar irradiated samples, using the ERDA technique with 1.8 MeV He$^+$ ions, confirm the previous results that there is a considerable loss of H atoms during the irradiation. The irradiation with 380 keV He and 760 keV Ar ions represents a loss of respectively 35% and 70% of the H atoms present in the sample as is shown in fig. 3a). The subsequent annealings of the photoresist film irradiated with 380 keV He ions, show a very small additional loss of H as displayed in fig. 3b). The ERDA experiment examined the H distribution in the first 2500 Å of the film.

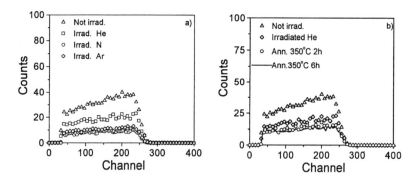

Fig. 3. Measured proton recoil spectra for H in the AZ-1350J photoresist. a) not irradiated, and He, N and Ar irradiated films (see table I); b) not irradiated, and He irradiated and annealed films.

Similar thermal treatments of the AZ-1350J films submitted to shallow implantations with N^{++} and Ar^{++} ions (see table I) show very different behaviors. The shallow implanted samples do not maintain the stability of the films under isothermal annealings and after 2 or 3 hours at 300°C blisters are observed at the surface of the film and a consequent very large loss of material. Fig. 4 shows a scanning electron micrograph of the films irradiated with 380 keV He$^+$ and 760 keV Ar^{++} after annealing. The He irradiated sample has a smooth surface in contrast to the destroyed Ar^{++} irradiated film, showing blisters after annealing.

Finally, we also investigated via FTIR, the modifications of the chemical structure of these samples after irradiation and annealing. Preliminary results of FTIR spectroscopy on the virgin, He (380 keV), N (760 keV) and Ar (760 keV) irradiated films are shown in fig. 5. The FTIR spectra illustrate the differences in the optical absorption between the irradiated photoresist films and the pristine film. The most important chemical transformations are:
i) after irradiation with N and Ar we see the disappearance of the O-H bonds corresponding to the large decrease in the band around 3600 cm^{-1}; ii) the He irradiated film shows the decrease of the symmetric and antisymmetric C-H stretching absorption peaks at 2850 and

2920 cm^{-2}; iii) the expected $C \equiv C$ bands in the Ar^{++} and N^{++} irradiated films around 2200 cm^{-3} can not be identified because of the superimposed very intense silicon substrate bands; iv) the irradiation with He ions produces the smaller affect on the chemical structure of the photoresist.

The large loss of H observed in the ERDA experiment after irradiation with Ar ions is confirmed by the above FTIR results.

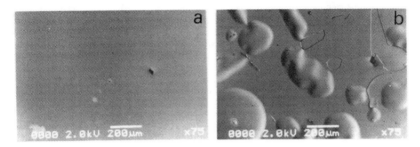

Fig. 4. SEM micrograph of AZ-1350J films irradiated with He (a) and Ar (b) and annealed at 300°C.

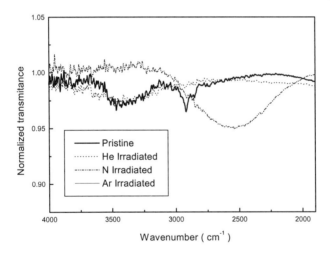

Wavenumber (cm^{-1})

Fig. 5. FTIR spectra of the AZ-1350J photoresist for pristine and ion irradiated films.

DISCUSSION AND CONCLUSIONS

Our data show that the irradiation of the AZ-1350J film with 380 keV He ions at a fluence of 10^{16} He cm^{-2} improves considerably the thermal stability of the photoresist as measured after an isothermal annealing of 15 hours, at a temperature up to 400°C. Similar thermal treatments of the photoresist submitted to shallow implantation of 760 keV Ar ions with a range R$_p$ = 9500 Å, reaching only half the film thickness show a large number of blisters produced by the gases evolved from the non-irradiated deeper half of the film indicating that the irradiated top layer is less permeable to gas release. This means that similarly as irradiations with γ-rays and fast-electrons, the energy deposition due to the ion has to occur homogeneously over the whole volume of the photoresist in order to produce a beneficial thermal effect.

The consensus of the previous works related to the improvement of mechanical properties like hardness and wear, of irradiated polymers, is that the ions cause ionization resulting in the production of free radicals and ultimately in an interconnected, cross-linked network. We can invoke such a scenario to explain the enhanced thermal stability of the AZ-1350J photoresist irradiated with the He ions. The large electronic energy density of 28 eV/Å^3 transferred by 380 keV 10^{16} He cm^{-2}, is very important, because as suggested in [9], the polystyrene and novolac polymers require simultaneous activation of sites in adjacent chains in order to increase the cross-linking. From this point of view, ions are much more efficient. A He ion of 1 MeV may ionize nearly all the molecules lying along its path, whereas a 1 MeV electron will ionize only about 1 in 2500 of the molecules it traverses.

Probably part of the lost H during the irradiation is related to the evolved atoms as a consequence of the cross-linking process at hydrogen sites between adjacent chains. However, the large H loss certainly produces a carbon-enriched irradiated region, having different properties from those of the pristine film. This is specially true for irradiation with heavier and higher energy ions as is confirmed by the FTIR results. Our data show that there is a loss of 70% of the H atoms in AZ-1350J after the irradiation with 760 keV Ar ions at a fluence of 2×10^{15} Ar cm^{-2}. In this circumstance the observed enhancement in thermal stability or hardness may be, at least partially attributed to the transformation of the polymer surface into a carbon layer.

The above results demand additional work with a careful analysis of the chemical structure modifications produced by the irradiation and annealing. Residual gas analysis during irradiation and annealing and detailed FTIR and gel permeation chromatography experiments are in progress in order to better understand the observed effects.

ACKNOWLEDGMENTS

The work was supported in part by the agencies FINEP, CNPq and CAPES.

REFERENCES

1. T. Venkatesan, L. Calcagno, B.S. Elman and G. Foti, in Ion Beam Modification of Polymers, edited by P. Mazzoldi and G.W. Arnold (Elsevier, N.Y. 1987), p. 301.
2. E.H. Lee, M.B. Lewis, P.J. Blau, and L.K. Mansur, J. Mater. Res. 6, p. 610 (1991).
3. G.R. Rao and E.H. Lee, J. Mater. Res. 11, p. 2661 (1996).
4. M. Behar, L. Amaral, F.C. Zawislak, R.B. Guimarães and D. Fink, Nucl. Intrum. and Meth. B 46, p. 350 (1990).
5. M. Behar, R.B. Guimarães, P.L. Grande, L. Amaral, J.P. Biersack, D. Fink, J.R. Kaschny and F.C. Zawislak, Phys. Rev. B 41, p. 6145 (1990).
6. R.L. Maltez, L. Amaral, M. Behar and F.C. Zawislak, Nucl. Intrum. and Meth. B 80/81, p. 1316 (1993).
7. J.P. Biersack and L.G. Haggmark, Nucl. Instrum. and Meth. 174, p. 257 (1980).
8. Yongqiang Wang, S.S. Mohite, L.B. Bridwell, R.E. Giedd and C.J. Sofield, J. Mater. Res. 8, p. 388 (1993).
9. T.M. Hall, A. Wayner and L.F. Thompson, J. Appl. Phys. 53, p. 3997 (1982).

STUDY OF THE EFFECTS OF MeV IONS ON PS AND PES

A. L. EVELYN*, D. ILA*, R. L. ZIMMERMAN*, K. BHAT*, D. B. POKER**, D. K. HENSLEY** AND N. JUST***
*Department of Physical and Natural Sciences, Alabama A&M University, P.O. Box 1447, Normal, AL 35762-1447
**Solid State Division, Oak Ridge National Laboratory, Oak Ridge, TN
***Université Claude Bernard de Lyon, France

ABSTRACT

The electronic and nuclear stopping effects produced by MeV ion bombardment in polyethylene (PE) and polyvinylidene chloride (PVDC) have been previously studied and reported. We have subsequently selected two other insulators: polystyrene (PS) and polyethersulfone (PES) which contains sulfur as a crosslinking agent, and irradiated them with MeV alpha particles. The electronic and nuclear effects of the incident ions were separated by stacking thin films of the polymers. A layered system was selected such that the first layers experienced most of the effects of the electronic energy deposited and the last layers received most of the effects of the nuclear stopping. The changes in the chemical structure were measured by residual gas analysis (RGA), Raman microprobe analysis, RBS and FTIR. The post-irradiation characterization resolved the effects of the stopping powers on the PS and PES and the results were compared with those from PE and PVDC.

INTRODUCTION

Incident ions moving through material lose energy as a result of their interactions with the surrounding electrons and nuclei of the host atoms. These stopping interactions are termed the electronic (ε_e) and the nuclear (ε_n) stopping powers, respectively [1]. The stopping power profiles for MeV ions show that ε_e dominates at shallower penetration depths compared with ε_n which is maximum at the end of the ion track. Each stopping power induces different effects in the target material due to the particular mechanisms and quantities of energy deposited. The effects of both ε_e and ε_n for MeV ions can be mapped, throughout the ion tracks using a thin polymer film stacking technique [2]. The deposited energies can be used to change properties such as the electrical conductivity, hardness and the optical absorption in polymers by changing the chemical structures [3-6]. Studies have shown increases in electrical conductivity as much as eight orders of magnitude [6, 7, 9]. The transfer of energy from the ions to the surrounding material results in a complex combination of polymer chain scissoring, covalent bond breaking and crosslinking.

Effects of ion beam irradiation on polyvinylidene chloride (PVDC) $[CH_2-CCl_2]_n$, a copolymer, and polyethylene (PE) $[CH_2]_n$, a homopolymer have been investigated and reported on [2, 4-5]. These polymers were initially chosen because of their simple linear chain chemical structures which kept the number of variables in the study as low as possible. The further study of ion interaction with more complex polymer structures is the focus of this paper which reports on the effects of the interaction of MeV ion beams with polyethersulfone (PES) $[OC_6H_4-SO_2C_6H_4]_n$ and polystyrene (PS) $[CH_2-CH-C_6H_4]_n$. PES contains sulfur which is a known crosslinking agent and PS has a pendant styrene group which also promotes crosslinking resulting in network polymerization. The effects of the additional functional groups (sulfone and styrene) in the chemical composition provide further information on the ion polymer interaction processes and are compared with the previous results from ion irradiated PE and PVDC. The objective was to further understand of the effects of MeV ions on these polymers by using the thin film stacking technique.

449

EXPERIMENTAL PROCEDURE

Self supporting thin films of PES were fabricated from pellets of Radel A 200 polyethersulfone, obtained from Amoco Performance Products of Alpharetta, GA. The pellets were dissolved in methylene chloride and spun coated to produce films with thicknesses on the order of 4 microns. The PS samples were prepared by spin coating a solution of PS onto silicon wafers creating ~ 1.5 μm thick supported films. PES films were stacked and bombarded with 5 MeV alpha particles. The SRIM (1996 version of TRIM) computer simulation code [1] was used to predict the penetration depth of the bombarding ions, This allowed a determination to be made of the required number of layers needed to stop the ions in the PES, to ensure that the ion tracks would not extend beyond the last laminae of the film stacks. SRIM predicted an ion penetration depth of 30 μm in the PES for the 5.0 MeV alpha particles. This required eight layers of the thin polymer films. The film stacking technique resulted in the first layers experiencing most of the effects of the deposited electronic energy and the last layer receiving almost all of the effects of the nuclear stopping power.

A rastered incident ion beam delivered 5.0 MeV alpha particles, at fluences of 10^{13}, 10^{14}, 10^{15} and 10^{16} ions/cm^2, to the target. The beam current density was kept below 35 nA/cm^2 to allow enough time for volatile gases, produced by the decomposing polymer, to diffuse out of the films and to allow shrinking without cracking. The films were attached to aluminum holders using silicone grease which allowed the films to slide along the surface of the holders during shrinking. Ion ranges for the bombarding alpha particles extended to 7 layers.

Changes in the chemical composition and structures of the PES and PS were analyzed by Raman microprobe analysis, Fourier transform infrared spectroscopy (FTIR), residual gas analysis (RGA) and Rutherford backscattering spectrometry (RBS).

RESULTS AND DISCUSSION

Raman spectra from the irradiated PES films show the presence of two peaks around 1595 cm^{-1} (G-line) which is attributed to graphene structure formation and around 1350 cm^{-1} (D-line) which is attributed to amorphous structure or "disorder" in the material. Figure 1 shows the Raman spectra for three layers of the 5 MeV alpha irradiated PES films to a fluence of 1 x 10^{16} ions/cm^2. It shows that the D/G ratio decreases from the first layer to the fourth then increases in the last layer. The electronic stopping power which dominates in the uppermost layers and which is effected by the greater ion energies is responsible for causes the sulfone group in the PES polymer chain to enhance the crosslinking of polymers. As the ions lose energy in the last layers the nuclear stopping, which accounts for most of the molecular dissociation and bond breakage, results in increasing D/G ratios as the crosslink enhancing group (sulfone) becomes disrupted. This effect was also seen in the films irradiated to 5×10^{15} cm^{-2}. Figure 2 compares the Raman spectra from the first layers of the irradiated PES films at the 1×10^{16} and 5×10^{15} cm^{-2} fluences. The D/G ratios are seen to increase as the fluence is increased, which is a result of increasing the number of violent, high energy reaction regions lying adjacent to the incident ion's track.

Figures 3 and 4 show the FTIR spectra from the irradiated PES films and compare them with the virgin material. Absorption bands for the CH$_2$ bonds appear in the 1460-1370 and 740-720 cm^{-1} regions [10]. Figure 3 shows a reduction in the number of the C–H bonds in the deeper irradiated layers while Figure 4 indicates that they completely disappear in the corresponding layers. The absorption bands corresponding to the C-O stretching vibrations at 1110 cm^{-1} and the sulfone stretching vibrations at 1160-1140 cm^{-1} [8] are also reduced in the deeper layers in Figure 3 and completely removed in Figure 4, along with other absorption bands. the disappearance of the and a complete removal of the C–Cl bonds around the 866 cm^{-1} region. Figure 3 shows that the bonds are reduced more in layers 1 through 6 than in the last one where the 5.0 MeV alpha particles are stopped. The remaining virgin material contributes to

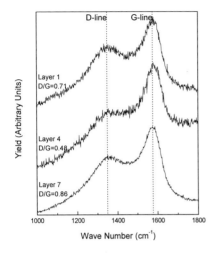

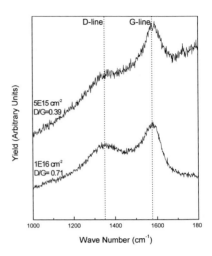

Fig. 1. Raman spectra obtained from the 5.0 MeV α-irradiated PES films at a fluence of 1×10^{16} ions/cm^2.

Fig. 2. Raman spectra obtained from the first layers of the 5.0 MeV α-irradiated PES films at 5×10^{15} & 1×10^{16} ions/cm^2.

the larger absorption bands in this layer. The spectra from the 3×10^{16} cm^{-1} sample (Figure 4) shows the appearance of a peak around 1700 cm-1, which corresponds to the -C=C- bond formation.

Figure 5 shows RBS spectra obtained from various layers of the 5 MeV helium irradiated PES films (5×10^{15} cm^{-2}). The RBS analysis indicated that the concentration ratios of sulfur to carbon increased from layers 3 through 5 then decreased in layer 6. This complements the observed effect in the Raman analysis and suggests that the presence of the sulfur enhanced the G/D (decreased D/G) ratios. The decreased sulfur to carbon concentration ratio in the last layer corresponds with the higher D/G ratios observed in the Raman spectra. The RBS spectra indicates that the ratios of oxygen to carbon correspondingly increased with the sulfur to carbon ratios suggesting that the sulfur maintains the bonds with the oxygen in the sulfone groups which enhances the polymer crosslinking.

Residual gas analysis (RGA) was done *in situ* during 3 MeV alpha particle irradiations on PES and PE films and showed how the polymer degraded with ion fluence over time. Figures 6 and 7 show the RGA spectra obtained from PES and PS, plotted as a function of received ion fluence. The usual peaks associated with water appear at mass number positions 16, 17 and 18. As the fluence increases, the amount of liberated hydrogen (1 and 2 AMU) increases, and so does carbon dioxide (44 AMU). Nitrogen appears at 28 AMU because of the nitrogen used to purge the implantation chamber as well as residual quantities from the stripper gas in the Pelletron accelerator. The growth in the peak at 72 AMU suggests the liberation of the aromatic benzene rings (C_6). The RGA spectra from the PS similarly show the rise in the C_6 peak as the fluence increases.

CONCLUSION

The Raman microprobe analysis verified that the film stacking technique resolved the effects of the MeV ions (attributed to the influence of ε_e and ε_n) while the FTIR analysis revealed a complex reorganization of the chemical structure in the polymers. RBS and RGA established the effects of the

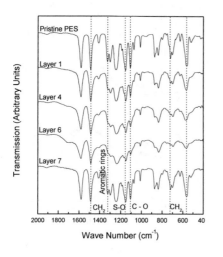

Fig. 3. FTIR spectra obtained from various layers of the 5 MeV α-irradiated PES Films at 5x10^{14} cm^{-2}.

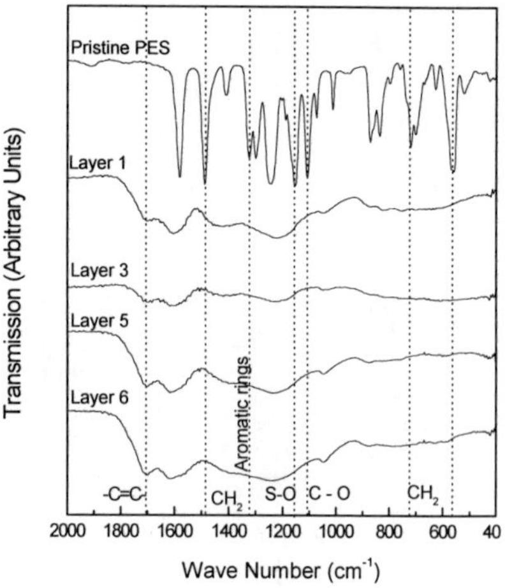

Fig. 4. FTIR spectra obtained from various layers of the 5 MeV α-irradiated PES Films at 3x10^{16} cm^{-2}.

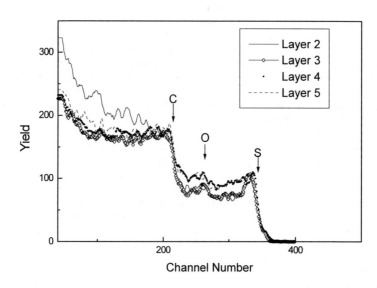

Fig. 5. RBS spectra obtained from various layers of 5 MeV α-irradiated PES at 5x10^{15} cm^{-2}.

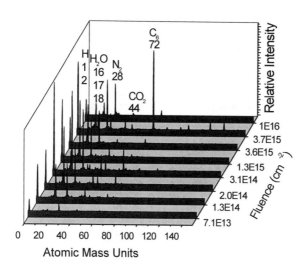

Fig. 6. RGA spectra obtained from PES acquired *in situ* during a 3 MeV alpha irradiation.

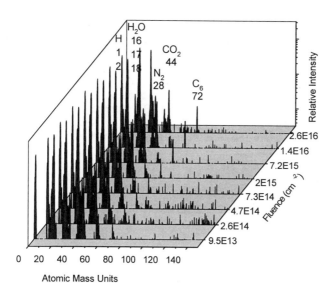

Fig. 7. RGA spectra obtained from PS acquired *in situ* during a 3 MeV alpha irradiation.

MeV ions on the polymer degradation. Comparisons with PVDC & PE demonstrate that the sulfone group in the PES polymer and the pendant styrene group in the PS enhanced the polymer chain crosslinking. Previous results from alpha particle irradiated polyethylene (PE) and polyvinylidene chloride (PVDC) showed that increasing the irradiating fluences resulted in corresponding increases in the disorder to graphene formation ratios. It was also shown that the increase in the disorder in the deeper irradiated regions resulted from the effects of the nuclear stopping powers. This general trend was also observed in the PES and PS polymer materials. However, the differences that were observed in the analysis of the PES and PS are attributed to the presence of the crosslink enhancing groups in these polymer chains.

The Raman spectra from the last layer shows an increase in the disorder as indicated by an increase in the D/G ratio. This disorder can be caused by breakage of the C-C or C=C bonds and the ratio increases as the disorder increases. The only component of the ion beam interaction which increases in the deeper layer is the nuclear stopping power. This is supported by our previous work on the interaction of low energy ions with polymeric glassy carbon.

ACKNOWLEDGMENTS

This project was supported by the Howard J. Foster Center for Irradiation of Materials at Alabama A&M University and Alabama-EPSCoR/ASIP Grant No. OSR-9559480. The work at ORNL was sponsored by the Division of Materials Sciences, U.S. Department of Energy, under contract DE-AC05-96OR22464 with Lockheed Martin Energy Research Corp.

The authors would like to especially thank Dan Nisen and La Sonya Cephus at CIM-AAMU for providing technical assistance during the ion beam experiments.

REFERENCES

1.   J. F. Ziegler, J. P. Biersack and U. Littmark, The Stopping and Range of Ions in Solids, New York, Pergamon Press Inc., 1985.
2.   A. L. Evelyn, D. Ila, R. L. Zimmerman, K. Bhat, D. B. Poker and D. K. Hensley, Nucl. Instr. and Meth. B 127/128 (1997) 694.
3.   D. Ila, A. L. Evelyn, and G. M. Jenkins, Nucl. Instr. and Meth. B 91 (1994) 580.
4.   A. L. Evelyn, D. Ila, and G. M. Jenkins, Nucl. Instr. & Meth. B 85 (1994) 861.
5.   A. L. Evelyn, D. Ila, J. Fisher and D. B. Poker in Ion-Solid Interactions for Materials Modification and Processing, edited by D. B. Poker, D. Ila, Y-T. Cheng, L. R. Harriott and T. W. Sigmon (Mat. Res. Soc. Symp. Proc. 396, Pittsburgh, PA 1996) 323.
6.   A. L. Evelyn, D. Ila, R. L. Zimmerman, K. Bhat, D. B. Poker and D. K. Hensley in Materials Modification and Synthesis by Ion Beam Processing, edited by D. E. Alexander, N. W. Cheung, B. Park and W. Skorupa (Mat. Res. Soc. Symp. Proc. 438, Pittsburgh, PA 1997) 499-504.
7.   D. Ila, A. L. Evelyn, and G. M. Jenkins in Crystallization and Related Phenomena in Amorphous Materials, edited by M. Libera, T. E. Haynes, P. Cebe and J. E. Dickinson Jr. (Mater. Res. Soc. Proc. 321, Pittsburgh, PA 1994) 441.
8.   J. R. Dyer, Applications of Absorption Spectroscopy of Organic Compounds, New Jersey, Prentice-Hall, Inc. 1965.
9.   D. Ila, A. L. Evelyn, and Y. Qian in Materials Reliability in Microelectronics, edited by P. Børgesen, J. C. Coburn, J. E. Sanchez, Jr., K. P. Rodbell and W. F. Filter (Mater. Res. Soc. Proc. 338, Pittsburgh, PA 1994) 613.
10. J. Davenas, and X. L. Xu, Nucl. Instr. and Meth. B 39 (1989) 754.

AUTHOR INDEX

455

SUBJECT INDEX

radiation(-)
  annealing, 221
  damage, 159
  induced
    amorphization, 227
    defects, 227
    segregation, 129
Raman
  microprobe, 449
  spectroscopy, 339
RBS, 123, 381, 443, 449
RDF analysis, 147
reactive ion-assisted deposition, 313

sapphire, 81, 283, 381, 393
scanning tunneling microscopy, 93
scratch test, 265
segregation, 153
self-interstitial atoms, 63
SEM, 443
semiconductor nanocluster, 405
shallow implantation, 69
Si, 411
SIA clusters, 63
SiC, 39, 45, 135, 141
silica, 345, 351, 405
silicon, 3, 15, 27, 33
SIMS, 69, 141, 153
simulations, 51
Si_3N_4, 39, 153
SiO_2, 39, 111, 123, 153, 327, 339, 345, 351, 405
Si-O-Si bond angle, 21
SnO_2, 313
solid phase epitaxy, 27
stacking fault tetrahedral, 189
stainless steel, 277
STEM, 147
STM, 93
storage gas, 437
surface
  damage, 425
  free energy, 341
  modification, 227, 437
  morphology, 51, 313
  topology, 63
swelling effects, 123

TEM, 183, 189, 363, 381, 393, 411
  HRTEM, 177, 183, 417
tetrahedral bonds, 39
textured films, 301, 319
thermal
  conductivity, 327
  demixing, 197
  spike, 111
  stability, 443
  stress, 283
third-order optical susceptibility, 327, 357
three-dimensional ion implantation, 69
Ti, 231
Tl-Ba-Ca-Cu-O alloy, 177
topological, 39
tracks, 99, 111, 171, 177
transient thermal model, 99
tribology, 241, 259

vacancy-type defects, 57, 135
vanadium oxide, 393
velocity effect, 111
viscosity, 117
vitreous silica, 99
VO_2, 393
V_2O_3, 393
V_7O_{13}, 393
voids, 393

wear, 265
  rate, 259
  resistance, 253
wettability, 425
wetting
  angle, 431
  property, 437

Xe nanocrystals, 417
X-ray
  diffraction, 231, 319, 339
  reflectivity, 197

$Y_3Fe_5O_{12}$, 111

ZnS, 399
ZrNi alloy, 227